THE FOUNDATIONS OF MECHANICS AND THERMODYNAMICS

SELECTED PAPERS

BY

W. NOLL

WITH A PREFACE BY C. TRUESDELL

SPRINGER-VERLAG

BERLIN HEIDELBERG NEW YORK 1974

WALTER NOLL

Professor of Mathematics
Carnegie-Mellon University, Pittsburgh, Pennsylvania, USA

CLIFFORD AMBROSE TRUESDELL, III

Professor of Rational Mechanics
The Johns Hopkins University, Baltimore, Maryland, USA

ISBN-13: 978-3-642-65819-8 e-ISBN-13: 978-3-642-65817-4
DOI: 10.1007/978-3-642-65817-4

Library of Congress Catalog Card Number 74-1651. Softcover reprint of the hardcover 1st edition 1974

Preface

WALTER NOLL's influence upon research into the foundations of mechanics and thermodynamics in the past twenty years is plain, everywhere acknowledged. Less obvious is the wide effect his writings have exerted upon those who apply mechanics to special cases, but it is witnessed by the now common use of terms, concepts, and styles of argument he introduced, use sometimes by young engineers who have learnt them in some recent textbook and hence take them for granted, often with no idea whence they come.

The purpose of this volume of reprints is to put into general hands those of NOLL's works which presently promise broadest and most fertile service to students of the thermomechanics of deformable bodies. This purpose explains the selection of the contents, about half of the pages he has published in periodicals or proceedings. The complete list of his works, which follows this preface, serves also as Table, the articles reprinted here being indicated by their respective page numbers at the right.

Had influence already manifest been the basis of the choice, it would have dictated inclusion of No. 2, which was NOLL's thesis, but by now that work is in part obsolete, in part available through the intermediacy of a dozen books and a hundred papers. As a single example, I may adduce the fact that in Volume 1 of *A Supplement to the Oxford English Dictionary,* 1972, that thesis is the earliest source quoted in the article "constitutive equation", although, with a lapse of accuracy scarcely to be expected, NOLL's name is misspelled there. Equally certainly, on the basis of influence I should have had to exclude his latest development of the foundations, presented in No. 35, since its impact is not yet widely felt. In the past NOLL's major works have often seemed forbiddingly abstract and remote at first sight, and five to ten years have gone by before they came to be widely understood and applied. Had quality been the criterion, it would have forbidden me to exclude No. 3, his paper on statistical mechanics, or No. 33, his study of GIBBS' phase rule, but their subjects are peripheral to the field of this volume. Instead, the reader will find below two largely expository papers, Nos. 24 and 25, which present in typical simplicity and elegance propositions in algebra and geometry now frequently applied in continuum mechanics.

This preface is no place for an obituary of a man not yet fifty and in full power of work, but I will record the circumstances in which NOLL began to study the foundations of mechanics.

Students well trained in mathematics, as mathematicians understand that discipline, rarely take any interest in mechanics. Students of mechanics, all too often aping their teachers, neglect the prime lesson the history of that discipline affords, namely, that "the paradise of the mathematical sciences" flowers and bears its best fruit when mathematicians cultivate it with their most powerful tools – indeed, design those tools for that purpose. After the war the relict

German scholars, against odds now not only forgotten but also hard to imagine, were striving to revivify the life of the mind which the mental and physical barbarity preached and practised by the -isms and -acies of 1933–1946 had all but eradicated. Thinking that among the disciples of these elders, restorers rather than progressives, I might find a student or two who would wish to master new mathematics but grasp it and use it with the wholeness of earlier times, in 1952 I wrote to Mr. HAMEL, one of the few then remaining mathematicians from the classical mould, to ask him to name some young men fit to study for the doctorate in The Graduate Institute for Applied Mathematics at Indiana University, flourishing at that time though soon to be destroyed by the jealous ambition of the local, stereotyped pure. Having just retired from the Technische Universität in Charlottenburg, he passed my inquiry on to Mr. SZABÒ, in whose institute there NOLL was then an assistant. Although Mr. NOLL informs me that he had attended only one course by HAMEL, and that "there were very few students, and none of us understood what he was talking about", nevertheless I like to think of WALTER NOLL as our link with the great Göttingen school, with the HILBERT who wrote:

> What a vital nerve would be cut off from mathematics by rooting out geometry and mathematical physics! On the contrary, I think that wherever mathematical ideas come up, whether from the theory of knowledge or in geometry, or out of the theories of natural science, mathematics ought to investigate the principles underlying these ideas and by means of a simple and complete system of axioms establish them in such a way that in deduction the new ideas shall be no whit inferior to the old arithmetic concepts.

In Berlin NOLL had followed courses by E. SCHMIDT and HASSE, the former, like HAMEL, once a doctorand under HILBERT and the latter once a colleague. The volume of the *Journal of Rational Mechanics and Analysis* in which NOLL's thesis is published is dedicated *in memoriam* to HAMEL. The present volume of reprints will mark the twentieth anniversary of the defense of that thesis on August 9, 1954, in my office. The room itself, like all that we valued of our circumstances then, has since been destroyed.

Nowadays, when the common student seeks a secure berth by grafting himself upon some modest little professor whom he regards as prone to foster painlessly his limaceous glide toward a dissertation not too strenuous or, even better, to draught it for him, tradition is moribund, and we lightly disregard it. Mathematics once was transmitted almost like a priesthood, through novitiate, trial, and the laying on of hands. The burden lay upon the candidate, not upon his professor. Before NOLL came to Indiana, a year at the Sorbonne had given him some of the abstract, direct mathematics of BOURBAKI, which then seemed to him unrelated to natural science, but from which he afterward created what is now become the common dialect of continuum mechanics. That year brought him into contact with living exponents of another tradition, even more powerful than HILBERT's: the didactic, systematic rationalism of DESCARTES.

C. TRUESDELL

Publications of Walter Noll

The numbers prefixed to the papers reprinted in this volume are set in **bold** type in the list below. The pages on which they appear here are indicated at the right.

In the texts reprinted, both such errata as have been published before and also some further ones are included. Whenever it was feasible to do so, the text itself has been corrected, but in some few cases errata are listed at the ends of the works to which they pertain.

BOOKS

I. *The Non-Linear Field Theories of Mechanics, Encyclopedia of Physics,* Volume III/3, 602 pages, Berlin-Heidelberg-New York, Springer-Verlag, 1965 (with C. Truesdell)

II. *Viscometric Flows of Non-Newtonian Fluids, Theory and Experiment,* Springer Tracts in Natural Philosophy, Volume 5, 130 pages, Berlin-Heidelberg-New York, Springer-Verlag, 1966 (with B. D. Coleman and H. Markovitz)

PAPERS AND NOTES

1. Eine Bemerkung zur Schwarzschen Ungleichheit, *Mathematische Nachrichten,* Volume 7, pp. 55–59 (1952) (with E. Mohr)
2. On the Continuity of the Solid and Fluid States, *Journal of Rational Mechanics and Analysis,* Volume 4, pp. 3–81 (1955)
3. Die Herleitung der Grundgleichungen der Thermomechanik der Kontinua aus der statistischen Mechanik, *Journal of Rational Mechanics and Analysis,* Volume 4, pp. 627–646 (1955)
4. Verschiebungsfunktionen für elastische Schwingungsprobleme, *Zeitschrift für angewandte Mathematik und Mechanik,* Volume 37, pp. 81–87 (1957)
5. On the Uniqueness and Non-existence of Stokes Flow, *Archive for Rational Mechanics and Analysis,* Volume 1, pp. 97–106 (1957) (with R. Finn)
6. On the Rotation of an Incompressible Continuous Medium in Plane Motion, *Quarterly of Applied Mathematics,* Volume 15, pp. 317–319 (1957)
7. On Exterior Boundary Value Problems in Linear Elasticity, *Archive for Rational Mechanics and Analysis,* Volume 2, pp. 191–196 (1958) (with R. J. Duffin)
8. **8.** A Mathematical Theory of the Mechanical Behavior of Continuous Media, *Archive for Rational Mechanics and Analysis,* Volume 2, pp. 197–226 (1958) 1–30

9. The Foundations of Classical Mechanics in the Light of Recent Advances in Continuum Mechanics, pp. 266–281 of *The Axiomatic Method, with Special Reference to Geometry and Physics* (Symposium at Berkeley, 1957), Amsterdam, North-Holland Publishing Co., 1959 32–47

10. On Certain Steady Flows of General Fluids, *Archive for Rational Mechanics and Analysis,* Volume 3, pp. 289–303 (1959) (with B. D. COLEMAN) 49–63

11. Helical Flow of General Fluids, *Journal of Applied Physics,* Volume 30, pp. 1508–1512 (1959) (with B. D. COLEMAN)

12. Conditions for Equilibrium at Negative Absolute Temperatures, *The Physical Review,* Volume 115, pp. 262–265 (1959) (with B. D. COLEMAN)

13. On the Thermostatics of Continuous Media, *Archive for Rational Mechanics and Analysis,* Volume 4, pp. 97–128 (1959) (with B. D. COLEMAN) 65–96

14. An Approximation Theorem for Functionals with Applications in Continuum Mechanics, *Archive for Rational Mechanics and Analysis,* Volume 6, pp. 355–370 (1960) (with B. D. COLEMAN) 97–112

15. Recent Results in the Continuum Theory of Viscoelastic Fluids, *Annals of the New York Academy of Science,* Volume 89, pp. 672–714 (1961) (with B. D. COLEMAN)

16. Foundations of Linear Viscoelasticity, *Reviews of Modern Physics,* Volume 33, pp. 239–249 (1961) (with B. D. COLEMAN) 113–123

17. Normal Stresses in Second-order Viscoelasticity, *Transactions of the Society of Rheology,* Volume 5, pp. 41–46 (1961) (with B. D. COLEMAN)

18. Steady Extension of Incompressible Simple Fluids, *The Physics of Fluids,* Volume 5, pp. 840–843 (1962) (with B. D. COLEMAN)

19. Simple Fluids with Fading Memory, pp. 530–552 of *Second-Order Effects in Elasticity, Plasticity, and Fluid Dynamics* (Symposium at Haifa, 1962), Oxford etc., Pergamon Press, 1964 (with B. D. COLEMAN)

20. Motions with Constant Stretch History, *Archive for Rational Mechanics and Analysis,* Volume 11, pp. 97–105 (1962) 125–133

21. La Mécanique Classique, Basée sur un Axiome d'Objectivité, pp. 47–56 of *La Méthode Axiomatique dans les Mécaniques Classiques et Nouvelles* (Colloque International, Paris, 1959), Paris, Gauthier-Villars, 1963 135–144

22. The Thermodynamics of Elastic Materials with Heat Conduction and Viscosity, *Archive for Rational Mechanics and Analysis,* Volume 13, pp. 167–178 (1963) (with B. D. COLEMAN) 145–156

23. Material Symmetry and Thermostatic Inequalities in Finite Elastic Deformations, *Archive for Rational Mechanics and Analysis,* Volume 15, pp. 87–111 (1964) (with B. D. COLEMAN) 157–181

24. Euclidean Geometry and Minkowskian Chronometry, *American Mathematical Monthly,* Volume 71, pp. 129–144 (1964) 183–198

25. Proof of the Maximality of the Orthogonal Group in the Unimodular Group, *Archive for Rational Mechanics and Analysis,* Volume 18, pp. 100–102 (1965) 200-202

26. The Equations of Finite Elasticity, pp. 93–101 of *Symposium on Applications of Nonlinear Partial Differential Equations in Mathematical Physics* (1964), Providence, American Mathematical Society, 1965

27. The Foundations of Mechanics, pp. 159–200 of *Non-Linear Continuum Theories* (C. I. M. E. Lectures, 1965), Rome, Cremonesi, 1966

28. Space-Time Structures in Classical Mechanics, pp. 28–34 of *Delaware Seminar in the Foundations of Physics,* Berlin-Heidelberg-New York, Springer-Verlag, 1967 204–210

29. Materially Uniform Simple Bodies with Inhomogeneities, *Archive for Rational Mechanics and Analysis,* Volume 27, pp. 1–32 (1967) 211–242

30. Inhomogeneities in Materially Uniform Simple Bodies, pp. 239–246 of *IUTAM Symposium on Mechanics of Generalized Continua* (1967), Berlin-Heidelberg-New York, Springer-Verlag, 1968

31. Quasi-invertibility in a Staircase Diagram, *Proceedings of the American Mathematical Society,* Volume 23, pp. 1–4 (1969)

32. Representations of Certain Isotropic Tensor Functions, *Archiv der Mathematik,* Volume 21, pp. 87–90 (1970)

33. On certain Convex Sets of Measures and on Phases of Reacting Mixtures, *Archive for Rational Mechanics and Analysis,* Volume 38, pp. 1–12 (1970)

34. Annihilators of Linear Differential Operators, *Journal d'Analyse,* Volume 24, pp. 205–284 (1971) (with H. D. DOMBROWSKI)

35. A New Mathematical Theory of Simple Materials, *Archive for Rational Mechanics and Analysis,* Volume 48, pp. 1–50 (1972) 243–292

36. Lectures on the Foundations of Continuum Mechanics and Thermodynamics, *Archive for Rational Mechanics and Analysis,* Volume 52, pp. 62–92 (1973) 294–324

Several of the above-listed papers have been reprinted already in collective volumes, as follows:

Nos. 2, 8, 10, 11, and 14 in

Continuum Mechanics II, The Rational Mechanics of Materials, International Science Review Series, Volume VIII, Part 2, New York etc., Gordon & Breach, 1965.

Nos. 13 and 16 in

Continuum Mechanics III, Foundations of Elasticity Theory, International Science Review Series, Volume VIII, Part 3, New York etc., Gordon & Breach, 1965.

Nos. 8, 29, and 30 in

Continuum Theory of Inhomogeneities in Simple Bodies, New York, Springer-Verlag, 1968.

A Mathematical Theory of the Mechanical Behavior of Continuous Media

WALTER NOLL

Communicated by C. TRUESDELL

Contents

1. Introduction

Until not long ago continuum mechanics meant to most people the theories of inviscid and linearly viscous fluids and of linearly elastic solids. However, the behavior of only few real materials can be described adequately by these classical theories. Experimental scientists, who had to deal with real materials, developed a science of non-classical materials called rheology. But they did not succeed in fitting their experimental results into a general mathematical framework. Most of the rheological theories are either one-dimensional, and hence appropriate at best for particular experimental situations, or are confined to infinitesimal deformations, in which case they are only of limited use, because large deformations occur easily in the materials these theories are intended to describe.

In the last few years a number of mathematicians have succeeded in devising general three-dimensional theories which are valid for large deformations*. Guided directly by physical experience or by the one-dimensional theories of the rheologists, these authors have proposed various special hypotheses intended to characterize the mechanical behavior of large classes of materials in arbitrary motions. Using the principle of objectivity of material properties (*cf.* Section 11 of this paper) and usually also the assumption of material isotropy** they then derived the general constitutive equation (also called rheological equation of state or stress-strain relation) compatible with the original hypothesis. In many cases they also solved particular problems in order to make possible a comparison with experimental results***.

This paper does not start with a special hypothesis. It starts (Chapter II) with a general principle, called *the principle of determinism for the stress*, which is implicit in all physical experience and applies to any material whatsoever as long as only its mechanical behavior is considered and as long as there are no constraints. From this principle and the principle of objectivity the most general constitutive equation for all materials is derived.

Constitutive equations characterize material properties. But material properties are of a local nature and may change from particle to particle in a body. It turned out to be necessary, therefore, to develop concepts which describe the local behavior of a motion at a particular particle. This is done in Chapter I. Some of the concepts introduced there are similar to those of the theory of "jets" introduced by EHRESMANN [*2*] into modern differential geometry. The treatment of strain and related notions given in Chapter I is, I believe, more concise and direct than previous treatments. It is based on the well known polar decomposition theorem for linear transformations.

With only one exception**** the ideal materials previously considered in the literature are special cases of what is called a *simple material* in Chapter III of

* A review of the work done up to 1953 was given by TRUESDELL [*12*]. For more recent researches we refer to [*1*], [*4*], [*5*], [*8*], [*9*], [*13*].

** These two things were not clearly distinguished for some time. A clarification was given by the author in [*5*], where the term "principle of isotropy of space" was used for the principle of objectivity.

*** *Cf.* [*3*], [*5*], [*10*], [*12*], [*13*], [*14*].

**** TRUESDELL's theory of Maxwellian fluids (*cf.* [*12*], Chapter V D).

the present paper*. There is reason to believe that most real materials are adequately covered by this category. A detailed discussion of the theory of these simple materials is given. It includes new and very general definitions of isotropic and anisotropic solids and of fluids in terms of the special invariances of the corresponding constitutive equations. Fluids are always isotropic in the sense in which the term is used in this paper. I believe that physically observed anisotropies in fluids are not really intrinsic in these fluids but depend on the past history of the substance.

In Chapter IV it is shown how two large classes of ideal materials fit into the general framework. In the isotropic case, these two classes were introduced by RIVLIN & ERICKSEN [*8*] and COTTER & RIVLIN [*1*]. The general non-isotropic case is discussed here, too.

The theory of simple materials presented in Chapter III generalizes the memory theory of linear visco-elasticity. The theory of materials of the rate type (Section 24) generalizes the differential operator theory of linear visco-elasticity. The materials of the differential type (Section 23) correspond to the special case when only the stress and not the stress rates occur in the basic equation.

Notation and terminology. Physicists customarily use symbols to denote physical quantities. They do not use symbols for the functions that relate these quantities, except sometimes the symbol f, which then stands for any function. Pure mathematicians, on the other hand, distinguish carefully between functions and their values, and they use different symbols for different functions. Both methods have advantages and disadvantages. I have found it necessary here to follow the pure mathematicians. Thus, in this paper, F means something entirely different in nature from $F(t)$. The first is a function, and the second a value of that function and hence a constant.

The terms "function" and "functional" are used here as synonyms for "mapping". A mapping is an operation that assigns to each element of one set, called the *domain* of the mapping, a value in another set. A mapping that has the same values as a given mapping α on a subset of the domain of α but is not defined outside this subset is called the *restriction* of α to the subset in question. Sometimes, if no confusion can arise, we use the same symbol α for such a restriction.

If α and β are two suitable mappings, then $\alpha\circ\beta$ denotes their *composition* defined by $(\alpha\circ\beta)(X)=\alpha(\beta(X))$. The inverse of a one-to-one mapping α is denoted by $\overset{-1}{\alpha}$. If $\mathscr{F}$ and $\mathscr{G}$ are two sets, then $\mathscr{F}\times\mathscr{G}$ means the set of all ordered pairs of elements of $\mathscr{F}$ and $\mathscr{G}$.

In the case of linear transformations F, G of a vector-space into itself we omit the symbol $\circ$ and write simply FG for their composition. The inverse of a linear transformation G is denoted by G^{-1}, its transpose by G^T. The identity transformation is denoted by I. A linear transformation Q is orthogonal if and only if $QQ^T=I$, *i.e.*, $Q^T=Q^{-1}$. The term *tensor* is used as a synonym for linear transformation. Tensors of order higher than two do not occur in this paper.

* More general materials are those "of order n", discussed by the author in [*7*]. They include TRUESDELL's theory of Maxwellian fluids. Simple materials are those of order one.

Points and vectors of a Euclidean space are denoted by boldface letters, except when they are values of mappings. If $\boldsymbol{x}$ and $\boldsymbol{y}$ are two points, then $\boldsymbol{y}-\boldsymbol{x}$ denotes the vector leading from $\boldsymbol{x}$ to $\boldsymbol{y}$. If $\boldsymbol{x}$ is a point and $\boldsymbol{u}$ a vector, then $\boldsymbol{x}+\boldsymbol{u}$ denotes the point $\boldsymbol{y}$ uniquely determined by $\boldsymbol{u}=\boldsymbol{y}-\boldsymbol{x}$.

I. Local kinematics

2. *Basic concepts*

Precise axiomatic definitions of the basic concepts used in this paper are given in [*6*]. Here is a brief summary:

A *body* $\mathscr{B}$ is a smooth manifold of elements $X, Z, \ldots$, called *particles*. The *configurations* $\varphi, \vartheta, \ldots$ of $\mathscr{B}$ are the elements of a set of one-to-one mappings of $\mathscr{B}$ into a three-dimensional Euclidean point space $\mathscr{E}$. The vector space associated with $\mathscr{E}$ will be denoted by $\mathscr{V}$. The *mass-distribution* m of $\mathscr{B}$ is a measure defined on all Borel subsets of $\mathscr{B}$. If φ is a configuration of $\mathscr{B}$, then there is a corresponding *mass density* ϱ_φ such that

$$m(\mathscr{C}) = \int_{\varphi(\mathscr{C})} \varrho_\varphi(\boldsymbol{X})\, dV \tag{2.1}$$

for all Borel subsets $\mathscr{C}$ of $\mathscr{B}$. A *motion* is a one-parameter family $\{\vartheta_\tau\}$, $-\infty<\tau<\infty$, of configurations. The parameter τ is called the *time*. A motion will often simply be denoted by ϑ, which is then regarded as a point-valued function of two variables, a particle and a time. The mass density corresponding to the configuration ϑ_τ is denoted by $\varrho(\tau)$.

A body, its configurations, its mass-distribution, and its possible motions are subject to the axioms given in [*6*]. These axioms insure that the customary notions and operations are meaningful. However, in some of the considerations of this paper, continuity and differentiability conditions stronger than those implied by the axioms of [*6*] have to be assumed. It will be clear from the context when this is the case, and we shall not mention these conditions explicitly.

3. *Deformations and linear transformations*

A smooth homeomorphism δ which maps a neighborhood of the null-vector $\boldsymbol{O}$ of the vector space $\mathscr{V}$ onto another such neighborhood and which maps $\boldsymbol{O}$ into itself,

$$\delta(\boldsymbol{O}) = \boldsymbol{O}, \tag{3.1}$$

shall be called a *local homeomorphism* of $\mathscr{V}$. We define an equivalence relation among all local homeomorphisms by local identity: $\delta \sim \tilde{\delta}$ if and only if $\delta(\boldsymbol{P}) = \tilde{\delta}(\boldsymbol{P})$ for all $\boldsymbol{P}$ in some neighborhood of $\boldsymbol{O}$, however small. The resulting equivalence classes Δ will be called *deformations*. The equivalence class of $\delta\circ\hat{\delta}$ depends only on the equivalence classes Δ and $\hat{\Delta}$ of δ and $\hat{\delta}$, and it will therefore be denoted by $\Delta\circ\hat{\Delta}$. With the law of composition thus defined, the deformations form a group denoted by $\mathscr{D}$.

A *linear transformation* G, *i.e.*, a mapping of $\mathscr{V}$ onto itself such that

$$G(\boldsymbol{P}_1 + \boldsymbol{P}_2) = G(\boldsymbol{P}_1) + G(\boldsymbol{P}_2), \qquad G(a\boldsymbol{P}) = a\,G(\boldsymbol{P}), \tag{3.2}$$

is a local homeomorphism provided that it is regular, *i.e.*, provided that $\boldsymbol{G}\boldsymbol{P}=\boldsymbol{O}$ is possible only for $\boldsymbol{P}=\boldsymbol{O}$. Two different linear transformations are never equivalent in the sense defined above. Hence a regular linear transformation defines a unique deformation and may thus be regarded as a special deformation. The regular linear transformations thus form a subgroup of the group of deformations $\mathscr{D}$, called the *linear group* and denoted by $\mathscr{L}$.

The gradient at $\boldsymbol{P}=\boldsymbol{O}$ of a local homeomorphism δ is a local property of δ and hence depends only on the equivalence class Δ to which δ belongs. We can, therefore, define the *gradient* $\nabla\Delta$ of a deformation Δ by

$$\nabla\Delta=\nabla\delta(\boldsymbol{O}) \quad \text{if} \quad \delta\in\Delta. \tag{3.3}$$

The gradient $\nabla\Delta$ is a regular linear transformation and hence itself a deformation. Since, by the chain rule,

$$\nabla(\Delta\circ\hat{\Delta})=(\nabla\Delta)(\nabla\hat{\Delta}), \tag{3.4}$$

it follows that the gradient operation ∇ is an endomorphism of the group $\mathscr{D}$. Its kernel is the group $\mathscr{N}$ of all *null-deformations, i.e.*, all deformations whose gradient is the identity I. The quotient group $\mathscr{D}/\mathscr{N}$ consists of the equivalence classes of deformations defined by the equivalence relation

$$\Delta\sim\tilde{\Delta} \quad \text{if and only if} \quad \nabla\Delta=\nabla\tilde{\Delta}. \tag{3.5}$$

The deformations in each class have the same gradient and any representing local homeomorphisms differ only by small terms of order one. The relation (3.5) may be read: Δ and $\tilde{\Delta}$ are equal up to a small term of order one. Each class has a unique representative which is a linear transformation. We have

$$\Delta\sim\nabla\Delta, \tag{3.6}$$

i.e. any deformation differs from its gradient only by a small term of order one. In the case when $\Delta=G$ is a linear transformation, we have

$$\nabla G=G. \tag{3.7}$$

The quotient group $\mathscr{D}/\mathscr{N}$ is isomorphic to the linear group $\mathscr{L}$.

A deformation Δ will be called *isochoric* if there is a volume preserving local homeomorphism δ in the class Δ. The isochoric deformations form a subgroup $\mathscr{I}$ of $\mathscr{D}$. The null-deformations are isochoric, and $\mathscr{N}$ is a subgroup of $\mathscr{I}$. A linear transformation G is isochoric if and only if

$$|\det G|=1. \tag{3.8}$$

The isochoric linear transformations form a subgroup $\mathscr{U}$ of $\mathscr{L}$, called the *unimodular group*. They are also called *unimodular transformations*. The group of *orthogonal transformations* is a subgroup of $\mathscr{U}$; it will be denoted by $\mathscr{O}$. The space of all *symmetric linear transformations* will be denoted by $\mathscr{S}$. The set of all *positive definite* and symmetric linear transformations will be denoted by $\mathscr{S}_+$. It is a subset of the linear group $\mathscr{L}$.

4. *Local configurations*

Consider a neighborhood $\mathscr{N}(X)$ of a particle X in a body $\mathscr{B}$, *i.e.*, a part of $\mathscr{B}$ containing X. Let ψ be a smooth homeomorphism of $\mathscr{N}(X)$ into the vector space $\mathscr{V}$ mapping X itself into the null-vector,

$$\psi(X) = \boldsymbol{O}. \tag{4.1}$$

We define an equivalence relation among all such homeomorphisms by the condition that $\psi \sim \tilde{\psi}$ if and only if $\psi(Z) = \tilde{\psi}(Z)$ for all Z in some neighborhood of X. The resulting equivalence classes $\boldsymbol{\Phi}$ will be called the *local configurations* of X. The set of all local configurations of X will be denoted by $\mathscr{C}_X$.

Let δ be a local homeomorphism of $\mathscr{V}$. The equivalence class of the mapping $\delta \circ \psi$ depends only on the equivalence class $\boldsymbol{\Phi}$ of ψ and on the deformation Δ of δ. Therefore, it is meaningful to speak about the local configuration $\Delta \circ \boldsymbol{\Phi}$ of X, which is the composite of the local configuration $\boldsymbol{\Phi}$ of X and the deformation Δ. If $\boldsymbol{\Phi}$ and $\hat{\boldsymbol{\Phi}}$ are two local configurations of X, we define

$$\Delta = \hat{\boldsymbol{\Phi}} \circ \overset{-1}{\boldsymbol{\Phi}} \quad \text{if} \quad \hat{\boldsymbol{\Phi}} = \Delta \circ \boldsymbol{\Phi}. \tag{4.2}$$

This Δ will be called the *deformation from the configuration* $\boldsymbol{\Phi}$ *to the configuration* $\hat{\boldsymbol{\Phi}}$.

Let φ be a configuration of the body $\mathscr{B}$. Then $\psi = \varphi - \varphi(X)$ maps $\mathscr{B}$ into the vector space $\mathscr{V}$ and has the property (4.1). Its equivalence class will be denoted by $\boldsymbol{\Phi}(\varphi, X)$, and it will be called the *localization* at X of φ. Assume that a motion ϑ of $\mathscr{B}$ is given. We then use the notation

$$\Theta(\tau) = \boldsymbol{\Phi}(\vartheta_\tau, X), \tag{4.3}$$

and we call Θ the *localization* at X of the motion ϑ. We note that Θ is a function of a real variable whose values are in $\mathscr{C}_X$. Any sufficiently smooth function of this type will be called a *local motion* of X.

Let Θ be a local motion and $\boldsymbol{\Phi}$ a local configuration of X. We then write

$$\Omega(\tau) = \Theta(\tau) \circ \overset{-1}{\boldsymbol{\Phi}} \tag{4.4}$$

and call Ω the *deformation function* of X in the local motion Θ relative to the *local reference configuration* $\boldsymbol{\Phi}$. Ω is a function of a real variable with values in $\mathscr{D}$. Any sufficiently smooth function of this type will be called a deformation function.

In the special case when $\boldsymbol{\Phi} = \Theta(t)$ in (4.4), where t is a particular time, we use the notation

$$\Omega_t(\tau) = \Theta(\tau) \circ \overset{-1}{\Theta}(t) \tag{4.5}$$

and call Ω_t the deformation function of X in Θ relative to the time t.

Let φ be a configuration of the body $\mathscr{B}$ and ϱ_φ the corresponding mass density. The value $\varrho_\varphi(\varphi(X))$ depends only on the localization $\boldsymbol{\Phi}(\varphi, X)$ of φ at X. We can, therefore, define

$$\varrho_{\boldsymbol{\Phi}} = \varrho_\varphi(\varphi(X)), \qquad \boldsymbol{\Phi} = \boldsymbol{\Phi}(\varphi, X), \tag{4.6}$$

and call $\varrho_{\boldsymbol{\Phi}}$ the *mass density* of X in the local configuration $\boldsymbol{\Phi}$.

5. *Gradients*

Assume that two local configurations Φ and $\hat{\Phi}$ of a particle X are given. The regular linear transformation

$$G = G(\hat{\Phi}, \Phi) = \nabla \Delta, \quad \Delta = \hat{\Phi} \circ \overset{-1}{\Phi} \tag{5.1}$$

will then be called the *deformation gradient* from Φ to $\hat{\Phi}$.

In the case when a local motion Θ and a local reference configuration Φ are given, we use the notation

$$F(\tau) = \nabla \Omega(\tau) = G(\Theta(\tau), \Phi), \tag{5.2}$$

$$F_t(\tau) = \nabla \Omega_t(\tau) = G(\Theta(\tau), \vartheta(t)). \tag{5.3}$$

Assume that $\hat{\Phi}$ is another local reference configuration, and let

$$\hat{F}(\tau) = \nabla \hat{\Omega}(\tau) = G(\Theta(\tau), \hat{\Phi}). \tag{5.4}$$

Then, by the chain rule for gradients, we have

$$F(\tau) = \hat{F}(\tau)\, G, \tag{5.5}$$

where $G = G(\hat{\Phi}, \Phi)$ is defined by (5.1). We note the following formulas:

$$G(\hat{\Phi}, \Phi) = G(\Phi, \hat{\Phi})^{-1}, \tag{5.6}$$

$$F(\tau) = F_t(\tau)\, F(t), \quad F_t(t) = I. \tag{5.7}$$

We define an equivalence relation among all local configurations $\Phi \in \mathscr{C}_X$ by

$$\Phi \sim \tilde{\Phi} \quad \text{if and only if} \quad G(\tilde{\Phi}, \Phi) = I. \tag{5.8}$$

The corresponding equivalence classes M will be called the *configuration gradients* at X. The class of local configurations equivalent to a given $\Phi \in \mathscr{C}_X$ is denoted by $\nabla \Phi$, so that

$$M = \nabla \Phi \quad \text{if and only if} \quad \Phi \in M. \tag{5.9}$$

The equivalence class of the local configuration $\Delta \circ \Phi$, where Δ is a deformation, depends only on the equivalence class of Φ and on the gradient $\nabla \Delta$ of Δ. Therefore, it is meaningful to speak about the configuration gradient GM, which is the product of the configuration gradient M and the regular linear transformation G. This product is characterized by the property

$$\nabla(\Delta \circ \Phi) = (\nabla \Delta)\,(\nabla \Phi). \tag{5.10}$$

If M and $\hat{M}$ are two configuration gradients, we define

$$G = \hat{M} M^{-1} \quad \text{if} \quad \hat{M} = GM. \tag{5.11}$$

We have

$$G(\hat{\Phi}, \Phi) = \hat{M} M^{-1}, \quad M = \nabla \Phi, \quad \hat{M} = \nabla \hat{\Phi}, \tag{5.12}$$

which shows that $G(\hat{\Phi}, \Phi)$ depends only on the gradients of Φ and $\hat{\Phi}$. In particular, the deformation gradient function F in (5.2) depends only on $\nabla \Phi$, and we may call

$$F(\tau) = (\nabla \Theta(\tau))\, M^{-1} \tag{5.13}$$

the deformation gradient at time τ of the local motion Θ relative to the configuration gradient M as a reference.

If Φ and $\hat{\Phi}$ are two local configuration at X, then the corresponding densities ϱ_Φ and $\varrho_{\hat{\Phi}}$ are related by

$$\varrho_{\hat{\Phi}} = |\det G(\hat{\Phi}, \Phi)|\, \varrho_\Phi. \tag{5.14}$$

It follows from (5.8) that $\varrho_{\hat{\Phi}} = \varrho_\Phi$ if $\hat{\Phi}$ and Φ have the same configuration gradient. Hence we can define

$$\varrho_M = \varrho_\Phi \quad \text{if} \quad M = \nabla \Phi, \tag{5.15}$$

and we call ϱ_M the mass density for the configuration gradient M at X. Equation (5.12) then implies

$$\varrho_{GM} = |\det G|\, \varrho_M. \tag{5.16}$$

It follows that

$$\varrho_M = \varrho_{\hat{M}} \quad \text{if and only if} \quad G = \hat{M} M^{-1} \quad \text{is unimodular.} \tag{5.17}$$

6. *Rotation and strain tensors*

Of fundamental importance for the local kinematics of continuous media is the well known *polar decomposition theorem:*

Let F be any regular linear transformation. Then there are unique decompositions

$$F = R\,U, \quad F = V R \tag{6.1}$$

in which R is orthogonal and U and V are symmetric and positive definite, *i.e.*, $R \in \mathcal{O}$, $U, V \in \mathcal{S}_+$. The following relations are valid:

$$U^2 = F^T F, \qquad V^2 = F F^T, \tag{6.2}$$

$$V = R\,U R^T, \quad V^2 = R\,U^2 R^T. \tag{6.3}$$

This theorem, when applied to a deformation gradient $F = G(\hat{\Phi}, \Phi)$, gives rise to the following terminology: We call R the *rotation tensor*, U the *right strain tensor*, and V the *left strain tensor* of the deformation from Φ to $\hat{\Phi}$. The squares

$$C = U^2 = F^T F, \quad B = V^2 = F F^T = R\,C\,R^T \tag{6.4}$$

will be called the *right* and the *left Cauchy-Green tensors* of the deformation from Φ to $\hat{\Phi}$.

The strain tensors U and V describe adequately what is meant physically by strain because their eigenvalues are the principal extension ratios of the deformation from Φ to $\hat{\Phi}$. However, it is often of advantage to use the Cauchy-Green tensors rather than the strain tensors as measures of strain because their components are rational functions of the components of the deformation gradient F, while the components of U and V are complicated irrational functions of the components of F. Both C and B are, of course, symmetric and positive definite. Their eigenvalues are the squares of the principal extension ratios of the deformation.

In the special case when F is the tensor function defined by (5.2) R, U, V, C, and B will also be tensor functions of a real variable. If F is replaced by the tensor function F_t defined in (5.3), we use the analogous notation R_t, U_t, V_t, C_t, and B_t. These functions have the same smoothness properties as F or F_t, respectively. We note that

$$F_t(t) = R_t(t) = U_t(t) = V_t(t) = C_t(t) = B_t(t) = I. \tag{6.5}$$

7. *Histories*

Let $\mathscr{F}$ be an arbitrary set. The class of all functions with values in $\mathscr{F}$ whose domain is the negative real axis will then be denoted by $\mathscr{F}^*$.

Let α be any function of a real variable with values in $\mathscr{F}$. We then define $\alpha^t \in \mathscr{F}^*$ by

$$\alpha^t(s) = \alpha(t+s) \quad \text{for} \quad s \leqq 0, \tag{7.1}$$

and we call it the *history up to time t* of the function α.

For functions $\alpha^* \in \mathscr{F}^*$ we generally use the notation

$$\alpha_0 = \alpha^*(0). \tag{7.2}$$

Then

$$\alpha_0^t = \alpha^t(0) = \alpha(t). \tag{7.3}$$

Let Θ be a local motion of a particle X, so that $\Theta(\tau) \in \mathscr{C}_X$. The corresponding history $\Theta^t \in \mathscr{C}_X^*$, defined by

$$\Theta^t(s) = \Theta(t+s) \quad \text{for} \quad s \leqq 0, \tag{7.4}$$

will then be called the *kinematical history* up to t of X. If, in addition, a local reference configuration Φ is given and if $\Omega = \Theta \circ \overset{-1}{\Phi}$ is the corresponding deformation function, we call $\Omega^t \in \mathscr{D}^*$, defined by

$$\Omega^t(s) = \Omega(t+s) \quad \text{for} \quad s \leqq 0, \tag{7.5}$$

the *deformation history* up to t of X.

8. *Rate of strain and spin*

Consider a local motion Θ of a particle X. Let F_t be the corresponding deformation gradient function defined by (5.3) and F_t^t the history of F_t up to time t defined according to (7.1).

We then call*

$$E(t) = \dot{F}_t(t) = \dot{F}_t^t(0) \tag{8.1}$$

the *velocity gradient* at time t of the local motion Θ. Similarly*

$$E_n(t) = \overset{(n)}{F_t}(t) = \overset{(n)}{F_t^t}(0), \qquad n = 0, 1, 2, \ldots \tag{8.2}$$

---

* A superposed dot denotes the operation of differentiation and a superposed (n) its nth iteration.

is called the nth *acceleration gradient* at time t of Θ. We have, by (6.5) and (8.1),

$$E_0 = I, \quad E_1 = E. \tag{8.3}$$

If Θ is the localization at X of a global motion ϑ, then $E(t)$ $(E_n(t))$ actually coincides with the gradient with respect to position of the velocity (n^{th} acceleration) at $\boldsymbol{x} = \vartheta_t(X)$ and at time t.

The polar decomposition

$$F_t(\tau) = R_t(\tau)\, U_t(\tau) \tag{8.4}$$

defines the rotation tensor $R_t(\tau)$ and the right strain tensor $U_t(\tau)$ of the deformation from the configuration at time t to the configuration at time τ. The histories R_t^t and U_t^t up to time t of R_t and U_t are defined according to (7.1). The tensor

$$W(t) = \dot{R}_t(t) = \dot{R}_t^t(0) \tag{8.5}$$

is called the *spin* at time t, and the tensor

$$D(t) = \dot{U}_t(t) = \dot{U}_t^t(0) \tag{8.6}$$

the *rate of strain* at time t. Similarly, we define the nth *spin* W_n and the nth *rate of strain* D_n by

$$W_n(t) = \overset{(n)}{R}_t(t) = \overset{(n)}{R_t^t}(0), \tag{8.7}$$

$$D_n(t) = \overset{(n)}{U}_t(t) = \overset{(n)}{U_t^t}(0). \tag{8.8}$$

Replacing U_t in (8.8) by the right Cauchy-Green tensor $C_t = U_t^2$, we get a tensor

$$A_n(t) = \overset{(n)}{C}_t(t) = \overset{(n)}{C_t^t}(0), \tag{8.9}$$

which we shall call the nth *Rivlin-Ericksen tensor**. By (6.5), (8.5), and (8.6) we have

$$R_0 = D_0 = A_0 = I, \quad R_1 = R, \quad D_1 = D. \tag{8.10}$$

One could give definitions similar to (8.8) and (8.9) by replacing the right strain and Cauchy-Green tensors by their left counterparts. But the tensors defined in this way are of little interest. Moreover, for $n = 1$ one would get nothing new, because

$$D(t) = \dot{U}_t(t) = \dot{V}_t(t), \tag{8.11}$$

as is not hard to prove.

9. *Rational expressions for the rates*

It is possible to find explicit expressions for W_n, D_n, and A_n as polynomials in the acceleration gradients E_k and their transpositions E_k^T, $k = 1, 2, \ldots, n$. Differentiating $C_t = F_t^T F_t$ [*cf.* (6.4)] n times, by the product rule we obtain

$$\overset{(n)}{C}_t(\tau) = \sum_{k=0}^{n} \binom{n}{k} \overset{(k)}{F_t^T}(\tau)\, \overset{(n-k)}{F_t}(\tau).$$

---

* RIVLIN & ERICKSEN [8] recognized the importance of these tensors and used them extensively.

Substituting $\tau=t$ yields, according to (8.2) and (8.9),

$$A_n=\sum_{k=0}^{n}\binom{n}{k}E_k^T E_{n-k}. \tag{9.1}$$

Since $E_0=I$ by (8.3), this may be written in the form

$$A_n=E_n+E_n^T+\sum_{k=1}^{n-1}\binom{n}{k}E_k^T E_{n-k}. \tag{9.2}$$

Differentiating $C_t^{(\tau)}=U_t^2(\tau)$ and then substituting $\tau=t$ we get

$$A_n=2D_n+\sum_{k=1}^{n-1}\binom{n}{k}D_k D_{n-k}, \tag{9.3}$$

and hence

$$D_n=\frac{1}{2}\left\{A_n-\sum_{k=1}^{n-1}\binom{n}{k}D_k D_{n-k}\right\}. \tag{9.4}$$

This is a recursion formula which can be used to find explicit expressions for the D_n as polynomials in the A_k, $k=1, 2, \dots, n$. Hence, after substitution of (9.2), one would also obtain explicit expressions for the D_n as polynomials in the E_k and E_k^T, $k=1, 2, \dots, n$.

Differentiating $F_t(\tau)=R_t(\tau)\,U_t(\tau)$ and then putting $\tau=t$, we find

$$E_n=W_n+D_n+\sum_{k=1}^{n-1}\binom{n}{k}W_k D_{n-k}, \tag{9.5}$$

and hence

$$W_n=E_n-D_n-\sum_{k=1}^{n-1}\binom{n}{k}W_k D_{n-k}. \tag{9.6}$$

This is again a recursive formula which permits us to express W_n as a polynomial in the E_k and D_k. Since the D_k are polynomials in the E_k and E_k^T, we can also find expressions for the W_n as polynomials in the E_k and E_k^T, $k=1, 2, \dots, n$.

In the special case $n=1$ we find

$$D=\tfrac{1}{2}A_1=\tfrac{1}{2}(E+E^T), \qquad W=\tfrac{1}{2}(E-E^T). \tag{9.7}$$

It follows that the spin W is skew ($W^T=-W$). The higher spins W_n, $n>1$ are not necessarily skew. Of course, the D_n and A_n are all symmetric. For $n=2$ we find

$$\begin{aligned} A_2&=E_2+E_2^T+2E^T E,\\ D_2&=\tfrac{1}{4}(A_2-\tfrac{1}{2}A_1^2)=\tfrac{1}{2}[E_2+E_2^T-(E+E^T)^2]+E^T E,\\ W_2&=\tfrac{1}{2}[E_2-E_2^T-E^T(E-E^T)]. \end{aligned} \tag{9.8}$$

II. The general constitutive equation

10. Basic concepts

A *dynamical process* for a body $\mathscr{B}$ is defined as a motion ϑ of $\mathscr{B}$ coupled with a system of forces for $\mathscr{B}$ at each time τ, subject to the principle of linear momentum and the principle of angular momentum (*cf.* [*6*], Section 5). A system of forces can be split into body forces and contact forces. If sufficient continuity assumptions are made, the contact forces are determined, for each time τ, by a field of symmetric *stress tensors* $S(\tau)$. If a motion ϑ and smooth fields of symmetric tensors $S(\tau)$ are arbitrarily prescribed, one can always find a dynamical process such that $S(\tau)$ is the corresponding stress tensor field at time τ. Appropriate body forces can be chosen, for example, by putting the mutual body forces equal to zero and by letting the density $\boldsymbol{b}$ of the external body forces be given by

$$\boldsymbol{b} = \dot{\boldsymbol{v}} - \frac{1}{\varrho} \operatorname{div} S, \tag{10.1}$$

where $\dot{\boldsymbol{v}}$ is the acceleration of the motion ϑ. Then CAUCHY's law of motion ([*6*], (5.25)) holds, and the principles of dynamics are satisfied.

If we subsequently talk about a *process* for a body $\mathscr{B}$ we mean a pair $\{\vartheta, S\}$, where ϑ is an arbitrary motion of $\mathscr{B}$ and S an arbitrary time family of symmetric stress fields. We can always adjoin suitable body forces to a given process so as to make it a dynamical process.

A *local process* for a particle $X \in \mathscr{B}$ is defined as a pair $\{\Theta, S\}$, where Θ is a local motion of X and S a symmetric tensor function of a real variable. The *localization* at X of a process $\{\vartheta, S\}$ is defined by replacing ϑ by its localization Θ at X and S by its values for X. We shall use the same symbol S for these values and write $S(\tau) = S(\tau, X)$. No confusion can arise because, in all local considerations, the particle X will be fixed.

A *constitutive assumption* is a restrictive condition on the possible dynamical processes a body can undergo, and it characterizes the material properties of the body. Restrictive conditions on the possible motions alone are particular constitutive assumptions; they are called *constraints*. Examples of constraints are rigidity (every possible motion is a rigid motion) and incompressibility (every possible motion is isochoric). In this paper, we shall assume that there are no constraints, *i.e.* that all motions ϑ are possible. We shall investigate constitutive assumptions in the form of functional relations between the stress S and the motion ϑ of a process $\{\vartheta, S\}$. Such relations will be called *constitutive equations*.

11. The principle of objectivity of material properties

A *change of frame* is a transformation of space and time specified by a point-valued function $\boldsymbol{c}$ of a real variable, a function Q of a real variable whose values are orthogonal transformations, and a real constant a. It transforms a pair $\{\boldsymbol{x}, \tau\}$ consisting in a point $\boldsymbol{x}$ and a time τ into the pair $\{\boldsymbol{x}', \tau'\}$ given by*

$$\boldsymbol{x}' = \boldsymbol{c}(\tau) + Q(\tau)(\boldsymbol{x} - \boldsymbol{O}), \tag{11.1}$$

$$\tau' = \tau - a, \tag{11.2}$$

where $\boldsymbol{O}$ is an arbitrary point, the same for all transformations. Vectors $\boldsymbol{u} \in \mathscr{V}$ transform according to

$$\boldsymbol{u}' = Q(\tau)\boldsymbol{u}. \tag{11.3}$$

* TOUPIN has analyzed these transformations recently [*11*]; he calls them "Euclidean transformations of space-time".

If G is a linear transformation and $\boldsymbol{u} \in \mathscr{V}$, then $G\boldsymbol{u} \in \mathscr{V}$. Hence, by (11.3), $(G\boldsymbol{u})' = Q(\tau)\, G\boldsymbol{u}$. Defining the transform G' of G by $(G\boldsymbol{u})' = G'\boldsymbol{u}'$, we get $Q(\tau)\, G\boldsymbol{u} = G'\, Q(\tau)\boldsymbol{u}$, and hence the following law of transformation for second order tensors:

$$G' = Q(\tau)\, G\, Q(\tau)^T. \tag{11.4}$$

Two dynamical processes are called *equivalent* if they are related by a change of frame as made precise in Definition 7 of [*6*]. Under such a change of frame the contact forces transform in an objective manner, *i.e.*, according to the law (11.3). Hence the stress tensor must transform according to the law (11.4). The motion ϑ determines the position of the particles and hence transforms according to (11.1). Disregarding the body forces, we can say that *two processes* $\{\vartheta, S\}$ *and* $\{\vartheta', S'\}$ *are equivalent if they are related by a change of frame in the form*

$$\begin{aligned} \vartheta'(Z, \tau') &= \boldsymbol{c}(\tau) + Q(\tau)\,[\vartheta(Z, \tau) - \boldsymbol{O}], \\ S'(Z, \tau') &= Q(\tau)\, S(Z, \tau)\, Q^T(\tau), \\ \tau' &= \tau - a. \end{aligned} \tag{11.5}$$

Constitutive equations are subject to the following invariance requirement:

Principle of objectivity of material properties. *If a process* $\{\vartheta, S\}$ *is compatible with a constitutive equation, then also all processes* $\{\vartheta', S'\}$ *equivalent to it must be compatible with the same constitutive equation.*

This is a special case of the general principle of objectivity stated in [*6*]. Its physical meaning is simply that the material properties of a body should not depend on the observer, no matter how he moves.

12. The principle of determinism for the stress

We ask for guiding principles which will enable us to find the most general form of a constitutive equation. The following two are implied by physical experience:

(*i*) The stress at a particle X should depend only on the physical state of an arbitrarily small neighborhood of X. The state of parts distant from X should have no direct influence on the stress at X. This condition is implicit in the concept of a contact force.

(*ii*) The physical state of a body at a time t should depend only on its past history, *i.e.*, on what happened to it at times $\tau \leqq t$, and not on its future, *i.e.*, on what will happen to it at times $\tau > t$. This condition expresses the causality of natural processes.

The physical history of a body consists of several components: its kinematical history, its thermodynamic history, its electromagnetic history, its chemical history, etc. In reality, each of these components may influence the stress. But, in this investigation, we shall disregard all non-mechanical influences.

Principle of determinism. *The stress* $S(t)$ *at a particle* X *and at time* t *is determined by the past history of the motion of an arbitrarily small neighborhood of* X.

In precise terms, this principle states that

(a) the stress $S(t)$ at X is a *functional* $\mathfrak{F}_t$ of the motion ϑ,

$$S(t) = \mathfrak{F}_t(\vartheta) \tag{12.1}$$

the domain of $\mathfrak{F}_t$ being the class of all possible motions and its values being symmetric tensors,

(b) for any two motions ϑ and $\bar{\vartheta}$ which coincide in some neighborhood of X for times $\tau \leqq t$ the value of the functional $\mathfrak{F}_t$ is the same, *i.e.*,

$$\mathfrak{F}_t(\vartheta) = \mathfrak{F}_t(\bar{\vartheta}) \tag{12.2}$$

whenever

$$\vartheta(Z, \tau) = \bar{\vartheta}(Z, \tau) \tag{12.3}$$

for $\tau \leqq t$ and Z in some neighborhood $\mathcal{N}(X)$ of X, however small.

13. The general constitutive equation

Not any arbitrary functional $\mathfrak{F}_t$ with the Property (b) is permissible in (12.1), because (12.1), as a constitutive equation, is subject to the principle of objectivity stated in Section 11. This principle states that (12.1) must hold equally if the process $\{\vartheta, S\}$ is replaced by any equivalent process $\{\vartheta', S'\}$, *i.e.*, that

$$S'(t') = \mathfrak{F}_{t'}(\vartheta') \tag{13.1}$$

must be valid for any process $\{\vartheta', S'\}$ related to $\{\vartheta, S\}$ by a transformation of the form (11.5). Choosing $Q = I$, $a = 0$, and $\boldsymbol{c}(\tau) = \boldsymbol{O} - [\vartheta_\tau(X) - \boldsymbol{O}]$ in (11.5), we get $S' = S$, $\tau' = \tau$, and $\vartheta'_\tau = \boldsymbol{O} + [\vartheta_\tau - \vartheta_\tau(X)]$. It follows that $S(t) = S'(t) = \mathfrak{F}_t(\vartheta')$ holds and hence that $S(t)$ depends only on the vector-valued function ψ defined by $\psi(Z, \tau) = \vartheta(Z, \tau) - \vartheta(X, \tau)$. Moreover, the Property (b) of $\mathfrak{F}_t$ implies that $S(t)$ can depend only on the local behavior of ψ in an arbitrarily small neighborhood of X. This means that $S(t)$ is determined by the localization Θ at X of the motion ϑ as defined by (4.3). Thus, (12.1) reduces to the form

$$S(t) = \mathfrak{F}_t(\Theta) \tag{13.2}$$

where $\mathfrak{F}_t$ is a functional with the property that

$$\mathfrak{F}_t(\Theta) = \mathfrak{F}_t(\bar{\Theta}) \tag{13.3}$$

whenever

$$\Theta(\tau) = \bar{\Theta}(\tau) \quad \text{for} \quad \tau \leqq t. \tag{13.4}$$

Now we consider another equivalent process by choosing $\boldsymbol{c}(\tau) = \boldsymbol{O}$, $Q = I$, and $a = t$ in (11.5). We then have

$$\tau' = \tau - t, \qquad t' = t - t = 0, \qquad S'(t') = S'(0) = S(t),$$

and, for the corresponding local motion Θ,

$$\Theta'(\tau') = \Theta(\tau) = \Theta(t + \tau').$$

It follows from (13.1) that

$$S(t) = S^t(0) = \mathfrak{F}_0(\Theta^t). \tag{13.5}$$

By (13.4) we have $\mathfrak{F}_0(\Theta^t) = \mathfrak{F}_0(\overline{\Theta}^t)$ whenever $\Theta^t(\tau') = \overline{\Theta}^t(\tau')$ for $\tau' \leqq 0$, which shows that the value $\mathfrak{F}_0(\Theta^t)$ depends only on the restriction of the function $\Theta^t(\tau')$ to $\tau' \leqq 0$. But this restriction is nothing but the kinematical history $\Theta^t \in \mathscr{C}_X^*$ up to time t as defined by (7.4). Omitting the index 0 in (13.5), we see that the constitutive equation (13.2) reduces to the form

$$S(t) = \mathfrak{F}(\Theta^t), \tag{13.6}$$

where Θ^t is the kinematical history of X for the local motion Θ. The form of the functional $\mathfrak{F}$ is independent of t and depends only on the particle X. Its domain is $\mathscr{C}_X^*$.

We finally consider an equivalent process by choosing $\boldsymbol{c}(\tau) = \boldsymbol{O}$ and $a = 0$ in (11.5), leaving the orthogonal tensor function Q arbitrary. The principle of objectivity then implies that the functional $\mathfrak{F}$ of (13.6) must satisfy the relation

$$Q_0 \mathfrak{F}(\Theta^*) Q_0^T = \mathfrak{F}(Q^* \circ \Theta^*), \quad Q_0 = Q^*(0), \tag{13.7}$$

for all kinematical histories $\Theta^* \in \mathscr{C}_X^*$ and all orthogonal tensor functions $Q^* \in \mathscr{O}^*$.

It is not hard to see that, conversely, the principle of determinism and the principle of objectivity are automatically satisfied for any equation of the form (13.6) provided the functional $\mathfrak{F}$ has the property (13.7). Hence (13.6) with (13.7) is the most general constitutive equation. An equation of this form restricts the class of all possible local processes $\{\Theta, S\}$ for the particle X and characterizes the local material properties of X.

If we subsequently speak about a particle X, we always assume that a functional of the type discussed above is associated with it. We call it the *functional of the particle X*.

14. Material isomorphisms

The nature of the domain $\mathscr{C}_X^*$ of the functional $\mathfrak{F}$ of the particle X varies with the particle X. Hence there is no direct way to compare the functionals of different particles. It is desirable to render such a comparison possible, because only then can a precise meaning be given to the statement that two different particles consist of the same material.

A body was defined in [6] as a certain mathematical structure. As in the case with any such structure, it is meaningful to talk about isomorphisms between bodies. An *isomorphism* of a body $\mathscr{B}$ onto a body $\overline{\mathscr{B}}$ is a one-to-one mapping γ of $\mathscr{B}$ onto $\overline{\mathscr{B}}$ such that

(*a*) the configurations $\overline{\varphi}$ of $\overline{\mathscr{B}}$ are of the form

$$\overline{\varphi} = \varphi \circ \overset{-1}{\gamma}, \tag{14.1}$$

(*b*) the mass distributions m and $\overline{m}$ of $\mathscr{B}$ and $\overline{\mathscr{B}}$ are related so that

$$m(\mathscr{C}) = \overline{m}(\gamma(\mathscr{C})) \tag{14.2}$$

for all Borel subsets $\mathscr{C}$ in $\mathscr{B}$.

Assume that X and $\bar{X}$ are two particles and consider the isomorphisms γ, if any, of neighborhoods $\mathcal{N}(X)$ onto neighborhoods $\bar{\mathcal{N}}(\bar{X})$ which map X into $\bar{X}=\gamma(X)$. We define an equivalence relation among all such isomorphisms by the condition that $\gamma\sim\tilde{\gamma}$ if and only if $\gamma(Z)=\tilde{\gamma}(Z)$ for Z in some neighborhood of X, however small. The resulting equivalence classes Γ will be called the *local isomorphisms* of X onto $\bar{X}$. If Γ is such a local isomorphism then the local configurations $\bar{\Phi}$ of $\bar{X}$ are of the form

$$\bar{\Phi}=\Phi\circ\overset{-1}{\Gamma}, \tag{14.3}$$

where Φ is a local configuration of X.

Definition 1. *A local isomorphism Γ of a particle X onto a particle $\bar{X}$ will be called a* material isomorphism *of X onto $\bar{X}$ provided the functionals $\mathfrak{F}$ and $\bar{\mathfrak{F}}$ of X and $\bar{X}$ are related by*

$$\mathfrak{F}(\Theta^*)=\bar{\mathfrak{F}}(\Theta^*\circ\overset{-1}{\Gamma}) \tag{14.4}$$

for all kinematical histories $\Theta^*\in\mathscr{C}_X^*$.

We shall say that two particles consist of the same material if they are materially isomorphic to each other.

15. Constitutive functionals

Assume that a local motion Θ and a local reference configuration Φ of a particle X are given. Let Θ^t and Ω^t be the corresponding kinematical history and deformation history, respectively, as defined in Section 7. By (4.4) they are related by

$$\Theta^t=\Omega^t\circ\Phi. \tag{15.1}$$

Hence the general constitutive equation (13.6) may be rewritten in the form

$$S(t)=\mathfrak{F}(\Omega^t\circ\Phi). \tag{15.2}$$

Definition 2. *A functional $\mathfrak{G}$ whose domain is the set $\mathscr{D}^*$ of all deformation histories and whose values are symmetric tensors is called a* constitutive functional *provided it has the following property: For all deformation histories $\Omega^*\in\mathscr{D}^*$ and all orthogonal tensor functions $Q^*\in\mathscr{O}^*$ the relation*

$$Q_0\,\mathfrak{G}(\Omega^*)\,Q_0^T=\mathfrak{G}(Q^*\circ\Omega^*),\qquad Q_0=Q^*(0), \tag{15.3}$$

holds.

It is clear that the functional $\mathfrak{G}_\Phi$ defined by

$$\mathfrak{G}_\Phi(\Omega^*)=\mathfrak{G}(\Omega^*;\Phi)=\mathfrak{F}(\Omega^*\circ\Phi) \tag{15.4}$$

is a constitutive functional, because the property (15.3) for $\mathfrak{G}=\mathfrak{G}_\Phi$ is equivalent to the property (13.7) for $\mathfrak{F}$. Hence the most general constitutive equation may be formulated in this way:

Theorem 1. *Given a particle X and a local reference configuration Φ of X, there is a constitutive functional $\mathfrak{G}_\Phi$ such that, for any local process $\{\Theta, S\}$ of X, the stress S is related to the local motion Θ by*

$$S(t) = \mathfrak{G}_\Phi(\Omega^t) = \mathfrak{G}(\Omega^t; \Phi), \tag{15.5}$$

where Ω^t is the deformation history of X defined in Section 7. The constitutive functionals $\mathfrak{G}_\Phi$ and $\mathfrak{G}_{\hat{\Phi}}$ corresponding to the local reference configurations Φ and $\hat{\Phi}$ are related, for all deformation histories $\Omega^* \in \mathscr{D}^*$, by

$$\mathfrak{G}(\Omega^* \circ \Delta; \Phi) = \mathfrak{G}(\Omega^*; \hat{\Phi}), \tag{15.6}$$

where $\Delta = \hat{\Phi} \circ \overset{-1}{\Phi}$ is the deformation from Φ to $\hat{\Phi}$.

The relation (15.6) is an immediate consequence of (15.4). We call $\mathfrak{G}_\Phi$ the constitutive functional of the particle X relative to the local reference configuration Φ.

Since the nature of the domain $\mathscr{D}^*$ of constitutive functionals is not related to particles, as was the case with the functionals $\mathfrak{F}$ in (13.6), they characterize materials in a manner that is independent of the particular particle. The following theorem is a consequence of Definition 1 and (15.4):

Theorem 2. *Two particles X and $\overline{X}$ are materially isomorphic,* i.e., *consist of the same material, if and only if there is a local configuration Φ of X and a local isomorphism Γ of X onto $\overline{X}$ such that*

$$\mathfrak{G}_\Phi(\Omega^*) = \mathfrak{G}_{\overline{\Phi}}(\Omega^*), \quad \overline{\Phi} = \Phi \circ \overset{-1}{\Gamma}, \tag{15.7}$$

for all $\Omega^* \in \mathscr{D}^*$.

16. The local isotropy group

Let Γ be a *material automorphism* of X, *i.e.*, a material isomorphism of X onto itself. By Theorem 2, there is a local configuration Φ such that

$$\mathfrak{G}(\Omega^*; \Phi) = \mathfrak{G}(\Omega^*; \hat{\Phi}), \quad \hat{\Phi} = \Phi \circ \overset{-1}{\Gamma}. \tag{16.1}$$

The material automorphism Γ and the deformation $\Delta = \hat{\Phi} \circ \overset{-1}{\Phi}$ from Φ to $\hat{\Phi}$ are related by

$$\Delta = \Phi \circ \overset{-1}{\Gamma} \circ \overset{-1}{\Phi}, \quad \Gamma = \overset{-1}{\Phi} \circ \overset{-1}{\Delta} \circ \Phi. \tag{16.2}$$

Since Γ, an isomorphism, preserves the mass distribution, it follows that Δ must be an isochoric deformation.

In addition to (16.1) we have the relation (15.6) between the two functionals $\mathfrak{G}_\Phi$ and $\mathfrak{G}_{\hat{\Phi}}$. Combining these two relations we obtain

$$\mathfrak{G}_\Phi(\Omega^* \circ \Delta) = \mathfrak{G}_\Phi(\Omega^*). \tag{16.3}$$

Conversely, if (16.3) holds for an isochoric deformation $\Delta \in \mathscr{I}$, then $\Gamma = \overset{-1}{\Phi} \circ \overset{-1}{\Delta} \circ \Phi$ can easily be seen to be a material automorphism of X.

Definition 3. *Let $\mathfrak{G}$ be a constitutive functional. The group $\mathcal{G}^l$ of all isochoric deformations $\Delta \in \mathcal{I}$ with the property that*

$$\mathfrak{G}(\Omega^* \circ \Delta) = \mathfrak{G}(\Omega^*) \tag{16.4}$$

holds for all $\Omega^* \in \mathcal{D}^*$ is called the local isotropy group *of* $\mathfrak{G}$.

$\mathcal{G}^l$ is a subgroup of the group $\mathcal{I}$ of all isochoric deformations. The material automorphisms of a particle also form a group. The relations (16.2) establish an isomorphism of this group with the local isotropy group $\mathcal{G}^l_\Phi$ of $\mathfrak{G}_\Phi$. The local isotropy groups $\mathcal{G}^l_\Phi$ and $\mathcal{G}^l_{\hat{\Phi}}$ corresponding to different local configurations Φ and $\hat{\Phi}$ will in general be different. But they are all isomorphic to each other because they are isomorphic to the group of material automorphisms of X. In fact, it is not hard to see that $\mathcal{G}^l_{\hat{\Phi}}$ is conjugate to $\mathcal{G}^l_\Phi$:

$$\mathcal{G}^l_{\hat{\Phi}} = \Delta \circ \mathcal{G}^l_\Phi \circ \Delta^{-1}, \qquad \Delta = \hat{\Phi} \circ \overset{-1}{\Phi}. \tag{16.5}$$

III. Simple materials

17. Simple constitutive functionals

It may happen that the values $\mathfrak{G}(\Omega^*)$ of a constitutive functional $\mathfrak{G}$ are not affected if $\Omega^*(s)$, for each $s \leqq 0$, is changed by a small term of order one. By (3.6), $\Omega^*(s)$ differs from $\nabla\Omega^*(s)$ only by such a small term.

Definition 4. *A constitutive functional $\mathfrak{G}$ is said to be* simple *if*

$$\mathfrak{G}(\Omega^*) = \mathfrak{G}(\nabla\Omega^*) \tag{17.1}$$

for all deformation histories $\Omega^* \in \mathcal{D}^*$.

The condition (17.1) is equivalent to the following: If $\nabla\Omega^*(s) = \nabla\tilde{\Omega}^*(s)$ for all $s \leqq 0$, then

$$\mathfrak{G}(\Omega^*) = \mathfrak{G}(\tilde{\Omega}^*). \tag{17.2}$$

The values of a simple constitutive functional $\mathfrak{G}$ are determined for all $\Omega^* \in \mathcal{D}^*$ if they are known for the histories of linear transformations $F^* \in \mathcal{L}^*$. In other words, $\mathfrak{G}$ is determined by its restriction to $\mathcal{L}^*$.

From now on we assume that $\mathfrak{G}$ is simple, and we use the same symbol $\mathfrak{G}$ for its restriction to $\mathcal{L}^*$. By (15.3) $\mathfrak{G}$ satisfies the relation

$$Q_0 \mathfrak{G}(F^*) Q_0^T = \mathfrak{G}(Q^* F^*), \qquad Q_0 = Q^*(0), \tag{17.3}$$

for all $F^* \in \mathcal{L}^*$ and all $Q^* \in \mathcal{O}^*$.

Let $G \in \mathcal{L}$ be an arbitrary regular linear transformation with the polar decomposition

$$G = PT, \qquad P \in \mathcal{O}, \qquad T \in \mathcal{S}_+. \tag{17.4}$$

For any $F^* \in \mathcal{L}^*$ we define

$$\hat{F}^* = F^* G^{-1} = \hat{R}^* \hat{U}^*, \qquad \hat{R}^* \in \mathcal{O}^*, \qquad \hat{U}^* \in \mathcal{S}_+^*, \tag{17.5}$$

where $\hat{R}^*$ and $\hat{U}^*$ are determined by the polar decomposition of $\hat{F}^*$. We then have

$$F^* = \hat{F}^* G = \hat{R}^* \hat{U}^* P T = (\hat{R}^* P)(\hat{P}^T U^* P) T.$$

Substituting $\hat{R}^* P$ for Q^* and $(P^T \hat{U}^* P)\, T$ for F^* in (17.3), we see that

$$\mathfrak{G}(F^*) = \hat{R}_0 P\, \mathfrak{G}\big((P^T \hat{U}^* P)\, T\big)\, P^T \hat{R}_0^T, \qquad \hat{R}_0 = \hat{R}^*(0). \tag{17.6}$$

We now define a functional $\mathfrak{R}$ with domain $\mathscr{S}_+^* \times \mathscr{S}_+$ by

$$\mathfrak{G}(U^* T) = \mathfrak{R}(U^*; T) \tag{17.7}$$

for all $U^* \in \mathscr{S}_+^*$ and $T \in \mathscr{S}_+$. Equation (17.6) then shows that

$$\mathfrak{G}(F^*) = \hat{R}_0 P\, \mathfrak{R}(P^T \hat{U}^* P; T)\, P^T \hat{R}_0^T. \tag{17.8}$$

The functional $\mathfrak{R}$ has the property that

$$\mathfrak{R}(U^*; T) = \mathfrak{R}(\tilde{U}^*; \tilde{T}) \quad \text{if} \quad U^* T = \tilde{U}^* \tilde{T}. \tag{17.9}$$

Conversely, if $\mathfrak{R}$ is any functional with domain $\mathscr{S}_+^* \times \mathscr{S}_+$ and with the property (17.9), then (17.8) defines a simple constitutive functional $\mathfrak{G}$. The variables on the right side of (17.8) are defined in terms of F^* and an arbitrary $G \in \mathscr{L}$. For the special choice $G = I$ we get

$$\mathfrak{G}(F^*) = R_0\, \mathfrak{R}(U^*; I)\, R_0^T, \qquad R_0 = R^*(0), \tag{17.10}$$

where U^* and R^* are defined by the polar decomposition

$$F^* = R^* U^*, \qquad R^* \in \mathscr{O}^*, \qquad U^* \in \mathscr{S}_+^*. \tag{17.11}$$

In the special case when $G = F_0 = F^*(0)$ (17.8) becomes

$$\mathfrak{G}(F^*) = R_0\, \mathfrak{R}(R_0^T U_*^* R_0; U_0)\, R_0^T, \tag{17.12}$$

where $R_0 = R^*(0)$, $U_0 = U^*(0)$, and where $U_*^* \in \mathscr{S}_+^*$ is determined by the polar decomposition

$$F^* F_0^{-1} = R_*^* U_*^*, \qquad R_*^* \in \mathscr{O}^*, \qquad U_*^* \in \mathscr{S}_+^*. \tag{17.13}$$

We define a functional $\mathfrak{R}_1$ with domain $\mathscr{S}_+^*$ by

$$\mathfrak{R}_1(U^*) = \mathfrak{R}(U^*; I) \quad \text{for} \quad U^* \in \mathscr{S}_+^*. \tag{17.14}$$

Then (17.10) takes the form

$$\mathfrak{G}(F^*) = R_0\, \mathfrak{R}_1(U^*)\, R_0^T. \tag{17.15}$$

Conversely, if $\mathfrak{R}_1$ is an arbitrary functional with domain $\mathscr{S}_+^*$, then (17.15) defines a simple constitutive functional.

We also define a functional $\mathfrak{R}_2$ as the restriction of $\mathfrak{R}$ obtained by allowing for its first variable only those functions $U_*^* \in \mathscr{S}_+^*$ whose value for $s = 0$ is the identity,

$$U_*^*(0) = I. \tag{17.16}$$

We denote the set of all U_*^* with the property (17.16) by $\mathscr{S}_{+*}^*$. The domain of $\mathfrak{R}_2$ is then $\mathscr{S}_{+*}^* \times \mathscr{S}_+$, and it is identical with $\mathfrak{R}$ in this domain. The function

U_*^* defined by (17.13) has the property (17.16), and the same is true for $R_0^T U_*^* R_0$. Hence (17.12) has the form

$$\mathfrak{G}(F^*) = R_0 \mathfrak{K}_2(R_0^T U_*^* R_0; U_0) R_0^T. \tag{17.17}$$

Conversely, if $\mathfrak{K}_2$ is an arbitrary functional with domain $\mathscr{S}_{+*}^* \times \mathscr{S}_+$, then (17.17) defines a simple constitutive functional $\mathfrak{G}$.

18. Simple materials

We say that the material at a particle X is *simple* or, briefly, that X is simple if the constitutive functional $\mathfrak{G}_\Phi$ of X, for some local configuration Φ of X, is simple. We assume from now on that this is the case. Let $\hat{\Phi}$ be another local configuration of X and let $\Delta = \hat{\Phi} \circ \overset{-1}{\Phi}$ be the deformation from Φ to $\hat{\Phi}$. For every deformation history $\Omega^* \in \mathscr{D}^*$, by (3.7) and the chain rule (3.4), we have

$$\nabla(\Omega^* \circ \Delta) = \nabla[(\nabla\Omega^*) \circ \Delta]. \tag{18.1}$$

It follows from (17.2) that

$$\mathfrak{G}(\Omega^* \circ \Delta; \Phi) = \mathfrak{G}(\nabla\Omega^* \circ \Delta; \Phi),$$

and hence from (15.6) that

$$\mathfrak{G}(\Omega^*; \hat{\Phi}) = \mathfrak{G}(\nabla\Omega^*; \hat{\Phi}),$$

which shows that $\mathfrak{G}_{\hat{\Phi}}$ is also simple. Therefore, if the material at a particle is simple then the constitutive functional $\mathfrak{G}_\Phi$ is simple for all local reference configurations Φ.

It follows from (15.6) that the functionals $\mathfrak{G}_\Phi$ and $\mathfrak{G}_{\hat{\Phi}}$ corresponding to two local configurations Φ and $\hat{\Phi}$ are related by

$$\mathfrak{G}(F^* G; \Phi) = \mathfrak{G}(F^*; \hat{\Phi}), \tag{18.2}$$

where $G = G(\hat{\Phi}, \Phi)$ is the deformation gradient from Φ to $\hat{\Phi}$ defined by (5.1). It follows from (18.2) that

$$\mathfrak{G}(F^*; \Phi) = \mathfrak{G}(F^*; \hat{\Phi}) \quad \text{if} \quad G(\hat{\Phi}, \Phi) = I.$$

This means, according to (5.8), that $\mathfrak{G}(F^*; \Phi)$ depends only on the gradient $M = \nabla\Phi$ of the local configuration Φ. We can therefore define

$$\mathfrak{G}_M(F^*) = \mathfrak{G}(F^*; M) = \mathfrak{G}(F^*; \Phi) \quad \text{if} \quad M = \nabla\Phi. \tag{18.3}$$

The following theorem is a corollary of Theorem 1.

Theorem 3. *Let X be a simple particle and let M be a configuration gradient of X. Then there is a simple constitutive functional $\mathfrak{G}_M$ such that, for any local process $\{\Theta, S\}$, the stress is related to the local motion by the constitutive equation*

$$S(t) = \mathfrak{G}_M(F^t) = \mathfrak{G}(F^t; M), \tag{18.4}$$

where $F^t \in \mathscr{L}^*$ is the history of the deformation gradient of the local motion Θ relative to M as a reference. The functionals $\mathfrak{G}_M$ and $\mathfrak{G}_{\hat{M}}$ corresponding to two configuration *gradients are related by*

$$\mathfrak{G}(F^* G; M) = \mathfrak{G}(F^*; \hat{M}), \quad G = \hat{M} M^{-1}, \tag{18.5}$$

for all $F^* \in \mathcal{L}^*$.

The results of the previous section enable us to put the constitutive equation into various other forms:

Theorem 4. *The constitutive equation for a simple particle X may be written, with reference to a fixed configuration gradient M of X, in one of the following four forms:*

(a) There is a functional $\mathfrak{K}_1$ *with domain* $\mathcal{S}_+^*$ *such that the stress is given by*

$$S(t) = R(t)\, \mathfrak{K}_1(U^t)\, R(t)^T, \tag{18.6}$$

where $R(t)$ *is the rotation tensor and* U^t *the history of the right strain tensor relative to M as a reference.*

(b) There is a functional $\mathfrak{K}$ *with domain* $\mathcal{S}_+^* \times \mathcal{S}$ *and with the property* (17.9) *such that, for any configuration gradient* $\hat{M}$ *as a reference, the stress is given by*

$$S(t) = \hat{R}(t)\, P\, \mathfrak{K}(P^T \hat{U}^t P; T_r)\, P^T \hat{R}(t)^T, \tag{18.7}$$

where $\hat{R}(t)$ *is the rotation tensor and* $\hat{U}^t$ *the history of the right strain tensor relative to* $\hat{M}$ *as a reference, and where P is the rotation tensor and* T_r *the right strain tensor of the deformation from M to* $\hat{M}$, $\hat{M} M^{-1} = P\, T$.

(c) In the special case when $\hat{M}$ *is the gradient of the configuration at some time* t_0 *the equation* (18.7) *takes the form*

$$S(t) = R_{t_0}(t)\, R(t_0)\, \mathfrak{K}\left(R(t_0)^T\, U_{t_0}^t\, R(t_0);\, U(t_0)\right) R(t_0)^T R_{t_0}(t)^T, \tag{18.8}$$

where the rotation tensors R and R_{t_0} *and the right strain tensors U and* U_{t_0} *are taken relative to M and the configuration at time* t_0 *as a reference, respectively.*

(d) There is a functional $\mathfrak{K}_2$ *with domain* $\mathcal{S}_{+*}^* \times \mathcal{S}_+$ *such that the stress is given by*

$$S(t) = R(t)\, \mathfrak{K}_2\left(R^T(t)\, U_t^t\, R(t);\, U(t)\right) R(t)^T, \tag{18.9}$$

where R, U, and U_t *are defined as before.*

The forms (18.6) and (18.9) have the advantage that the functionals $\mathfrak{K}_1$ and $\mathfrak{K}_2$ are not subject to any restrictive condition.

19. The isotropy group

If $\mathfrak{G}$ is a simple constitutive functional, then it follows from (17.2) that $\mathfrak{G}(\Omega^*) = \mathfrak{G}(\Omega^* \circ \Delta)$ holds for all null-deformations $\Delta \in \mathcal{N}$, *i.e.*, whenever $\nabla \Delta = I$. Hence, by definition 3, the local isotropy group $\mathcal{G}^l$ of $\mathfrak{G}$ contains the group $\mathcal{N}$ of all null-deformations as a normal subgroup. The quotient group $\mathcal{G}^l / \mathcal{N}$ is isomorphic to a group of unimodular transformations.

Definition 5. *The group* $\mathcal{G}$ *of all unimodular transformations H with the property that*

$$\mathfrak{G}(F^* H) = \mathfrak{G}(F^*) \tag{19.1}$$

holds for all $F^* \in \mathcal{L}^*$ *is called the* isotropy group *of the simple constitutive functional* $\mathfrak{G}$.

$\mathcal{G}$ is a subgroup of the group $\mathcal{U}$ of all unimodular transformations and it is isomorphic to $\mathcal{G}^l/\mathcal{N}$.

Theorem 5. *An orthogonal transformation* Q *is an element of the isotropy group* $\mathcal{G}$ *of a constitutive functional* $\mathfrak{G}$ *if and only if one of the following conditions is satisfied:*

(a) For all $F^* \in \mathcal{L}^*$

$$Q\,\mathfrak{G}(F^*)\,Q^T = \mathfrak{G}(Q F^* Q^T). \tag{19.2}$$

(b) For all $U^* \in \mathcal{S}_+^*$ *and all* $T \in \mathcal{S}_+$

$$Q\,\mathfrak{R}(U^*;T)\,Q^T = \mathfrak{R}(Q\,U^*\,Q^T; Q\,T\,Q^T), \tag{19.3}$$

where $\mathfrak{R}$ *is defined by* (17.7).

(c) For all $U^* \in \mathcal{S}_+^*$

$$Q\,\mathfrak{R}_1(U^*)\,Q^T = \mathfrak{R}_1(Q\,U^*\,Q^T), \tag{19.4}$$

where $\mathfrak{R}_1$ *is defined by* (17.14).

(d) For all $U_*^* \in \mathcal{S}_{+*}^*$ *and all* $T \in \mathcal{S}_+$

$$Q\,\mathfrak{R}_2(U_*^*;T)\,Q^T = \mathfrak{R}_2(Q\,U_*^*\,Q^T; Q\,T\,Q^T) \tag{19.5}$$

where $\mathfrak{R}_2$ *is defined in Section* 17.

Proof. If we substitute the constant tensor Q for the tensor function Q^* in (17.3), and QF^* for F^* and Q for H in (19.1), we see that (19.2) holds if and only if $Q \in \mathcal{G}$. The equivalence of the condition (a) with (b), (c), and (d) follows directly from the definitions of $\mathfrak{R}$, $\mathfrak{R}_1$, and $\mathfrak{R}_2$.

If $\mathcal{G}$ contains the full orthogonal group $\mathcal{O}$, then (19.2), (19.3), (19.4), and (19.5) are valid for all orthogonal Q.

The isotropy group $\mathcal{G}_M$ of the constitutive functional $\mathfrak{G}_M$ of a particle will, in general, depend on the choice of the configuration gradient M. But, as in the case of local isotropy, the groups $\mathcal{G}_M$ and $\mathcal{G}_{\hat{M}}$ corresponding to two configuration gradients M and $\hat{M}$ are conjugate:

$$\mathcal{G}_{\hat{M}} = G\,\mathcal{G}_M\,G^{-1}, \qquad G = \hat{M}\,M^{-1}. \tag{19.6}$$

It follows from (18.5) that $H \in \mathcal{G}_M$ if and only if

$$\mathfrak{G}(F^*; M) = \mathfrak{G}(F^*; H\,M) \tag{19.7}$$

for all $F^* \in \mathcal{L}^*$.

20. Isotropic and anisotropic solids

We say that a constitutive functional $\mathfrak{G}$ defines a *solid* if its isotropy group is a subgroup of the orthogonal group, *i.e.*, if $\mathcal{G} < \mathcal{O}$. A particle X is said to be a solid particle if there is a configuration gradient M of X such that $\mathfrak{G}_M$ defines a solid. A solid is called an *isotropic solid* if the isotropy group of its defining functional is the full orthogonal group, *i.e.*, if $\mathscr{G}=\mathcal{O}$. Let X be an isotropic solid particle. Any configuration gradient M such that $\mathscr{G}_M=\mathcal{O}$ is then called an *undistorted state* of X.

The following theorem follows immediately from Theorems 4 and 5.

Theorem 6. *Let X be an isotropic solid particle. Its constitutive equation may be written, with reference to an undistorted state M of X, in one of the following forms:*

(20.1) $$S(t)=\mathfrak{K}_1\big(R(t)\,U^t R^T(t)\big),$$

(20.2) $$S(t)=\hat{R}(t)\,\mathfrak{K}(\hat{U}^t;T_l)\,\hat{R}(t)^T,$$

(20.3) $$S(t)=R_{t_0}(t)\,\mathfrak{K}\big(U_{t_0}^t;V(t_0)\big)\,R_{t_0}(t),$$

(20.4) $$S(t)=\mathfrak{K}_2\big(U_t^t;V(t)\big).$$

The notation of Theorem 4 applies here. In addition, $V=RUR^T$ is the left strain tensor relative to M as a reference and T_l is the left strain tensor of the deformation from M to $\hat{M}$. The functionals $\mathfrak{K}$, $\mathfrak{K}_1$ and $\mathfrak{K}_2$ satisfy the conditions (b), (c), and (d) of Theorem 5 for all orthogonal transformations Q.

A solid is called *anisotropic* if the isotropy group of its defining functional is a proper subgroup of the orthogonal group, *i.e.*, if $\mathscr{G}<\mathcal{O}$ and $\mathscr{G}\neq\mathcal{O}$. Material symmetries in anisotropic solids, such as *orthotropy*, *transverse isotropy*, and the various types of *crystal symmetry* are defined according to the special nature of the isotropy group $\mathscr{G}$.

21. Fluids

We say that a constitutive functional defines a *fluid* if its isotropy group $\mathscr{G}$ is the full unimodular group $\mathscr{U}$, *i.e.*, if $\mathscr{G}=\mathscr{U}$. A particle X is said to be a fluid particle if, for some configuration gradient M of X, the corresponding constitutive functional defines a fluid, *i.e.*, if $\mathscr{G}_M=\mathscr{U}$. Let $\hat{M}$ be any other configuration gradient. By (19.6) $\mathscr{G}_{\hat{M}}$ is conjugate to $\mathscr{G}_M=\mathscr{U}$. But $\mathscr{U}$ is a normal subgroup of $\mathscr{L}$ and hence coincides with all its conjugates. It follows that, if X is a fluid particle, then $\mathscr{G}_M=\mathscr{U}$ for all configuration gradients M of X.

For a fluid it follows from (19.7) that

(21.1) $$\mathfrak{G}(F^*;M)=\mathfrak{G}(F^*;\hat{M})$$

whenever $H=\hat{M}\overset{-1}{M}$ is unimodular. But, by (5.17), this is the case if and only if the densities ϱ_M and $\varrho_{\hat{M}}$ coincide. If follows that the value $\mathfrak{G}(F^*;M)$ can depend only on the density ϱ_M. Therefore, we can define a functional $\mathfrak{H}$ with domain $\mathscr{L}^*\times\mathscr{R}_+$ ($\mathscr{R}_+=$ set of positive real numbers) such that

(21.2) $$\mathfrak{G}(F^*;M)=\mathfrak{H}(F^*;\varrho_M)$$

for all $F^*\in\mathscr{L}^*$ and all configuration gradients M of X. Let $F^*=R^*U^*$ be the polar decomposition of F^*. Substituting R^* for Q^* and U^* for F^* in (17.3) and using (21.2) we see that

(21.3) $$\mathfrak{G}(F^*;M)=R_0\,\mathfrak{H}(U^*;\varrho_M)\,R_0^T,\qquad R_0=R^*(0),$$

which shows that $\mathfrak{H}$ is determined by its restriction to $\mathscr{S}_+^* \times \mathscr{R}_+$. We use the same symbol $\mathfrak{H}$ for this restriction. Since $\mathscr{G}_M = \mathscr{U}$ contains the orthogonal group it follows from (19.2) that $\mathfrak{H}$ satisfies the relation

$$Q\,\mathfrak{H}(U^*; d)\,Q^T = \mathfrak{H}(Q\,U^*\,Q^T; d) \tag{21.4}$$

for all $U^* \in \mathscr{S}_+^*$, all $Q \in \mathscr{O}$, and all $d > 0$.

Theorem 7. *The constitutive equation for a fluid particle X may be written in one of the following forms:*

(a) There is a functional $\mathfrak{H}$ with domain $\mathscr{S}_+^* \times \mathscr{R}_+$ and with the property (21.4) *such that the stress is given by*

$$S(t) = R(t)\,\mathfrak{H}(U^t; \varrho_M)\,R(t)^T, \tag{21.5}$$

where $R(t)$ is the rotation tensor and U^t the history of the right strain tensor relative to an arbitrary configuration gradient M with mass density ϱ_M.

(b) In the special case when M is the configuration gradient at some time t_0, the equation (21.5) *takes the form*

$$S(t) = R_{t_0}(t)\,\mathfrak{H}\big(U_{t_0}^t; \varrho(t_0)\big)\,R_{t_0}(t)^T. \tag{21.6}$$

(c) There is a functional $\mathfrak{H}_1$ with domain $\mathscr{S}_{+*}^* \times \mathscr{R}_+$ and with the property that

$$Q\,\mathfrak{H}_1(U_*^*; d)\,Q^T = \mathfrak{H}_1(Q\,U_*^*\,Q^T; d) \tag{21.7}$$

for all $U_*^* \in \mathscr{S}_{+*}^*$, all $Q \in \mathscr{O}$, and all $d > 0$, such that the stress is given by

$$S(t) = \mathfrak{H}_1\big(U_t^t; \varrho(t)\big). \tag{21.8}$$

Proof. The part (*a*) and its special case (*b*) follow from (21.3) and Theorem 3. The part (*c*) follows from (*b*) by choosing $t_0 = t$ and by defining $\mathfrak{H}_1$ to be the restriction of $\mathfrak{H}$ to $\mathscr{S}_{+*}^* \times \mathscr{R}_+$.

We note that an arbitrary functional $\mathfrak{H}_1$ with domain $\mathscr{S}_{+*}^* \times \mathscr{R}_+$ and with the property (21.7) may define a fluid. The constitutive equation (21.8) is intrinsic in the sense that it does not depend on the choice of a reference configuration. The constitutive equation of a solid cannot be put into such an intrinsic form.

22. Constitutive equations involving the Cauchy-Green tensors

As we pointed out in Section 6, it is often better to use the Cauchy-Green tensors C and B instead of the strain tensors U and V. In order to do so, we define new functionals $\overline{\mathfrak{R}}$, $\overline{\mathfrak{R}}_1$, $\overline{\mathfrak{R}}_2$, $\overline{\mathfrak{H}}$ and $\overline{\mathfrak{H}}_1$, of the same nature as the corresponding functionals without the superposed bars defined in Sections 17 and 20, by the following formulas:

$$\overline{\mathfrak{R}}(U^{*2}; T^2) = U_0\,\mathfrak{R}(U^*; T)\,U_0, \tag{22.1}$$

$$\overline{\mathfrak{R}}_1(U^{*2}) = U_0\,\mathfrak{R}_1(U^*)\,U_0, \tag{22.2}$$

$$\overline{\mathfrak{R}}_2(U_*^{*2}; T^2) = \mathfrak{R}_2(U_*^*; T), \tag{22.3}$$

$$\overline{\mathfrak{H}}(U^{*2}; d) = U_0\,\mathfrak{H}(U^*; d)\,U_0, \tag{22.4}$$

$$\overline{\mathfrak{H}}_1(U_*^{*2}; d) = \mathfrak{H}_1(U_*^*; d), \tag{22.5}$$

valid for all $U^* \in \mathscr{S}_+^*$ $(U_0 = U^*(0))$, all $U_*^* \in \mathscr{S}_{+*}^*$ $(U_*^*(0) = I)$, all $T \in \mathscr{S}_+$, and all $d > 0$.

It is not hard to see that the general constitutive equations (18.6), (18.8), and (18.9) of Theorem 4 for simple materials then take the form

$$F(t)^T S(t) F(t) = \overline{\mathfrak{K}}_1(C^t), \text{*} \tag{22.6}$$

$$F_{t_0}(t)^T S(t) F_{t_0}(t) = R(t_0)\, \overline{\mathfrak{K}}\big(R(t_0)^T C_{t_0}^t R(t_0); C(t_0)\big)\, R(t_0)^T, \tag{22.7}$$

$$S(t) = R(t)\, \overline{\mathfrak{K}}_2\big(R(t)^T C_t^t R(t); C(t)\big)\, R(t)^T. \tag{22.8}$$

In the case of isotropic solids, the functionals $\overline{\mathfrak{K}}$, $\overline{\mathfrak{K}}_1$, and $\overline{\mathfrak{K}}_2$ satisfy, for all orthogonal Q, the functional relations obtained from (19.3), (19.4), and (19.5) by superposing bars. Moreover, the simplified constitutive equations (20.3) and (20.4) for isotropic solids take the form

$$F_{t_0}(t)^T S(t) F_{t_0}(t) = \overline{\mathfrak{K}}\big(C_{t_0}^t; B(t_0)\big), \tag{22.9}$$

$$S(t) = \overline{\mathfrak{K}}_2\big(C_t^t; B(t)\big). \tag{22.10}$$

The functionals $\overline{\mathfrak{H}}$ and $\overline{\mathfrak{H}}_1$ have the same properties (21.4) and (21.7) as the corresponding functionals without the superposed bars. The constitutive equations (21.6) and (21.8) for fluids take the form

$$F_{t_0}(t)^T S(t) F_{t_0}(t) = \overline{\mathfrak{H}}\big(C_{t_0}^t; \varrho(t_0)\big), \tag{22.11}$$

$$S(t) = \overline{\mathfrak{H}}_1\big(C_t^t; \varrho(t)\big). \tag{22.12}$$

IV. Special classes of materials

23. *Materials of the differential type*

The value $\mathfrak{G}(F^*)$ of a simple constitutive functional is determined by the values $F^*(s)$ of the tensor function F^* for $s \leq 0$. It may happen that $\mathfrak{G}(F^*)$ depends only on the values $F^*(s)$ for s very near to zero. If F^* has sufficiently many continuous derivatives then $F^*(s)$ may be approximated, for small values of s, by its Taylor expansion up to some order n. This Taylor expansion is determined by the value of F^* and its derivatives up to the order n at $s = 0$, *i.e.*, by

$$F_0 = F^*(0), \quad \dot{F}_0 = \dot{F}^*(0), \ldots, \overset{(n)}{F}_0 = \overset{(n)}{F}{}^*(0). \tag{23.1}$$

Definition 6. *A simple constitutive functional $\mathfrak{G}$ is said to be of the differential type if*

$$\mathfrak{G}(F^*) = \mathfrak{G}(\tilde{F}^*) \tag{23.2}$$

whenever

$$\overset{(k)}{F}{}^*(0) = \overset{(k)}{\tilde{F}}{}^*(0), \qquad k = 0, 1, \ldots, n. \tag{23.3}$$

We have seen that every simple constitutive functional $\mathfrak{G}$ has a representation of the form (17.17) in terms of a functional $\mathfrak{K}_2$. Since $\mathfrak{K}_2$ is just a restriction of $\mathfrak{K}$

* This form, in other notation, has been proposed independently by GREEN & RIVLIN [4].

which is defined by (17.7) in terms of $\mathfrak{G}$, we have

$$\mathfrak{R}_2(U_*^*; T) = \mathfrak{G}(U_*^* T) \tag{23.4}$$

for all $U_*^* \in \mathcal{S}_{+*}^*$ and all $T \in \mathcal{S}_+$. If $\mathfrak{G}$ is of the differential type then its value depends only on the values at $s=0$ of its argument and its first n derivatives. Hence, by (23.4), the value $\mathfrak{R}_2(U_*^*; T)$ depends only on the first n derivatives of U_*^* at $s=0$, since $U_*^*(0)=I$. It follows that there is a function $\mathfrak{k}$ of $n+1$ symmetric tensor variables such that

$$\mathfrak{R}_2(U_*^*; T) = \mathfrak{k}\big(\dot{U}_*^*(0), \ddot{U}_*^*(0), \ldots, \overset{(n)}{U_*^*}(0); T\big) \tag{23.5}$$

for all $U_*^* \in \mathcal{S}_{+*}^*$ and all $T \in \mathcal{S}_+$. In the case when $U_*^* = U_t^t$ is the history up to time t of the right strain tensor relative to the configuration at time t of a local motion, the derivative $\overset{(k)}{U_t^t}(0)$ coincides by (8.8) with the kth rate of strain $D_k(t)$. For simplicity we use the notation

$$\mathfrak{k}(D_1, D_2, \ldots, D_n; T) = \mathfrak{k}(D_k; T). \tag{23.6}$$

We say that a particle X is of differential type if its constitutive functional G_M, for some configuration gradient M, is of the differential type. It is not hard to see that $G_{\hat{M}}$, for any other configuration gradient $\hat{M}$, is then also of the differential type. The following theorem is a consequence of (18.9) and the remarks made above:

Theorem 8. *The constitutive equation of a particle X of differential type may be written, with reference to a configuration gradient M of X, in the following form:*

There is a symmetric-tensor-valued function $\mathfrak{k}$ of $n+1$ symmetric tensor variables such that the stress is given by

$$S(t) = R(t)\, \mathfrak{k}\big(R^T(t)\, D_k(t)\, R(t); U(t)\big)\, R(t)^T, \tag{23.7}$$

where $D_k(t)$ is the kth rate of strain and where the rotation tensor $R(t)$ and the right strain tensor $U(t)$ are taken relative to M as a reference.

In a material of the differential type the stress depends only on the immediate past of the motion and not on its course at times long ago.

Since the rates of strain D_k can be expressed as polynomials in the acceleration gradients E_k and E_k^T as shown in Section 9, it follows that the constitutive equation (23.7) is a relation involving the displacement gradient $G = RU$, the acceleration gradients, and the stress.

In the case of *isotropic solids*, it follows from Theorem 6 and (23.5) that $\mathfrak{k}$ must satisfy the relation

$$Q\, \mathfrak{k}(D_k; T)\, Q^T = \mathfrak{k}(Q D_k Q^T; Q T Q^T) \tag{23.8}$$

for all $D_k \in \mathcal{S}$, $k = 1, 2, \ldots, n$, all $Q \in \mathcal{O}$, and all $T \in \mathcal{S}_+$. A function with this property is called an *isotropic tensor function*. By (20.4) the constitutive equation (23.7) reduces for isotropic solids to

$$S(t) = \mathfrak{k}\big(D_k(t); V(t)\big). \tag{23.9}$$

The constitutive equation for a *fluid* of the differential type has the form

(23.10) $$S(t) = \mathfrak{h}\left(D_k(t);\, \varrho(t)\right)$$

where $\mathfrak{h}$ is an isotropic function of n symmetric tensor variables and one positive scalar variable, *i.e.*, it satisfies the relation

(23.11) $$Q\,\mathfrak{h}(D_k;\, d)\, Q^T = \mathfrak{h}(Q\, D_k\, Q^T;\, d)$$

for all $D_k \in \mathscr{S}$, $k = 1, 2, \ldots, n$, all $d > 0$, and all $Q \in \mathcal{O}$. This is an immediate consequence of Theorem 7, (21.8).

If we use the alternate forms (22.8), (22.10), and (22.12) of the general constitutive equations we arrive at the following forms for materials of the differential type: In the general case,

(23.12) $$S(t) = R(t)\, \bar{\mathfrak{k}}\left(R(t)^T A_k(t)\, R(t);\, C(t)\right) R(t)^T.$$

For isotropic solids*,

(23.13) $$S(t) = \bar{\mathfrak{k}}\left(A_k(t);\, B(t)\right).$$

For fluids,

(23.14) $$S(t) = \bar{\mathfrak{h}}\left(A_k(t);\, \varrho(t)\right).$$

In these equations $\bar{\mathfrak{k}}$ and $\bar{\mathfrak{h}}$ are of the same type as $\mathfrak{k}$ and $\mathfrak{h}$; in (23.12) $\bar{\mathfrak{k}}$ may be arbitrary but in (23.13) it must be isotropic. Of course, $\bar{\mathfrak{h}}$ is also isotropic. $A_k(t)$ is the kth Rivlin-Ericksen tensor, defined by (8.9) and related to the acceleration gradients by (9.2).

24. *Materials of the rate type*

The general constitutive equation of a simple material in the form (18.8) may be rewritten as

(24.1) $$\bar{S}_{t_0}(t) = \mathfrak{R}\left(\bar{U}^t_{t_0};\, U(t_0)\right)$$

where

(24.2) $$\bar{S}_{t_0}(t) = R(t_0)^T R_{t_0}(t)^T S(t)\, R_{t_0}(t)\, R(t_0)$$

and

(24.3) $$\bar{U}^t_{t_0}(s) = \bar{U}_{t_0}(t+s) = R(t_0)^T U_{t_0}(t+s)\, R(t_0).$$

Keeping t_0 and $U(t_0)$ fixed, we may interpret (24.1) in the following manner: Assuming that the function $\bar{U}_{t_0}$ defined by

(24.4) $$\bar{U}_{t_0}(t) = R(t_0)\, U_{t_0}(t)\, R(t_0)^T$$

is given, the function $\bar{S}_{t_0}$ defined by (24.2) is completely determined. In other words, (24.1) defines an operation on functions $\bar{U}_{t_0}$ with values in $\mathscr{S}_+$ which gives functions $\bar{S}_{t_0}$ with values in $\mathscr{S}$. It may happen that this operation is defined by the process of solution of a differential equation for $\bar{S}_{t_0}$ in the form

(24.5) $$\mathfrak{f}\left(\bar{S}_{t_0}(t),\, \dot{\bar{S}}_{t_0}(t), \ldots, \overset{(m)}{\bar{S}}_{t_0}(t);\, \bar{U}_{t_0}(t),\, \dot{\bar{U}}_{t_0}(t), \ldots, \overset{(n)}{\bar{U}}_{t_0}(t);\, U(t_0)\right) = 0,$$

* This form was first derived by Rivlin & Ericksen [8].

where $\mathfrak{f}$ is a symmetric-tensor-valued function of $m+n+3$ symmetric tensor variables. For simplicity we use a notation similar to (23.6), so that (24.5) becomes

$$\mathfrak{f}\left(\overset{(l)}{\overline{S}}_{t_0}(t);\ \overset{(k)}{\overline{U}}_{t_0}(t);\ U(t_0)\right) = 0. \tag{24.6}$$

If the function $\overline{U}_{t_0}$ is given, then (24.6) is a differential equation of order m for the function $\overline{S}_{t_0}$. We assume that the form of $\mathfrak{f}$ is such that there is a unique solution $\overline{S}_{t_0}$ which assumes given initial values $\overset{(l)}{\overline{S}}_{t_0}(t_1)$, $l=0, 1, \ldots, m-1$, no matter how we choose $U(t_0)$, $\overset{(k)}{\overline{U}}_{t_0}$, t_1, and $\overset{(l)}{\overline{S}}_{t_0}(t_1)$.

Since t_0 is arbitrary, in (24.6) we may make the special choice $t_0=t$ obtaining

$$\mathfrak{f}\left(\overset{(l)}{\overline{S}}_{t}(t);\ \overset{(k)}{\overline{U}}_{t}(t);\ U(t)\right) = 0. \tag{24.7}$$

By (24.4) and (8.8) we have

$$\overset{(k)}{\overline{U}}_{t}(t) = R^T(t)\, D_k(t)\, R(t). \tag{24.8}$$

The tensor function $\hat{S}_l$ defined by

$$\hat{S}_l(t) = \overline{R_{t_0}(t)^T \overset{(l)}{S}(t)\, R_{t_0}(t)}\Big|_{t_0=t} \tag{24.9}$$

will be called the lth *invariant stress rate*. If we carry out the differentiation in (24.9) according to the product rule, and if we observe (8.7), we see that

$$\hat{S}_l = \sum_{\substack{p,\,q,\,r=0,\ldots,l\\ p+q+r=l}} \frac{l!}{p!\,q!\,r!}\, W_p^T \overset{(q)}{S} W_r; \tag{24.10}$$

thus $\hat{S}_l$ can be expressed explicitly in terms of the stress S, its time derivatives $\overset{(q)}{S}$ up to the order l, and the spins W_p up to the order l. We have $\hat{S}_0=S$. For $l=1$, we get the *invariant stress rate**

$$\hat{S}_1 = \hat{S} = \dot{S} - W S + S W. \tag{24.11}$$

Observing (24.2), (24.9), and (24.8), we see that (24.7) has the form

$$\mathfrak{f}\left(R(t)^T \hat{S}_l(t)\, R(t);\ R(t)^T D_k(t)\, R(t);\ U(t)\right) = 0. \tag{24.12}$$

A material with a constitutive equation of this form will be called a *material of the rate type.*

It must be noted that (24.12) is not really a complete constitutive equation. The stress is not determined by the local motion alone but only when, in addition, initial values $\hat{S}_l(t_1)$, $l=0, 1, \ldots, m-1$, for some initial time t_1, are given. These initial values, on the other hand, should be determined by the history of the local motion up to the time t_1. A constitutive equation of the type (24.12) characterizes not a single material but a family of materials depending on m symmetric tensor parameters.

* It was introduced by ZAREMBA [*15*].

In the case of *isotropic solids* it follows from Theorem 6 that the tensor function $\mathfrak{f}$ in (24.12) must be isotropic and that the constitutive equation reduces to

(24.13) $$\mathfrak{f}\left(\hat{S}_l(t); D_k(t); V(t)\right) = 0.$$

Fluids of the rate type are described by an equation of the form

(24.14) $$\mathfrak{g}\left(\hat{S}_l(t); D_k(t); \varrho(t)\right) = 0,$$

where $\mathfrak{g}$ is an isotropic tensor function of $m+n+1$ tensor variables and one positive scalar variable. However, not every equation of the form (24.14) defines a class of fluids of the rate type. It may also define a class of isotropic solids, because the stress may depend on the right strain tensor relative to some reference configuration through the initial values.

Starting from the general constitutive equations (22.7), (22.9), and (22.11), one can easily derive alternate forms for the constitutive equations of the rate type:

In the general case we obtain

(24.15) $$\bar{\mathfrak{f}}\left(R(t)^T \tilde{S}_l(t)\, R(t); R(t)^T A_k(t)\, R(t); C(t)\right) = 0$$

where $A_k(t)$ is the kth Rivlin-Ericksen tensor (9.2) and where $\tilde{S}_l(t)$ is the lth *stress flux* defined by

(24.16) $$\tilde{S}_l(t) = \overline{F_{t_0}(t)^T S(t) F_{t_0}(t)}^{(l)}\Big|_{t_0=t}.$$

We find that $\tilde{S}_l$ may be expressed explicitly in terms of the stress S, its derivatives up to the order m, and the acceleration gradients E_l up to the order m by the formula

(24.17) $$\tilde{S}_l = \sum_{\substack{q,p,r=0,\ldots,l\\ p+q+r=l}} \frac{l!}{p!\,q!\,r!} E_p^T \overset{(q)}{S} E_r.$$

For $l=0$ we have $\tilde{S}_0 = S$, and for $l=1$ we get the *stress flux**

(24.18) $$\tilde{S}_1 = \tilde{S} = \dot{S} + E^T S + S E.$$

In the case of isotropic solids $\bar{\mathfrak{f}}$ is isotropic and (24.15) reduces to

(24.19) $$\bar{\mathfrak{f}}\left(\tilde{S}_l(t); A_k(t); B(t)\right) = 0^{**}.$$

For fluids we get

(24.20) $$\bar{\mathfrak{h}}\left(\tilde{S}_l(t); A_k(t); \varrho(t)\right) = 0,$$

where $\bar{\mathfrak{h}}$ is an isotropic tensor function.

Acknowledgement. The research leading to this paper was sponsored in part by the National Science Foudation under Grant NSF-G 5250 and in part by the Air Force Office of Scientific Research under Contract AF 18(600)-1138 with Carnegie Institute of Technology.

---

* It differs from the flux introduced by CAUCHY (*cf.* the discussion given by TRUESDELL in [*12*], Section 55 *bis*), in which there are minus signs on the right.

** This form was first derived by COTTER & RIVLIN [*1*].

References

[1] COTTER, B., & R. S. RIVLIN: Tensors associated with time-dependent stress. Quart. Appl. Math. **13**, 177—182 (1955).

[2] EHRESMANN, C.: Introduction à la théorie des structures infinitésimales et des pseudo-groupes de Lie. Colloques internationaux du centre national de la recherche scientifique. Géométrie différentielle, 97—110. Paris 1953.

[3] GREEN, A. E.: Simple extension of a hypo-elastic body of grade zero. J. Rational Mech. Anal. **5**, 637—642 (1956).

[4] GREEN, A. E., & R. S. RIVLIN: The mechanics of non-linear materials with memory. Arch. Rational Mech. Anal. **1**, 1—34 (1957).

[5] NOLL, W.: On the continuity of the solid and fluid states. J. Rational Mech. Anal. **4**, 3—81 (1955).

[6] NOLL, W.: The foundations of classical mechanics in the light of recent advances in continuum mechanics. Proceedings of the Berkeley Symposium on the Axiomatic Method, 1958.

[7] NOLL, W.: On the foundations of the mechanics of continuous media. Carnegie Institute of Technology, Technical Report no. 17, Air Force Office of Scientific Research, 1957.

[8] RIVLIN, R. S., & J. L. ERICKSEN: Stress-deformation relations for isotropic materials. J. Rational Mech. Anal. **4**, 323—425 (1955).

[9] RIVLIN, R. S.: Further remarks on stress-deformation relations for isotropic materials. J. Rational Mech. Anal. **4**, 681—702 (1955).

[10] RIVLIN, R. S.: Solution of some problems in the exact theory of viscoelasticity. J. Rational Mech. Anal. **5**, 179—188 (1956).

[11] TOUPIN, R. A.: World invariant kinematics. Arch. Rational Mech. Anal. **1**, 181—211 (1958).

[12] TRUESDELL, C.: The mechanical foundations of elasticity and fluid dynamics. J. Rational Mech. Anal. **1**, 125—300 (1952) and **2**, 593—616 (1953).

[13] TRUESDELL, C.: Hypo-elasticity. J. Rational Mech. Anal. **4**, 83—133 (1955).

[14] TRUESDELL, C.: Hypo-elastic shear. J. Appl. Physics **27**, 441—447 (1956).

[15] ZAREMBA, S.: Sur une forme perfectionnée de la théorie de la relaxation. Bull. Int. Acad. Sci. Cracovie **1903**, 594—614.

Carnegie Institute of Technology
Pittsburgh, Pennsylvania

(Received July 31, 1958)

THE FOUNDATIONS OF CLASSICAL MECHANICS IN THE LIGHT OF RECENT ADVANCES IN CONTINUUM MECHANICS

WALTER NOLL

THE FOUNDATIONS OF CLASSICAL MECHANICS IN THE LIGHT OF RECENT ADVANCES IN CONTINUUM MECHANICS [1]

WALTER NOLL

1. Introduction. It is a widespread belief even today that classical mechanics is a dead subject, that its foundations were made clear long ago, and that all that remains to be done is to solve special problems. This is not so. It is true that the mechanics of systems of a finite number of mass points has been on a sufficiently rigorous basis since Newton. Many textbooks on theoretical mechanics dismiss continuous bodies with the remark that they can be regarded as the limiting case of a particle system with an increasing number of particles. They cannot. The erroneous belief that they can had the unfortunate effect that no serious attempt was made for a long period to put classical continuum mechanics on a rigorous axiomatic basis. Only the recent advances in the theory of materials other than perfect fluids and linearly elastic solids have revived the interest in the foundations of classical mechanics. A clarification of these foundations is of importance also for the following reason. It is known that continuous matter is really made up of elementary particles. The basic laws governing the elementary particles are those of quantum mechanics. The science that provides the link between these basic laws and the laws describing the behavior of gross matter is statistical mechanics. At the present time this link is quite weak, partly because the mathematical difficulties are formidable, and partly because the basic laws themselves are not yet completely clear. A rigorous theory of continuum mechanics would give at least some precise information on what kind of gross behavior the basic laws ought to predict.

I want to give here a brief outline of an axiomatic scheme for continuum mechanics, and I shall attempt to introduce the same level of rigor and clarity as is now customary in pure mathematics. The mathematical

[1] The results presented in this paper were obtained in the course of research sponsored by the U.S. Air Force Office of Scientific Research under contract no. AF 18 (600)–1138 with Carnegie Institute of Technology.

structures involved are quite complex, and some fine details have to be omitted in order not to overburden the paper with technicalities.

Notation: Points and vectors in Euclidean space will be indicated by bold face letters. If $\boldsymbol{x}$ and $\boldsymbol{y}$ are two points, then $\boldsymbol{x} - \boldsymbol{y}$ denotes the vector determined by the ordered pair $(\boldsymbol{y}, \boldsymbol{x})$. If $\boldsymbol{x}$ is a point and $\boldsymbol{v}$ a vector, then $\boldsymbol{x} + \boldsymbol{v}$ denotes the point uniquely determined by $(\boldsymbol{x} + \boldsymbol{v}) - \boldsymbol{x} = \boldsymbol{v}$. The word "smooth" will be used instead of "continuously differentiable". Some equations will be valid only up to a set of measure zero. It will be clear from the context when this is the case.

2. Bodies.

DEFINITION 1: *A* BODY *is a set* $\mathfrak{B}$ *endowed with a structure defined by*
(*a*) *a set* Φ *of mappings of* $\mathfrak{B}$ *into a three-dimensional Euclidean point space* E, *and*
(*b*) *a real valued set function* m *defined for a set of subsets of* $\mathfrak{B}$

subject to seven axioms as follows:

(*S*.1) *Every mapping* $\varphi \in \Phi$ *is one-to-one.*

(*S*.2) *For each* $\varphi \in \Phi$, *the image* $B = \varphi(\mathfrak{B})$ *is a region in the space* E, *a region being defined as a compact set with piecewise smooth boundaries.*

(*S*.3) *If* $\varphi \in \Phi$ *and* $\psi \in \Phi$ *then the mapping* $\chi = \psi \circ \overset{-1}{\varphi}$ [2] *of* $\varphi(\mathfrak{B})$ *onto* $\psi(\mathfrak{B})$ *can be extended to a smooth homeomorphism of* E *onto itself.*

(*S*.4) *If* χ *is a smooth homeomorphism of* E *onto itself and if* $\varphi \in \Phi$, *then also* $\chi \circ \varphi \in \Phi$.

These four axioms give $\mathfrak{B}$ the structure of a piece of a differentiable manifold that is isomorphic to a region in Euclidean three-space. The following three axioms give $\mathfrak{B}$ the structure of a measure space.

(*M*.1) m *is a non-negative measure, defined for all Borel subsets* $\mathfrak{C}$ *of* $\mathfrak{B}$.

(*M*.2) *For each* $\varphi \in \Phi$, *the measure* $\mu_\varphi = m \circ \overset{-1}{\varphi}$ *induced by* m *on the region* $B = \varphi(\mathfrak{B})$ *in space is absolutely continuous relative to the Lebèsgue measure in* B. *Hence it has a density* ρ_φ *so that*

$$m(\mathfrak{C}) = \int_{\varphi(\mathfrak{C})} \rho_\varphi(\boldsymbol{x})\,\mathrm{d}V. \tag{2.1}$$

(*M*.3) *For each* $\varphi \in \Phi$ *the density* ρ_φ *is positive and bounded.*

[2] The symbol $\circ$ denotes the composition of mappings and a superposed -1 denotes the inverse of a mapping.

We use the following terminology: The elements $X, Y, \ldots$ of $\mathfrak{B}$ are the PARTICLES of the body. The mappings $\varphi \in \Phi$ are the CONFIGURATIONS of the body. The point $\boldsymbol{x} = \varphi(X)$ is the POSITION of the particle X in the configuration φ. The set function m is the MASS DISTRIBUTION of the body. The number $m(\mathfrak{C})$ is the MASS of the set $\mathfrak{C}$. Here and subsequently we refer to Borel sets simply as sets. The density ρ_φ is the MASS DENSITY of $\mathfrak{B}$ in the configuration φ. Note that it would have been sufficient to require the existence of ρ_φ only for one particular configuration φ. It then follows that the mass density exists also for all other configurations.

A compact subset $\mathfrak{P}$ of $\mathfrak{B}$ with piecewise smooth boundaries will be called a PART of the body $\mathfrak{B}$. It may again be regarded as a body whose configurations are the restrictions to $\mathfrak{P}$ of the configurations of $\mathfrak{B}$ and whose mass distribution is the restriction of the mass distribution of $\mathfrak{B}$ to the subsets of $\mathfrak{P}$. Two parts $\mathfrak{P}$ and $\mathfrak{Q}$ will be called SEPARATE if

$$\mathfrak{P} \cap \mathfrak{Q} \subset \overline{\mathfrak{P}} \cap \overline{\mathfrak{Q}},$$

where $\overline{\mathfrak{P}}$ denotes the boundary of $\mathfrak{P}$.

3. Kinematics

DEFINITION 2: *A* MOTION *of a body $\mathfrak{B}$ is a one-parameter family $\{\theta_t\}$, $-\infty < t < \infty$, of configurations $\theta_t \in \Phi$ of $\mathfrak{B}$ such that*

(*K*.1) *The derivative*

$$\boldsymbol{v}(X, t) = \frac{d}{dt}\,\theta_t(X) \tag{3.1}$$

exists for all $X \in \mathfrak{B}$ and all t, it is a continuous function of X and t jointly, and it is a smooth function of X.

(K.2) *The derivative*

$$\dot{\boldsymbol{v}}(X, t) = \frac{d}{dt}\,\boldsymbol{v}(X, t) = \frac{d^2}{dt^2}\,\theta_t(X) \tag{3.2}$$

exists piecewise and is piecewise continuous in X and t jointly.

The parameter t is called the TIME. Derivatives with respect to t will be denoted by superposed dots. $\boldsymbol{v}(X, t)$ is called the VELOCITY of the particle X at time t. $\dot{\boldsymbol{v}}(X, t)$ is called the ACCELERATION of X at t.

Let h be any real, vector, or tensor valued function of X and t, and assume that $h(X, t)$ is smooth in X and t jointly. We may then associate with h the function $\hat{h}$ defined by

$$\hat{h}(\boldsymbol{x}, t) = h(\overset{-1}{\theta_t}(\boldsymbol{x}), t) \tag{3.3}$$

for $-\infty < t < \infty$ and $\boldsymbol{x} \in \theta_t(\mathfrak{B})$. By the chain rule of differentiation we have

$$\dot{h}(X, t) = \dot{\hat{h}}(\theta_t(X), t) + \nabla\hat{h}(\theta_t(X), t) \cdot \boldsymbol{v}(X, t), \tag{3.4}$$

where $\nabla\hat{h}$ denotes the gradient of $\hat{h}$ with respect to $\boldsymbol{x}$. It is customary in the literature to use the same symbol for h and $\hat{h}$, to omit the independent variables, and to distinguish $\dot{\hat{h}}$ from $\dot{h}$ by writing $\dot{\hat{h}} = \frac{\partial h}{\partial t}$. Equation (3.4) then takes the familiar form

$$\dot{h} = \frac{\partial h}{\partial t} + \boldsymbol{v} \cdot \operatorname{grad} h. \tag{3.5}$$

The LINEAR MOMENTUM at time t of a set $\mathfrak{C} \subset \mathfrak{B}$ is defined by

$$\boldsymbol{g}(\mathfrak{C}; t) = \int_{\mathfrak{C}} \boldsymbol{v}(X, t) \mathrm{d}m. \tag{3.6}$$

It follows from (K.1) and (K.2) that $\boldsymbol{g}(\mathfrak{C}, t)$ is piecewise smooth in t. As a function of $\mathfrak{C}$ it is a vector valued measure, absolutely continuous relative to m with density $\boldsymbol{v}$.

The ANGULAR MOMENTUM at time t of a set $\mathfrak{C} \subset \mathfrak{B}$, relative to a point $\boldsymbol{O} \in E$, is defined by

$$\boldsymbol{h}(\mathfrak{C}; t; \boldsymbol{O}) = \int_{\mathfrak{C}} [\theta_t(X) - \boldsymbol{O}] \times \boldsymbol{v}(X, t) \mathrm{d}m. \tag{3.7}$$

It is piecewise smooth in t, and, as a function of $\mathfrak{C}$, it is a vector valued measure.

4. Forces

DEFINITION 3: *A* SYSTEM OF BODY FORCES *for a body* $\mathfrak{B}$ *is a family* $\{\boldsymbol{B}_{\mathfrak{P}}\}$ *of vector valued set functions subject to the following axioms:*

(B.1) *For each part* $\mathfrak{P}$ *of* $\mathfrak{B}$, $\boldsymbol{B}_{\mathfrak{P}}$ *is a vector valued measure defined on the Borel subsets of* $\mathfrak{P}$.

(B.2) *For each* $\mathfrak{P}$, $\boldsymbol{B}_{\mathfrak{P}}$ *is absolutely continuous relative to the mass distribution* m *of* $\mathfrak{P}$. *Hence it has a density* $\boldsymbol{b}_{\mathfrak{P}}$ *so that*

$$\boldsymbol{B}_{\mathfrak{P}}(\mathfrak{C}) = \int_{\mathfrak{C}} \boldsymbol{b}_{\mathfrak{P}}(X) \mathrm{d}m. \tag{4.1}$$

(B.3) *The density* $\boldsymbol{b}_{\mathfrak{P}}$ *is bounded, i.e.*

$$|\boldsymbol{b}_{\mathfrak{P}}(X)| < k < \infty,$$

where k *is independent of* $\mathfrak{P}$ *and* $X \in \mathfrak{P}$.

DEFINITION 4: *A* SYSTEM OF CONTACT FORCES *for a body* $\mathfrak{B}$ *is a family* $\{\boldsymbol{C}_{\mathfrak{P}}\}$ *of vector valued set functions subject to the following axioms:*

(C.1) *For each part* $\mathfrak{P}$ *of* $\mathfrak{B}$, $\boldsymbol{C}_{\mathfrak{P}}$ *is a vector valued measure defined on the Borel subsets of* $\mathfrak{P}$.

(C.2) $\boldsymbol{C}_{\mathfrak{P}}(\mathfrak{C}) = \boldsymbol{C}_{\mathfrak{P}}(\mathfrak{C} \cap \overline{\mathfrak{P}})$.

(C.3) *If* $\mathfrak{c} \subset \overline{\mathfrak{Q}}$, $\mathfrak{c} \subset \overline{\mathfrak{P}}$, *and* $\mathfrak{P} \subset \mathfrak{Q}$, *then*

$$\boldsymbol{C}_{\mathfrak{P}}(\mathfrak{c}) = \boldsymbol{C}_{\mathfrak{Q}}(\mathfrak{c}).$$

(C.4) *If* $\varphi \in \Phi$ *is any configuration of* $\mathfrak{B}$ *and if* $\overline{P} = \varphi(\overline{\mathfrak{P}})$, *then the induced measure* $\boldsymbol{C}_{\mathfrak{P}} \circ \overset{-1}{\varphi}$, *when restricted to the Borel subsets of the boundary surface* $\overline{P}$ *of* $P = \varphi(\mathfrak{P})$, *is absolutely continuous relative to the Lebèsgue surface measure on* $\overline{P}$. *Hence it has a density* $\boldsymbol{s}(\mathfrak{P}, \varphi)$ *so that*

$$\boldsymbol{C}_{\mathfrak{P}}(\mathfrak{c}) = \int_{\varphi(\mathfrak{c})} \boldsymbol{s}(\mathfrak{P}, \varphi; \boldsymbol{x}) \mathrm{d}A \tag{4.2}$$

for all Borel subsets $\mathfrak{c} \subset \overline{\mathfrak{P}}$.

(C.5) *The density* $\boldsymbol{s}(\mathfrak{P}, \varphi)$ *is bounded, i.e.*

$$|\boldsymbol{s}(\mathfrak{P}, \varphi; \boldsymbol{x})| < l < \infty,$$

where l *does not depend on* $\mathfrak{P}$ *or* $\boldsymbol{x} \in \varphi(\overline{\mathfrak{P}})$.

As in the case of a mass distribution, it would, have been sufficient in (C.4) to require the existence of $\boldsymbol{s}(\mathfrak{P}, \varphi)$ only for a particular $\varphi \in \Phi$. The existence of $\boldsymbol{s}$ for all other configurations is then an automatic consequence. The axiom (C.2) means that $\boldsymbol{C}_{\mathfrak{P}}$ is essentially a vector measure on the boundary $\overline{\mathfrak{P}}$.

It is useful to consider surfaces in $\mathfrak{B}$ as being *oriented*, and to employ the operation of addition of oriented surfaces in the sense of algebraic topology. The boundary $\overline{\mathfrak{P}}$ of a part $\mathfrak{P}$ of $\mathfrak{B}$ will be regarded as oriented in such a way that the positive side of $\overline{\mathfrak{P}}$ is exterior to $\mathfrak{P}$. If $\mathfrak{P}$ and $\mathfrak{Q}$ are two separate parts of $\mathfrak{B}$, then

$$\overline{\mathfrak{P} \cup \mathfrak{Q}} = \overline{\mathfrak{P}} + \overline{\mathfrak{Q}}. \tag{4.3}$$

This is true because the common boundary of $\mathfrak{P}$ and $\mathfrak{Q}$, if any, appears twice with opposite orientation on the right side of (4.3) and hence cancels. We shall say that the surface $\mathfrak{c}$ is a PIECE of the surface $\mathfrak{d}$ if $\mathfrak{c}$ is a subset of $\mathfrak{d}$ and if the orientation of $\mathfrak{c}$ is induced by $\mathfrak{d}$. The significance of the axiom (C.3) is brought out by the following theorem:

THEOREM I: *There is a vector valued function $\boldsymbol{S}$, defined for all oriented surfaces $\mathfrak{c}$ in $\mathfrak{B}$, such that*

$$C_{\mathfrak{P}}(\mathfrak{c}) = \boldsymbol{S}(\mathfrak{c}) \tag{4.4}$$

whenever $\mathfrak{c}$ is a piece of the boundary $\overline{\mathfrak{P}}$ of $\mathfrak{P}$. We say that $\boldsymbol{S}(\mathfrak{c})$ is the CONTACT FORCE ACTING ACROSS THE ORIENTED SURFACE $\mathfrak{c}$.

Proof: For each $\mathfrak{c}$ which is not a piece of $-\overline{\mathfrak{B}}$ we can find a part $\mathfrak{Q}(\mathfrak{c})$ of $\mathfrak{B}$ such that $\mathfrak{c}$ is a piece of $\overline{\mathfrak{Q}(\mathfrak{c})}$. We then define $\boldsymbol{S}(\mathfrak{c}) = \boldsymbol{C}_{\mathfrak{Q}(\mathfrak{c})}(\mathfrak{c})$. Now let $\mathfrak{P}$ be an arbitrary part of $\mathfrak{B}$ and let $\mathfrak{c}$ be a piece of $\overline{\mathfrak{P}}$. We then have

$$\mathfrak{c} \subset \overline{\mathfrak{P}}, \quad \mathfrak{c} \subset \overline{\mathfrak{Q}(\mathfrak{c})}, \quad \mathfrak{c} \subset \overline{\mathfrak{Q}(\mathfrak{c}) \cap \mathfrak{P}},$$
$$\mathfrak{P} \cap \mathfrak{Q}(\mathfrak{c}) \subset \mathfrak{P}, \quad \mathfrak{P} \cap \mathfrak{Q}(\mathfrak{c}) \subset \mathfrak{Q}(\mathfrak{c}).$$

Applying axiom (C.3) twice, we get

$$\boldsymbol{C}_{\mathfrak{P}}(\mathfrak{c}) = \boldsymbol{C}_{\mathfrak{P}\cap\mathfrak{Q}(\mathfrak{c})}(\mathfrak{c}), \quad \boldsymbol{C}_{\mathfrak{Q}(\mathfrak{c})}(\mathfrak{c}) = \boldsymbol{C}_{\mathfrak{P}\cap\mathfrak{Q}(\mathfrak{c})}(\mathfrak{c}).$$

Hence

$$\boldsymbol{C}_{\mathfrak{P}}(\mathfrak{c}) = \boldsymbol{C}_{\mathfrak{Q}(\mathfrak{c})}(\mathfrak{c}) = \boldsymbol{S}(\mathfrak{c}).$$

If $\mathfrak{c}$ is a part of $-\overline{\mathfrak{B}}$ we define

$$\boldsymbol{S}(\mathfrak{c}) = -\boldsymbol{S}(-\mathfrak{c}). \tag{4.5}$$

It follows from theorem I and axiom (C.4) that there is a vector valued function $\boldsymbol{s}(\mathfrak{c}, \varphi; \boldsymbol{x})$ such that

$$\boldsymbol{S}(\mathfrak{c}) = \int_{\varphi(\mathfrak{c})} \boldsymbol{s}(\mathfrak{c}, \varphi; \boldsymbol{x}) \mathrm{d}A. \tag{4.6}$$

Also, if $\boldsymbol{x} \in \varphi(\mathfrak{d}) \subset \varphi(\mathfrak{c})$ and if $\mathfrak{d}$ is a piece of $\mathfrak{c}$, then

$$\boldsymbol{s}(\mathfrak{c}, \varphi; \boldsymbol{x}) = \boldsymbol{s}(\mathfrak{d}, \varphi; \boldsymbol{x}). \tag{4.7}$$

If $\mathfrak{c}_1$ and $\mathfrak{c}_2$ are two pieces of a surface $\mathfrak{c}$ and if $\mathfrak{c} = \mathfrak{c}_1 + \mathfrak{c}_2$, then

$$\boldsymbol{S}(\mathfrak{c}) = \boldsymbol{S}(\mathfrak{c}_1) + \boldsymbol{S}(\mathfrak{c}_2). \tag{4.8}$$

This is true because $\boldsymbol{C}_{\mathfrak{P}}$, as a measure, is additive and because, by axiom (C.4) the value of $\boldsymbol{C}_{\mathfrak{P}}$ for the common boundary curve of $\mathfrak{c}_1$ and $\mathfrak{c}_2$ is zero.

Definition 5: *A* system of forces *for a body* $\mathfrak{B}$ *is a family of vector valued measures* $\{\boldsymbol{F}_{\mathfrak{P}}\}$ *such that, for each part* $\mathfrak{P}$ *of* $\mathfrak{B}$, $\boldsymbol{F}_{\mathfrak{P}}$ *is defined on the subsets of* $\mathfrak{P}$ *and such that the* $\boldsymbol{F}_{\mathfrak{P}}$ *have decompositions*

$$\boldsymbol{F}_{\mathfrak{P}} = \boldsymbol{B}_{\mathfrak{P}} + \boldsymbol{C}_{\mathfrak{P}}, \tag{4.9}$$

where $\{\boldsymbol{B}_{\mathfrak{P}}\}$ *is a system of body forces and* $\{\boldsymbol{C}_{\mathfrak{P}}\}$ *is a system of contact forces.*

It is not hard to see that the decomposition (4.9), if it exists, is automatically unique.

We use the following terminology: The measure $\boldsymbol{F}_{\mathfrak{B}}$ is the force acting on the part $\mathfrak{P}$ of $\mathfrak{B}$. The vector $\boldsymbol{F}_{\mathfrak{P}}(\mathfrak{P})$ is the resultant force acting on $\mathfrak{P}$. Let $\mathfrak{P}$ and $\mathfrak{Q}$ be two separate parts of $\mathfrak{B}$. The vector measure

$$\boldsymbol{F}_{\mathfrak{P},\mathfrak{Q}} = \boldsymbol{F}_{\mathfrak{P}} - \boldsymbol{F}_{\mathfrak{P}\cup\mathfrak{Q}} \tag{4.10}$$

defined on the subsets of $\mathfrak{P}$, is the mutual force exerted on $\mathfrak{P}$ by $\mathfrak{Q}$. The mutual force exerted on a part $\mathfrak{P}$ of $\mathfrak{B}$ by the closure of its complement is denoted by $\boldsymbol{F}_{(\mathfrak{P})}$ and it is called the internal force acting on $\mathfrak{P}$. The restriction of $\boldsymbol{F}_{\mathfrak{B}}$ to a part $\mathfrak{P}$ of $\mathfrak{B}$ is the external force acting on $\mathfrak{P}$. A similar terminology and notation will be used when "force" is replaced by "body force" or by "contact force".

Let $\{\boldsymbol{F}_{\mathfrak{P}}\}$ be a system of forces for a body $\mathfrak{B}$, $\varphi \in \boldsymbol{\Phi}$ a configuration of $\mathfrak{B}$, and $\boldsymbol{O} \in E$ a point in space. The moment about $\boldsymbol{O}$ of the force $\boldsymbol{F}_{\mathfrak{P}}$ acting on the part $\mathfrak{P}$ of $\mathfrak{B}$ in the configuration φ is the vector valued measure $\boldsymbol{M}(\boldsymbol{F}_{\mathfrak{P}}, \varphi, \boldsymbol{O})$ defined by

$$\boldsymbol{M}(\boldsymbol{F}_{\mathfrak{P}}, \varphi, \boldsymbol{O}; \mathfrak{C}) = \int_{\mathfrak{C}} [\varphi(X) - \boldsymbol{O}] \times \mathrm{d}\boldsymbol{F}_{\mathfrak{P}} \tag{4.11}$$

for the subsets $\mathfrak{C}$ of $\mathfrak{P}$. The vector $\boldsymbol{M}(\boldsymbol{F}_{\mathfrak{P}}, \varphi, \boldsymbol{O}; \mathfrak{P})$ is the resultant moment about $\boldsymbol{O}$ acting on $\mathfrak{P}$.

5. Dynamical processes

Definition 6: *A* dynamical process *is a triple* $\{\mathfrak{B}, \theta_t, \boldsymbol{F}_{\mathfrak{P},t}\}$, *where* $\mathfrak{B}$ *is a body,* θ_t *is a motion of* $\mathfrak{B}$, *and* $\boldsymbol{F}_{\mathfrak{P},t}$ *is a one-parameter family of systems of forces for* $\mathfrak{B}$, *subject to the following two axioms:*

(D.1) *Principle of linear momentum: For all parts* $\mathfrak{P}$ *of* $\mathfrak{B}$ *and all times t,*

$$\boldsymbol{F}_{\mathfrak{P},t}(\mathfrak{P}) = \dot{\boldsymbol{g}}(\mathfrak{P}; t), \tag{5.1}$$

where $\boldsymbol{g}$ *is defined by* (3.6). *In words: The resultant force acting on the part* $\mathfrak{P}$ *is equal to the rate of change of the linear momentum of* $\mathfrak{P}$.

(D.2) *Principle of angular momentum: Let* $\boldsymbol{O} \in E$ *be any point in space.*

Then for all parts $\mathfrak{P}$ *of* $\mathfrak{B}$ *and all times t,*

$$M(F_{\mathfrak{P},t}, \theta_t, O; \mathfrak{P}) = \dot{h}(\mathfrak{P}; t; O), \tag{5.2}$$

where h *and* M *are defined by* (3.7) *and* (4.11), *respectively. In words: The resultant moment about* O *acting on a part* $\mathfrak{P}$ *is equal to the rate of change of the angular momentum of* $\mathfrak{P}$ *relative to* O.

It would have been sufficient to require that (5.2) holds for a particular $O \in E$. It is then automatically valid for all points in space. Also, (5.2) remains valid if the fixed point O is replaced by the variable mass center

$$c(\mathfrak{P}, t) = O + \frac{1}{m(\mathfrak{P})} \int_{\mathfrak{P}} (\theta_t(X) - O) dm \tag{5.3}$$

of the part $\mathfrak{P}$. These statements can be proved in the classical manner.

We now prove a number of important theorems. For simplicity we omit the variable t; we write

$$s(\mathfrak{c}; x) = s(\mathfrak{c}, \theta_t; x) \tag{5.4}$$

for the density of the contact force as defined by (4.6).

THEOREM II: *For any two separate parts* $\mathfrak{P}$ *and* $\mathfrak{Q}$ *of* $\mathfrak{B}$ *we have*

$$F_{\mathfrak{P},\mathfrak{Q}}(\mathfrak{P}) = - F_{\mathfrak{Q},\mathfrak{P}}(\mathfrak{Q}) \tag{5.5}$$

i.e. the resultant mutual force exerted on $\mathfrak{P}$ *by* $\mathfrak{Q}$ *is equal and opposite to the resultant mutual force exerted on* $\mathfrak{Q}$ *by* $\mathfrak{P}$.

Proof: We apply axiom (D.1) to $\mathfrak{P}$, $\mathfrak{Q}$, and $\mathfrak{P} \cup \mathfrak{Q}$:

$$F_{\mathfrak{P}}(\mathfrak{P}) = \dot{g}(\mathfrak{P}), \quad F_{\mathfrak{Q}}(\mathfrak{Q}) = \dot{g}(\mathfrak{Q}), \quad F_{\mathfrak{P}\cup\mathfrak{Q}}(\mathfrak{P} \cup \mathfrak{Q}) = \dot{g}(\mathfrak{P} \cup \mathfrak{Q}). \tag{5.6}$$

Since $\mathfrak{P} \cap \mathfrak{Q}$ has no mass by (M.2), it follows from (3.6) that

$$g(\mathfrak{P} \cup \mathfrak{Q}) = g(\mathfrak{P}) + g(\mathfrak{Q});$$

hence, by (5.6),

$$F_{\mathfrak{P}}(\mathfrak{P}) + F_{\mathfrak{Q}}(\mathfrak{Q}) = F_{\mathfrak{P}\cup\mathfrak{Q}}(\mathfrak{P} \cup \mathfrak{Q}).$$

It is not hard to see that $F_{\mathfrak{P}\cup\mathfrak{Q}}(\mathfrak{P} \cap \mathfrak{Q}) = 0$. Hence

$$F_{\mathfrak{P}\cup\mathfrak{Q}}(\mathfrak{P} \cup \mathfrak{Q}) = F_{\mathfrak{P}\cup\mathfrak{Q}}(\mathfrak{P}) + F_{\mathfrak{P}\cup\mathfrak{Q}}(\mathfrak{Q}).$$

The assertion follows now from the definition (4.10).

THEOREM III (*reaction principle*) [3]: *The contact force* $\boldsymbol{S}(\mathfrak{c})$ *acting cross* $\mathfrak{c}$ *is opposite to the contact force acting across* $-\mathfrak{c}$, *i.e.*

$$\boldsymbol{S}(\mathfrak{c}) = -\boldsymbol{S}(-\mathfrak{c}) \tag{5.7}$$

Proof: If $\mathfrak{c}$ is a piece of $-\overline{\mathfrak{B}}$, then (5.7) is true by the definition (4.5). If not, it is possible to find two separate parts $\mathfrak{P}$ and $\mathfrak{Q}$ such that $\mathfrak{P} \cap \mathfrak{Q} = \mathfrak{c}$ (see Fig. 1). We orient $\mathfrak{c}$ such that it is a piece of $\overline{\mathfrak{P}}$. Then $-\mathfrak{c}$

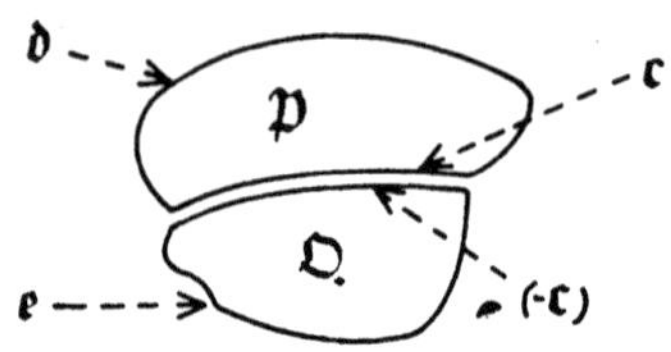

Fig. 1

will be a piece of $\overline{\mathfrak{Q}}$. The surfaces $\overline{\mathfrak{P}}$, $\overline{\mathfrak{Q}}$, and $\overline{\mathfrak{P} \cup \mathfrak{Q}}$ have decompositions

$$\overline{\mathfrak{P}} = \mathfrak{c} + \mathfrak{d}, \quad \overline{\mathfrak{Q}} = (-\mathfrak{c}) + \mathfrak{e}, \quad \overline{\mathfrak{P} \cup \mathfrak{Q}} = \mathfrak{d} + \mathfrak{e}.$$

It follows from theorem I and (4.8) that

$$\boldsymbol{C}_{\mathfrak{P}}(\mathfrak{P}) = \boldsymbol{S}(\mathfrak{c}) + \boldsymbol{S}(\mathfrak{d}), \quad \boldsymbol{C}_{\mathfrak{P} \cup \mathfrak{Q}}(\mathfrak{P}) = \boldsymbol{S}(\mathfrak{d})$$

and hence that

$$\boldsymbol{C}_{\mathfrak{P},\mathfrak{Q}}(\mathfrak{P}) = \boldsymbol{C}_{\mathfrak{P}}(\mathfrak{P}) - \boldsymbol{C}_{\mathfrak{P} \cup \mathfrak{Q}}(\mathfrak{P}) = \boldsymbol{S}(\mathfrak{c}).$$

Similiarly, we obtain

$$\boldsymbol{C}_{\mathfrak{Q},\mathfrak{P}}(\mathfrak{Q}) = \boldsymbol{S}(-\mathfrak{c}).$$

For the total resultant mutual forces, we get

$$\begin{aligned} \boldsymbol{F}_{\mathfrak{Q},\mathfrak{P}}(\mathfrak{P}) &= \boldsymbol{B}_{\mathfrak{P},\mathfrak{Q}}(\mathfrak{P}) + \boldsymbol{S}(\mathfrak{c}) \\ \boldsymbol{F}_{\mathfrak{Q},\mathfrak{P}}(\mathfrak{Q}) &= \boldsymbol{B}_{\mathfrak{Q},\mathfrak{P}}(\mathfrak{Q}) + \boldsymbol{S}(-\mathfrak{c}). \end{aligned} \tag{5.8}$$

Application of theorem II gives

$$\boldsymbol{S}(\mathfrak{c}) + \boldsymbol{S}(-\mathfrak{c}) = -[\boldsymbol{B}_{\mathfrak{P},\mathfrak{Q}}(\mathfrak{P}) + \boldsymbol{B}_{\mathfrak{Q},\mathfrak{P}}(\mathfrak{Q})]. \tag{5.9}$$

Using axiom (M.3) one can show that the parts $\mathfrak{P}$ and $\mathfrak{Q}$ can be chosen

[3] Various statements, mostly quite vague, pass under the title "principle of action and reaction" in the literature. All of these statements, when made precise, are provable theorems in the theory presented here.

such that their masses $m(\mathfrak{P})$ and $m(\mathfrak{Q})$ are arbitrarily small. Axiom (B.3) then implies that the right side of (5.9) can be made arbitrarily small in absolute value. It follows that the left side of (5.9) must vanish. Q.e.d.

As a corollary, it follows that

$$S(\mathfrak{c}_1 + \mathfrak{c}_2) = S(\mathfrak{c}_1) + S(\mathfrak{c}_2), \tag{5.10}$$

no matter whether $\mathfrak{c}_1$ and $\mathfrak{c}_2$ are pieces of $\mathfrak{c} = \mathfrak{c}_1 + \mathfrak{c}_2$, as in (4.8), or not. Hence S may be regarded as an additive vector valued function of oriented surfaces in $\mathfrak{B}$. Another corollary is that the statement of theorem II remains true if "mutual force" there is replaced by "mutual contact force" or by "mutual body force".

THEOREM IV (*stress principle*) [4]: *There is a vector valued function* $s(x, n)$, *where* $x \in \theta_t(\mathfrak{B})$ *and where* n *is a unit vector, such that*

$$s(\mathfrak{c}; x) = s(x, n) \tag{5.11}$$

whenever $\theta_t(\mathfrak{c})$ *has the unit normal* n *at* $x \in \theta_t(\mathfrak{c})$, *directed towards the positive side of the oriented surface* $\theta_t(\mathfrak{c})$, *the orientation of* $\theta_t(\mathfrak{c})$ *being induced by the orientation of* $\mathfrak{c}$.

Proof: Let $\mathfrak{c}_1$ and $\mathfrak{c}_2$ be two surfaces in $\mathfrak{B}$ tangent to each other at $X = \overset{-1}{\theta_t}(x)$. The surfaces $c_1 = \theta_t(\mathfrak{c}_1)$ and $c_2 = \theta_t(\mathfrak{c}_2)$ in space E are then tangent to each other at the point x. We assume that n is their unit normal at x and that c_1 and c_2 are oriented in such a way that n is directed

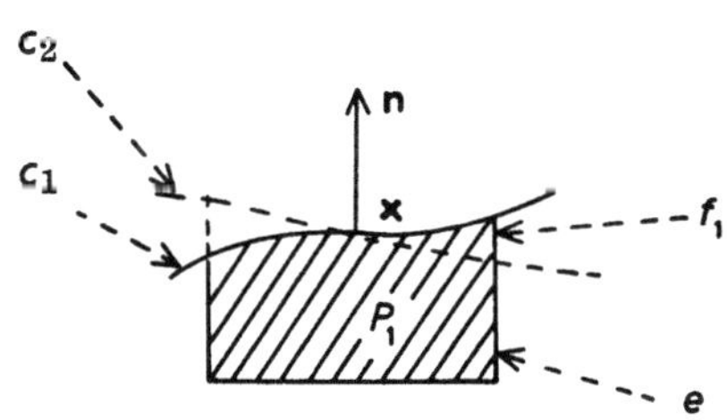

Fig. 2

toward the positive side of c_1 and c_2. Consider the region P_1 bounded by a piece d_1 of c_1, a piece of a circular cylinder f of radius r whose axis is n

[4] The assertion of this theorem appears in all of the past literature as an assumption. It has been proposed occasionally that one should weaken this assumption and allow the stress to depend not only on the tangent plane at x, but also on the curvature of the surface c at x. The theorem given here shows that such dependence on the curvature, or on any other local property of the surface at x, is impossible.

and by a plane perpendicular to $\boldsymbol{n}$ at a distance r from $\boldsymbol{x}$ as shown in Fig. 2. The region P_2 is defined in a similiar manner. Denote the common boundary of P_1 and P_2 on the cylinder and the plane by e. The boundaries $\bar{P}_1$ and $\bar{P}_2$ then have decompositions into separate pieces of the form

$$\bar{P}_1 = d_1 + e + f_1, \quad \bar{P}_2 = d_2 + e + f_2, \tag{5.12}$$

where f_1 and f_2 are pieces of the cylinder f. We denote the surface area of a surface c by $A(c)$ and the volume of a region P by $V(P)$. It is not hard to see that then

$$A(d_i) = \pi r^2 + o(r^2), \tag{5.13}$$

$$A(f_i) = o(r^2), \tag{5.14}$$

$$V(P_i) = o(r^2) \tag{5.15}$$

for $i = 1, 2$. $\mathfrak{P}_1 = \overset{-1}{\theta}_t(P_1)$ and $\mathfrak{P}_2 = \overset{-1}{\theta}_t(P_2)$ will be parts of $\mathfrak{B}$ for small enough r, except when $\boldsymbol{x} \in \theta_t(\bar{\mathfrak{B}})$, and $\boldsymbol{n}$ is directed toward the interior of $\theta_t(\mathfrak{B})$. Applying axiom (D.1) to $\mathfrak{P}_1$ and $\mathfrak{P}_2$ gives

$$\boldsymbol{F}_{\mathfrak{P}_i}(\mathfrak{P}_i) = \boldsymbol{B}_{\mathfrak{P}_i}(\mathfrak{P}_i) + \boldsymbol{C}_{\mathfrak{P}_i}(\mathfrak{P}_i) = \int_{\mathfrak{P}_i} \dot{\boldsymbol{v}}\,\mathrm{d}m, \quad i = 1, 2. \tag{5.16}$$

By (4.1) and (4.4) this may be written in the form

$$\boldsymbol{S}(\bar{\mathfrak{P}}_i) = \int_{\mathfrak{P}_i} (\dot{\boldsymbol{v}} - \boldsymbol{b}_{\mathfrak{P}_i})\mathrm{d}m. \tag{5.17}$$

By (4.6), (4.7), (4.8), and (5.12) we have

$$\boldsymbol{S}(\bar{\mathfrak{P}}_i) = \int_{d_i} \boldsymbol{s}(\mathfrak{c}_i)\mathrm{d}A + \int_{f_i} \boldsymbol{s}(\mathfrak{f}_i)\mathrm{d}A + \int_{e} \boldsymbol{s}(\mathfrak{e})\mathrm{d}A; \quad i = 1, 2, \tag{5.18}$$

where $\mathfrak{f}_i = \overset{-1}{\theta}_t(f_i)$, $\mathfrak{e} = \overset{-1}{\theta}_t(e)$. Subtracting the two equations (5.18) and using (5.17), we get

$$\int_{d_1} \boldsymbol{s}(\mathfrak{c}_1)\mathrm{d}A - \int_{d_2} \boldsymbol{s}(\mathfrak{c}_2)dA = \tag{5.19}$$

$$= \int_{\mathfrak{P}_1} (\dot{\boldsymbol{v}} - \boldsymbol{b}_{\mathfrak{P}_1})\mathrm{d}m - \int_{\mathfrak{P}_2} (\dot{\boldsymbol{v}} - \boldsymbol{b}_{\mathfrak{P}_2})\mathrm{d}m - \int_{f_1} \boldsymbol{s}(\mathfrak{f}_1)\mathrm{d}A + \int_{f_2} \boldsymbol{s}(\mathfrak{f}_2)\mathrm{d}A.$$

Since $\dot{\boldsymbol{v}}$, $\boldsymbol{b}_{\mathfrak{P}_i}$ and the mass density are bounded by constants independent of $\mathfrak{P}$, according to the axioms (K.2), (B.3), and (M.3), it follows from (5.15) that

$$\int_{\mathfrak{P}_i} (\dot{\boldsymbol{v}} - \boldsymbol{b}_{\mathfrak{P}_i})\mathrm{d}m = o(r^2), \quad i = 1, 2.$$

Similarly, it follows from axiom (C.5) and from (5.14) that

$$\int_{f_i} \boldsymbol{s}(\mathfrak{f}_i)\mathrm{d}A = o(r^2), \quad i = 1, 2.$$

Hence, by (5.19),

$$\int_{d_1} \boldsymbol{s}(\mathfrak{c})\mathrm{d}A = \int_{d_2} \boldsymbol{s}(\mathfrak{c}_2)\mathrm{d}A + o(r^2).$$

Dividing by πr^2 and using (5.13), we get

$$\frac{\int_{d_1} \boldsymbol{s}(\mathfrak{c}_1)\mathrm{d}A}{A(d_1)} = \frac{\int_{d_2} \boldsymbol{s}(\mathfrak{c}_2)\mathrm{d}A}{A(d_2)} + \frac{o(r^2)}{\pi r^2}. \tag{5.20}$$

By a theorem on measures with density, we have

$$\lim_{r\to 0} \frac{\int_{d_i} \boldsymbol{s}(\mathfrak{c}_i)\mathrm{d}A}{A(d_i)} = \boldsymbol{s}(\mathfrak{c}_i; \boldsymbol{x}), \quad i = 1, 2.$$

Thus, letting r go to zero in (5.20), we finally obtain

$$\boldsymbol{s}(\mathfrak{c}_1; \boldsymbol{x}) = \boldsymbol{s}(\mathfrak{c}_2; \boldsymbol{x}),$$

which shows that $\boldsymbol{s}(\mathfrak{c}; \boldsymbol{x})$ has the same value for all surfaces $\mathfrak{c}$ with the same unit normal $\boldsymbol{n}$. The exceptional case when $\boldsymbol{x} \in \overset{-1}{\theta_t}(\overline{\mathfrak{B}})$ and $\boldsymbol{n}$ is directed toward the interior of $\theta_t(\mathfrak{B})$ is taken care of by the definition (4.5).

The vector $\boldsymbol{s}(\boldsymbol{x}, \boldsymbol{n})$ is called the STRESS acting at $\boldsymbol{x}$ across the surface element with unit normal $\boldsymbol{n}$. By (4.6) the contact force $\boldsymbol{S}(\mathfrak{c})$ acting across $\mathfrak{c}$ is given by

$$\boldsymbol{S}(\mathfrak{c}) = \int_{\theta_t(\mathfrak{c})} \boldsymbol{s}(\boldsymbol{x}, \boldsymbol{n})\mathrm{d}A, \tag{5.21}$$

where $\boldsymbol{n}$ is the unit normal at $\boldsymbol{x}$ to the oriented surface $\theta_t(\mathfrak{c})$. By theorem II we have

$$\boldsymbol{s}(\boldsymbol{x}, \boldsymbol{n}) = -\boldsymbol{s}(\boldsymbol{x}, -\boldsymbol{n}). \tag{5.22}$$

The following two additional assumptions suffice to ensure the validity of the classical theorems of continuum mechanics:

(a) The stress $\boldsymbol{s}(\boldsymbol{x}, \boldsymbol{n})$, for each $\boldsymbol{n}$, is a smooth function of $\boldsymbol{x} \in \theta_t(\mathfrak{B})$.

(b) For almost all $X \in \mathfrak{B}$, the limit

$$\boldsymbol{b}(X) = \lim_{\mathfrak{P}\to X} \frac{1}{m(\mathfrak{P})} \boldsymbol{B}_{\mathfrak{B}}(\mathfrak{P}), \tag{5.23}$$

where $\mathfrak{P}$ is a neighborhood of X contracting to X, exists.

Under these assumptions, one can prove the following theorems in the classical manner:

(1) There is a field of linear transformations $S(\boldsymbol{x})$, $\boldsymbol{x} \in \theta_t(\mathfrak{B})$, such that

$$\boldsymbol{s}(\boldsymbol{x}, \boldsymbol{n}) = S(\boldsymbol{x})\boldsymbol{n}. \tag{5.24}$$

$S(\boldsymbol{x})$ is called the STRESS TENSOR at $\boldsymbol{x}$.

(2) The stress tensor $S(\boldsymbol{x})$ is symmetric.

(3) Cauchy's equation of motion

$$\operatorname{div} S + \rho\boldsymbol{b} = \rho\dot{\boldsymbol{v}} \tag{5.25}$$

holds, where S is the stress tensor, ρ is the mass density, $\dot{\boldsymbol{v}}$ is the acceleration, and $\boldsymbol{b}$ is defined by (5.23).

6. Equivalence of dynamical processes. The position of a particle can be specified physically not in an absolute sense but only relative to a given *frame of reference*. Such a frame is a set of objects whose mutual distances change very little in time, like the walls of a laboratory, the fixed stars, or the wooden horses on a merry-go-round. In classical physics, a change of frame corresponds to a transformation of space and time which preserves distances and time intervals. It is well known that the most general such transformation is of the form

$$\begin{aligned} \boldsymbol{x}^* &= \boldsymbol{c}(t) + Q(t)(\boldsymbol{x} - \boldsymbol{O}), \\ t^* &= t + a, \end{aligned} \tag{6.1}$$

where $\boldsymbol{c}(t)$ is a point valued function of t, $Q(t)$ is a function of t whose values are orthogonal transformations, a is a real constant, and $\boldsymbol{O}$ is a point, which may be fixed once and for all. We assume that $\boldsymbol{c}(t)$ and $Q(t)$ are twice continuously differentiable. A change of frame (6.1) also induces a transformation on vectors and tensors. A vector $\boldsymbol{u}$, for example, is transformed into

$$\boldsymbol{u}^* = Q(t)\boldsymbol{u}. \tag{6.2}$$

Let $\{\mathfrak{B}, \theta_t, \boldsymbol{F}_{\mathfrak{B},t}\}$ be a dynamical process. A change of frame $\{\boldsymbol{c}, Q, a\}$

will transform the motion θ_t into a new motion θ_t' defined by

$$\theta_t'(X) = \boldsymbol{c}(t-a) + Q(t-a)[\theta_{t-a}(X) - \boldsymbol{O}]. \tag{6.3}$$

The velocities and the accelerations of the two motions θ_t and θ_t' are, in general, not related by the transformation formula (6.2) for vectors. They depend on the choice of the frame of reference. We say that they are not *objective*. However, there are objective kinematical quantities, for example the rate of deformation tensor.

If we wish to assume that forces have an objective meaning we would have to require that $\boldsymbol{F}_{\mathfrak{P},t}(\mathfrak{C})$ transforms according to the law (6.2) under a change of frame. However, when this assumption is made, a dynamical process does not transform into a dynamical process because the axioms (D.1) and (D.2) are not preserved, except when $\boldsymbol{c}$ is linear in t and Q is constant. It is this difficulty which has led to the concept of absolute space and which has caused much controversy in the history of mechanics. A clarification was finally given by Einstein in his general theory of relativity, in which gravitational forces and inertial forces cannot be separated from each other in an objective manner. If we wish to stay in the realm of classical mechanics we may resolve the paradox by sacrificing the objectivity of the external body forces while retaining the objectivity of the essential types of forces, the contact forces and the mutual body forces. This can be done by assuming that the forces transform according to a law of the form

$$\boldsymbol{F}'_{\mathfrak{P},t}(\mathfrak{C}) = Q(t-a)\boldsymbol{F}_{\mathfrak{P},t-a}(\mathfrak{C}) + \boldsymbol{I}(\mathfrak{C}, t). \tag{6.4}$$

Here $\boldsymbol{I}(\mathfrak{C}, t)$ will be called the INERTIAL FORCE acting on $\mathfrak{C}$ due to the change of frame $\{\boldsymbol{c}, Q, a\}$.

DEFINITION 7: *Two dynamical processes* $\{\mathfrak{B}, \theta_t, \boldsymbol{F}_{\mathfrak{P},t}\}$ *and* $\{\mathfrak{B}, \theta_t', \boldsymbol{F}'_{\mathfrak{P},t}\}$ *are called* EQUIVALENT *if there is a change of frame* $\{\boldsymbol{c}, Q, a\}$ *such that* θ_t' *and* $\boldsymbol{F}'_{\mathfrak{P},t}$ *are related to* θ_t *and* $\boldsymbol{F}_{\mathfrak{P},t}$ *by* (6.3) *and* (6.4).

The classical analysis of relative motion shows that the inertial force $\boldsymbol{I}(\mathfrak{C}, t)$ is necessarily of the form

$$\boldsymbol{I}(\mathfrak{C}, t) = \int_{\mathfrak{C}} \boldsymbol{i}(X, t)\,dm \tag{6.5}$$

with

$$\boldsymbol{i}(X, t) = \ddot{\boldsymbol{c}}(t-a) + 2V(t-a)[\boldsymbol{v}'(X, t) - \dot{\boldsymbol{c}}(t-a)] + [V^2(t-a) - \dot{V}(t-a)][\theta_t'(X) - \boldsymbol{c}(t-a)] \tag{6.6}$$

where $\boldsymbol{v}'$ is the velocity of the motion θ_t', and where $V(t)$ is defined by

$$V(t) = \dot{Q}(t)Q(t)^{-1}. \tag{6.7}$$

It is not hard to see that the inertial force $\boldsymbol{I}$ gives a contribution only to the external body forces and that the contact forces and the mutual forces transform according to (6.2) and hence are objective. The external body forces and the inertial forces cannot be separated from each other in an objective manner. Experience shows that, for the body consisting of the entire solar system, there are frames relative to which the external body forces nearly vanish. These are the classical Galilean frames. Two equivalent dynamical processes really correspond to the same physical process, viewed only from two different frames of reference.

7. Constitutive assumptions. An axiom that characterizes the particular material properties of a body is called a CONSTITUTIVE ASSUMPTION. It restricts the class of dynamical processes the body can undergo. A familiar example is the assumption that the body is rigid. It restricts the possible motions to those in which the distance between any two particles remains unchanged in time. More important for modern continuum mechanics are constitutive assumptions in the form of functional relations between the stress tensor S and the motion θ_t. Such relations are called CONSTITUTIVE EQUATIONS (sometimes also rheological equations of state or stress-strain relations). A classical example is the constitutive equation for linear viscous fluids

$$S = (-p + \lambda \operatorname{tr} D)I + 2\mu D, \tag{7.1}$$

where D is the rate of deformation tensor, I is the unit tensor, p is the pressure, and λ and μ are viscosity constants. A wide variety of constitutive equations have been investigated in recent years [5], and a general theory of such equations has been developed [2].

Constitutive assumptions are subject to a general restriction:

PRINCIPLE OF OBJECTIVITY: *If a dynamical process is compatible with a constitutive assumption then all processes equivalent to it must also be compatible with this constitutive assumption. In other words, constitutive assumptions must be invariant under changes of frame.*

This principle, although implicitly used by many scientists in the history of mechanics, was stated explicitly first by Oldroyd [3] and was

[5] A review of the literature and a bibliography is given in [1].

clarified further by the author [4]. It is of great importance in the theory of constitutive equations.

8. Unsolved problems. The axiomatic treatment given here is still too special. It does not cover concentrated forces, contact couples and body couples, sliding, impact, rupture, and other discontinuities, singularities, and degeneracies. It would be desirable to have a universal scheme which covers any conceivable situation.

A more fundamental physical problem is to find a rigorous unified theory of continuum mechanics and thermodynamics. Classical thermodynamics deals only with equilibrium states and hence is not adequate for processes with fast changes of state in time. Such a unified theory should lead to further restrictive conditions on the form of constitutive equations and hence to more definite and realistic theories for special materials. Also, a satisfactory connection with statistical mechanics can be expected only after such a theory has been developed.

References

[1] NOLL, W., ERICKSEN, J. L. and TRUESDELL, C., *The Non-linear Field Theories of Mechanics*. Article to appear in the Encyclopedia of Physics.

[2] ——, *A general theory of constitutive equations*. To appear in Archive for Rational Mechanics and Analysis.

[3] OLDROYD, J. G., *On the formulation of rheological equations of state*. Proceedings of the Royal Society of London (A) 200 (1950), 523–541.

[4] NOLL, W., *On the Continuity of the Solid and Fluid States*. Journal of Rational Mechanics and Analysis 4 (1955), 3–81.

On Certain Steady Flows of General Fluids

BERNARD D. COLEMAN & WALTER NOLL

Contents

Introduction

In a recent article on the mechanical behavior of continuous media [*1*], the physical concept of a *fluid* was given a mathematical definition. We believe that this definition covers almost all real fluids (whenever thermal and other non-mechanical effects can be disregarded) and is more general than any proposed previously because it accounts for all hereditary effects including stress relaxation. The general fluids covered by this definition may even exhibit, in some situations, physical phenomena which are usually attributed to solids.

Here we show that certain steady flow problems can be solved assuming only the definition of a general fluid and incompressibility. The class of problems considered includes most of the classical flows: simple shearing flow, channel flow (*i.e.* flow between fixed plates), Poiseuille flow, and Couette flow. It also includes torsional and other flows for which we do not work out the details, because, under reasonable body forces, such flows are not compatible with the dynamical equations unless inertia is neglected.

The solutions obtained here are in terms of three unspecified real functions of one variable which depend on the particular material; we call them *material functions*. Our solutions are simply related. If experimental measurements are combined with any one of the solutions to determine the three material functions, then complete stress and velocity profiles can be predicted for the other flows. Thus, our work really establishes correspondence principles for the flows considered. It is hoped that our results may help experimental rheologists to rationalize the treatment of their data by giving precise meaning to such concepts as "shear-dependent viscosity".

If one uses a special theory, either molecular or phenomenological, to solve a particular flow problem of our class (say, simple shearing flow), then our three material functions can be readily identified*. The solutions to the other flow problems can then be obtained from the formulae given here without further reference to the original special theory**.

It should be pointed out that our three material functions do not determine the behavior of the fluid in all experimental situations, since the functional that predicts the response of a material to all situations may contain vastly more information than the three material functions introduced here.

Some of the solutions and relations obtained here have appeared in the literature, but the derivations given have rested upon special assumptions about the nature of the fluid considered. Of these previous investigations, the most general is RIVLIN'S study of flows in fluids of the differential type [4]. Such fluids do not have long range memory and hence constitute only a special class of the general fluids considered here. It turns out, however, that RIVLIN'S solutions (including those for helical flow, which will be discussed in a future note) remain valid for general fluids.

This paper can be read independently of reference [1] if the reader is willing to accept equations (1.1)—(1.3) below.

Notation. We denote vectors and points by bold face Latin or Greek minuscules: $\boldsymbol{v}$, $\boldsymbol{x}$, $\boldsymbol{\xi}$.

Second order tensors are always denoted by light face Latin majuscules: A, B, I, Q. The transpose of a tensor Q is written Q^T. The unit or "metric" tensor is denoted by I. The symbol t_{ij} denotes the *physical* components of T relative to an orthogonal, but not necessarily Cartesian, coordinate system, except when a statement is made to the contrary. We use the summation convention throughout. The matrix of the physical components of T is denoted by (T) or $\|t_{ij}\|$, and the trace of T is written $\operatorname{tr} T$.

Partial differentiation is indicated as follows: $\partial_y = \partial/\partial y$ or $\partial_i = \partial/\partial x_i$.

1. Basic concepts

In a general fluid ([1], §§ 21 and 22) the stress tensor S at time t depends on the history of the motion up to t and is given by a constitutive equation which may be stated in the form (22.12) of [1]. In this paper we discuss only isochoric motions; we do not make the dependence of S on the density $\varrho(t)$ explicit, and we write

$$S(t) = \mathfrak{H}(C_t^t). \tag{1.1}$$

Here $\mathfrak{H}$ is a functional of the tensor-valued function C_t^t and satisfies the relation

$$Q\,\mathfrak{H}(C_t^t)\,Q^T = \mathfrak{H}(Q\,C_t^t\,Q^T) \tag{1.2}$$

for any constant orthogonal tensor Q (*cf.* [1], (21.7)). The function C_t^t is the history of the right Cauchy-Green tensor taken with respect to the configuration

* A summary of results pertaining to the normal stress effect for special theories has been presented by MARKOVITZ [2]. The present investigation gives a proof of his equations (12)—(16), which he offered as conjectures.

** For example, much of the work in Chapter IV of [3] could have been avoided if the present results had been known to the author at the time.

at time t and is defined by

$$C_t^t(s) = C_t(t+s) = F_t^T(t+s)\, F_t(t+s), \qquad s \leqq 0, \tag{1.3}$$

where $F_t(t+s)$ is the displacement gradient at time $t+s$ relative to the configuration at time t (*cf.* [1], §§ 6 and 7).

In this paper we consider homogeneous *incompressible* fluids. The stress is then determined by the history of the motion only up to a hydrostatic pressure p, and (1.1) must be replaced by

$$T(t) = \mathfrak{H}(C_t^t), \tag{1.4}$$

where

$$T = S + pI \tag{1.5}$$

is called the *extra-stress*. Moreover, the tensor-valued functional $\mathfrak{H}$ is determined only up to an arbitrary scalar-valued functional of C_t^t. We remove this indeterminacy by the normalization

$$\operatorname{tr} T = \operatorname{tr}[\mathfrak{H}(C_t^t)] = 0. \tag{1.6}$$

The pressure p then reduces to the mean pressure

$$p = -\tfrac{1}{3}\operatorname{tr} S, \tag{1.7}$$

and T is the deviatoric part of S. We note that the extra stress T must vanish if the motion reduces to a state of rest, in which case we have $C_t^t(s) = I = \text{const.}$ Hence

$$0 = \mathfrak{H}(I). \tag{1.8}$$

Consider a particle which at time t finds itself at the position $\boldsymbol{x}$, with general coordinates x^i, and which was at time $t+s$, $s \leqq 0$, at the point $\boldsymbol{\xi}$, with coordinates ξ^i. For the dependence of $\boldsymbol{\xi}$ on $\boldsymbol{x}$, t, and $t+s$, we write

$$\boldsymbol{\xi} = \boldsymbol{\vartheta}_t(\boldsymbol{x}, t+s). \tag{1.9}$$

The function $\boldsymbol{\vartheta}_t$, with coordinates $\vartheta^i_{(t)}$, is the displacement function relative to the configuration at time t.

It follows from (1.3) that the covariant components of $C_t^t(s)$, relative to the natural basis at $\boldsymbol{x}$ of the coordinate system, are

$$c^{(t)}_{(t)\,ij}(s) = f^k_{(t)\,i}(t+s)\, f^l_{(t)\,j}(t+s)\, g_{kl}(t+s), \tag{1.10}$$

where

$$f^i_{(t)\,j}(t+s) = \partial_j[\vartheta^i_{(t)}(\boldsymbol{x}, t+s)], \tag{1.11}$$

and $g_{kl}(t+s)$ denotes the covariant components of the unit tensor I at $\boldsymbol{\xi} = \boldsymbol{\vartheta}_t(\boldsymbol{x}, t+s)$. If the velocity $\boldsymbol{v}(\boldsymbol{\xi}, t+s)$ is given as a function of the position $\boldsymbol{\xi}$ and the time $t+s$, the displacement function

$$\boldsymbol{\xi}(s) = \boldsymbol{\vartheta}_t(\boldsymbol{x}, t+s) \tag{1.12}$$

is that solution of the differential equation

$$\frac{d\boldsymbol{\xi}}{ds} = \boldsymbol{v}(\boldsymbol{\xi}, t+s) \tag{1.13}$$

which satisfies the initial condition

$$\boldsymbol{\xi}(0) = \boldsymbol{\vartheta}_t(\boldsymbol{x}, t) = \boldsymbol{x}. \tag{1.14}$$

If the motion is a steady flow, the velocity $\boldsymbol{v}(\boldsymbol{\xi})$ depends on only the position $\boldsymbol{\xi}$ and not on the time $t+s$. In this case $\boldsymbol{\vartheta}_t(\boldsymbol{x}, t+s)$, and hence also $F_t(t+s)$ and $C_t^t(s)$, are independent of the reference time t. It follows from (1.4) that the extra-stress T must be independent of time.

We assume that the body forces $\boldsymbol{g}$ (per unit mass) have a single-valued potential ψ:

$$\boldsymbol{g} = -\operatorname{grad}\psi. \tag{1.15}$$

We introduce the *modified pressure*:

$$\varphi = p + \varrho\psi. \tag{1.16}$$

It coincides with the ordinary pressure p if there are no body forces. The dynamical equations take the form

$$\operatorname{div} T - \operatorname{grad}\varphi = \varrho\dot{\boldsymbol{v}}. \tag{1.17}$$

2. A class of steady flows

We consider here steady flows for which the velocity field $\boldsymbol{v}(\boldsymbol{x})$ has the contravariant components

$$\{v^1, v^2, v^3\} = \{0, v(x^1), 0\} \tag{2.1}$$

in some appropriate *orthogonal* coordinate system x^1, x^2, x^3. The matrix $\|g_{ij}\|$ of the covariant components of the unit tensor I is then diagonal. We further assume that the g_{ij} are independent of x^2, so that $\|g_{ij}\|$ has the form

$$\|g_{ij}(\boldsymbol{x})\| = \begin{Vmatrix} [g_1(x^1, x^3)]^2 & 0 & 0 \\ 0 & [g_2(x^1, x^3)]^2 & 0 \\ 0 & 0 & [g_3(x^1, x^3)]^2 \end{Vmatrix}, \tag{2.2}$$

where g_1, g_2, and g_3 are three positive functions of x^1 and x^3. The flows defined in this manner are isochoric, because the equation

$$\operatorname{div}\boldsymbol{v} = 0, \qquad \text{i.e.,} \qquad (\sqrt{g})^{-1}\partial_i(\sqrt{g}\,v^i) = 0, \tag{2.3}$$

where $\sqrt{g} = g_1 g_2 g_3$, is obviously satisfied.

In the special case of the velocity field (2.1), integration of the differential equation (1.13) with the initial condition (1.14) yields the simple result

$$\xi^1 = x^1, \qquad \xi^2 = x^2 + s\,v(x^1), \qquad \xi^3 = x^3. \tag{2.4}$$

From (2.4), (1.12) and (1.11) we obtain

$$\|f^i_{(t)j}(t+s)\| = \begin{Vmatrix} 1 & 0 & 0 \\ s\,v'(x^1) & 1 & 0 \\ 0 & 0 & 1 \end{Vmatrix}, \tag{2.5}$$

where v' denotes the derivative of v. Equations (2.4) and (2.2) imply that

$$g_i(\xi^1, \xi^3) = g_i(x^1, x^3), \qquad i = 1, 2, 3; \tag{2.6}$$

i.e., that the covariant components of the unit tensor remain constant along the paths of the particles. Using equations (1.10), (2.5), (2.2), and (2.6) we obtain the following expression for the covariant components of the tensor $C_t^t(s)$:

$$\|c^{(t)}_{(t)\,ij}(s)\| = \left\| \begin{matrix} [g_1(x^1,x^3)]^2 + [s\,g_2(x^1,x^3)\,v'(x^1)]^2 & s\,[g_2(x^1,x^3)]^2\,v'(x^1) & 0 \\ s\,[g_2(x^1,x^3)]^2\,v'(x^1) & [g_2(x^1,x^3)]^2 & 0 \\ 0 & 0 & [g_3(x^1,x^3)]^2 \end{matrix} \right\|. \qquad (2.7)$$

The matrix $(C_t^t(s))$ of the physical components of $C_t^t(s)$ is

$$(C_t^t(s)) = \left\| \frac{c^{(t)}_{(t)\,ij}(s)}{g_i(x^1,x^3)\,g_j(x^1,x^3)} \right\| = \left\| \begin{matrix} 1+\varkappa^2 s^2 & \varkappa s & 0 \\ \varkappa s & 1 & 0 \\ 0 & 0 & 1 \end{matrix} \right\|, \qquad (2.8)$$

where

$$\varkappa = \frac{g_2(x^1,x^3)}{g_1(x^1,x^3)}\,v'(x^1). \qquad (2.9)$$

In the remainder of this paper we shall always use the physical components of second order tensors.

3. Reduction of the general constitutive equation

The results of the previous section may be summarized in this way: If the velocity field has the form (2.1) in an orthogonal coordinate system obeying (2.2), the function C_t^t has the form

$$C_t^t(s) = I + s\,A + s^2\,B, \qquad (3.1)$$

where the matrices (A), (B) of the physical components of A and B are given by

$$(A) = \varkappa \left\| \begin{matrix} 0 & 1 & 0 \\ 1 & 0 & 0 \\ 0 & 0 & 0 \end{matrix} \right\|, \qquad (B) = \varkappa^2 \left\| \begin{matrix} 1 & 0 & 0 \\ 0 & 0 & 0 \\ 0 & 0 & 0 \end{matrix} \right\|. \qquad (3.2)$$

When (3.1) holds, the function C_t^t is completely determined by A and B, and the functional $\mathfrak{H}$ of (1.4) reduces to a function $\mathfrak{h}$ of the two tensors A and B, so that

$$T = \mathfrak{h}(A, B). \qquad (3.3)^\star$$

It follows from (1.2) that the tensor function $\mathfrak{h}$ must be isotropic; *i.e.*,

$$\mathfrak{h}(Q\,A\,Q^T,\, Q\,B\,Q^T) = Q\,\mathfrak{h}(A,B)\,Q^T \qquad (3.4)$$

must hold for all orthogonal tensors Q. It is a consequence of this fact that the component functions $\mathfrak{h}_{ij}$ of $\mathfrak{h}$ are the same for all *orthonormal* bases. Since physical components are components relative to orthonormal bases, the functions $\mathfrak{h}_{ij}$ do not depend on the coordinate system or the point under consideration, if physical components are used.

The normalization (1.6) of the functional $\mathfrak{H}$ implies that the function $\mathfrak{h}$ is normalized by

$$\operatorname{tr} T = \operatorname{tr}[\mathfrak{h}(A,B)] = 0. \qquad (3.5)$$

⋆ We note that, if (3.1) holds, A and $2B$ coincide with the first two Rivlin-Ericksen tensors A_1 and A_2 as defined in [*1*], (8.9). Equation (3.3) has the form of a constitutive equation of the differential type (*cf.* [*1*] (23.14)).

Furthermore, it follows from (1.8) that

$$0 = \mathfrak{h}(0, 0). \tag{3.6}$$

We now consider the orthogonal tensor Q whose physical components are given by the matrix

$$(Q) = \begin{Vmatrix} 1 & 0 & 0 \\ 0 & 1 & 0 \\ 0 & 0 & -1 \end{Vmatrix}. \tag{3.7}$$

Then, by (3.2), we have

$$Q A Q^T = A, \qquad Q B Q^T = B. \tag{3.8}$$

Hence, from (3.4) we get

$$Q T Q^T = T \tag{3.9}$$

for $T = \mathfrak{h}(A, B)$. Denoting the *physical components* of T by t_{ij}, we find that the matrix of the physical components of $Q T Q^T$ is given by

$$(Q T Q^T) = \begin{Vmatrix} t_{11} & t_{12} & -t_{13} \\ t_{21} & t_{22} & -t_{23} \\ -t_{31} & -t_{32} & t_{33} \end{Vmatrix}. \tag{3.10}$$

According to (3.9), this matrix must coincide with the matrix $(T) = \|t_{ij}\|$, which is possible only if $t_{13} = t_{31} = t_{23} = t_{32} = 0$. Hence (T) must have the form

$$(T) = \begin{Vmatrix} t_{11} & t_{12} & 0 \\ t_{21} & t_{22} & 0 \\ 0 & 0 & t_{33} \end{Vmatrix}. \tag{3.11}$$

By (3.5), we have

$$t_{11} + t_{22} + t_{33} = 0. \tag{3.12}$$

From the fact that the component functions $\mathfrak{h}_{ij}$ are independent of the coordinate system, and from (3.2) and (3.3), it follows that the physical components t_{ij} of T are functions of $\varkappa$ alone. It is clear from (3.11) and (3.12) that these components t_{ij} may be expressed in terms of three independent functions of $\varkappa$. We choose

$$t_{12} = \tau(\varkappa), \qquad t_{11} - t_{33} = \sigma_1(\varkappa), \qquad t_{22} - t_{33} = \sigma_2(\varkappa). \tag{3.13}$$

For the physical components $s_{ij} = t_{ij} - p\,\delta_{ij}$ of the stress tensor we also have

$$s_{12} = s_{21} = \tau(\varkappa), \tag{3.14}$$

$$s_{11} - s_{33} = \sigma_1(\varkappa), \qquad s_{22} - s_{33} = \sigma_2(\varkappa). \tag{3.15}$$

The three functions τ, σ_1, and σ_2 depend only on the material, not on the particular flow; they are *material functions*.

The functions τ, σ_1, and σ_2 are not completely arbitrary, as may be seen from the following argument: For a second choice of Q in (3.4) we take

$$(Q) = \begin{Vmatrix} 1 & 0 & 0 \\ 0 & -1 & 0 \\ 0 & 0 & 1 \end{Vmatrix}. \tag{3.16}$$

Equation (3.2) then yields

$$Q A Q^T = -A, \qquad Q B Q^T = B, \tag{3.17}$$

and, from (3.11), we get

$$(Q T Q^T) = \begin{Vmatrix} t_{11} & -t_{12} & 0 \\ -t_{21} & t_{22} & 0 \\ 0 & 0 & t_{33} \end{Vmatrix} \tag{3.18}$$

It is clear from (3.2) that the conjugation operation (3.17) is equivalent to replacing $\varkappa$ by $-\varkappa$. Thus, by (3.13), equation (3.18) states that

$$\tau(-\varkappa) = -\tau(\varkappa), \qquad \sigma_i(\varkappa) = \sigma_i(-\varkappa); \tag{3.19}$$

i.e., that τ must be an *odd* function, while σ_1 and σ_2 must be *even* functions. No further restrictions are imposed on τ, σ_1, and σ_2 by the isotropy condition (3.4).

The condition (3.6) implies that the material functions τ, σ_1, and σ_2 must vanish for $\varkappa = 0$:

$$\tau(0) = \sigma_1(0) = \sigma_2(0) = 0. \tag{3.20}$$

A further restrictive condition on τ follows from the principle that the dissipation of energy must be positive. This condition is

$$\varkappa\, \tau(\varkappa) > 0, \qquad \varkappa \neq 0; \tag{3.21}$$

i.e. $\tau(\varkappa)$ must have the same sign as $\varkappa$. If it is assumed that τ is twice continuously differentiable, then (3.20) and (3.21) imply that $\tau(\varkappa)$ must be a strictly increasing function of $\varkappa$ in some interval $-\varkappa_0 \leqq \varkappa \leqq +\varkappa_0$ around $\varkappa = 0$. In this interval τ will have a strictly increasing and odd inverse, $\overset{-1}{\tau}$. In the remainder of this paper, whenever $\overset{-1}{\tau}$ occurs, we assume that we are in the range in which $\overset{-1}{\tau}$ is defined. If $\overset{-1}{\tau}$ is not defined over the entire real axis then the flows which we discuss below will not occur at high speeds.

We can give a precise meaning to the "shear-dependent viscosity" used in the rheological literature by identifying it with the material function η:

$$\eta(\varkappa) = \frac{\tau(\varkappa)}{\varkappa}. \tag{3.22}$$

Clearly η must be an *even* function. Since $\tau(0) = 0$ and τ is twice differentiable, η is differentiable, even at $\varkappa = 0$, and

$$\eta'(0) = 0. \tag{3.23}$$

If we assume that σ_1 and σ_2 are differentiable, we also have

$$\sigma_1'(0) = \sigma_2'(0) = 0. \tag{3.24}$$

If a polynomial approximation for η, σ_1, and σ_2, in a neighborhood of $\varkappa = 0$, is used to fit experimental data, only even powers of $\varkappa$ can occur.

4. Rectilinear flow

If the velocity field has the form (2.1) in a Cartesian coordinate system

$$x^1 = x, \qquad x^2 = y, \qquad x^3 = z, \tag{4.1}$$

then we call the motion *rectilinear flow*. In this case

$$g_1 = g_2 = g_3 = 1, \tag{4.2}$$

and hence, by (2.9),

$$\varkappa = v'(x). \tag{4.3}$$

By (3.13), the extra stresses t_{ij} are then functions of x only, and the dynamical equations (1.17) become

$$\partial_x t_{xx} - \partial_x \varphi = 0, \quad \partial_x t_{xy} - \partial_y \varphi = 0, \quad \partial_z \varphi = 0. \tag{4.4}$$

A simple analysis shows that these equations are satisfied only if

$$t_{xy}(x) = -a\,x + b, \tag{4.5}$$

$$\varphi = \varphi(x, y) = -a\,y + t_{xx}(x) - c, \tag{4.6}$$

where a, b, and c are constants.

There are two important special cases of rectilinear flow which we shall discuss in the next two sections.

5. Simple shearing flow

This is a rectilinear flow between an infinite plate I at rest and an infinite plate II moving with a constant speed V parallel to plate I. We denote the distance between the plates by d. We choose the y and z axes to be in the plane of plate I and such that plate II moves in the y direction. The x axis is then perpendicular to the plates with $x = 0$ at plate I. Assuming that the fluid adheres to the walls, we have the boundary conditions

$$v(0) = 0, \quad v(d) = V. \tag{5.1}$$

We assume now that the modified pressure φ, defined by (1.16), has no gradient in direction of the flow; *i.e.*, that $\partial_y \varphi = 0$. We then have $a = 0$ in (4.6). Hence, by (3.13), (4.3) and (4.5), the shearing stress

$$t_{xy}(x) = \tau(v'(x)) = b \tag{5.2}$$

is constant. It follows that $v'(x) = \overset{-1}{\tau}(b)$ must be constant and hence that $v(x)$ is linear. The conditions (5.1) then imply that

$$v(x) = \frac{V}{d}\,x, \tag{5.3}$$

and (4.3) becomes

$$\varkappa = \frac{V}{d} = \text{const}. \tag{5.4}$$

According to (3.14), the shearing stress is

$$s_{xy} = \tau\left(\frac{V}{d}\right). \tag{5.5}$$

Since s_{xy} is equal to the tangential force per unit area that must be applied to the moving plate II in order to produce the flow, equation (5.5) can be used to determine the material function τ experimentally.

It follows from (5.4) and (3.13) that the t_{ij} must actually be constant. Hence, by (4.6), since $a=0$,

$$\varphi = t_{xx} - c = \text{const}. \tag{5.6}$$

Using (1.16) and $s_{xx} = t_{xx} - p$, we find

$$s_{xx} = c - \varrho\psi, \tag{5.7}$$

and (3.15) becomes

$$s_{xx} - s_{zz} = \sigma_1\left(\frac{V}{d}\right), \qquad s_{yy} - s_{zz} = \sigma_2\left(\frac{V}{d}\right). \tag{5.8}$$

Hence, the matrix of the stress tensor is

$$(S) = \left\| \begin{matrix} c-\varrho\psi & \tau\left(\frac{V}{d}\right) & 0 \\ \tau\left(\frac{V}{d}\right) & c-\varrho\psi+\sigma_2\left(\frac{V}{d}\right)-\sigma_1\left(\frac{V}{d}\right) & 0 \\ 0 & 0 & c-\varrho\psi-\sigma_1\left(\frac{V}{d}\right) \end{matrix} \right\|. \tag{5.9}$$

Since the normal stresses can be measured, at least in principle, equations (5.8) can be used to determine the material functions σ_1 and σ_2.

6. Flow through a channel

This is a rectilinear flow between two parallel infinite plates which are both at rest. We now place the y and z axes in a plane half-way between the plates; the x axis is then perpendicular to the plates. Let the distance between the plates be d; it follows that the plates are in the planes $x = \pm\frac{1}{2}d$.

The fluid adheres to the plates. For rectilinear flow, to get any motion at all we must have $a \neq 0$ in (4.6). When a is positive, there is a flow in the positive y-direction. From (3.13), (4.3) and (4.5) we get $v'(x) = \overset{-1}{\tau}(-ax+b)$. The boundary conditions are $v(\pm\frac{1}{2}d) = 0$. The fact that $\overset{-1}{\tau}$ is an odd function implies $b=0$; *i.e.*,

$$v'(x) = -\overset{-1}{\tau}(ax). \tag{6.1}$$

After integration we obtain the velocity profile

$$v(x) = -\int_{-\frac{1}{2}d}^{x} \overset{-1}{\tau}(a\xi)\, d\xi. \tag{6.2}$$

The total applied force f in the direction of the flow, exerted on a column of fluid with cross-section $\mathscr{A}$ and lying between two planes $y = y_{\text{I}}$ and $y = y_{\text{II}}$, is given by

$$f = \Big(\int_{\mathscr{A}} s_{yy}\, dA\Big)_{y=y_{\text{II}}} - \Big(\int_{\mathscr{A}} s_{yy}\, dA\Big)_{y=y_{\text{I}}} - \int_{y_{\text{I}}}^{y_{\text{II}}} \int_{\mathscr{A}} (\partial_y \psi)\, \varrho\, dA\, dy. \tag{6.3}$$

This reduces to

$$f = \int_{\mathscr{A}} \{(s_{yy} - \varrho\psi)_{y=y_{\text{II}}} - (s_{yy} - \varrho\psi)_{y=y_{\text{I}}}\}\, dA. \tag{6.4}$$

By (1.16) and (4.6) we have

$$s_{yy} - \varrho\psi = t_{yy} - p - \varrho\psi = t_{yy} - \varphi = ay + t_{yy}(x) - t_{xx}(x) + c,$$

hence

$$s_{yy} - \varrho\psi = a\,y + h(x). \tag{6.5}$$

Substitution into (6.4) yields

$$f = a\,(y_{II} - y_I)\,A\,,$$

where A is the area of the cross-section of the column. We thus have

$$a = \frac{f}{A\,(y_{II} - y_I)}\,, \tag{6.6}$$

which gives a physical meaning to the parameter a. It is the *applied force per unit volume in the direction of the flow*. It is the driving force that produces the flow. If there are no body forces, a reduces to what is usually called the *pressure head* per unit length. It must be produced by a pump or similar device. If the only body force is gravity, acting in the direction of flow, and if there is no pressure head, then a is simply the *specific weight* of the fluid. In any case, a is a measurable quantity. We call it the *driving force*.

The volume discharge per unit time through a cross-section of the channel one unit deep is

$$Q = \int_{-\frac{1}{2}d}^{\frac{1}{2}d} v(x)\,dx. \tag{6.7}$$

Integrating by parts, remembering the boundary conditions $v(\pm\frac{1}{2}d) = 0$, and using (6.1), we find

$$Q = -\int_{-\frac{1}{2}d}^{\frac{1}{2}d} x\,v'(x)\,dx = \int_{-\frac{1}{2}d}^{+\frac{1}{2}d} x\,\overset{-1}{\tau}(a\,x)\,dx. \tag{6.8}$$

We finally obtain

$$Q = \frac{2}{a^2}\int_0^{\frac{1}{2}ad} \xi\,\overset{-1}{\tau}(\xi)\,d\xi. \tag{6.9}$$

Thus, the experimentally measurable rate of discharge Q is related to the measurable driving force a through the material function τ which can be determined by independent measurements.

Equation (6.9) is easily solved for the function $\overset{-1}{\tau}$ in terms of the function $Q = Q(a)$:

$$\overset{-1}{\tau}\left(\frac{a\,d}{2}\right) = \frac{2}{a\,d^2}\,\frac{\mathrm{d}}{\mathrm{d}a}\left[a^2\,Q(a)\right]. \tag{6.10}$$

This formula may be used to calculate the material function τ when Q has been measured as a function of a.

The normal stresses are obtained from (4.6), (3.15), (4.3) and (6.1):

$$\begin{aligned} s_{xx} &= a\,y + \varrho\psi - c\,, \\ s_{xx} - s_{zz} &= \sigma_1\left(\overset{-1}{\tau}(a\,x)\right), \\ s_{yy} - s_{zz} &= \sigma_2\left(\overset{-1}{\tau}(a\,x)\right). \end{aligned} \tag{6.11}$$

If the material functions τ, σ_1, and σ_2 are known from independent measurements, these formulae may be used to predict the measurable normal stresses. It will be noticed that the normal stresses are determined only up to a constant hydrostatic pressure of magnitude c.

If only τ is known, then (6.11) may be used to determine σ_1 and σ_2 by measuring $s_{xx}-s_{zz}$ and $s_{yy}-s_{zz}$. If the potential ψ of the body forces is known, then a measurement of s_{xx} just determines the value of the constant c.

7. Poiseuille flow

We consider a steady flow through an infinite circular pipe of radius R and use cylindrical coordinates with the z-axis coincident with the axis of the pipe. We label the coordinates as follows:

$$r = x^1, \qquad z = x^2, \qquad \vartheta = x^3. \tag{7.1}$$

We say that we have Poiseuille flow if in this coordinate system the velocity field has the form (2.1) and if the fluid adheres to the walls of the pipe; *i.e.*

$$v(R) = 0. \tag{7.2}$$

The coordinate system r, z, ϑ is orthogonal and obeys (2.2) with

$$g_1 = g_2 = 1, \qquad g_3 = r. \tag{7.3}$$

Therefore, we find that (2.9) becomes

$$\varkappa = v'(r). \tag{7.4}$$

By (3.13), the extra stresses t_{ij} are then functions of r only, and the dynamical equations (1.17) become

$$\begin{aligned} \partial_r t_{rr} + r^{-1}(t_{rr} - t_{\vartheta\vartheta}) - \partial_r \varphi &= 0, \\ \partial_r t_{rz} + r^{-1} t_{rz} - \partial_z \varphi &= 0, \\ - \partial_\vartheta \varphi &= 0. \end{aligned} \tag{7.5}$$

An elementary analysis shows that these equations are satisfied only if

$$t_{rz} = -\tfrac{1}{2} a r + b r^{-1}, \tag{7.6}$$

$$\varphi = \varphi(r, z) = -a z + h(r), \tag{7.7}$$

$$h'(r) = \partial_r t_{rr} + r^{-1}(t_{rr} - t_{\vartheta\vartheta}), \tag{7.8}$$

where a and b are constants and h is an arbitrary function of r. Since the shearing stress $s_{rz} = t_{rz}$ must be continuous at the center of the pipe (*i.e.* for $r=0$), we must put $b=0$ in (7.6). According to (3.14) and (7.4), we thus have

$$s_{rz} = -\tfrac{1}{2} a r = \tau\big(v'(r)\big);$$

hence, since τ is an odd function,

$$v'(r) = -\overset{-1}{\tau}(\tfrac{1}{2} a r). \tag{7.9}$$

This equation, together with the boundary condition (7.2), determines the velocity profile:

$$v(r) = \int_r^R \overset{-1}{\tau}(\tfrac{1}{2} a \xi)\, d\xi. \tag{7.10}$$

An investigation analogous to that given in § 6 for the case of a flow through a channel shows that the parameter a is again the applied force per unit volume in the direction of the flow. We again call a the *driving force*. It reduces to the pressure head per unit length if there are no body forces. If the flow takes place in a vertical pipe under the action of gravity alone, a is just the specific weight of the fluid.

The volume discharge per unit time through a cross-section of the pipe is

$$Q = 2\pi \int_0^R v(r)\, r\, dr. \tag{7.11}$$

After integration by parts and use of (7.2) and (7.9), we find*

$$Q = \frac{8\pi}{a^3} \int_0^{\frac{1}{2}aR} \xi^2 \overset{-1}{\tau}(\xi)\, d\xi. \tag{7.12}$$

Thus, if the material function τ is known, one can predict, for pipe flow, the dependence of the rate of discharge Q on the driving force a.

As in the case of channel flow, (7.12) is easily inverted:

$$\overset{-1}{\tau}(\tfrac{1}{2} a R) = \frac{1}{\pi a^2 R^3} \frac{\mathrm{d}}{\mathrm{d}a} [a^3 Q(a)]. \tag{7.13}$$

If $Q = Q(a)$ has been measured, this formula may be used to calculate the material function τ.

To calculate the normal stresses, we observe that (3.13), (7.4), (7.9) and the fact that σ_1 is an even function imply

$$t_{rr} - t_{\vartheta\vartheta} = \sigma_1(\overset{-1}{\tau}(\tfrac{1}{2} a r)). \tag{7.14}$$

Integration of (7.8) yields

$$h(r) = t_{rr} - \int_r^R \xi^{-1} \sigma_1(\overset{-1}{\tau}(\tfrac{1}{2} a \xi))\, d\xi - c, \tag{7.15}$$

where c is a constant. Substituting (7.15) into (7.7) and using $\varphi = p + \varrho\psi$ and $s_{rr} = t_{rr} - p$, we obtain

$$s_{rr} = a z + \varrho\psi + \int_r^R \xi^{-1} \sigma_1(\overset{-1}{\tau}(\tfrac{1}{2} a \xi))\, d\xi + c. \tag{7.16}$$

By (3.15), we have

$$s_{rr} - s_{\vartheta\vartheta} = \sigma_1(\overset{-1}{\tau}(\tfrac{1}{2} a r)), \qquad s_{zz} - s_{\vartheta\vartheta} = \sigma_2(\overset{-1}{\tau}(\tfrac{1}{2} a r)). \tag{7.17}$$

The remarks made in § 6 about the prediction of measurable normal stresses in channel flow also apply here.

8. Couette flow

Here we consider a steady flow between two infinite coaxial cylinders; the inner cylinder has radius R_1, and the outer cylinder has radius R_2. Using cylindrical coordinates with the z axis coincident with the axis of the cylinders, we

* Equivalent relations have appeared previously in the literature (*cf.* [5], pp. 249 and 250). The derivation given here, we believe, is the first that is not based on special assumptions.

now label the coordinates

$$r = x^1, \quad \vartheta = x^2, \quad z = x^3. \tag{8.1}*$$

If, in these coordinates, the velocity field has the form (2.1), we speak about Couette flow. We than have

$$g_1 = g_3 = 1, \quad g_2 = r, \tag{8.2}$$

so that (2.9) becomes

$$\varkappa = r v'(r). \tag{8.3}$$

Note that $v(r)$ has now the physical dimension of an angular velocity. By (3.13) and (8.3), the extra stresses t_{ij} depend only on r. The dynamical equations (1.17) become

$$\begin{aligned} \partial_r t_{rr} + r^{-1}(t_{rr} - t_{\vartheta\vartheta}) - \partial_r \varphi &= -\varrho r v^2, \\ \partial_r t_{r\vartheta} + 2r^{-1} t_{r\vartheta} - r^{-1}\partial_\vartheta \varphi &= 0, \\ -\partial_z \varphi &= 0. \end{aligned} \tag{8.4}$$

An analysis of these equations, using the fact that φ must be single valued and hence periodic in ϑ, shows that φ can depend only on r and that $t_{r\vartheta} = M/(2\pi r^2)$, where M is a constant. From (3.13) and (8.3), it follows that

$$t_{r\vartheta} = \tau\big(r v'(r)\big) = \frac{M}{2\pi r^2}. \tag{8.5}$$

The moment per unit height exerted on the fluid inside the cylindrical surface $r = \text{const.}$ is, by (8.5), equal to

$$r(2\pi r)\, t_{r\vartheta} = M \tag{8.6}$$

and is independent of r. From (8.5) we get

$$v'(r) = \frac{1}{r}\,\overset{-1}{\tau}\left(\frac{M}{2\pi r^2}\right). \tag{8.7}$$

We consider further the special case in which the inner cylinder remains at rest while the outer cylinder rotates with constant angular velocity Ω. Assuming that the fluid adheres to the walls, we then have

$$v(R_1) = 0, \quad v(R_2) = \Omega. \tag{8.8}$$

M is the moment per unit height that must be applied to the outer cylinder in order to produce the flow.

Integration of (8.7) with the boundary condition (8.8) yields

$$\Omega = \frac{1}{2}\int_{M/(2\pi R_2^2)}^{M/(2\pi R_1^2)} \frac{1}{\xi}\,\overset{-1}{\tau}(\xi)\, d\xi. \tag{8.9}**$$

Thus, a knowledge of the material function τ makes it possible to predict the angular velocity Ω of the outer cylinder as a function of the moment per unit height M. If the gap between the two cylinders is very small $[(R_2 - R_1)/R_1 \ll 1]$,

* Note that this differs from the labeling (7.1) used for Poiseuille flow.

** A derivation, based on special assumptions, of a formula equivalent to (8.9) is given in [6], § 3.

the formula (8.9) may be approximated by

$$\Omega \approx \frac{R_2 - R_1}{R_1} \overset{-1}{\tau}\left(\frac{M}{2\pi R_2^2}\right). \tag{8.10}$$

We wish to invert (8.9) so as to be able to calculate the material function τ from an experimental determination of $\Omega = \Omega(M)$. Differentiation of (8.9) with respect to M gives

$$2M \frac{\mathrm{d}\Omega(M)}{\mathrm{d}M} = \overset{-1}{\tau}\left(\frac{M}{2\pi R_1^2}\right) - \overset{-1}{\tau}\left(\frac{M}{2\pi R_2^2}\right). \tag{8.11}$$

With the abbreviations

$$F(M) = 2M \frac{\mathrm{d}\Omega(M)}{\mathrm{d}M}, \qquad \alpha = \frac{R_1^2}{R_2^2}, \tag{8.12}$$

equation (8.11) becomes

$$F(M) = \overset{-1}{\tau}\left(\frac{M^M}{2\pi R_1^2}\right) - \overset{-1}{\tau}\left(\alpha \frac{M}{2\pi R_1^2}\right). \tag{8.13}$$

If we replace here M by $\alpha^n M$ and sum from $n = 0$ to $n = N$, we get

$$\sum_{n=0}^{N} F(\alpha^n M) = \overset{-1}{\tau}\left(\frac{M}{2\pi R_1^2}\right) - \overset{-1}{\tau}\left(\alpha^{N+1} \frac{M}{2\pi R_1^2}\right). \tag{8.14}$$

Since $R_1 < R_2$, we have $\alpha < 1$ and hence $\lim_{n\to\infty} \alpha^{n+1} = 0$. The continuity of $\overset{-1}{\tau}$ at zero and the fact that $\overset{-1}{\tau}(0) = 0$ (*cf.* (3.20)) insure that

$$\lim_{n\to\infty} \overset{-1}{\tau}\left(\alpha^{n+1} \frac{M}{2\pi R_1^2}\right) = 0.$$

Hence, (8.14) yields*

$$\overset{-1}{\tau}\left(\frac{M}{2\pi R_1^2}\right) = \sum_{n=0}^{\infty} F(\alpha^n M). \tag{8.15}$$

Equations (8.12) and (8.15) make it possible to calculate τ if $\Omega = \Omega(M)$ is known. When the gap between the cylinders is small, τ can be calculated directly from (8.10).

We now determine the normal stresses. From (3.13), (8.3), and (8.7), we get

$$t_{rr} - t_{\vartheta\vartheta} = \sigma_1\left(\overset{-1}{\tau}\left(\frac{M}{2\pi r^2}\right)\right) - \sigma_2\left(\overset{-1}{\tau}\left(\frac{M}{2\pi r^2}\right)\right). \tag{8.16}$$

Substitution of (8.16) into the first equation of (8.4) and integration yields, after using $s_{rr} = t_{rr} - p$ and $\varphi = p + \varrho\psi$,

$$s_{rr} = \varrho\psi - \int_{R_1}^{r} \left\{\varrho\xi[v(\xi)]^2 + \frac{1}{\xi}\sigma_1\left(\overset{-1}{\tau}\left(\frac{M}{2\pi\xi^2}\right)\right) - \frac{1}{\xi}\sigma_2\left(\overset{-1}{\tau}\left(\frac{M}{2\pi\xi^2}\right)\right)\right\} d\xi + c, \tag{8.17}$$

where $v(\xi)$, by (8.7) and (8.8), is

$$v(\xi) = \int_{R_1}^{\xi} \frac{1}{\zeta} \overset{-1}{\tau}\left(\frac{M}{2\pi\zeta^2}\right) d\zeta. \tag{8.18}$$

---

* Note added in proof: It has been brought to our attention that an equivalent inversion of (8.9) was published by KRIEGER & ELROD [7].

By (3.15), we have

$$s_{rr} - s_{zz} \doteq \sigma_1\left(\overset{-1}{\tau}\left(\frac{M}{2\pi r^2}\right)\right), \qquad s_{\vartheta\vartheta} - s_{zz} = \sigma_2\left(\overset{-1}{\tau}\left(\frac{M}{2\pi r^2}\right)\right), \tag{8.19}$$

The remarks made at the end of § 6 apply here also.

When the potential ψ of the body forces is independent of r, (8.17) yields the following expression for the difference of the normal stresses at the outer and inner cylinders:

$$s_{rr}(R_2) - s_{rr}(R_1) = -\int_{R_1}^{R_2} \left\{\varrho\, r\,[v(r)]^2 + \frac{1}{r}\sigma_1\left(\overset{-1}{\tau}\left(\frac{M}{2\pi r^2}\right)\right) - \frac{1}{r}\sigma_2\left(\overset{-1}{\tau}\left(\frac{M}{2\pi r^2}\right)\right)\right\} dr. \tag{8.20}$$

When $(R_2 - R_1)/R_1 \ll 1$, this formula may be approximated by

$$s_{rr}(R_2) - s_{rr}(R_1) \approx \frac{R_2 - R_1}{R_1}\left[\sigma_2\left(\overset{-1}{\tau}\left(\frac{M}{2\pi R_2^2}\right)\right) - \sigma_1\left(\overset{-1}{\tau}\left(\frac{M}{2\pi R_2^2}\right)\right)\right]. \tag{8.21}$$

The error in both the approximate formulae (8.10) and (8.21) is of the order of $[(R_2 - R_1)/R_1]^2$. The inertial term $\varrho r[v(r)]^2$ in the integral on the right side of (8.20) makes a contribution to $s_{rr}(R_2) - s_{rr}(R_1)$ of the order of $[(R_2 - R_1)/R_1]^3$ and hence does not appear in (8.21).

Acknowledgement. The research leading to this paper was partly supported by the National Science Foundation under Grant NSF-G 5250 to Carnegie Institute of Technology.

References

[1] NOLL, W.: A mathematical theory of the mechanical behavior of continuous media. Arch. Rational Mech. Anal. **2**, 197–226 (1958).

[2] MARKOVITZ, H.: Normal stress effect in polyisobutylene solutions. II. Classification and application of rheological theories. Trans. Society of Rheology **1**, 37–52 (1957).

[3] NOLL, W.: On the continuity of the solid and fluid states. J. Rational Mech. Anal. **4**, 3–81 (1955).

[4] RIVLIN, R. S.: Solution of some problems in the exact theory of viscoelasticity. J. Rational Mech. Anal. **5**, 179–188 (1956).

[5] HERMANS, J. J.: Flow properties of disperse systems, Chapter 5. Amsterdam: North-Holland Publishing Co. 1953.

[6] MOONEY, M.: Explicit formulas for slip and fluidity. J. Rheology **2**, 210–222 (1931).

[7] KRIEGER, I. M., & H. ELROD: Direct determination of the flow curves of non-Newtonian fluids. II. Shearing rate in the concentric cylinder viscometer. J. Appl. Phys. **27**, 134–136 (1953).

Mellon Institute
Pittsburgh, Pennsylvania
and
Carnegie Institute of Technology
Pittsburgh, Pennsylvania

(Received March 13, 1959)

On the Thermostatics of Continuous Media

BERNARD D. COLEMAN & WALTER NOLL

Contents

Introduction

In this article we regard thermostatics as being that branch of thermodynamics which deals with bodies which are at rest at the present time and which, for all practical purposes, may be regarded as having been at rest at all times in the past.

We attempt to develop here a rigorous theory of thermostatics for continuous bodies in arbitrary states of strain. The thermodynamics of chemical reactions, phase transitions, and capillarity is not discussed. Our aim is to derive some of the fundamental laws of hydrostatics and elastostatics from thermodynamic principles. Among these laws are the existence of elastic potentials for stress-strain relations, the known inequalities of hydrostatics, and some new inequalities for hydrostatics and elastostatics.

In his classic work, "On the Equilibrium of Heterogeneous Substances", J. W. GIBBS [*1*] laid down criteria for determining whether a given (global) state of a body is thermodynamically stable. He used these criteria to derive particular equations and inequalities which represent conditions (in some cases necessary and other cases sufficient) for various special states to be stable. The equations GIBBS obtained as necessary conditions for thermodynamic stability are now recognized as fundamental laws in physical chemistry. GIBBS also derived inequalities which, apparently because they are in obvious accord with everyday experience and thus might be mistakenly called trivial, have attracted relatively little attention and are sometimes not even mentioned in modern thermodynamics courses. For example, in his treatment of homogeneous systems at rest under uniform hydrostatic pressure, GIBBS showed that a necessary condition for such a system to be in a stable state is that both its heat capacity at constant volume and its adiabatic modulus of compression be non-negative. It is inequalities of this type which are emphasized in the present paper. We take, however, a point of view different from that of GIBBS.

In the classical treatments of thermostatics (*e.g.*, [*1*], [*2*]) the adjective *stable* is used in two senses. It is sometimes used as a modifier for the word *equilibrium*; *i.e.* one refers to "states of stable equilibrium"; or it is used as a modifier for the word *state*; *i.e.* one refers to "stable states". In this paper we never use the word stable in the former sense. The theory which we develop here makes a careful distinction between *local* states, referring to a material point in a body, and *global* states, referring to the body as a whole. A local thermomechanic state is specified by giving the entropy density and the local configuration at a material point. A global thermomechanic state, on the other hand, is specified only when the entropy field and the complete configuration are specified for the entire body. We regard *thermal equilibrium* to be a property of *local states*. We consider just one type of thermal equilibrium. We define a state of thermal equilibrium as a local thermomechanic state which minimizes an appropriate potential rather than as a state at which a first variation vanishes. We regard *stability* as a property of only *global states*. We consider several types of stable states, defined as global thermomechanic states which minimize certain energy integrals subject to different constraints.

Our theory is based on two physical postulates. The first asserts that, at a material point, any local thermomechanic state can be an equilibrium state provided the local temperature and local forces have appropriate values. The second postulate is essentially the assumption that, at least in continuum mechanics, absolute temperatures are never negative. We believe that these physical postulates, which are stated in terms of our definition of equilibrium, contain the physical content for the statics of continuous media of the First and Second Laws of Thermodynamics. From our postulates we prove relationships between the stress-strain equation and the caloric equation of state, and we derive various inequalities restricting the form of the caloric equation of state. We should like to propose that the inequalities which we obtain for the finite theory of elasticity answer some of the questions raised by C. TRUESDELL [*3*] in his recent article, "Das ungelöste Hauptproblem der endlichen Elastizitätstheorie".

Although our definition of thermal equilibrium is new, some of the definitions of the stability of global states which we propose for study are similar to stability definitions considered by GIBBS [*1*] and J. HADAMARD [*4*]. In particular, our concepts of isothermal and adiabatic stability *at fixed boundary* are closely related to, but not identical to, *Hadamard stability**. We briefly discuss GIBBS' theory of the stability of fluid phases in § 16. In a future article we hope to give a discussion of GIBBS' theory of the stability of fluid mixtures.

We regard the main tasks of the science of thermostatics to be, first, the exploration of the consequences for the caloric equation of state of the existence of local states of thermal equilibrium and, second, the derivation of useful necessary and sufficient criteria for global states to be stable. In the present paper, § 6—§ 13 are devoted to the first task and § 14—§ 16 deal briefly with the second. From our present point of view, we should say that the great classical thermodynamicists, GIBBS and DUHEM, devoted their main efforts to the second task.

It will be noticed that in this paper we never mention such notions as "reversible processes" and "quasi-static processes"; in fact, our theory of thermostatics, being a truly statical theory, has no need of "processes" at all.

In writing the present paper we have striven for a level of mathematical rigor comparable to that of works in pure mathematics rather than to that customary in physics.

Notation and basic mathematical concepts. We often find it convenient to distinguish between functions and their values. The basic local thermodynamic variables are denoted by light face Greek minuscules: $\varepsilon, \psi, \eta, \vartheta, \ldots$. Symbols such as $\hat{\varepsilon}, \bar{\varepsilon}, \bar{\bar{\varepsilon}} \ldots$ and $\hat{\psi}, \tilde{\psi}, \bar{\psi} \ldots$ represent real valued *functions* whose values are the thermodynamic variables ε and ψ.

We denote vectors and points of the three-dimensional Euclidean space $\mathscr{E}$ by bold face Latin minuscules: $\boldsymbol{v}, \boldsymbol{x}, \boldsymbol{y} \ldots$.

Second order tensors are denoted by light face Latin majuscules: F, U, Q, R, I. However, we reserve the symbols X and Z to represent material points of a physical body. The term tensor is used as a synonym for linear transformation. Tensors of order higher than two do not occur in this paper. For the trace of the tensor F we write $\operatorname{tr} F$ and for the determinant of F we write $\det F$. We say that F is invertible if F has an inverse F^{-1}; *i.e.* if $\det F \neq 0$. The transpose of F is denoted by F^T. The identity transformation is written I. For the composition, or product, of two linear transformations A and B we write simply AB.

* Hadamard stability requires (roughly) that the first variation of the integral of the elastic potential vanish, and that the second variation be non-negative, for all smooth variations in the state of strain which are compatible with a fixed boundary. This sort of stability is necessary but not sufficient for stability at fixed boundary as we define it here. In the theory of the propagation of waves in a perfectly elastic solid, Hadamard stability of a particular rest state implies the reality of all roots of the wave velocity equation for acceleration waves of arbitrary direction which might impinge on an object in that state. J. L. ERICKSEN & R. A. TOUPIN [*5*] have recently considered a modification of Hadamard stability in which they require that the second variation of the integral of the elastic potential be strictly positive. They use their definition of stability to prove uniqueness theorems in the theory of small deformations superimposed on large. R. HILL [*15*] also has recently discussed relationships between uniqueness and stability. In the third article of his "Recherches sur l'élasticité" P. DUHEM [*6*] formulated several definitions of stability which are applicable to bodies with fixed and partially free surfaces; he also derived several necessary conditions on the equation of state for particular states of strain to be stable.

Let $\boldsymbol{h}(\boldsymbol{x})$ be a function for which both the range and the domain consist of either vectors or points in Euclidean space $\mathscr{E}$. Assume that for $\boldsymbol{x}$ in some region the derivative

$$\frac{d}{ds}\boldsymbol{h}(\boldsymbol{x}+s\,\boldsymbol{v})\Big|_{s=0}=\nabla\boldsymbol{h}(\boldsymbol{x};\boldsymbol{v}) \tag{1}$$

exists for all $\boldsymbol{v}$ and is continuous in $\boldsymbol{x}$. It is the content of a fundamental theorem of analysis that $\nabla\boldsymbol{h}(\boldsymbol{x};\boldsymbol{v})$ is then a linear function of $\boldsymbol{v}$, and hence we can write

$$\nabla\boldsymbol{h}(\boldsymbol{x};\boldsymbol{v})=[\nabla\boldsymbol{h}(\boldsymbol{x})]\,\boldsymbol{v}, \tag{2}$$

where $\nabla\boldsymbol{h}(\boldsymbol{x})$ is a linear transformation (tensor), called the *gradient* of $\boldsymbol{h}$ at $\boldsymbol{x}$.

Similarly, the gradient of a real valued function $\zeta(F)$ of a tensor variable F is a tensor valued function $\zeta_F(F)$ defined by the relation

$$\frac{d}{ds}\zeta(F+s\,A)\Big|_{s=0}=\operatorname{tr}[\zeta_F(F)\,A], \tag{3}$$

where A is an arbitrary tensor. If Cartesian coordinates are used, and if $\|f_{ij}\|$ is the matrix of F, then the matrix of ζ_F is given by

$$\|\zeta_F(F)\|=\left\|\frac{\partial\zeta}{\partial f_{ji}}\right\|,$$

where i is the row and j the column index.

We make frequent use of the following theorem, called the *polar decomposition theorem: Any invertible tensor F has unique decompositions*

$$F=RU=VR \tag{4}$$

where R is orthogonal (i.e., $RR^T=I$) and U, V are positive definite symmetric tensors (i.e., $U=U^T$, $V=V^T$, and the proper numbers of U and V are all real and greater than zero). In addition, we have

$$U=R^T V R,\qquad U^2=F^T F,\qquad V^2=FF^T. \tag{5}$$

Consider a smooth (*i.e.*, continuously differentiable) real valued function $\zeta(\boldsymbol{w})$ whose domain $\mathscr{W}$ is a region in a finite dimensional vector space. The function ζ is called *strictly convex* if either of the following two equivalent conditions are satisfied:

(a) For all $\boldsymbol{w}_1$ and $\boldsymbol{w}_2\neq\boldsymbol{w}_1$ in $\mathscr{W}$ and all positive α,β with $\alpha+\beta=1$, the inequality

$$\zeta(\alpha\,\boldsymbol{w}_1+\beta\,\boldsymbol{w}_2)<\alpha\,\zeta(\boldsymbol{w}_1)+\beta\,\zeta(\boldsymbol{w}_2) \tag{6}$$

holds.

(b) For all $\boldsymbol{w}$ and $\boldsymbol{w}^*\neq\boldsymbol{w}$ in $\mathscr{W}$ the inequality

$$\zeta(\boldsymbol{w}^*)-\zeta(\boldsymbol{w})-(\boldsymbol{w}^*-\boldsymbol{w})\cdot\nabla\zeta(\boldsymbol{w})>0 \tag{7}$$

is satisfied.

When $\mathscr{W}$ is a region in the space of all tensors, we use the notation of (3) and the convexity inequality (7) becomes

$$\zeta(F^*)-\zeta(F)-\operatorname{tr}[(F^*-F)\,\zeta_F(F)]>0. \tag{8}$$

For a twice continuously differentiable function $\zeta(\boldsymbol{w})$ to be strictly convex in $\mathscr{W}$, it is sufficient that the second gradient $\nabla\nabla\zeta(\boldsymbol{w})$ be positive definite for $\boldsymbol{w}$ in $\mathscr{W}$. This condition is not necessary, however: if $\zeta(\boldsymbol{w})$ is convex, it follows only that $\nabla\nabla\zeta(\boldsymbol{w})$ is positive *semi*definite.

1. Mechanical preliminaries

We give a brief summary of those concepts from the mechanics of continuous media that are relevant to the present investigation. For a detailed discussion we refer to [*7*] and [*8*].

A *body* $\mathscr{B}$ is a smooth manifold of elements $X, Z, \ldots$, called *material points**. A *configuration* $\boldsymbol{f}$ of $\mathscr{B}$ is a smooth one-to-one mapping of $\mathscr{B}$ onto a region in a three-dimensional Euclidean point space $\mathscr{E}$. The point $\boldsymbol{x}=\boldsymbol{f}(X)$ is the *position* of the material point X in the configuration $\boldsymbol{f}$. The *mass distribution* m of $\mathscr{B}$ is a measure defined on all Borel subsets of $\mathscr{B}$. For the total mass of $\mathscr{B}$ we write $m(\mathscr{B})$. To each configuration $\boldsymbol{f}$ of $\mathscr{B}$ corresponds a mass density ϱ.

Consider a neighborhood $\mathscr{N}(X)$ of a material point in a body; *i.e.*, a part of the body containing X in its interior. Let $\boldsymbol{g}$ be a smooth homeomorphism of $\mathscr{N}(X)$ into the three-dimensional vector space $\mathscr{V}$ such that X itself is mapped into the zero vector $\mathbf{0}$. The inverse mapping of $\boldsymbol{g}$ is denoted by $\overset{-1}{\boldsymbol{g}}$. Let $\boldsymbol{g}_1$ and $\boldsymbol{g}_2$ be two such homeomorphisms. The composition $\boldsymbol{g}_2\circ\overset{-1}{\boldsymbol{g}_1}$ of $\boldsymbol{g}_2$ and $\overset{-1}{\boldsymbol{g}_1}$ is defined by

$$\big(\boldsymbol{g}_2\circ\overset{-1}{\boldsymbol{g}_1}\big)(\boldsymbol{x})=\boldsymbol{g}_2\big(\overset{-1}{\boldsymbol{g}_1}(\boldsymbol{x})\big).$$

It is a mapping of a neighborhood of $\mathbf{0}$ onto another neighborhood of $\mathbf{0}$. We define an equivalence relation "$\sim$" among all these homeomorphisms by the condition that $\boldsymbol{g}_1\sim\boldsymbol{g}_2$ if the gradient of the mapping $\boldsymbol{g}_2\circ\overset{-1}{\boldsymbol{g}_1}$ at $\mathbf{0}$ is the identity I. The resulting equivalence classes will be called the *local configurations*** M of X. If M_1 is the equivalence class of $\boldsymbol{g}_1$ and M_2 the equivalence class of $\boldsymbol{g}_2$ then the gradient at $\mathbf{0}$ of $\boldsymbol{g}_2\circ\overset{-1}{\boldsymbol{g}_1}$, *i.e.*

$$G=\nabla\big(\boldsymbol{g}_2\circ\overset{-1}{\boldsymbol{g}_1}\big)(\mathbf{0}), \tag{1.1}$$

depends only on M_1 and M_2. We write

$$G=M_2M_1^{-1}, \quad M_2=GM_1, \tag{1.2}$$

and call G the *deformation gradient* from M_1 to M_2; G is an invertible linear transformation.

It is often convenient to employ a *local reference configuration* M_r and to characterize the other local configurations

$$M=FM_r \tag{1.3}$$

by their deformation gradients F from the local reference configuration M_r. If, in this way, two local configurations M_1 and M_2 correspond, respectively, to F_1 and F_2 then the deformation gradient G from M_1 to M_2 is given by

$$G=F_2F_1^{-1}, \quad F_2=GF_1. \tag{1.4}$$

---

* The term "particle" is often used. We prefer "material point" to avoid confusion with molecules and other physical particles.

** The term "configuration gradient" was used in [*7*].

The *rotation tensor* R, the *right stretch tensor** U, and the *left stretch tensor* V of a deformation gradient F are defined by the unique polar decompositions

$$F = RU = VR, \tag{1.5}$$

where R is orthogonal, while U and $V = RUR^T$ are symmetric and positive definite. We note that U and V have the same proper numbers; these proper numbers are called the *principal stretches* v_1, v_2, v_3. A deformation gradient G is called a *pure stretch* if its rotation tensor reduces to the identity I; *i.e.*, if G is symmetric and positive definite and hence coincides with its own right and left stretch tensors.

The mass densities at X corresponding to the local configurations M_1 and M_2 are denoted, respectively, by ϱ_1 and ϱ_2. We have

$$\varrho_2 = \frac{1}{|\det G|}\,\varrho_1, \tag{1.6}$$

where G is related to M_1 and M_2 by (1.2).

2. Thermomechanic states

A global thermomechanic state, or simply a *state*, of a body $\mathscr{B}$ is a pair $\{\boldsymbol{f}, \eta\}$ consisting of a configuration $\boldsymbol{f}$ of $\mathscr{B}$ and a scalar field η defined on $\mathscr{B}$; η is called the *entropy distribution* of the state.

A local thermomechanic state, or simply a *local state*, of a material point X is defined as a pair (M, η) consisting of a local configuration M of X and a real number η, called the *entropy density* (per unit mass) of the local state**.

In the following we often use a local reference configuration M_r and, according to (1.3), characterize the other local configurations M by the deformation gradients F from M_r. We then use the pair (F, η) to characterize the local states.

Two local states (F, η) and (F', η') will be called *equivalent* if they differ only by a change of frame of reference. The local configuration transforms under a change of frame according to the law $F' = QF$ where Q is orthogonal. We assume that the entropy density η is *objective*; *i.e.*, it remains invariant under a change of frame. Thus, the local states (F, η) and (F', η') are equivalent if and only if

$$F' = QF, \quad \eta' = \eta \tag{2.1}$$

for some orthogonal Q.

We say that two global states $\{\boldsymbol{f}, \eta\}$ and $\{\boldsymbol{f}', \eta'\}$ are equivalent if they differ by only a change of frame. This is the case if and only if

$$\eta'(X) = \eta(X), \quad F'(X) = QF(X) \tag{2.2}$$

for all X in the body and some orthogonal tensor Q independent of X. Here, $F(X)$ and $F'(X)$ are the deformation gradients at X corresponding to $\boldsymbol{f}$ and $\boldsymbol{f}'$ respectively.

---

* The term "strain tensor" was used in [7].

** In this article, pairs in braces, { }, always refer to *global* properties; the elements of such pairs are fields over $\mathscr{B}$. On the other hand, pairs in brackets, ( ), always refer to *local* properties and have elements which are either real numbers or tensors. Note that the symbol η in $\{\boldsymbol{f}, \eta\}$ and (M, η) denotes different entities; in the first case η denotes a field while in the second case it denotes a number. No confusion should arise, however.

3. The caloric equation of state

A *material* is characterized by a real valued function of local states, whose values ε are called the *energy densities* (per unit mass) of the local states. We pick a fixed local reference configuration M_r and characterize the state (M, η) by the pair (F, η) where $F = M M_r^{-1}$. We write

$$\varepsilon = \hat{\varepsilon}(F, \eta). \tag{3.1}$$

It is assumed here that the function $\hat{\varepsilon}$ has continuous derivatives with respect to F and η*.

We assume that the energy density is objective; *i.e.* invariant under a change of frame. It follows from (2.1) that the function $\hat{\varepsilon}$ must satisfy the relation

$$\hat{\varepsilon}(QF, \eta) = \hat{\varepsilon}(F, \eta) \tag{3.2}$$

for all orthogonal Q. Using the polar decomposition (1.5) and putting $Q = R^T$ in (3.2) we see that

$$\varepsilon = \hat{\varepsilon}(F, \eta) = \hat{\varepsilon}(U, \eta); \tag{3.3}$$

i.e., that the energy density is determined by the right stretch tensor U and the entropy η.

The function $\hat{\varepsilon}$ in (3.3) depends on the choice of the local reference configuration M_r. The function $\hat{\varepsilon}'$ corresponding to some other local reference configuration M_r' is related to $\hat{\varepsilon}$ by

$$\hat{\varepsilon}'(F, \eta) = \hat{\varepsilon}(FG, \eta), \tag{3.4}$$

where $G = M_r' M_r^{-1}$ is the deformation gradient from M_r to M_r'.

The equation (3.3) characterizes the thermal and mechanical properties of a material in statics. It is called the *caloric equation of state* of the material.

4. The isotropy group

It may happen that the energy function $\hat{\varepsilon}$ remains the same function if the local reference configuration M_r is changed to another local reference configuration $M_r' = H M_r$ with the same density. It follows from (3.4) that $\hat{\varepsilon}$ then satisfies the relation

$$\hat{\varepsilon}(F, \eta) = \hat{\varepsilon}(FH, \eta). \tag{4.1}$$

Since M_r' and M_r have the same density, it is clear from (1.6) that $|\det H| = 1$; *i.e.*, H is a unimodular transformation. The unimodular transformations H for which (4.1) holds form a group, called the *isotropy group* $\mathscr{G}$ of $\hat{\varepsilon}$ or of the material defined by $\hat{\varepsilon}$. This group depends, in general, on the choice of the local reference configuration, but it can be shown that the groups corresponding to two different local configurations are always conjugate and hence isomorphic.

We say that the energy function $\hat{\varepsilon}$ defines a *simple fluid* if its isotropy group $\mathscr{G}$ is the full unimodular group $\mathscr{U}$. If $\mathscr{G} = \mathscr{U}$ for one reference configuration, then $\mathscr{G} = \mathscr{U}$ for all reference configurations. A material point is called a *fluid material point* if its energy function defines a simple fluid. The caloric equation

---

* For the application to physical situations it is necessary to limit the domain of $\hat{\varepsilon}$ to a region in the space of local configurations and an interval on the η-axis. We do not supply the mathematical details which may arise in the consideration of limitations of this kind.

of state (3.3) then reduces to the form

$$\hat{\varepsilon} = \hat{\varepsilon}(F, \eta) = \bar{\varepsilon}(v, \eta), \tag{4.2}$$

where

$$v = \frac{1}{\varrho} = |\det F| \frac{1}{\varrho_r} \tag{4.3}$$

is the *specific volume* of the local configuration $M = F M_r$; ϱ and ϱ_r are the mass densities corresponding to M and M_r. The function $\bar{\varepsilon}$ in (4.2) does not depend on the choice of the reference configuration.

We say that a material point is an *isotropic material point* if the isotropy group of its energy function $\hat{\varepsilon}$, relative to some local reference configuration, contains the orthogonal group $\mathcal{O}$. Those local reference configurations of the material point for which $\mathcal{G}$ contains $\mathcal{O}$ are said to be *undistorted*. A simple fluid is isotropic, and all of its local configurations are undistorted. For any isotropic material, it follows from (3.2) and (4.1) that $\hat{\varepsilon}$ satisfies the relation

$$\hat{\varepsilon}(Q U Q^T, \eta) = \hat{\varepsilon}(U, \eta) \tag{4.4}$$

for all symmetric and positive definite U and all orthogonal Q, provided the local reference configuration for $\hat{\varepsilon}$ is undistorted. Taking $Q = R$, so that $V = R U R^T$ is the left stretch tensor, we see that for isotropic material points the caloric equation of state (3.3) may be written in the form

$$\varepsilon = \hat{\varepsilon}(F, \eta) = \hat{\varepsilon}(V, \eta). \tag{4.5}$$

It is a further consequence of (4.4) that for each fixed value of η, ε may be expressed as a symmetric function of the three principal stretches v_1, v_2, v_3:

$$\varepsilon = \hat{\varepsilon}(F, \eta) = \hat{\varepsilon}(V, \eta) = \bar{\varepsilon}(v_1, v_2, v_3; \eta) = \bar{\varepsilon}(v_j, \eta). \tag{4.6}$$

It may also be expressed as a function of the three principal invariants I_V, II_V, III_V of V and U:

$$\varepsilon = \hat{\varepsilon}(V, \eta) = \bar{\bar{\varepsilon}}(I_V, II_V, III_V; \eta). \tag{4.7}$$

We say that the energy function $\hat{\varepsilon}$ defines a *simple solid* if its isotropy group $\mathcal{G}$ is contained as a subgroup in the orthogonal group $\mathcal{O}$. A material point is called a *solid material point* if its energy function $\hat{\varepsilon}$, relative to some local configuration as a reference, defines a simple solid. The local reference configurations with this property are again called the *undistorted* states of the solid material point. For an *isotropic simple solid*, the isotropy group $\mathcal{G}$ is identical to the orthogonal group $\mathcal{O}$.

Throughout the rest of this paper, whenever we discuss isotropic materials it is to be understood that the local reference configuration for the energy density function is undistorted, unless the reference configuration is explicitly specified.

5. Forces, stresses, and work

A *system of forces* is a system of vector valued measures, one for each part of the body $\mathcal{B}$ under consideration*. One must distinguish between contact and body forces. The contact force acting across an oriented surface element in $\mathcal{B}$ will be denoted by $d\boldsymbol{c}$.

* For a detailed axiomatic treatment *cf.* [8].

Definition of mechanical equilibrium. *In order that a body $\mathscr{B}$ be in* ***mechanical equilibrium*** *under a given system of forces, two conditions must be fulfilled for each part $\mathscr{P}$ of $\mathscr{B}$:* (a) *the sum of the forces acting on $\mathscr{P}$ must vanish, and* (b) *the sum of the moments, about any point, of the forces acting on $\mathscr{P}$ must vanish.*

The condition (a), called the *force condition,* depends only on the body and the force system, not on the configuration of the body. The condition (b), called the *moment condition,* does depend on the configuration of the body; *i.e.,* for a given force system, the moment condition may be satisfied for one configuration but not for another.

The force condition alone implies that, for each configuration, the contact forces $d\boldsymbol{c}$ arise from a stress-tensor S, so that

$$d\boldsymbol{c} = S\boldsymbol{n}\,dA\,, \tag{5.1}$$

where $\boldsymbol{n}$ is the unit normal vector of the oriented surface element and dA its area in the configuration under consideration. For fixed contact forces $d\boldsymbol{c}$, the stress tensor S will be different for different configurations.

We consider now a neighborhood $\mathscr{N}(X)$ of a material point X and assume that a system of contact forces $d\boldsymbol{c}$ is given for $\mathscr{N}(X)$. Let $\boldsymbol{f}_r$ be a fixed reference configuration and $\boldsymbol{f}$ some other configuration of $\mathscr{N}(X)$. If $d\boldsymbol{c}$ is such that the force condition is satisfied, then (5.1) is valid for all configurations; we can write for the reference configuration $\boldsymbol{f}_r$, in particular,

$$d\boldsymbol{c} = S_r \boldsymbol{n}_r\,dA_r\,, \tag{5.2}$$

where $\boldsymbol{n}_r$ is the unit normal of the oriented surface element in the reference configuration $\boldsymbol{f}_r$, and dA_r is the area of the surface element in $\boldsymbol{f}_r$. We denote the position vector, in the configuration $\boldsymbol{f}$, of a typical material point Z in $\mathscr{N}(X)$, relative to the position of X as origin, by $\boldsymbol{p}$, and we consider the tensor K defined by

$$K = \frac{1}{v(\mathscr{N}(X))} \int\limits_{\overline{\mathscr{N}(X)}} \boldsymbol{p} \otimes d\boldsymbol{c}\,, \tag{5.3}$$

where $\overline{\mathscr{N}(X)}$ denotes the boundary surface of $\mathscr{N}(X)$ and $v(\mathscr{N}(X))$ the volume of $\mathscr{N}(X)$ in the configuration $\boldsymbol{f}$, and where $\otimes$ denotes a tensor product. If the force condition is satisfied, the relation (5.1) is valid, and we have

$$K^T = \frac{1}{v(\mathscr{N}(X))} \int\limits_{\overline{\mathscr{N}(X)}} S\boldsymbol{n} \otimes \boldsymbol{p}\,dA\,.$$

In the limit as $\mathscr{N}(X)$ shrinks to X, we obtain, after using Green's theorem,

$$S^T = \lim_{\mathscr{N}(X)\to X} K\,. \tag{5.4}$$

The same argument, with the configuration $\boldsymbol{f}$ replaced by the reference configuration $\boldsymbol{f}_r$, gives

$$S_r^T = \lim_{\mathscr{N}(X)\to X} \frac{1}{v_r(\mathscr{N}(X))} \int\limits_{\overline{\mathscr{N}(X)}} \boldsymbol{p}_r \otimes d\boldsymbol{c}\,, \tag{5.5}$$

where $v_r(\mathscr{N}(X))$ is the volume of $\mathscr{N}(X)$ in the reference configuration and $\boldsymbol{p}_r$ the position vector, in the reference configuration, of a typical material point Z

in $\mathcal{N}(X)$, relative to the position of X as origin. The position vector $\boldsymbol{p}$ of Z in the configuration $\boldsymbol{f}$ is related to $\boldsymbol{p}_r$ by the relation

$$\boldsymbol{p} = F\boldsymbol{p}_r + o(|\boldsymbol{p}_r|) \tag{5.6}$$

where F is the gradient at X of the deformation from $\boldsymbol{f}_r$ to $\boldsymbol{f}$ and where

$$\lim_{d\to 0} \frac{o(d)}{d} = 0.$$

Substitution of (5.6) into (5.3) and use of (5.4) and (5.5) yields

$$S = \frac{\varrho}{\varrho_r} F S_r, \tag{5.7}$$

where ϱ and ϱ_r are, respectively, the mass densities at X in the configurations $\boldsymbol{f}$ and $\boldsymbol{f}_r$.

The skew part of K, defined by (5.3), is the moment about X, per unit volume, of the contact forces $d\boldsymbol{c}$ acting on $\mathcal{N}(X)$ in the configuration $\boldsymbol{f}$. If the moment condition is satisfied for the configuration $\boldsymbol{f}$, then the total moment (*i.e.* the moment of the contact forces *and* the body forces) about X in $\boldsymbol{f}$ must vanish. Since the moment per unit volume about X of the body forces on $\mathcal{N}(X)$ goes to zero as $\mathcal{N}(X)$ shrinks to X, it follows from (5.4) that S must be symmetric if the moment condition is satisfied in $\boldsymbol{f}$.

We say that a material point X is in *local mechanical equilibrium*, when the body is in a given configuration and under a given force system, if the stress tensor S exists at X and is symmetric.

The local behavior at X of a system of contact forces is completely determined by the tensor S_r defined by (5.2). It is called the *Kirchhoff tensor** of the system. For a given force system, the Kirchhoff tensor depends only on the choice of the reference configuration and remains the same if the actual configuration is changed. From (5.7) we see that the existence of the Kirchhoff tensor S_r and the symmetry of $F S_r$ are necessary and sufficient conditions for local mechanical equilibrium at a material point in the local configuration determined by F.

In order that a body $\mathcal{B}$ in a configuration $\boldsymbol{f}$ be in mechanical equilibrium, it is not sufficient that all its material points be in local mechanical equilibrium; *i.e.*, that the stress tensor exist and be symmetric at each material point. Global mechanical equilibrium will prevail only if, in addition, Cauchy's law

$$\operatorname{div} S + \varrho\, \boldsymbol{b} = \boldsymbol{0} \tag{5.8}$$

is satisfied. In this equation, S, ϱ, and the density $\boldsymbol{b}$ of the body forces are to be regarded as fields with domain $\boldsymbol{f}(\mathcal{B})$.

We consider now a smooth one-parameter family of configurations $\boldsymbol{f}(s)$ with deformation gradients $F(s)$ at X. The *work per unit mass* done on $\mathcal{N}(X)$ by the contact forces $d\boldsymbol{c}$ along the path of configurations $\boldsymbol{f}(s)$ from $s = s_1$ to $s = s_2$ is defined by

$$w = \frac{1}{m(\mathcal{N}(X))} \int_{s_1}^{s_2} \left[\int_{\mathcal{N}(X)} \frac{d\boldsymbol{p}}{ds} \cdot d\boldsymbol{c} \right] ds, \tag{5.9}$$

* *Cf.* TRUESDELL [9], (26.5).

where $m(\mathcal{N}(X))$ is the mass of $\mathcal{N}(X)$ and $\boldsymbol{p}(s)$ denotes the position vector, in the configuration $\boldsymbol{f}(s)$, of a typical material point in $\mathcal{N}(X)$. Assuming that the contact forces $\boldsymbol{dc}$ are independent of s, we obtain

$$w = \frac{1}{m(\mathcal{N}(X))}\left[\int_{\mathcal{N}(X)} \boldsymbol{p}(s_2)\cdot \boldsymbol{dc} - \int_{\mathcal{N}(X)} \boldsymbol{p}(s_1)\cdot \boldsymbol{dc}\right]. \tag{5.10}$$

Observing (5.3), (5.4) and (5.7), and taking the limit $\mathcal{N}(X)\to X$, we get

$$\varrho_r w = \mathrm{tr}\left[F(s_2)\, S_r\right] - \mathrm{tr}\left[F(s_1)\, S_r\right]. \tag{5.11}$$

This relation shows that $-\frac{1}{\varrho_r}\,\mathrm{tr}(F\,S_r)$ has the physical meaning of the potential energy, per unit mass, of the local contact forces.

6. Definition of thermal equilibrium

A *force temperature pair* for a material point X is a pair (S_r, ϑ) consisting of a tensor S_r, to be interpreted as the Kirchhoff tensor of a system of contact forces at X, and a real number ϑ, to be interpreted as the temperature at X.

Let a force temperature pair (S_r, ϑ) be given and consider the function

$$\hat{\lambda}(F,\eta) = \hat{\varepsilon}(F,\eta) - \frac{1}{\varrho_r}\,\mathrm{tr}(F\,S_r) - \eta\,\vartheta. \tag{6.1}$$

To help motivate the definition of thermal equilibrium given below, we make the following remarks. According to (5.11) the term $-\frac{1}{\varrho_r}\,\mathrm{tr}(F\,S_r)$ is the potential energy, per unit mass, of the local contact forces. The term $-\eta\vartheta$ may be interpreted as a thermal potential energy. Thus, the value $\lambda = \hat{\lambda}(F,\eta)$ gives a kind of free energy per unit mass of the local state (F,η) when under the action of the force temperature pair (S_r, ϑ).

Definition of thermal equilibrium. *The local state (F,η) is called a* ***state of thermal equilibrium under a given force temperature pair*** (S_r, ϑ) *if*

(a) *the stress tensor $S = (\varrho/\varrho_r)F\,S_r$ is symmetric,*

(b) *the inequality*

$$\hat{\lambda}(F^*,\eta^*) > \hat{\lambda}(F,\eta) \tag{6.2}$$

holds for all states $(F^*,\eta^*) \neq (F,\eta)$ such that

$$F^* = G F, \tag{6.3}$$

where G is symmetric and positive definite.

The condition (a) means that F corresponds to a local configuration in local mechanical equilibrium (*cf.* § 5). The condition (b) means that a change of state increases the free energy λ provided that the configuration of the changed state is related to the original configuration by a pure stretch G (*cf.* § 1).

7. Conditions for thermal equilibrium

In this section we show that, for a local state (F,η) to be a state of thermal equilibrium under the force temperature pair (S_r, ϑ), the following three conditions are necessary and sufficient:

(α) The stress tensor $S = \frac{\varrho}{\varrho_r} F S_r$ is given by the *stress relation**

$$S = \varrho F \hat{\varepsilon}_F(F, \eta). \tag{7.1}$$

(β) The temperature ϑ is given by the *temperature relation*

$$\vartheta = \hat{\varepsilon}_\eta(F, \eta). \tag{7.2}$$

(γ) The inequality

$$\hat{\varepsilon}(F^*, \eta^*) - \hat{\varepsilon}(F, \eta) - \operatorname{tr}\left[(F^* - F)\,\hat{\varepsilon}_F(F, \eta)\right] - (\eta^* - \eta)\,\hat{\varepsilon}_\eta(F, \eta) > 0 \tag{7.3}$$

holds if $(F^*, \eta^*) \neq (F, \eta)$ and F^* is related to F by $F^* = GF$, where G is positive definite and symmetric.

We assume first that (F, η) is a state of thermal equilibrium and prove the validity of (α), (β), and (γ). By (6.2) and (6.3), the function $\hat{\lambda}(GF, \eta^*)$ of the symmetric tensor variable G and the scalar variable η^* has a minimum for $G = I$ and $\eta^* = \eta$. By a theorem of calculus, it follows that the derivatives of $\hat{\lambda}(GF, \eta^*)$ with respect to G and η^* must vanish for $G = I$ and $\eta^* = \eta$. If we set the derivative of $\hat{\lambda}(GF, \eta^*)$ with respect to η^* equal to zero at $\eta^* = \eta$, we obtain the temperature relation (7.2). The gradient of $\hat{\lambda}(GF, \eta^*)$ with respect to G may be computed using the formula (3) of the mathematical preliminaries and (6.1); we obtain the equation

$$\operatorname{tr}\left\{\left[F\,\hat{\varepsilon}_F(F, \eta) - \frac{1}{\varrho_r} F S_r\right] A\right\} = 0, \tag{7.4}$$

which is valid for arbitrary symmetric tensors A. Using (5.7) the equation (7.4) may be rewritten in the form

$$\operatorname{tr}\{[\varrho F \hat{\varepsilon}_F(F, \eta) - S]\,A\} = 0. \tag{7.5}$$

By the condition (a) of the definition of thermal equilibrium, S is symmetric. It follows from (3.2) and Theorem I of reference [10], p. 42, that $\varrho F \hat{\varepsilon}_F(F, \eta)$ is also symmetric. Thus, the tensor $\varrho F \hat{\varepsilon}_F(F, \eta) - S$ is symmetric. On the other hand, (7.5) can be valid for arbitrary symmetric A only if $\varrho F \hat{\varepsilon}_F(F, \eta) - S$ is skew; whence it follows that $\varrho F \hat{\varepsilon}_F(F, \eta) - S$ must vanish, which proves (7.1). The inequality (7.3) is obtained simply by substitution of (7.1) and (7.2) into the inequality (6.2), after $\hat{\lambda}$ is replaced by its definition (6.1).

We assume now that the conditions (α), (β), and (γ) are satisfied. It then follows from (7.1), (3.2) and the theorem of reference [10] mentioned above that the stress tensor S must be symmetric, so that condition (a) of the definition of thermal equilibrium is satisfied. Furthermore, the Kirchhoff tensor is given by

$$S_r = \varrho_r\,\hat{\varepsilon}_F(F, \eta). \tag{7.6}$$

Substitution of (7.6) and (7.2) into the inequality (7.3) gives the inequality (6.2); hence condition (b) of the definition of equilibrium is also satisfied.

* This is the familiar stress-strain relation of finite elasticity theory (*cf.* [10], (16.4)).

8. The fundamental postulates

We are now able to lay down our two fundamental postulates:

Postulate I. *For every local state (F,η) for which $\hat{\varepsilon}(F,\eta)$ is defined there exists a force temperature pair (S_r,ϑ) such that (F,η) is a state of thermal equilibrium under (S_r,ϑ).*

Postulate II. *The energy function $\hat{\varepsilon}(F,\eta)$ is strictly increasing in η for each fixed F.*

Postulate I and the results of the previous section yield the following theorems:

Theorem 1. *The force temperature pair (S_r,ϑ) which makes the local state (F,η) a state of thermal equilibrium is given by*

$$S_r=\varrho_r\,\hat{\varepsilon}_F(F,\eta)\,, \tag{8.1}$$

$$\vartheta=\hat{\varepsilon}_\eta(F,\eta)\,. \tag{8.2}$$

Theorem 2. *The energy function $\hat{\varepsilon}$ obeys the inequality*

$$\hat{\varepsilon}(F^*,\eta^*)-\hat{\varepsilon}(F,\eta)-\mathrm{tr}\left[(F^*-F)\,\hat{\varepsilon}_F(F,\eta)\right]-(\eta^*-\eta)\,\hat{\varepsilon}_\eta(F,\eta)>0 \tag{8.3}$$

for any two local states (F,η) and (F^*,η^*), in the domain of definition of $\hat{\varepsilon}$, which are related by

$$F^*=GF\,, \tag{8.4}$$

where G is symmetric and positive definite.

The discussion of the previous section shows that Theorem 2 is equivalent to Postulate I. In fact, if we are given a state (F,η), we can define a force temperature pair (S_r,ϑ) according to (8.1) and (8.2) and then use the inequality (8.3) to prove that (F,η) is in equilibrium under (S_r,ϑ).

The inequality (8.3) of Theorem 2 is a restricted convexity condition on the function $\hat{\varepsilon}$. If we take, in particular, $F^*=F$, then (8.3) reduces to

$$\hat{\varepsilon}(F,\eta^*)-\hat{\varepsilon}(F,\eta)-(\eta^*-\eta)\,\hat{\varepsilon}_\eta(F,\eta)>0 \tag{8.5}$$

for $\eta^*\neq\eta$. This inequality is the content of the following corollary to Theorem 2:

Theorem 3. *For each fixed local configuration, the energy density is given by a strictly convex function of the entropy density.*

This theorem is equivalent to the statement that $\hat{\varepsilon}_\eta(F,\eta)$ must be a strictly increasing function of η for each fixed F. It follows that the equation (8.2) can be solved for η in a unique manner:

$$\eta=\tilde{\eta}(F,\vartheta)\,. \tag{8.6}$$

Here, $\tilde{\eta}$ is a strictly increasing function* of ϑ for each F. The fact that (8.6) is obtained by solving (8.2) for η is expressed by the identity

$$\hat{\varepsilon}_\eta\left[F,\tilde{\eta}(F,\vartheta)\right]=\vartheta\,. \tag{8.7}$$

* The specific heat c at fixed strain is given by $c=\vartheta\,\tilde{\eta}_\vartheta(F,\vartheta)$. Hence, it is a consequence of Theorem 3 that c/ϑ is never negative and, for each F, is strictly positive except possibly for a nowhere dense set of values of ϑ.

If we take $\eta^* = \eta$ in (8.3), we obtain

$$\hat{\varepsilon}(F^*, \eta) - \hat{\varepsilon}(F, \eta) - \operatorname{tr}[(F^* - F)\,\hat{\varepsilon}_F(F, \eta)] > 0; \tag{8.8}$$

this inequality holds whenever $F^* = GF$, where $G \neq I$ is symmetric and positive definite.

A local state (F, η) is called a ***natural state*** if the corresponding stress (8.1) vanishes. Keeping the entropy fixed, we may use the local configuration of the natural state as the reference configuration, so that $F = I$ and $\hat{\varepsilon}_F(I, \eta) = \frac{1}{\varrho} S = 0$. In this case, the inequality (8.8), by (8.4), reduces to

$$\hat{\varepsilon}(G, \eta) > \hat{\varepsilon}(I, \eta), \tag{8.9}$$

which is valid for arbitrary symmetric and positive definite $G \neq I$. Replacing G by the right stretch tensor U of an arbitrary deformation gradient F and using (3.3), we see that

$$\hat{\varepsilon}(F, \eta) \geqq \hat{\varepsilon}(I, \eta); \tag{8.10}$$

this expression becomes an equality only when F is orthogonal; *i.e.*, when (F, η) is equivalent to (I, η). Hence, the *energy density is smallest in a natural state.* It should be pointed out that this observation, though important for the theory of simple solids, is vacuous for fluids. For, we shall prove in § 11 that the stress on a fluid material point in thermal equilibrium is always a strictly positive pressure; thus, for a fluid there is no natural state.

We note that the restriction (8.4) on the inequality (8.3) of Theorem 2 is essential for application of the present theory to physical situations. This restriction means that the local configurations corresponding to F^* and F must be related by a pure stretch. If, for example, these local configurations were related by a rotation so that $F^* = QF$, with Q an orthogonal transformation, then the left side of (8.8) would reduce to $\operatorname{tr}[(Q - I)F\,\hat{\varepsilon}_F(F, \eta)]$, since $\hat{\varepsilon}(F^*)$ would equal $\hat{\varepsilon}(F)$ by (3.2). The stress relation (7.1) shows that the left side of (8.8) would then become $\frac{1}{\varrho}\operatorname{tr}[(Q - I)\,S]$. One can show that this expression can be made negative by an appropriate choice of Q if S has at least one negative proper number. Thus, the inequality (8.8), were it to hold for arbitrary pairs F, F^*, would exclude the possibility of thermal equilibrium under compression stresses, which is certainly not in accord with experience*.

9. An alternative axiomatization

In this section we hope to make clear our reasons for assuming Postulate II and to motivate further our definition of equilibrium.

It follows from Postulate II that, for each fixed F, the caloric equation of state has a unique solution for η:

$$\eta = \hat{\hat{\eta}}(F, \varepsilon), \tag{9.1}$$

and that the function $\hat{\hat{\eta}}$ is strictly increasing in ε for each F. This one-to-one correspondence between ε and η at each F makes it possible to give an alternative

* It has also been pointed out by HILL [15] that an assumption of unrestricted convexity of $\hat{\varepsilon}$ in the deformation gradient would lead to unacceptable physical behavior.

axiomatization of our present theory of thermostatics by taking ε and F as independent variables and defining thermal equilibrium in terms of the function $\hat{\hat{\eta}}$. In such a formulation a local thermomechanic state is characterized by a pair (F, ε), and thermal equilibrium is defined as follows:

Alternative definition of thermal equilibrium. *The local state* (F, ε) *is called a state of thermal equilibrium under the force temperature pair* (S_r, ϑ), *with* $\vartheta \neq 0$, *if*

(a) *the stress tensor* $S = (\varrho/\varrho_r) F S_r$ *is symmetric,*

(b) *the inequality*

$$\hat{\hat{\eta}}(F, \varepsilon) > \hat{\hat{\eta}}(F^*, \varepsilon^*) + \frac{1}{\vartheta \varrho_r} \operatorname{tr}\left[(F^* - F) S_r\right] - \frac{\varepsilon^* - \varepsilon}{\vartheta} \tag{9.2}$$

holds for all states $(F^*, \varepsilon^*) \neq (F, \varepsilon)$ *such that* $F^* = GF$, *where* G *is symmetric and positive definite.*

Theorem 4. *The definition of thermal equilibrium given in* § 6 *and the alternative definition of thermal equilibrium are equivalent* (*for* $\vartheta \neq 0$) *if Postulate II is assumed.*

Proof. In § 7 we showed that, under the original definition of § 6, in order for a state (F, η) to be a state of thermal equilibrium for the force temperature pair (S_r, ϑ) it is necessary that

$$\vartheta = \hat{\varepsilon}_\eta(F, \eta). \tag{9.3}$$

By a very similar argument it can be shown, using the alternative definition of thermal equilibrium, that in order for the state (F, ε) to be a state of thermal equilibrium it is necessary that

$$\frac{1}{\vartheta} = \hat{\hat{\eta}}_\varepsilon(F, \varepsilon). \tag{9.4}$$

Now, by Postulate II, the functions $\hat{\varepsilon}$ and $\hat{\hat{\eta}}$ are strictly increasing in η and ε, respectively, for fixed F. Hence, ϑ cannot be negative if (S_r, ϑ) is to be a force temperature pair for some state of thermal equilibrium, regardless of which of the two definitions is used. Since we here assume $\vartheta \neq 0$, we have $\vartheta > 0$, and (9.2) can be multiplied by ϑ and then rearranged to give

$$-\vartheta \hat{\hat{\eta}}(F^*, \varepsilon^*) - \frac{1}{\varrho_r} \operatorname{tr}(F^* S_r) + \varepsilon^* > -\vartheta \hat{\hat{\eta}}(F, \varepsilon) - \frac{1}{\varrho_r} \operatorname{tr}(F S_r) + \varepsilon. \tag{9.5}$$

Noting the relations

$$\begin{aligned} \eta^* &= \hat{\hat{\eta}}(F^*, \varepsilon^*), & \eta &= \hat{\hat{\eta}}(F, \varepsilon), \\ \varepsilon^* &= \hat{\varepsilon}(F^*, \eta^*), & \varepsilon &= \hat{\varepsilon}(F, \eta), \end{aligned} \tag{9.6}$$

and (6.1), we see that (9.5) is equivalent to (6.2). The requirement that $(F, \eta) \neq (F^*, \eta^*)$ and the requirements on $G = F^* F^{-1}$ are the same for (6.2) and (9.5). The condition (a) is obviously the same in both definitions; hence the definitions are equivalent, q.e.d.

From a certain point of view the alternative definition of thermal equilibrium given in this section is more fundamental than the original definition of § 6. The alternative definition is more closely related to the physical notion that, since entropy tends to increase, equilibrium states should be, in some sense, states of maximum entropy. The definition of § 6 is closely related to the idea, which is often used in mechanics, that equilibrium states should be, in some sense, states of minimum potential. It should be emphasized that the two definitions are equivalent only if Postulate II is assumed; *i.e.*, only if states of negative temperature are excluded. Of course, negative temperatures never occur in continuum mechanics, but there are subjects in which they do occur (*cf.* [*11*], [*12*]). Statistical mechanical considerations suggest that for systems capable of negative temperatures a practical definition of thermal equilibrium should be based on the idea of maximum entropy.

10. Infinitesimal deformations from an arbitrary state

Here we consider the classical theory of *infinitesimal deformations* from an arbitrary initial configuration. We make no attempt to justify the use of the theory of infinitesimal deformations as an approximation to the theory of finite deformations.

In the theory of infinitesimal deformations one considers cases in which $F^* = GF$ is obtained from F by superimposing an infinitesimal deformation. The *infinitesimal strain tensor* E is defined as the *symmetric part* of $G - I$.

In the special case in which G is positive definite and symmetric (*i.e.*, when F^* is related to F by a pure stretch) we have

$$E = G - I, \tag{10.1}$$

and the excess energy $\hat{\varepsilon}(GF, \eta) - \hat{\varepsilon}(F, \eta)$ is a function of E alone:

$$\sigma(E) = \hat{\varepsilon}(GF, \eta) - \hat{\varepsilon}(F, \eta). \tag{10.2}$$

Equation (10.2) is valid approximately even when G is not symmetric.

In the infinitesimal theory it is assumed (i) that (10.2) is valid exactly for all G and (ii) that the excess energy is exactly given by the sum,

$$\sigma(E) = \sigma_1(E) + \sigma_2(E), \tag{10.3}$$

of a term $\sigma_1(E)$ linear in E and a term $\sigma_2(E)$ quadratic in E.

By taking the gradient of (10.2) with respect to E and then putting $E = 0$, it is easily shown that the linear term $\sigma_1(E)$ must be given by

$$\sigma_1(E) = \operatorname{tr}[E F \hat{\varepsilon}_F(F, \eta)]. \tag{10.4}$$

Hence, using the stress relation (7.1), we have

$$\sigma_1(E) = \frac{1}{\varrho} \operatorname{tr}(E S), \tag{10.5}$$

where S is the stress of the original state (F, η).

Now, the fundamental inequality (8.8) may be written

$$\hat{\varepsilon}(GF, \eta) - \hat{\varepsilon}(F, \eta) - \operatorname{tr}[(G - I) F \hat{\varepsilon}_F(F, \eta)] > 0. \tag{10.6}$$

From (10.1), (10.2) we get

$$\sigma(E) - \operatorname{tr}[E F \hat{\varepsilon}_F(F, \eta)] > 0, \tag{10.7}$$

and it follows from (10.3), (10.4) and (10.5) that

$$\sigma_2(E) = \sigma(E) - \frac{1}{\varrho} \operatorname{tr}(E S) > 0. \tag{10.8}$$

This inequality is the content of the following theorem:

Theorem 5. *For an infinitesimal deformation superimposed, at fixed entropy, on an arbitrary state, the excess energy is the sum of a positive definite quadratic form in the infinitesimal strain tensor E of the superimposed strain and a linear term $\frac{1}{\varrho}\operatorname{tr}(E\,S)$, where ϱ is the density and S the stress corresponding to the original state.*

If the original state is a natural state in which the stress vanishes, the above theorem reduces to the familiar statement that the strain energy is a positive definite quadratic form in the infinitesimal strain tensor. For isotropic materials, this statement is equivalent to following well known inequalities for the Lamé constants:

$$\mu > 0, \quad 3\lambda + 2\mu > 0, \tag{10.9}$$

which state that the shear modulus and the compression modulus must be positive.

11. Simple fluids

For simple fluids we have, by (4.2) and (4.3),

$$\hat{\varepsilon}(F,\eta) = \bar{\varepsilon}(|\det F|\, v_r, \eta), \tag{11.1}$$

where $v_r = 1/\varrho_r$ is the specific volume in the reference configuration. Taking the gradient of (11.1) with respect to F, we obtain

$$\hat{\varepsilon}_F(F,\eta) = \bar{\varepsilon}_v(v,\eta)\, v\, F^{-1}, \tag{11.2}$$

where $v = |\det F|\, v_r$. On substituting (11.2) into the fundamental inequality (8.3) and using (8.4), we obtain

$$\bar{\varepsilon}(v^*,\eta^*) - \bar{\varepsilon}(v,\eta) - v\,\bar{\varepsilon}_v(v,\eta)\operatorname{tr}(G-I) - (\eta^*-\eta)\,\bar{\varepsilon}_\eta(v,\eta) > 0, \tag{11.3}$$

which must hold for all positive definite symmetric $G = FF^{*-1}$ whenever either $G \neq I$ or $\eta \neq \eta^*$.

We assume now that $v^* = v$; *i.e.*, that $|\det F^*| = |\det F|$, which means that G is unimodular. We also choose $\eta^* = \eta$. Then (11.3) reduces to

$$-\,v\,\bar{\varepsilon}_v(v,\eta)\operatorname{tr}(G-I) > 0, \tag{11.4}$$

which must be valid for all symmetric positive definite unimodular tensors $G \neq I$. Let g_1, g_2, g_3 be the proper numbers of G. We then have

$$g_i > 0, \quad g_1 g_2 g_3 = 1 \tag{11.5}$$

and

$$\operatorname{tr}(G-I) = g_1 + g_2 + g_3 - 3. \tag{11.6}$$

Using the fact that the arithmetic mean is greater than the geometric mean,

$$\frac{g_1+g_2+g_3}{3} > \sqrt[3]{g_1 g_2 g_3}, \tag{11.7}$$

we see that (11.5) and (11.6) imply

$$\operatorname{tr}(G-I) > 0. \tag{11.8}$$

Hence, it follows from (11.4) that

$$\bar{\varepsilon}_v(v,\eta) < 0 \tag{11.9}$$

for all v and η for which $\bar{\varepsilon}$ is defined. Thus, $\bar{\varepsilon}(v,\eta)$ must be a strictly decreasing function of v for each fixed η.

Substitution of (11.2) into (7.1) shows that the stress relation reduces to

$$S = -\bar{p}(v,\eta)\, I, \tag{11.10}$$

where

$$\bar{p}(v,\eta) = -\,\bar{\varepsilon}_v(v,\eta) \tag{11.11}$$

is the hydrostatic pressure. By (11.9) it is positive.

For further exploitation of (11.3) we choose $G = \alpha I$, $\alpha > 0$. Since

$$\alpha^3 = |\det G| = \left|\frac{\det F^*}{\det F}\right| = \frac{v^*}{v},$$

we have

$$G = \left(\sqrt[3]{\frac{v^*}{v}}\right) I. \tag{11.12}$$

Substitution of (11.12) into (11.3) yields the inequality

$$\bar{\varepsilon}(v^*,\eta^*) - \bar{\varepsilon}(v,\eta) - 3v\,\bar{\varepsilon}_v(v,\eta)\left(\sqrt[3]{\frac{v^*}{v}} - 1\right) - (\eta^* - \eta)\,\bar{\varepsilon}_\eta(v,\eta) > 0, \tag{11.13}$$

which must be valid for all v, v^*, η, η^* except, of course, when both $v = v^*$ and $\eta = \eta^*$. In order to understand the significance of this inequality we introduce the new variable

$$\nu = \sqrt[3]{v} \tag{11.14}$$

and define the function $\bar{\bar{\varepsilon}}$ by

$$\bar{\varepsilon}(v,\eta) = \bar{\bar{\varepsilon}}(\nu,\eta) = \bar{\bar{\varepsilon}}(\sqrt[3]{v},\eta). \tag{11.15}$$

A straightforward calculation shows that (11.13) is equivalent to the inequality

$$\bar{\bar{\varepsilon}}(\nu^*,\eta^*) - \bar{\bar{\varepsilon}}(\nu,\eta) - (\nu^* - \nu)\,\bar{\bar{\varepsilon}}_\nu(\nu,\eta) - (\eta^* - \eta)\,\bar{\bar{\varepsilon}}_\eta(\nu,\eta) > 0, \tag{11.16}$$

which states that $\bar{\bar{\varepsilon}}(\nu,\eta)$ is strictly convex in ν and η jointly. If $\bar{\bar{\varepsilon}}(\nu,\eta)$ happens to possess continuous second derivatives, it follows that the matrix

$$\left\|\begin{matrix} \bar{\bar{\varepsilon}}_{\nu\nu} & \bar{\bar{\varepsilon}}_{\nu\eta} \\ \bar{\bar{\varepsilon}}_{\eta\nu} & \bar{\bar{\varepsilon}}_{\eta\eta} \end{matrix}\right\| \tag{11.17}$$

must be positive semi-definite.

We summarize in the following theorem:

Theorem 6. *For a simple fluid in thermal equilibrium, the stress S reduces to a hydrostatic pressure $S = -\bar{p}(v,\eta)I$. The pressure $\bar{p}(v,\eta) = -\bar{\varepsilon}_v(v,\eta)$ is always positive. The energy density $\bar{\bar{\varepsilon}}(\nu,\eta)$ is a strictly convex function of the cube root ν of the specific volume and the entropy η jointly.*

It is not hard to show that, for simple fluids, the positivity of $\bar{p}(v,\eta)$ and the convexity of $\bar{\bar{\varepsilon}}(\nu,\eta)$ are not only necessary but also sufficient conditions for the validity of the fundamental inequality (11.3). Hence, these conditions are also sufficient conditions for the validity of Postulate I for simple fluids.

12. Isotropic materials

For isotropic materials in general, if we pick an undistorted state as reference, we have, by (4.5),

$$\hat{\varepsilon}(F,\eta) = \hat{\varepsilon}(VR,\eta) = \hat{\varepsilon}(V,\eta), \tag{12.1}$$

where V is the left stretch tensor, defined by the polar decomposition $F = VR$. On computing the gradient of (12.1) with respect to V, we find

$$\hat{\varepsilon}_F(F,\eta) = R^T \hat{\varepsilon}_V(V,\eta). \tag{12.2}$$

If we substitute (12.2) into (7.1) and again use $F = VR$, we see that the stress relation may be written in the form

$$S = \varrho V \hat{\varepsilon}_V(V,\eta) = \varrho \hat{\varepsilon}_V(V,\eta) V. \tag{12.3}$$

On substituting (8.4), (12.1) and (12.2) into the fundamental inequality (8.3) and observing that

$$\hat{\varepsilon}(F^*) = \hat{\varepsilon}(GF) = \hat{\varepsilon}(GVR) = \hat{\varepsilon}(GV),$$

we obtain

$$\hat{\varepsilon}(GV,\eta^*) - \hat{\varepsilon}(V,\eta) - \operatorname{tr}[(G-I)V\hat{\varepsilon}_V(V,\eta)] - (\eta^* - \eta)\hat{\varepsilon}_\eta(V,\eta) > 0. \tag{12.4}$$

This inequality must be valid for all η, η^* and all symmetric and positive definite G and V, except, of course, when both $\eta = \eta^*$ and $G = I$.

We consider now the special case when G and V commute; *i.e.*, when

$$V^* = GV \tag{12.5}$$

is symmetric. In this case the tensors V and V^* have an orthonormal basis of proper vectors in common. The matrices of V and V^*, relative to this basis, are

$$\|V\| = \begin{Vmatrix} v_1 & 0 & 0 \\ 0 & v_2 & 0 \\ 0 & 0 & v_3 \end{Vmatrix}, \qquad \|V^*\| = \begin{Vmatrix} v_1^* & 0 & 0 \\ 0 & v_2^* & 0 \\ 0 & 0 & v_3^* \end{Vmatrix}, \tag{12.6}$$

where the v_i and the v_i^* are the proper numbers of V and V^*, respectively. The matrix of $\hat{\varepsilon}_V(V,\eta)$ is

$$\|\hat{\varepsilon}_V(V,\eta)\| = \begin{Vmatrix} \varepsilon_1 & 0 & 0 \\ 0 & \varepsilon_2 & 0 \\ 0 & 0 & \varepsilon_3 \end{Vmatrix}, \tag{12.7}$$

where

$$\varepsilon_i = \bar{\varepsilon}_i(v_j,\eta) = \frac{\partial}{\partial v_i}\bar{\varepsilon}(v_j,\eta) \tag{12.8}$$

are the partial derivatives of the function (4.6). Substitution of (12.5), (4.6), (12.6), and (12.7) into (12.4) gives the inequality

$$\bar{\varepsilon}(v_j^*,\eta^*) - \bar{\varepsilon}(v_j,\eta) - \sum_{i=1}^{3}(v_i^* - v_i)\,\bar{\varepsilon}_i(v_j,\eta) - (\eta^* - \eta)\,\bar{\varepsilon}_\eta(v_j,\eta) > 0, \tag{12.9}$$

which is valid except when $\eta^* = \eta$ and $v_i = v_i^*$ for all i. We have thus proved*

* We have shown that for isotropic materials the inequality (12.9) is a *necessary* condition for validity of the fundamental inequality (8.3). At the present time, it is an open matter as to whether (12.9) is *sufficient* for the validity of (8.3) in the isotropic case, or whether further inequalities which are independent of (12.9) can be deduced from Postulate I for isotropic materials.

Theorem 7. *For an isotropic material, the energy density $\bar{\varepsilon}(v_j, \eta)$ is a strictly convex function of the principal stretches v_j and the entropy density η jointly.*

If $\bar{\varepsilon}$ happens to be twice continuously differentiable, it follows that the matrix

$$\begin{Vmatrix} \bar{\varepsilon}_{11} & \bar{\varepsilon}_{12} & \bar{\varepsilon}_{13} & \bar{\varepsilon}_{1\eta} \\ \bar{\varepsilon}_{21} & \bar{\varepsilon}_{22} & \bar{\varepsilon}_{23} & \bar{\varepsilon}_{2\eta} \\ \bar{\varepsilon}_{31} & \bar{\varepsilon}_{32} & \bar{\varepsilon}_{33} & \bar{\varepsilon}_{3\eta} \\ \bar{\varepsilon}_{\eta 1} & \bar{\varepsilon}_{\eta 2} & \bar{\varepsilon}_{\eta 3} & \bar{\varepsilon}_{\eta\eta} \end{Vmatrix} \tag{12.10}$$

is positive semidefinite. Here the indices 1, 2, 3, and η denote the derivatives of $\bar{\varepsilon}$ with respect to v_1, v_2, v_3, and η, respectively.

A corollary of the convexity inequality (12.9) is

Theorem 8. *For an isotropic material, the functions $\bar{\varepsilon}_i(v_j, \eta)$, defined by* (12.8), *have the property that $v_i > v_k$ implies $\bar{\varepsilon}_i(v_j, \eta) > \bar{\varepsilon}_k(v_j, \eta)$.*

Proof. Without loss of generality, we take $i=1$ and $k=2$. We then choose $v_1^* = v_2$, $v_2^* = v_1$, $v_3^* = v_3$, and $\eta^* = \eta$. Since $\bar{\varepsilon}(v_j, \eta)$ is a symmetric function of the principal stretches v_j (*cf.* § 4), and since the v_j^* differ from the v_j only by their order, we have

$$\bar{\varepsilon}(v_j^*, \eta) = \bar{\varepsilon}(v_j, \eta).$$

Hence (12.9) reduces to

$$-(v_2 - v_1)\,\bar{\varepsilon}_1(v_j, \eta) - (v_1 - v_2)\,\bar{\varepsilon}_2(v_j, \eta) > 0;$$

i.e.,

$$(v_1 - v_2)\left[\bar{\varepsilon}_1(v_j, \eta) - \bar{\varepsilon}_2(v_j, \eta)\right] > 0.$$

Thus, if $v_1 > v_2$, then $\bar{\varepsilon}_1(v_j, \eta) > \bar{\varepsilon}_2(v_j, \eta)$, q.e.d.

In an isotropic material, the left stretch tensor V and the stress tensor S have an orthonormal basis $\boldsymbol{e}_i$ of proper vectors in common. The $\boldsymbol{e}_i$ determine the principal axes of stress. It follows from (12.3), (12.6), and (12.7) that the principal stresses are given by

$$s_i = \varrho\, v_i\, \bar{\varepsilon}_i(v_j, \eta). \tag{12.11}$$

When measured per unit area in the undistorted reference state, these principal stresses must be replaced by

$$s_i' = \varrho_r\, \bar{\varepsilon}_i(v_j, \eta). \tag{12.12}$$

Hence Theorem 8 has the following simple physical interpretation:

Theorem 8a. *If, at a given value η, the principal stretch v_i is greater than the principal stretch v_k, then the principal stress, measured per unit area in the undistorted reference state, in the direction of v_i is greater than that in the direction of v_k.*

It should be noted that the statement of this theorem does not necessarily remain valid if the principal stresses are measured per unit area of the deformed state*, except when these stresses are all positive; *i.e.*, except in a state of pure tension.

* Such a statement was proposed as a postulate by M. BAKER & J. L. ERICKSEN [*13*]. In our theory, only the modification given by Theorem 8a is valid. Related inequalities have been studied by J. BARTA [*14*].

13. The free energy

It is often useful to employ the deformation gradient F and the temperature ϑ, rather than F and the entropy η, as the independent variables. This is possible because, by (8.2) and (8.6), there is a one-to-one correspondence between η and ϑ for each fixed F.

The *free energy function* $\tilde{\psi}$ is defined by

$$\tilde{\psi}(F,\vartheta) = \hat{\varepsilon}[F,\tilde{\eta}(F,\vartheta)] - \vartheta\tilde{\eta}(F,\vartheta), \tag{13.1}$$

where the entropy function $\tilde{\eta}$ is defined in (8.6) as the unique solution of the equation (8.2). The values ψ of the free energy function $\tilde{\psi}$ are called *free energy densities**.

Differentiation of (13.1) with respect to F, using the chain rule, gives

$$\tilde{\psi}_F(F,\vartheta) = \hat{\varepsilon}_F[F,\tilde{\eta}(F,\vartheta)] + \hat{\varepsilon}_\eta[F,\tilde{\eta}(F,\vartheta)]\,\tilde{\eta}_F(F,\vartheta) - \vartheta\,\tilde{\eta}_F(F,\vartheta).$$

It follows from (8.7) that the last two terms cancel, so that

$$\tilde{\psi}_F(F,\vartheta) = \hat{\varepsilon}_F[F,\tilde{\eta}(F,\vartheta)]. \tag{13.2}$$

Differentiation of (13.1) with respect to ϑ gives

$$\tilde{\psi}_\vartheta(F,\vartheta) = -\,\tilde{\eta}(F,\vartheta). \tag{13.3}$$

From Theorems 1 and 2, (13.2), and (13.3) we get

Theorem 9. *For the force temperature pair (S_r,ϑ) to make the local state (F,η) a state of thermal equilibrium, it is necessary and sufficient that S_r and η be given by*

$$S_r = \varrho_r\,\tilde{\psi}_F(F,\vartheta), \tag{13.4}$$

$$\eta = \quad \tilde{\psi}_\vartheta(F,\vartheta). \tag{13.5}$$

On multiplying (13.4) on the left by $(\varrho/\varrho_r)\,F$ and noting that $S = (\varrho/\varrho_r) F\,S_r$, we get the following form for the stress relation:

$$S = \varrho\,F\,\tilde{\psi}_F(F,\vartheta). \tag{13.6}$$

Assuming that two temperatures ϑ, ϑ^* and two deformation gradients F, F^* are given, we now put

$$\begin{aligned}\eta &= \tilde{\eta}(F,\vartheta) = -\,\tilde{\psi}_\vartheta(F,\vartheta)\\ \eta^* &= \tilde{\eta}(F^*,\vartheta^*) = -\,\tilde{\psi}_\vartheta(F^*,\vartheta^*).\end{aligned} \tag{13.7}$$

By substituting (13.1)—(13.7) into the fundamental inequality (8.3) of Theorem 2, we obtain

Theorem 10. *The free energy function $\tilde{\psi}$ obeys the inequality*

$$\tilde{\psi}(F^*,\vartheta^*) - \tilde{\psi}(F,\vartheta) - \operatorname{tr}[(F^*-F)\,\tilde{\psi}_F(F,\eta)] - (\vartheta^*-\vartheta)\,\tilde{\psi}_\vartheta(F^*,\vartheta^*) > 0 \tag{13.8}$$

for any two pairs (F,ϑ) and $(F^*,\vartheta^*) \neq (F,\vartheta)$ in the domain of definition of $\tilde{\psi}$ which are related by

$$F^* = GF, \tag{13.9}$$

where G is symmetric and positive definite.

* The term "Helmholtz free energy per unit mass" would also be in accord with common usage.

As Theorem 2, so also Theorem 10 is equivalent to Postulate I.

If in (13.8) we take the special case $F^*=F$ and interchange ϑ and ϑ^*, we obtain

$$\tilde{\psi}(F,\vartheta^*) - \tilde{\psi}(F,\vartheta) - (\vartheta^* - \vartheta)\,\tilde{\psi}_\vartheta(F,\vartheta) < 0. \tag{13.10}$$

This inequality, which is valid for all F and all $\vartheta^* \neq \vartheta$, states that the free energy function $\tilde{\psi}(F,\vartheta)$ is *strictly concave* in ϑ for each F.

Putting $\vartheta^*=\vartheta$ in (13.8) gives the following restricted convexity of $\tilde{\psi}$ in F:

$$\tilde{\psi}(F^*,\vartheta) - \tilde{\psi}(F,\vartheta) - \operatorname{tr}\left[(F^* - F)\,\tilde{\psi}_F(F,\vartheta)\right] > 0, \tag{13.11}$$

the restriction being the condition (13.9).

The considerations and results of § 11 and § 12 on simple fluids and isotropic materials remain valid if the energy function $\hat{\varepsilon}$ is replaced by the free energy function $\tilde{\psi}$, except that the convexity of $\hat{\varepsilon}(F,\eta)$ in η corresponds to the concavity of $\tilde{\psi}(F,\vartheta)$ in ϑ. We summarize the relevant results.

For a simple fluid, the free energy density reduces to a function of the specific volume v and the temperature ϑ only:

$$\psi = \tilde{\psi}(F,\vartheta) = \bar{\psi}(v,\vartheta). \tag{13.12}$$

The stress reduces to a hydrostatic pressure given by

$$S = -\bar{p}(v,\vartheta)\,I, \quad \bar{p}(v,\vartheta) = -\bar{\psi}_v(v,\vartheta). \tag{13.13}$$

The pressure is always positive. The function $\bar{\bar{\psi}}$, giving the free energy as a function of the cube root ν of the specific volume and the temperature,

$$\bar{\bar{\psi}}(\nu,\vartheta) = \bar{\psi}(\nu^3,\vartheta), \tag{13.14}$$

satisfies the inequality

$$\bar{\bar{\psi}}(\nu^*,\vartheta^*) - \bar{\bar{\psi}}(\nu,\vartheta) - (\nu^* - \nu)\,\bar{\bar{\psi}}_\nu(\nu,\vartheta) - (\vartheta^* - \vartheta)\,\bar{\bar{\psi}}_\vartheta(\nu^*,\vartheta^*) > 0. \tag{13.15}$$

This inequality implies that $\bar{\bar{\psi}}(\nu,\vartheta)$ is strictly convex in ν for each ϑ and strictly concave in ϑ for each ν.

For isotropic materials in general, the free energy reduces to a function of the temperature ϑ and the three principal stretches v_1, v_2, v_3, computed relative to an undistorted state:

$$\tilde{\psi}(F,\vartheta) = \bar{\psi}(v_1, v_2, v_3; \vartheta). \tag{13.16}$$

The function $\bar{\psi}$ is symmetric and strictly convex in the variables v_1, v_2, v_3; $\bar{\psi}$ is strictly concave in ϑ. Theorems 8 and 8a remain valid if $\bar{\varepsilon}$ is replaced by $\bar{\psi}$; *i.e.*, if the temperature, rather than the entropy is fixed at a given value. The stress relation may be written in the form

$$S = \varrho\, V\,\tilde{\psi}_V(V,\vartheta), \tag{13.17}$$

where V is the left stretch tensor.

The forms, (13.6), (13.13), and (13.17), of the stress relation are useful in discussing experiments involving equilibrium states for which the temperature is controlled, while the forms, (7.1), (11.10), and (12.3), are appropriate for discussing experiments involving equilibrium states for which the entropy is controlled.

14. Thermal stability

Consider a body $\mathscr{B}$ and a global thermomechanic state $\{\boldsymbol{f}, \eta\}$ of $\mathscr{B}$, defined by a configuration $\boldsymbol{f}$ of $\mathscr{B}$ and an entropy distribution η of $\mathscr{B}$; (*cf.* § 2). Let the caloric equation of state of the material point X of $\mathscr{B}$ be given by

$$\varepsilon(X) = \hat{\varepsilon}[F(X), \eta(X); X]. \tag{14.1}$$

Here $F(X)$ is the deformation gradient at X of the configuration $\boldsymbol{f}$ relative to some reference configuration $\boldsymbol{f}_r$. We do not assume that the body is homogeneous, and hence the function $\hat{\varepsilon}$ may depend explicitly on X as indicated in (14.1). The *total entropy* of $\mathscr{B}$ in the given state is defined by

$$\mathsf{H} = \int_{\mathscr{B}} \eta(X)\, dm \tag{14.2}$$

and the *total internal energy* of $\mathscr{B}$ by

$$\mathsf{E} = \int_{\mathscr{B}} \hat{\varepsilon}[F(X), \eta(X); X]\, dm. \tag{14.3}$$

In this section we shall deal with situations in which the deformation gradient $F(X)$ is kept fixed at each X while the entropy field $\eta = \eta(X)$ is varied. It will not be necessary to make the dependence of ε on F explicit, and the following abbreviated notation will be convenient:

$$\hat{\varepsilon}[F(X), \eta(X); X] = \varepsilon(X, \eta(X)). \tag{14.4}$$

Definition of thermal stability. *Let $\{\boldsymbol{f}, \eta\}$ be a state of $\mathscr{B}$ and let E and H be, respectively, the total internal energy and total entropy corresponding to the state $\{\boldsymbol{f}, \eta\}$. We say that $\{\boldsymbol{f}, \eta\}$ is a* **thermally stable** *state of $\mathscr{B}$ if every other state $\{\boldsymbol{f}, \eta^*\}$, with the same configuration as $\{\boldsymbol{f}, \eta\}$ and the same total entropy as $\{\boldsymbol{f}, \eta\}$,*

$$\mathsf{H}^* = \int_{\mathscr{B}} \eta^*(X)\, dm = \mathsf{H} = \int_{\mathscr{B}} \eta(X)\, dm, \tag{14.5}$$

has a greater total internal energy than the state $\{\boldsymbol{f}, \eta\}$; i.e.,

$$\mathsf{E}^* = \int_{\mathscr{B}} \varepsilon(X, \eta^*(X))\, dm > \mathsf{E} = \int_{\mathscr{B}} \varepsilon(X, \eta(X))\, dm. \tag{14.6}$$

We give another condition, equivalent to the one given above, which could also be used to define thermal stability.

Theorem 11. *A state $\{\boldsymbol{f}, \eta\}$ of $\mathscr{B}$ is thermally stable if and only if every other state $\{\boldsymbol{f}, \eta^*\}$ with the same configuration as $\{\boldsymbol{f}, \eta\}$ and the same total energy as $\{\boldsymbol{f}, \eta\}$,*

$$\mathsf{E}^* = \int_{\mathscr{B}} \varepsilon(X, \eta^*(X))\, dm = \mathsf{E} = \int_{\mathscr{B}} \varepsilon(X, \eta(X))\, dm, \tag{14.7}$$

has a lower total entropy than the state $\{\boldsymbol{f}, \eta\}$; i.e.,

$$\mathsf{H} = \int_{\mathscr{B}} \eta(X)\, dm > \mathsf{H}^* = \int_{\mathscr{B}} \eta^*(X)\, dm. \tag{14.8}$$

Proof. We show that the hypothesis of Theorem 11 is necessary for thermal stability by showing that if there exists a state $\{\boldsymbol{f}, \eta_1\}$ (with η_1 not identical to η) which obeys the equation (14.7) of Theorem 11 but violates (14.8), then there must exist a state $\{\boldsymbol{f}, \eta_2\}$ (with η_2 not identical to η) which obeys the equation

(14.5) of the definition of thermal stability but which does not obey (14.6). Let η_1 be the entropy density distribution which obeys (14.7) but not (14.8); we construct η_2 as follows:

$$\eta_2(X) = \eta_1(X) + \frac{\mathsf{H}-\mathsf{H}_1}{m(\mathscr{B})}, \tag{14.9}$$

where H_1 is the total entropy corresponding to η_1. The total entropy corresponding to η_2 is

$$\mathsf{H}_2 = \int_{\mathscr{B}} \eta_2(X)\, dm = \mathsf{H}. \tag{14.10}$$

Hence, the state $\{\boldsymbol{f}, \eta_2\}$ obeys the equation (14.5) of the definition. We have assumed that η_2 is not identical to η and that $\mathsf{H}_1 \geqq \mathsf{H}$. If $\mathsf{H}_1 = \mathsf{H}$, then η_2 is the same as η_1 and hence different from η. In this trivial case of $\eta_2 = \eta_1$, it follows from the fact that η_1 obeys (14.7) that

$$\int_{\mathscr{B}} \varepsilon(X, \eta_2(X))\, dm = \int_{\mathscr{B}} \varepsilon(X, \eta_1(X))\, dm = \int_{\mathscr{B}} \varepsilon(X, \eta(X))\, dm. \tag{14.11}$$

If $\mathsf{H}_1 > \mathsf{H}$, then $\eta_2(X) < \eta_1(X)$ for all X in $\mathscr{B}$. It then follows from Postulate II of § 8 and the assumption that η_1 obeys (14.7) that

$$\int_{\mathscr{B}} \varepsilon(X, \eta_2(X))\, dm < \int_{\mathscr{B}} \varepsilon(X, \eta_1(X))\, dm = \int_{\mathscr{B}} \varepsilon(X, \eta(X))\, dm. \tag{14.12}$$

It is clear from (14.12) that η_2 is not identical to η. Hence, whenever $\mathsf{H}_1 \geqq \mathsf{H}$, we have, by the construction (14.9), a state $\{\boldsymbol{f}, \eta_2\}$ with η_2 different from η but with $\mathsf{H}_2 = \mathsf{H}$ and

$$\mathsf{E}_2 = \int_{\mathscr{B}} \varepsilon(X, \eta_2(X))\, dm \leqq \int_{\mathscr{B}} \varepsilon(X, \eta(X))\, dm = \mathsf{E}. \tag{14.13}$$

Thus, a violation of the hypothesis of Theorem 11 implies the existence of a state different from $\{\boldsymbol{f}, \eta\}$ which obeys (14.5) yet violates (14.6).

The sufficiency of the hypothesis of Theorem 11 is proved analogously by starting with a state which obeys (14.5) of the definition, but not (14.6), and then using Postulate II to construct a state which obeys (14.7) of the theorem, but which violates (14.8).

The main result of the present section is the following theorem:

Theorem 12. *A state $\{\boldsymbol{f}, \eta\}$ of a body is thermally stable if and only if it is of uniform temperature;* i.e., *if and only if*

$$\vartheta = \varepsilon_\eta(X, \eta(X)) \tag{14.14}$$

is a constant, independent of the material point X.

Proof. To show the necessity of $\vartheta =$ constant, we observe that, by (14.6), the function $\eta(X)$ is the solution of the variational problem

$$\int_{\mathscr{B}} \varepsilon(X, \eta^*(X))\, dm = \text{Minimum} \tag{14.15}$$

subject to the constraint (14.5). It follows that the first variation of

$$\int_{\mathscr{B}} [\varepsilon(X, \eta^*(X)) - \alpha\, \eta^*(X)]\, dm \tag{14.16}$$

must vanish for $\eta^* = \eta$. Here α is a constant Lagrange parameter. We obtain

$$\alpha = \varepsilon_\eta(X, \eta(X)) = \vartheta = \text{constant}. \tag{14.17}$$

To prove the sufficiency of $\vartheta = \text{constant}$, we substitute the function values $F(X)$, $\eta(X)$ and $\eta^*(X)$ for F, η and η^* in the convexity inequality (8.5). Using the abbreviation (14.4) and the equation (14.14), we get

$$\varepsilon\big(X, \eta^*(X)\big) - \varepsilon\big(X, \eta(X)\big) - \left[\eta^*(X) - \eta(X)\right]\vartheta \geqq 0. \tag{14.18}$$

This inequality must be strict for some X if η^* and η are different continuous functions. If ϑ is a constant and if (14.5) holds, then integration of (14.18) over the body $\mathscr{B}$ gives the inequality (14.6), which proves that $\{\boldsymbol{f}, \eta\}$ is thermally stable, *q.e.d.*

15. Mechanical stability

Consider a state $\{\boldsymbol{f}, \eta\}$ of a body $\mathscr{B}$. According to Postulate I of § 8 it is possible to find a temperature field ϑ and a stress field S such that every material point of $\mathscr{B}$ is in *thermal equilibrium* for the force temperature field defined by S and ϑ. In fact, S and ϑ are given by the stress relation (7.1) and the temperature relation (7.2), respectively. If a field of body forces $\boldsymbol{b}$ is given, then the state $\{\boldsymbol{f}, \eta\}$ will be a state of *mechanical equilibrium* if Cauchy's condition

$$\operatorname{Div} S + \varrho\, \boldsymbol{b} = 0 \tag{15.1}$$

holds. If $\{\boldsymbol{f}, \eta\}$ is such that every material point is in thermal equilibrium, it is always possible to choose $\boldsymbol{b}$ such that the state $\{\boldsymbol{f}, \eta\}$ is a state of mechanical equilibrium. We need only to define $\boldsymbol{b}$ by (15.1). We say that the fields S, ϑ, and $\boldsymbol{b}$, given by (7.1), (7.2) and (15.1), make $\{\boldsymbol{f}, \eta\}$ a state of equilibrium. We call S, ϑ and $\boldsymbol{b}$, respectively, the stress, temperature, and body force fields of $\{\boldsymbol{f}, \eta\}$.

We investigate the possible meaning that can be given to the statement that an equilibrium state $\{\boldsymbol{f}, \eta\}$ is *stable*. First, we require that it be *thermally stable* which, according to Theorem 12, means that the temperature ϑ must be uniform. In addition, we require that some condition of *mechanical stability* be satisfied. One must distinguish between various types of *isothermal mechanical stability* and *adiabatic mechanical stability*.

In the case of isothermal mechanical stability, one compares the given equilibrium state $\{\boldsymbol{f}, \eta\}$ with a class of states $\{\boldsymbol{f}^*, \eta^*\}$ corresponding to the same uniform temperature $\vartheta = \hat{\vartheta}(F, \eta)$ as the given state. Each of these states is characterized by its configuration $\boldsymbol{f}^*$ alone, because the corresponding entropy distribution is then determined by

$$\eta = \tilde{\eta}(F^*, \vartheta). \tag{15.2}$$

External forces or boundary conditions must be prescribed for each of the comparison configurations $\boldsymbol{f}^*$. The configuration $\boldsymbol{f}$ is called stable if the increase in the total free energy would always be greater than the work done on the body by the external forces if the configuration were to be deformed into any of the comparison configurations $\boldsymbol{f}^*$. We give more precise definitions in two special cases.

Definition of isothermal stability at fixed boundary (IFB stability). *An equilibrium state $\{\boldsymbol{f}, \eta\}$ is called* **IFB stable** *if $\{\boldsymbol{f}, \eta\}$ has a uniform temperature ϑ*

and if for every state $\{\boldsymbol{f}^*, \eta\}$ *which satisfies the following conditions:*

(a) $\boldsymbol{f}^*$ *lies in a prescribed neighborhood* of* $\boldsymbol{f}$,

(b) $\boldsymbol{f}^*(X) = \boldsymbol{f}(X)$, *when* X *belongs to* $\overline{\mathscr{B}}$, (15.3)

(c) *the temperature corresponding to* $\{\boldsymbol{f}^*, \eta\}$ *is equal to* ϑ *for all* X *in* $\mathscr{B}$,

the following inequality holds:

$$\int_{\mathscr{B}} \{\psi(F^*) - \psi(F) - \boldsymbol{b}\cdot(\boldsymbol{f}^* - \boldsymbol{f})\}\, dm \geqq 0. \tag{15.4}$$

Here $\overline{\mathscr{B}}$ is the boundary of $\mathscr{B}$, and $\psi(F)$ is an abbreviation for

$$\psi(F) = \tilde{\psi}(F(X), \vartheta; X); \tag{15.5}$$

$F^*(X)$ and $F(X)$ are the deformation gradients at X for the configurations $\boldsymbol{f}^*$ and $\boldsymbol{f}$, respectively, both computed relative to the same fixed reference configuration. As in § 14, we do not assume that the body is homogeneous, and hence the function $\tilde{\psi}$ may depend explicitly on X.

We say that $\{\boldsymbol{f}, \eta\}$ is *strictly* IFB *stable* if the inequality (15.4) is strict whenever $\{\boldsymbol{f}^*, \eta^*\}$ obeys (a), (b) and (c) and is such that $f^* \neq f$.

Note that the surface tractions do no work if the boundary is fixed and that $-\int_{\mathscr{B}} \boldsymbol{f}^*\cdot\boldsymbol{b}\, dm$ is a potential of the work done by the body forces if these are held at their values $\boldsymbol{b}(X)$ in the equilibrium state $\{\boldsymbol{f}, \eta\}$.

The type of stability considered is affected by the prescription of the neighborhood in the requirement (a) of the definition of IFB stability. A global state may be stable with respect to some (small) neighborhood without being stable with respect to other (larger) neighborhoods.

Definition of isothermal stability at fixed surface tractions (IFT stability). *An equilibrium state* $\{\boldsymbol{f}, \eta\}$ *is called* ***IFT stable*** *if* $\{\boldsymbol{f}, \eta\}$ *has a uniform temperature* ϑ *and if for every state* $\{\boldsymbol{f}^*, \eta\}$ *which satisfies the following conditions:*

(a) $\boldsymbol{f}^*$ *lies in a prescribed neighborhood of* $\boldsymbol{f}$,

(b) *the temperature corresponding to* $\{\boldsymbol{f}^*, \eta\}$ *is equal to* ϑ *for all* X *in* $\mathscr{B}$,

the following inequality holds:

$$\Gamma = \int_{\mathscr{B}} \{\psi(F^*) - \psi(F) - \boldsymbol{b}\cdot(\boldsymbol{f}^* - \boldsymbol{f})\}\, dm - \int_{\overline{\mathscr{B}}} (\boldsymbol{f}^* - \boldsymbol{f})\cdot S\boldsymbol{n}\, dA \geqq 0. \tag{15.6}$$

Here $\overline{\mathscr{B}}$ is the boundary surface of the region occupied by $\mathscr{B}$ in the configuration $\boldsymbol{f}$; dA is the element of that surface; and $\boldsymbol{n}$ is the exterior unit normal.

Note that $-\int_{\overline{\mathscr{B}}} \boldsymbol{f}^*\cdot S\boldsymbol{n}\, dA$ is a potential of the work done by the surface tractions if they are held at their values in the equilibrium state $\{\boldsymbol{f}, \eta\}$.

An IFT stable state is always also IFB stable. This follows from the fact that the surface integral in (15.6) gives no contribution if the boundary condition (15.3) holds, so that the inequalities (15.4) and (15.6) become the same in this case.

* A neighborhood of a configuration is defined by the metric

$$\delta(\boldsymbol{f}, \boldsymbol{f}^*) = \sup_{X\in\mathscr{B}} \{|\boldsymbol{f}^*(X) - \boldsymbol{f}(X)| + |F^{*-1}(X)\,F(X) - I|\}$$

over the space of all configurations.

If the inequality (15.6) holds for all states which obey items (a) and (b) of the definition of IFT stability and is, furthermore, a strict inequality for all such states for which $F^*(X) \neq F(X)$ for at least one material point X, then we say that $\{\boldsymbol{f}, \eta\}$ is *strictly* IFT *stable against deformations and rotations.* For in that case (15.6) can reduce to an equality only if $\boldsymbol{f}^*$ is related to $\boldsymbol{f}$ by a simple rigid translation.

To investigate adiabatic mechanical stability, one compares the given equilibrium state $\{\boldsymbol{f}, \eta\}$ with a class of states which correspond to the same total entropy as $\{\boldsymbol{f}, \eta\}$. We again consider two special cases.

Definition of adiabatic stability at fixed boundary (AFB stability). *An equilibrium state $\{\boldsymbol{f}, \eta\}$ is called **AFB stable** if $\{\boldsymbol{f}\ \eta\}$ is thermally stable and if for every state $\{\boldsymbol{f}^*, \eta^*\}$ which satisfies the following conditions:*

(a) *$\boldsymbol{f}^*$ lies in a prescribed neighborhood of $\boldsymbol{f}$,*

(b) *$\boldsymbol{f}^*(X) = \boldsymbol{f}(X)$, when X belongs to $\bar{\mathscr{B}}$,*

(c) $\int_{\mathscr{B}} \eta^*(X)\, dm = \int_{\mathscr{B}} \eta(X)\, dm,$

the following inequality holds:

$$\int_{\mathscr{B}} \{\hat{\varepsilon}\,[F^*(X), \eta^*(X); X] - \hat{\varepsilon}\,[F(X), \eta(X); X] - \boldsymbol{b} \cdot (\boldsymbol{f}^* - \boldsymbol{f})\}\, dm \geqq 0. \tag{15.7}$$

If the inequality in (15.7) is strict for all $\{\boldsymbol{f}^*, \eta^*\}$ satisfying (a), (b) and (c) and for which $\boldsymbol{f}^* \neq \boldsymbol{f}$, then we say that $\{\boldsymbol{f}, \eta\}$ is *strictly* AFB *stable.*

Theorem 13. *A thermally stable equilibrium state $\{\boldsymbol{f}, \eta\}$ is AFB stable if and only if for every state $\{\boldsymbol{f}^*, \eta^*\}$ which satisfies the following conditions:*

(a) *$\boldsymbol{f}^*$ lies in a prescribed neighborhood of $\boldsymbol{f}$,*

(b) *$\boldsymbol{f}^*(X) = \boldsymbol{f}(X)$ when X belongs to $\bar{\mathscr{B}}$,*

(c) $$\int_{\mathscr{B}} \{\hat{\varepsilon}\,(F^*(X), \eta^*(X); X) - \boldsymbol{b} \cdot \boldsymbol{f}^*\}\, dm = \int_{\mathscr{B}} \{\hat{\varepsilon}\,(F(X), \eta(X); X) - \boldsymbol{b} \cdot \boldsymbol{f}\}\, dm, \tag{15.8}$$

the following inequality holds:

$$\int_{\mathscr{B}} \eta^*(X)\, dm \leqq \int_{\mathscr{B}} \eta(X)\, dm. \tag{15.9}$$

Furthermore, $\{\boldsymbol{f}, \eta\}$ is strictly AFB stable if and only if (15.9) *is a strict inequality for every state $\{\boldsymbol{f}^*, \eta^*\} \neq \{\boldsymbol{f}, \eta\}$ obeying* (a), (b) *and* (c).

We omit the proof of Theorem 13 because it is analogous to that of Theorem 11. Of course, the validity of Theorem 13 requires the assumption of Postulate II.

Definition of adiabatic stability at fixed surface tractions (AFT stability). *An equilibrium state $\{\boldsymbol{f}, \eta\}$ is called **AFT stable** if it is thermally stable and if for every state $\{\boldsymbol{f}^*, \eta^*\}$ which satisfies the following conditions:*

(a) *$\boldsymbol{f}^*$ is in a prescribed neighborhood of $\boldsymbol{f}$,*

(b) $\int_{\mathscr{B}} \eta^*(X)\, dm = \int_{\mathscr{B}} \eta(X)\, dm,$

the following inequality holds:

$$\begin{aligned}\int_{\mathscr{B}} \{\hat{\varepsilon}\,[F^*(X), \eta^*(X); X] - \hat{\varepsilon}\,[F(X), \eta(X); X] - \boldsymbol{b} \cdot (\boldsymbol{f}^* - \boldsymbol{f})\}\, dm\ - \\ - \int_{\bar{\mathscr{B}}} (\boldsymbol{f}^* - \boldsymbol{f}) \cdot S\, \boldsymbol{n}\, dA \geqq 0.\end{aligned} \tag{15.10}$$

It will be noticed that a state which is AFT stable is always AFB stable.

If the inequality (15.10) holds for all states which obey (a) and (b) and is a strict inequality for all such states for which $F^*(X) \neq F(X)$ for at least one X, then we say that $\{\boldsymbol{f}, \eta\}$ is *strictly* AFT *stable against deformations and rotations.*

It is clear that, in analogy to Theorem 13, an alternative, but equivalent, definition of AFT stability can be formulated in which a stable state is defined to be one of maximum entropy among all those states for which (15.10) reduces to an equality.

The definitions of IFB, IFT, AFB and AFT stability given above are applicable only to those physical situations in which the body force field $\boldsymbol{b} = \boldsymbol{b}(X)$ is independent of the comparison configuration $\boldsymbol{f}^*$. If one is interested in studying cases in which the body force on X depends on X and is also a functional of $\boldsymbol{f}^*$, one can modify the definitions of stability by connecting the comparison state $\boldsymbol{f}^*$ to $\boldsymbol{f}$ by means of a continuous one-parameter family $\boldsymbol{f}_s$, $0 \leq s \leq 1$, $\boldsymbol{f}_0 = \boldsymbol{f}$, $\boldsymbol{f}_1 = \boldsymbol{f}^*$ and replacing the term

$$-\int_{\mathscr{B}} \boldsymbol{b} \cdot (\boldsymbol{f}^* - \boldsymbol{f})\, dm$$

in (15.4), (15.6), (15.7), (15.8) and (15.10) by

$$-\int_{\mathscr{B}} \int_0^1 \boldsymbol{b}(X, \boldsymbol{f}_s) \cdot \frac{\partial \boldsymbol{f}_s(X)}{\partial s}\, ds\, dm .$$

If the body force on each material point is derivable from a single-valued potential, then the integral exhibited above is independent of the paramatization, and is simply the difference in the potentials at $\boldsymbol{f}$ and $\boldsymbol{f}^*$.

In the definitions of IFT and AFT stability, we assumed that not only the body forces but also the contact forces at the surface do not depend on the comparison configuration. One can also study, in a way analogous to that outlined above for the body forces, those cases in which the surface tractions depend on the comparison configuration.

Theorem 14. *A state which has isothermal stability of a certain type also has adiabatic stability of the corresponding type.*

Proof. Consider a state $\{\boldsymbol{f}, \eta\}$ which has a uniform temperature ϑ and which has isothermal stability of a particular type. Let $\boldsymbol{f}^*$ be a configuration which satisfies the boundary conditions, if any, for the appropriate comparison configurations. Define the entropy field η_1 by

$$\eta_1(X) = \tilde{\eta}\left(F^*(X), \vartheta\right), \tag{15.11}$$

where F^* is the deformation gradient field corresponding to the configuration $\boldsymbol{f}^*$. By (13.1) we have

$$\tilde{\psi}(F^*, \vartheta) - \tilde{\psi}(F, \vartheta) = \hat{\varepsilon}(F^*, \eta_1) - \hat{\varepsilon}(F, \eta) - (\eta_1 - \eta)\, \vartheta . \tag{15.12}$$

Here F corresponds to $\boldsymbol{f}$. Let η^* be any entropy distribution satisfying the condition

$$\int_{\mathscr{B}} \eta^*(X)\, dm = \int_{\mathscr{B}} \eta(X)\, dm , \tag{15.13}$$

which is required for comparison states in adiabatic stability. Define the field β by

$$\beta = \hat{\varepsilon}(F^*, \eta^*) - \hat{\varepsilon}(F^*, \eta_1) - (\eta^* - \eta_1)\, \vartheta . \tag{15.14}$$

From (8.5) we get

$$\beta(X) \geqq 0$$

for all X. From (15.12) we have

$$\tilde{\psi}(F^*, \vartheta) - \tilde{\psi}(F, \vartheta) = \hat{\varepsilon}(F^*, \eta^*) - \hat{\varepsilon}(F, \eta) - \beta - (\eta^* - \eta)\,\vartheta. \tag{15.15}$$

We integrate (15.15) over $\mathscr{B}$. According to (15.13) we get no contribution from the term $-(\eta^* - \eta)\vartheta$; hence, since β is non-negative,

$$\int_{\mathscr{B}} [\tilde{\psi}(F^*, \vartheta) - \tilde{\psi}(F, \vartheta)]\, dm \leqq \int_{\mathscr{B}} [\hat{\varepsilon}(F^*, \eta^*) - \hat{\varepsilon}(F, \eta)]\, dm. \tag{15.16}$$

Since the work W done by the external forces in going from $\boldsymbol{f}$ to $\boldsymbol{f}^*$ is the same in adiabatic and isothermal stability, it follows from (15.16) that if

$$\int_{\mathscr{B}} [\tilde{\psi}(F^*, \vartheta) - \tilde{\psi}(F, \vartheta)]\, dm - W \tag{15.17}$$

is non-negative, then

$$\int_{\mathscr{B}} [\hat{\varepsilon}(F^*, \eta^*) - \hat{\varepsilon}(F, \eta)]\, dm - W \tag{15.18}$$

is non-negative (and strictly positive when (15.17) is strictly positive). Hence, the isothermal stability of $\{\boldsymbol{f}, \eta\}$ implies the corresponding adiabatic stability for $\{\boldsymbol{f}, \eta\}$, *q.e.d.*

Although in writing our proof of Theorem 14 we have used a notation which implies that $\mathscr{B}$ is homogeneous, it is clear that the same argument is valid when $\mathscr{B}$ is not homogeneous.

It appears to us that the converse of Theorem 14 need not be true; *i.e.*, an equilibrium state may have adiabatic stability without being isothermally stable.

16. Gibbs' thermostatics of fluids

We now consider a type of stability which was proposed by GIBBS* for fluids free from body forces. GIBBS states** that he had in mind a physical situation in which the fluid is "enclosed in a rigid envelop which is non-conducting to heat and impermeable to all the components of the fluid". A body which may be regarded as being in such an envelop is usually called an "isolated system".

Definition of G stability*.** *An equilibrium state $\{\boldsymbol{f}, \eta\}$ of a fluid body $\mathscr{B}$ is called* **G stable** *if the following condition is satisfied. Let $\{\boldsymbol{f}^*, \eta^*\}$ be any other state with the same total volume and the same total entropy as $\{\boldsymbol{f}, \eta\}$,*

$$\int_{\mathscr{B}} v^*\, dm = \int_{\mathscr{B}} v\, dm, \qquad \int_{\mathscr{B}} \eta^*\, dm = \int_{\mathscr{B}} \eta\, dm, \tag{16.1}$$

* See the section of [1] which is entitled "Internal stability of homogeneous fluids as indicated by the fundamental equations", (b), pp. 100—115, particularly the subsection entitled "Stability with respect to continuous changes of phase" (b), pp. 105—111.

** [1] (b), p. 100.

*** In this definition we again restrict ourselves to those physical situations in which fluctuations in chemical composition are surpressed. We have in mind situations in which chemical reactions are prohibited and in which the fluid is either homogeneous or does not allow diffusion. For fluids the homogeneous case is the one of practical importance. Situations in which flow is permitted but diffusion is prohibited are rare.

then $\{\boldsymbol{f}, \eta\}$ *has a lower total internal energy than* $\{\boldsymbol{f}^*, \eta^*\}$,

$$\int_{\mathscr{B}} \bar{\varepsilon}[v^*(X), \eta^*(X); X]\, dm > \int_{\mathscr{B}} \bar{\varepsilon}[v(X), \eta(X); X]\, dm, \tag{16.2}$$

unless $v^*(X) = v(X)$ *and* $\eta^*(X) = \eta(X)$ *for all* X *in* $\mathscr{B}$.

In (16.1) v and v^* denote the specific volume fields for $\mathscr{B}$ corresponding to the configurations $\boldsymbol{f}$ and $\boldsymbol{f}^*$.

In the following alternative definition $\boldsymbol{f}$ and ε are taken as the independent variables, and the permitted comparison states are such that the total internal energy and total volume of the body are conserved during the variations. This alternative formulation may suggest to the reader why G stability is regarded as being appropriate for discussing the physics of *isolated systems* composed of fluids:

Alternative definition of G stability. *An equilibrium state* $\{\boldsymbol{f}, \varepsilon\}$ *of a fluid body* $\mathscr{B}$ *is called G stable if any other state* $\{\boldsymbol{f}^*, \varepsilon^*\}$ *with the same total volume and the same total internal energy as* $\{\boldsymbol{f}, \varepsilon\}$,

$$\int_{\mathscr{B}} v^*\, dm = \int_{\mathscr{B}} v\, dm, \qquad \int_{\mathscr{B}} \varepsilon^*\, dm = \int_{\mathscr{B}} \varepsilon\, dm, \tag{16.3}$$

has a higher total entropy,

$$\int_{\mathscr{B}} \hat{\bar{\eta}}[v^*(X), \varepsilon^*(X); X]\, dm < \int_{\mathscr{B}} \hat{\bar{\eta}}[v(X), \varepsilon(X); X]\, dm, \tag{16.4}$$

unless $v^*(X) = v(X)$ *and* $\varepsilon^*(X) = \varepsilon(X)$ *for all* X *in* $\mathscr{B}$.

The function $\hat{\bar{\eta}}$ in (16.4) is obtained by solving $\varepsilon = \bar{\varepsilon}(v, \eta; X)$ for η, which is possible in a unique way because $\bar{\varepsilon}$ is strictly increasing in η.

The proof of the equivalence of the two definitions of G stability is analogous to the one given for Theorem 11 of § 14 in the case of thermal stability; one must again use Postulate II of § 8.

The main result of this section is

Theorem 15. *An equilibrium state* $\{\boldsymbol{f}, \eta\}$ *of a fluid body is G stable if and only if its temperature and pressure are uniform.*

Proof. To prove that the condition is necessary we observe that the functions v, η are solutions of the variational problem

$$\int_{\mathscr{B}} \bar{\varepsilon}(v^*, \eta^*; X)\, dm = \text{Minimum} \tag{16.5}$$

subject to the constraints (16.1). Therefore, the first variation of

$$\int_{\mathscr{B}} [\bar{\varepsilon}(v^*, \eta^*; X) - \lambda \eta^* - \mu v^*]\, dm$$

must vanish for $v^* = v$ and $\eta^* = \eta$. Here λ and μ are constant Lagrange parameters. It follows that

$$\bar{\varepsilon}_\eta(v, \eta; X) = \lambda = \text{constant}, \qquad \bar{\varepsilon}_v(v, \eta; X) = \mu = \text{constant}. \tag{16.6}$$

Hence, by (8.2) and (11.11), both the temperature, $\vartheta = \bar{\varepsilon}_\eta(v, \eta; X)$, and the pressure, $p = -\bar{\varepsilon}_v(v, \eta; X)$ are uniform over $\mathscr{B}$.

To prove the sufficiency of the condition of the theorem, we assume that ϑ and p are uniform and that (16.1) holds. From the convexity inequality (11.16), the inequality (11.9), and the fact that $v=\nu^3$, $\nu>0$, is a convex function of ν, one can easily infer that $\bar{\varepsilon}(v,\eta)$ must be convex in v and η. Hence, the inequality

$$\bar{\varepsilon}(v^*,\eta^*;X)-\bar{\varepsilon}(v^*,\eta;X)-(v^*-v)\,\bar{\varepsilon}_v(v,\eta;X)-(\eta^*-\eta)\,\bar{\varepsilon}_\eta(v,\eta;X)\geqq 0 \quad (16.7)$$

is valid at all material points X in $\mathscr{B}$; (16.7) cannot reduce to an equality for all X unless $v(X)=v^*(X)$ and $\eta(X)=\eta^*(X)$ for all X. Since $p=-\varepsilon_v(v,\eta)$ and $\vartheta=\bar{\varepsilon}_\eta(v,\eta)$ are independent of X, integration of (16.7) over $\mathscr{B}$ gives

$$\int_{\mathscr{B}}\{\bar{\varepsilon}(v^*,\eta^*;X)-\bar{\varepsilon}(v,\eta;X)\}\,dm+p\int_{\mathscr{B}}(v^*-v)\,dm-\vartheta\int_{\mathscr{B}}(\eta^*-\eta)\,dm>0.$$

The condition (16.1) states that the last two terms vanish and hence that (16.2) holds, q.e.d.

In his discussion of the stability of homogeneous fluids, GIBBS used a definition of stability which is identical to what we have called G stability, except that he did not demand, as we do, that $\{f,\eta\}$ be an equilibrium state*. GIBBS was able to prove that uniform values of $\bar{\varepsilon}_\eta(v,\eta)$ and $\bar{\varepsilon}_v(v,\eta)$ are necessary for his stability and, furthermore, that the inequality (16.7) is also necessary. He also realized that the constancy of $\bar{\varepsilon}_\eta$ and $\bar{\varepsilon}_v$ over $\mathscr{B}$ and the validity of (16.7) are sufficient for his stability. If he had gone a step further and postulated that for homogeneous fluids stable states exist for every value of v and η for which $\bar{\varepsilon}$ is defined, he would have obtained (16.7) as a property of the function $\bar{\varepsilon}$. Such a procedure, however, cannot yield the statements, made in Theorem 6, that $-\bar{\varepsilon}_v$ is positive and that $\bar{\bar{\varepsilon}}$ is jointly and strictly convex in ν and η.

We conclude with

Theorem 16. *An equilibrium state $\{f,\eta\}$ of a fluid body $\mathscr{B}$ is G stable if and only if both of the following conditions hold:*

(a) *The temperature corresponding to $\{f,\eta\}$ is uniform.*

(b) *Any other state $\{f^*,\eta^*\}$ with the same total volume,*

$$\int_{\mathscr{B}} v^*\,dm=\int_{\mathscr{B}} v\,dm, \quad (16.8)$$

and the same uniform temperature ϑ has a higher total free energy,

$$\int_{\mathscr{B}}\bar{\psi}(v^*,\vartheta;X)\,dm>\int_{\mathscr{B}}\bar{\psi}(v,\vartheta;X)\,dm, \quad (16.9)$$

unless $v^*(X)=v(X)$ for all X in $\mathscr{B}$.

Proof. The proof that the conditions (a) and (b) are sufficient for the G stability of $\{f,\eta\}$ is completely analogous to the proof of Theorem 14 of § 15.

The necessity of the condition (a) for the G stability of $\{f,\eta\}$ follows from Theorem 15. To prove that (b) is necessary we assume that $\{f,\eta\}$ is stable. We consider another state $\{f^*,\eta^*\}$ which obeys (16.8) and which has the uniform temperature ϑ. Since $v=\nu^3$ is a convex function of ν for $\nu>0$, and $\bar{\psi}_v(v,\vartheta)<0$, the inequality (13.15) implies that

$$\bar{\psi}(v^*,\vartheta;X)-\bar{\psi}(v,\vartheta;X)-(v^*-v)\,\bar{\psi}_v(v,\vartheta;X)\geqq 0; \quad (16.10)$$

* GIBBS does not use either our Postulate I or our definition of (local) thermal equilibrium.

(16.10) cannot reduce to equality for all X unless $v(X) = v^*(X)$ for all X. Now, since we are assuming that $\{f, \eta\}$ is stable, it follows from Theorem 15 and (13.13) that $\bar{\bar{\psi}}_v(v, \vartheta; X)$ is independent of X. Thus, by (16.8), if we compute the mass integral of (16.10) over $\mathscr{B}$, the last term on the left makes no contribution, and we get (16.9). Hence, when $\{f, \eta\}$ is G stable, the condition (b) is valid, q.e.d.

This theorem shows that for G stability of fluids adiabatic and isothermal stability are equivalent.

Acknowledgement. This research was supported in part by the Air Force Office of Scientific Research under Contract AF 49(638)-541 with the Mellon Institute and by the National Science Foundation under Grant NSF-G 5250 to Carnegie Institute of Technology.

References

[1] (a) GIBBS, J. W.: On the equilibrium of heterogeneous substances. Trans. Conn. Acad. **3**, 108—248, 343—524 (1875—1878), or (b) The Scientific Papers of J. WILLARD GIBBS **1**, 55—372, particularly 55—62 and 100—115, Longmans, Green, 1906.

[2] DUHEM, P.: Dissolutions et Mélanges. Travaux et Mémoires des Facultés de Lille **3**, No. 2, 1—136 (1893).

[3] TRUESDELL, C.: Das ungelöste Hauptproblem der endlichen Elastizitätstheorie. Z. angew. Math. u. Mech. **36**, 97—103 (1956).

[4] HADAMARD, J.: Leçons sur la Propagation des Ondes et les Équations de l'Hydrodynamique. Paris 1903.

[5] ERICKSEN, J. L., & R. A. TOUPIN: Implications of Hadamard's condition for elastic stability with respect to uniqueness theorems. Can. J. Math. **8**, 432—436 (1956).

[6] DUHEM, P.: Recherches sur l'élasticité. Troisième partie. La stabilité des milieux élastiques. Ann. École Norm. (3) **22**, 143—217 (1905).

[7] NOLL, W.: A mathematical theory of the mechanical behavior of continuous media. Arch. Rat. Mech. Anal. **2**, 197—226 (1958).

[8] NOLL, W.: The fundations of classical mechanics in the light of recent advances in continuum mechanics. Proceedings of the Berkeley Symposium on the Axiomatic Method, 266—281, 1959.

[9] TRUESDELL, C.: The mechanical foundations of elasticity and fluid dynamics. J. Rat. Mech. Anal. **1**, 125—300 (1952); **2**, 593—616 (1953).

[10] NOLL, W.: On the continuity of the solid and fluid states. J. Rat. Mech. Anal. **4**, 3—81 (1955).

[11] RAMSEY, N. F.: Thermodynamics and statistical mechanics at negative absolute temperatures. Phys. Rev. **103**, 20—28 (1956).

[12] COLEMAN, B. D., & W. NOLL: Conditions for equilibrium at negative absolute temperatures. Phys. Rev. **115**, 262—265 (1959).

[13] BAKER, M., & J. L. ERICKSEN: Inequalities restricting the form of the stress-deformation relations for isotropic elastic solids and Reiner-Rivlin fluids. J. Washington Acad. Sciences **44**, 33—35 (1954).

[14] BARTA, J.: On the non-linear elasticity law. Acta Tech. Acad. Scient. Hungaricae **18**, 55—65 (1957).

[15] HILL, R.: On uniqueness and stability in the theory of finite elastic strain. J. Mech. Phys. Solids **5**, 229—241 (1957).

Mellon Institute
Pittsburgh, Pennsylvania
and
Carnegie Institute of Technology
Pittsburgh, Pennsylvania

(Received August 24, 1959)

An Approximation Theorem for Functionals, with Applications in Continuum Mechanics

BERNARD D. COLEMAN & WALTER NOLL

Contents

1. *Introduction*

We start with a brief discussion of the physical motivation behind the mathematical considerations to be presented in Part I of this paper.

Often in theoretical physics one is concerned with a functional relationship,

$$\boldsymbol{\sigma} = \underset{s=0}{\overset{\infty}{\mathfrak{F}}}\left(\boldsymbol{g}(s)\right), \tag{1.1}$$

which states that the present value $\boldsymbol{\sigma}$ of a physical quantity is determined by the values $\boldsymbol{g}(s)$ of a second quantity at all times s in the past. In (1.1), $\boldsymbol{g}(s)$ is the value of the second quantity s time units ago. If the functional $\mathfrak{F}$ is given, the relation (1.1) may be used to predict the present value $\boldsymbol{\sigma}$ of the first quantity from a knowledge of the "history" $\boldsymbol{g}$ of the second quantity. The relation (1.1) may be interpreted as expressing the causal nature of a class of physical processes. For definiteness we assume that the values of $\boldsymbol{\sigma}$ and $\boldsymbol{g}(s)$ lie in appropriate normed vector spaces, not necessarily finite-dimensional.

In many physical situations, the value $\boldsymbol{\sigma}$ in (1.1) will be, in some sense, more sensitive to the values of $\boldsymbol{g}$ for small s than for large s. Intuitively speaking, the "memory" of the system described by (1.1) will "fade away" in time. In order to make this idea precise, we shall introduce a *norm* $\|\boldsymbol{g}\|$ in the function space of the histories $\boldsymbol{g}$ for which (1.1) is meaningful. We first choose a number p,

$1 \leqq p \leqq \infty$, which will be kept fixed, and then define the norm of $\boldsymbol{g}$ to be

$$\|\boldsymbol{g}\| = \begin{cases} \sqrt[p]{\int\limits_0^\infty (|\boldsymbol{g}(s)|\, h(s))^p\, ds} & \text{if} \quad 1 \leqq p < \infty \\ \sup\limits_{s \geqq 0} |\boldsymbol{g}(s)|\, h(s) & \text{if} \quad p = \infty, \end{cases} \tag{1.2}$$

where $|\boldsymbol{g}(s)|$ is the magnitude of $\boldsymbol{g}(s)$ and where $h(s)$ is a real-valued function which approaches zero rapidly as $s \to \infty$. Thus, in computing the norm $\|\boldsymbol{g}\|$ we assign a greater influence to the values of $\boldsymbol{g}$ for small s (recent past) than for very large s (distant past). We call the function h in (1.2) an *influence function*. The physical idea that the memory of the system is fading corresponds to the mathematical assumption that the functional $\mathfrak{F}$ in (1.1) is *continuous* with respect to convergence in the norm (1.2) of the function space of histories. The influence function h characterizes the rapidity with which the memory is fading.

It is always possible to reduce the relation (1.1) to one which is *normalized* in the sense that the possible histories $\boldsymbol{g}$ have the value $\boldsymbol{g}(0) = \boldsymbol{0}$ at the present instant and that the value of $\mathfrak{F}$ for the zero history $\boldsymbol{g}(s) \equiv \boldsymbol{0}$ is zero. Now, physical experience indicates that phenomena which one would expect to be described exactly by a normalized functional relation (1.1) often follow a simpler relation of the form

$$\boldsymbol{\sigma} = \mathfrak{l}(\overset{(1)}{\boldsymbol{g}}), \qquad \overset{(1)}{\boldsymbol{g}} = \frac{d}{ds} \boldsymbol{g}(s)\Big|_{s=0}, \tag{1.3}$$

where $\mathfrak{l}$ is a linear transformation. In particular, it appears that (1.3) accounts well for the observed phenomena in the case of slow processes. This observation leads to the conjecture that, in some mathematically precise sense, the relation (1.3) approximates the normalized relation (1.1) for slow processes. It is the purpose of the present paper to prove an approximation theorem of this kind. Theorem 2 of §5 asserts not only that (1.3) is the first-order approximation to the normalized relation (1.1) for slow processes but shows also the form of the approximations of higher order. The theorem is based on the assumption that the functional $\mathfrak{F}$ of (1.1) is not only continuous but also Fréchet differentiable at the zero history $\boldsymbol{g}(0) \equiv \boldsymbol{0}$ in the function space with norm (1.2). A normalized functional $\mathfrak{F}$ satisfying this assumption will be called a *memory functional*. The continuity of $\mathfrak{F}$ for histories $\boldsymbol{g}$ different from the zero history is not needed to prove the approximation theorem.

In Part II of the paper we apply the approximation theorem to constitutive equations of continuum mechanics. Our main interest there is in the logical status of the theory of Newtonian fluids within the framework of a recently proposed general theory of fluids with memory effects*.

The theory of compressible Newtonian fluids is based on the following constitutive equation for the stress tensor $\boldsymbol{S}$:

$$\boldsymbol{S} = -p\,\boldsymbol{I} + 2\eta\,\boldsymbol{D} + \lambda(\operatorname{tr}\boldsymbol{D})\,\boldsymbol{I}; \tag{1.4}$$

here the rate of deformation tensor $\boldsymbol{D}$ is the symmetric part of the velocity gradient tensor; $p = p(\varrho)$ is the hydrostatic pressure the fluid would be supporting

* *Cf.* [*1*], [*2*] and [*3*].

if it were at rest at its present density ϱ; η and λ are functions of ϱ alone and are called coefficients of viscosity.

Experience shows that for some substances, such as water, the theory of Newtonian fluids accounts remarkably well for experimental measurements over a very wide range of conditions. Other substances, such as molten plastics, definitely do not obey (1.4) exactly, but yet their behavior appears to approximate that of Newtonian fluids in the limit of slow motions.

In §6 we review the mathematical definition of a *simple fluid*. As we have frequently remarked in the past, we believe this definition is capable of covering the behavior of nearly all real fluids including such substances as molten plastics which exhibit "hereditary", "non-Newtonian" and "viscoelastic" effects. Here we add to the definition of a simple fluid a new requirement: we require that the functionals occurring in the definition be memory functionals in the sense of the definition used in this paper. As we have indicated, this requirement is related to the physical assumption that simple fluids have a fading memory. If this assumption were not satisfied, the term "fluid" would hardly be appropriate.

In §7 we conclude that the theory of Newtonian fluids is indeed the complete first-order approximation to the theory of simple fluids for slow flows. We also indicate what an experimenter should expect to find as second-order corrections to the constitutive equation of a Newtonian fluid. We point out that several special flow problems for incompressible second-order fluids lead to third-order linear partial differential equations.

In §8 we apply our approximation theorem to the theory of the general simple materials defined in [*1*], Part III.

I. The Approximation Theorem

2. *Influence Functions and Histories*

An *influence function* h of order $r>0$ is a real-valued function of a real variable with the following properties:

(i) $h(s)$ is defined and continuous for $0\leqq s<\infty$.

(ii) $h(s)$ is positive, $h(s)>0$.

(iii) For each $\sigma>0$, there is a constant M_σ, independent of α, such that

$$\sup_{s\geqq\sigma}\frac{h(s/\alpha)}{\alpha^r h(s)}\leqq M_\sigma \quad \text{for} \quad 0<\alpha\leqq 1. \tag{2.1}$$

The last condition (iii) means that $h(s)$ must decay to zero at a fast enough rate as $s\to\infty$. In fact, we have

$$\sup_{s>0} s^r h(s) = N < \infty, \tag{2.2}$$

which follows from (2.1) by taking $\sigma=1$ and observing that

$$\left(\frac{1}{\alpha}\right)^r h\left(\frac{1}{\alpha}\right)\leqq h(1)\sup_{s\geqq 1}\frac{h(s/\alpha)}{\alpha^r h(s)}\leqq h(1)\,M_1.$$

Let a real Banach space $\mathscr{S}$ with norm $|\;|$ be given. We then define a *history* $\boldsymbol{g}$ to be a measurable function defined for $0\leqq s<\infty$ with values $\boldsymbol{g}(s)$ in $\mathscr{S}$. Two such functions will be regarded as the same if they differ only on a set of measure zero.

For a given influence function h and a given number p, $1 \leqq p \leqq \infty$, we define the $\mathscr{L}_{h,p}$-*norm*, $\|\boldsymbol{g}\|_{h,p}$, or simply $\|\boldsymbol{g}\|$, of a history $\boldsymbol{g}$ by

$$\|\boldsymbol{g}\|_{h,p} = \sqrt[p]{\int_0^\infty (|\boldsymbol{g}(s)|\, h(s))^p\, ds} \quad \text{if} \quad 1 \leqq p < \infty, \tag{2.3}$$

$$\|\boldsymbol{g}\|_{h,\infty} = \sup_{s \geqq 0} |\boldsymbol{g}(s)|\, h(s) \qquad \text{if} \quad p = \infty. \tag{2.4}$$

In (2.4) and subsequently $\sup f(s)$ stands for the essential supremum of $f(s)$, *i.e.* for the greatest lower bound of the suprema of all functions which differ from $f(s)$ only on a set of measure zero.

The set of all histories with finite $\mathscr{L}_{h,p}$-norm forms a Banach space, which we denote by $\mathscr{L}_{h,p}$.

We remark that (2.2) is a necessary but not a sufficient condition for the decay relation (2.1). However, if a function $h(s)$ satisfies (i), (ii), and the limit relation

$$\lim_{s \to \infty} s^r h(s) = 0 \quad \text{monotonically for large } s, \tag{2.5}$$

then it also satisfies the decay relation (2.1) and hence is an influence function of order r. To prove this fact we consider the expression

$$\frac{h(s/\alpha)}{\alpha^r h(s)} = \frac{(s/\alpha)^r h(s/\alpha)}{s^r h(s)}, \qquad 0 < \alpha \leqq 1. \tag{2.6}$$

Since $s^r h(s)$ is monotonically decreasing for s larger than some value s_1, it follows that (2.6) is not greater than 1 for all $s > s_1$. For $0 < \sigma \leqq s \leqq s_1$ the denominator $s^r h(s)$ of (2.6) has a positive minimum because h satisfies the conditions (i) and (ii). The numerator of (2.6) is bounded by the maximum of $s^r h(s)$, which exists and is finite because of (2.5). It follows that (2.6) is bounded, for $\sigma \leqq s < \infty$, by a constant M_σ independent of α, which is the content of (2.1).

The function

$$h(s) = \frac{1}{(s+1)^r}$$

is an influence function of order r. An exponential,

$$h(s) = e^{-\beta s}, \qquad \beta > 0,$$

is an influence function of any order.

3. Retardation

The *retardation* Γ_α with *retardation factor* α, $0 < \alpha \leqq 1$, is the linear transformation $\boldsymbol{g} \to \boldsymbol{g}_\alpha$ defined, for all histories $\boldsymbol{g}$, by

$$(\Gamma_\alpha \boldsymbol{g})(s) = \boldsymbol{g}_\alpha(s) = \boldsymbol{g}(\alpha s). \tag{3.1}$$

We show that Γ_α maps the space $\mathscr{L}_{h,p}$ into itself. For $p \neq \infty$, we have*

$$\begin{aligned} \|\boldsymbol{g}_\alpha\|^p &= \int_0^\infty (|\boldsymbol{g}(\alpha s)|\, h(s))^p\, ds \\ &= \frac{1}{\alpha} \int_0^\infty (|\boldsymbol{g}(s)|\, h(s))^p \left(\frac{h(s/\alpha)}{h(s)}\right)^p ds \leqq \|\boldsymbol{g}\|^p \frac{1}{\alpha} \left(\sup_{s>0} \frac{h(s/\alpha)}{h(s)}\right)^p. \end{aligned} \tag{3.2}$$

* Since no confusion can arise, we omit the indices h, p on the norm.

Since h is continuous and positive,

$$\sup_{\sigma \geqq s \geqq 0} \frac{h(s/\alpha)}{h(s)} < \infty$$

for any $\sigma > 0$. The decay condition (2.1) implies that

$$\sup_{s \geqq \sigma} \frac{h(s/\alpha)}{h(s)} \leqq \alpha^r M_\sigma < \infty.$$

Hence,

$$\sup_{s \geqq 0} \frac{h(s/\alpha)}{h(s)} = K_\alpha < \infty. \tag{3.3}$$

Combining (3.2) and (3.3) gives

$$\|\boldsymbol{g}_\alpha\| \leqq \alpha^{-\frac{1}{p}} K_\alpha \|\boldsymbol{g}\|. \tag{3.4}$$

It is easily verified that this inequality remains valid for $p = \infty$ when we put $\sqrt[\infty]{\alpha} = 1$. It follows from (3.4) that when $\boldsymbol{g}$ has a finite norm $\boldsymbol{g}_\alpha$ has a finite norm and is, therefore, in $\mathscr{L}_{h,p}$. Intuitively, retardation replaces a given history by one which is essentially the same, but slower.

If, possibly after a suitable alteration of $\boldsymbol{g}(s)$ on a set of measure zero, the limits

$$\overset{(0)}{\boldsymbol{g}} = \lim_{s \to 0} \boldsymbol{g}(s), \qquad \overset{(k)}{\boldsymbol{g}} = \lim_{s \to 0} \frac{k!}{s^k} \left(\boldsymbol{g}(s) - \sum_{j=0}^{k-1} \frac{s^j}{j!} \overset{(j)}{\boldsymbol{g}} \right) \tag{3.5}$$

exist for $k = 0, 1, \ldots, n$, then we shall say that the history $\boldsymbol{g}$ has n *generalized derivatives* at $s = 0$. Of course, the existence of ordinary derivatives implies the existence of the corresponding generalized derivatives. Here we shall use the term "derivative" always in the generalized sense of (3.5). Generalized derivatives at $s = 0$ of any order may exist even though $\boldsymbol{g}(s)$ is not continuous near $s = 0$.

The *Taylor transformation* Π_n is the linear transformation $\boldsymbol{g} \to \Pi_n \boldsymbol{g}$ defined, for all histories $\boldsymbol{g}$ which are n times differentiable at $s = 0$, by

$$(\Pi_n \boldsymbol{g})(s) = \sum_{j=0}^{n} \frac{s^j}{j!} \overset{(j)}{\boldsymbol{g}}, \tag{3.6}$$

where the $\overset{(j)}{\boldsymbol{g}}$ are the derivatives (3.5).

This Taylor transformation replaces $\boldsymbol{g}$ by its nth "Taylor approximation" $\Pi_n \boldsymbol{g}$. The history $\Pi_n \boldsymbol{g}$ is a polynomial of degree $\leqq n$.

We note that retardation Γ_α preserves the differentiability of a history and that Π_n and Γ_α commute; *i.e.*, for all histories $\boldsymbol{g}$ having n derivatives at $s = 0$, we have

$$\Pi_n \Gamma_\alpha \boldsymbol{g} = \Gamma_\alpha \Pi_n \boldsymbol{g}. \tag{3.7}$$

The set of all histories which have n derivatives at $s = 0$ and which also belong to the Banach space $\mathscr{L}_{h,p}$ will be denoted by $\mathscr{D}_n$. This set $\mathscr{D}_n$ is a linear subspace of $\mathscr{L}_{h,p}$, but it is not closed in $\mathscr{L}_{h,p}$ and hence not a Banach space.

The following theorem is an analogue of the classical Taylor approximation theorem.

Theorem 1. *Assume that n, p, and the order r of the influence function h satisfy the inequality*

$$n < r - \frac{1}{p} \qquad \left(\frac{1}{p} = 0 \quad \text{if} \quad p = \infty \right). \tag{3.8}$$

Then the Taylor transformation Π_n maps the subspace $\mathscr{D}_n$ of $\mathscr{L}_{h,p}$ into itself, and, for all $\boldsymbol{g}$ in $\mathscr{D}_n$,

$$\lim_{\alpha\to 0} \frac{1}{\alpha^n} \|\Gamma_\alpha(\boldsymbol{g} - \Pi_n \boldsymbol{g})\|_{h,p} = 0. \tag{3.9}$$

We can also write (3.9), using (3.7), in the form

$$\Gamma_\alpha \boldsymbol{g} - \Pi_n \Gamma_\alpha \boldsymbol{g} = \boldsymbol{g}_\alpha - \Pi_n \boldsymbol{g}_\alpha = o(\alpha^n), \tag{3.10}$$

where the order symbol o must be understood in terms of the norm of the function space $\mathscr{L}_{h,p}$. Roughly speaking, the theorem states that in the space $\mathscr{D}_n \in \mathscr{L}_{h,p}$ a slow history is close to its Taylor approximation and that the distance between them is $o(\alpha^n)$.

Proof of Theorem 1. We consider only the case when p is finite. The case when $p=\infty$ can be treated analogously.

First, we show that every polynomial of degree $\leqq n$ has a finite norm. For this purpose it is sufficient to prove that

$$\int_0^\infty (s^k h(s))^p \, ds = L_k < \infty \quad \text{for} \quad k \leqq n. \tag{3.11}$$

It follows from (2.2) that

$$\int_1^\infty (s^k h(s))^p \, ds = \int_1^\infty (s^r h(s))^p s^{p(k-r)} \, ds \leqq N^p \int_1^\infty s^{-p(r-k)} \, ds. \tag{3.12}$$

The inequalities (3.8) and $k \leqq n$ imply that $p(r-k) > 1$ and hence that the integral

$$\int_1^\infty s^{-p(r-k)} \, ds = \frac{1}{p(r-k)-1}$$

is finite. Since $s^k h(s)$ is continuous, it then follows from (3.12) that the integral (3.11) is finite.

We have shown that any polynomial of degree $\leqq n$ belongs to $\mathscr{L}_{h,p}$ and hence to $\mathscr{D}_n$, because it has n derivatives at $s=0$. Since a Taylor approximation $\Pi_n \boldsymbol{g}$ is a polynomial of degree $\leqq n$, it follows that Π_n maps $\mathscr{D}_n$ into itself.

The definitions (3.5) and (3.6) imply that the history

$$\boldsymbol{f} = \boldsymbol{g} - \Pi_n \boldsymbol{g} \in \mathscr{D}_n \tag{3.13}$$

satisfies the limit relation

$$\lim_{s\to 0} \frac{|\boldsymbol{f}(s)|}{s^n} = 0. \tag{3.14}$$

The definitions (2.3), (3.1), and (3.13) show that the assertion (3.9) of the theorem is equivalent to

$$\lim_{\alpha\to 0} \frac{1}{\alpha^{np}} \int_0^\infty (|\boldsymbol{f}(\alpha s)| \, h(s))^p \, ds = 0. \tag{3.15}$$

We observe that

$$\frac{1}{\alpha^{np}} \int_0^\infty (|\boldsymbol{f}(\alpha s)| \, h(s))^p \, ds = \frac{1}{\alpha^{np+1}} \int_0^\infty \left(|\boldsymbol{f}(s)| \, h\left(\frac{s}{\alpha}\right)\right)^p ds \tag{3.16}$$

and investigate the latter expression. Let $\varepsilon>0$ be given. It follows from (3.14) that we can find a $\sigma(\varepsilon)>0$ such that

$$\frac{|\boldsymbol{f}(s)|}{s^n} \leqq \varepsilon \quad \text{for} \quad 0<s\leqq\sigma(\varepsilon).$$

Hence,

$$\begin{aligned}\frac{1}{\alpha^{n p+1}} \int_0^{\sigma(\varepsilon)} \left(|\boldsymbol{f}(s)|\, h\left(\frac{s}{\alpha}\right)\right)^p ds &\leqq \frac{1}{\alpha^{n p+1}}\, \varepsilon^p \int_0^{\sigma(\varepsilon)} \left(s^n h\left(\frac{s}{\alpha}\right)\right)^p ds \\ &\leqq \varepsilon^p \int_0^{\infty} \left(\frac{s^n}{\alpha^n} h\left(\frac{s}{\alpha}\right)\right)^p d\left(\frac{s}{\alpha}\right) = \varepsilon^p L_n,\end{aligned} \tag{3.17}$$

where L_n is given by (3.11). On the other hand, we have

$$\frac{1}{\alpha^{n p+1}} \int_{\sigma(\varepsilon)}^{\infty} \left(|\boldsymbol{f}(s)|\, h\left(\frac{s}{\alpha}\right)\right)^p ds = \alpha^w \int_{\sigma(\varepsilon)}^{\infty} \left(|\boldsymbol{f}(s)|\, h(s)\right)^p \left(\frac{h(s/\alpha)}{\alpha^r h(s)}\right)^p ds, \tag{3.18}$$

where $w=p(r-n)-1$. Applying the decay condition (2.1) to (3.18), we find that

$$\frac{1}{\alpha^{n p+1}} \int_{\sigma(\varepsilon)}^{\infty} \left(|\boldsymbol{f}(s)|\, h\left(\frac{s}{\alpha}\right)\right)^p ds \leqq \alpha^w M_{\sigma(\varepsilon)}^p \|\boldsymbol{f}\|^p. \tag{3.19}$$

On combining (3.17) and (3.19), we see that

$$\frac{1}{\alpha^{n p+1}} \int_0^{\infty} \left(|\boldsymbol{f}(s)|\, h\left(\frac{s}{\alpha}\right)\right)^p ds \leqq \varepsilon^p L_n + \alpha^w M_{\sigma(\varepsilon)}^p \|\boldsymbol{f}\|^p \tag{3.20}$$

holds for all $\varepsilon>0$ and all $0<\alpha\leqq 1$. The assumed inequality (3.8) insures that $w=p(r-n)-1$ is positive. Therefore, by choosing first ε and then α sufficiently small, we see that the right side of (3.20) can be made as small as desired. It then follows from (3.16) that the limit relation (3.15) holds, which completes the proof of the theorem.

In the special case $n=0$, $r>\frac{1}{p}$, $\overset{(0)}{\boldsymbol{g}}=\boldsymbol{0}$ Theorem 1 states the following. If $\boldsymbol{g}\in\mathscr{L}_{h,p}$ is continuous at $s=0$ with value $\boldsymbol{0}$, then

$$\lim_{\alpha\to 0} \|\Gamma_\alpha \boldsymbol{g}\| = 0. \tag{3.21}$$

If $1\leqq n<r-\frac{1}{p}$ and $\overset{(0)}{\boldsymbol{g}}=\boldsymbol{0}$, instead of (3.21), we have the stronger result

$$\boldsymbol{g}_\alpha = \Gamma_\alpha \boldsymbol{g} = O(\alpha). \tag{3.22}$$

4. *Memory Functionals*

We first recall some definitions* from the theory of functions defined on a real vector space $\mathscr{H}$ with norm $\|\ \|$ and having values in another real vector space $\mathscr{T}$ with norm $|\ |$.

(1) Suppose $\boldsymbol{g}_1, \ldots, \boldsymbol{g}_k$ are variables in $\mathscr{H}$. Then a function $\mathfrak{P}(\boldsymbol{g}_1, \ldots, \boldsymbol{g}_k)$, defined for all values of the variables $\boldsymbol{g}_i$ in $\mathscr{H}$ and having values in $\mathscr{T}$, is called

* These definitions are analogous to those given in Chap. XXVI of [4].

a *bounded k-linear form* if it is linear in each variable $\boldsymbol{g}_i$ separately and if there is a constant M, independent of $\boldsymbol{g}_i$, such that

$$|\mathfrak{P}(\boldsymbol{g}_1, \ldots, \boldsymbol{g}_k)| \leqq M \|\boldsymbol{g}_1\| \cdots \|\boldsymbol{g}_k\|. \tag{4.1}$$

The form $\mathfrak{P}(\boldsymbol{g}_1, \ldots, \boldsymbol{g}_k)$ is said to be *symmetric* if any permutation of the variables leaves the value unchanged.

(2) A function $\mathfrak{P}(\boldsymbol{g})$, defined for all $\boldsymbol{g} \in \mathscr{H}$ and having values in $\mathscr{T}$, is called a *bounded homogeneous polynomial of degree k* if there is a bounded symmetric k-linear form $\mathfrak{P}(\boldsymbol{g}_1, \ldots, \boldsymbol{g}_k)$ such that

$$\mathfrak{P}(\boldsymbol{g}) = \mathfrak{P}(\boldsymbol{g}, \ldots, \boldsymbol{g}). \tag{4.2}$$

The symmetric n-linear form $\mathfrak{P}(\boldsymbol{g}_1, \ldots, \boldsymbol{g}_k)$ is uniquely determined by the homogeneous polynomial $\mathfrak{P}(\boldsymbol{g})$ and is called the *polar form* of the polynomial.

(3) A function $\mathfrak{F}$, defined on a neighborhood of $\mathbf{0} \in \mathscr{H}$ and having values in $\mathscr{T}$ is said to be *n times Fréchet-differentiable* at $\mathbf{0} \in \mathscr{H}$ if there are bounded homogeneous polynomials $\delta^k \mathfrak{F}(\boldsymbol{g})$ of degree $k = 0, 1, \ldots, n$ such that

$$\mathfrak{F}(\boldsymbol{g}) = \sum_{k=0}^{n} \frac{1}{k!} \delta^k \mathfrak{F}(\boldsymbol{g}) + \|\boldsymbol{g}\|^n \mathfrak{R}(\boldsymbol{g}), \tag{4.3}$$

where

$$\lim_{\|\boldsymbol{g}\| \to 0} |\mathfrak{R}(\boldsymbol{g})| = 0. \tag{4.4}$$

The polynomial $\delta^k \mathfrak{F}(\boldsymbol{g})$ or its polar form $\delta^k \mathfrak{F}(\boldsymbol{g}_1, \ldots, \boldsymbol{g}_k)$ is called the kth *Fréchet-differential* or the kth *variation* of $\mathfrak{F}$ at $\mathbf{0} \in \mathscr{H}$. The differentials may be obtained recursively by

$$\delta^0 \mathfrak{F}(\boldsymbol{g}) = \mathfrak{F}(0), \qquad \delta^k \mathfrak{F}(\boldsymbol{g}) = k! \lim_{\lambda \to 0} \frac{1}{\lambda^k} \left[\mathfrak{F}(\lambda \boldsymbol{g}) - \sum_{j=0}^{k-1} \frac{\lambda^j}{j!} \delta^j \mathfrak{F}(\boldsymbol{g}) \right]. \tag{4.5}$$

We here consider the case in which $\mathscr{H}$ is the function space of all histories $\boldsymbol{g}$ with the following properties:

(α) $\boldsymbol{g}$ has a finite $\mathscr{L}_{h,p}$-norm,

(β) $\boldsymbol{g}$ has n generalized derivatives at $s = 0$,

(γ) $\boldsymbol{g}$ has a zero limit at $s = 0$:

$$\lim_{s \to 0} \boldsymbol{g}(s) = \overset{(0)}{\boldsymbol{g}} = \mathbf{0}, \tag{4.6}$$

(δ) n, p, and the order r of the influence function h obey the inequality (3.8).

The conditions (α) and (β) state that $\mathscr{H} < \mathscr{D}_n$. Condition ($\delta$) insures that the conclusion (3.9) of Theorem 1 is valid for all $\boldsymbol{g} \in \mathscr{H}$.

A function $\mathfrak{F}$ defined on a neighborhood in $\mathscr{H}$ of the zero function $\mathbf{0} \in \mathscr{H} < \mathscr{L}_{h,p}$ and having values in a real Banach space $\mathscr{T}$ will be called a *memory functional of type* (h, n) if it is n times Fréchet-differentiable at $\mathbf{0} \in \mathscr{H}$ and if it is normalized by

$$\delta^0 \mathfrak{F}(\boldsymbol{g}) = \mathfrak{F}(0) = \mathbf{0}. \tag{4.7}$$

In some applications it may be more natural to assume that $\mathfrak{F}$ is defined and Fréchet-differentiable on a neighborhood of zero in the entire space $\mathscr{L}_{h,p}$. However, the approximation theorem of the following section applies only to histories which belong to the subspace $\mathscr{H}$ of $\mathscr{L}_{h,p}$.

5. *The Approximation Theorem*

Theorem 2. *Let* $\mathfrak{F}$ *be a memory functional of type* (h, n)*; the histories* $\boldsymbol{g}$ *in the domain space* $\mathscr{H}$ *of* $\mathfrak{F}$ *have values in the space* $\mathscr{S}$*, and* $\mathfrak{F}$ *itself has values in* $\mathscr{T}$*. Then, for each k-tuple of indices* $(j_1, j_2, \ldots, j_k)$ *such that*

$$1 \leqq j_1 \leqq j_2 \leqq \cdots \leqq j_k \leqq n, \qquad j_1 + j_2 + \cdots + j_n \leqq n, \tag{5.1}$$

there exists a bounded k-linear form $\mathfrak{l}_{j_1 \ldots j_k}$ *with variables in* $\mathscr{S}$ *and values in* $\mathscr{T}$ *such that, for all* $\boldsymbol{g} \in \mathscr{H}$,

$$\mathfrak{F}(\boldsymbol{g}_\alpha) = \sum_{(j_1, \ldots, j_k)} \mathfrak{l}_{j_1 \ldots j_k}(\overset{(j_1)}{\boldsymbol{g}_\alpha}, \ldots, \overset{(j_k)}{\boldsymbol{g}_\alpha}) + o(\alpha^n), \tag{5.2}$$

where

$$\boldsymbol{g} = \Gamma_\alpha \boldsymbol{g}, \quad \text{i.e.,} \quad \boldsymbol{g}_\alpha(s) = \boldsymbol{g}(\alpha s), \tag{5.3}$$

and where

$$\overset{(j)}{\boldsymbol{g}_\alpha} = \alpha^j \overset{(j)}{\boldsymbol{g}} \tag{5.4}$$

is the generalized j^{th} *derivative at* $s=0$ *of* $\boldsymbol{g}_\alpha$*, defined according to* (3.5)*. The sum in* (5.2) *is to be extended over all sets* $(j_1, \ldots, j_k)$ *of indices satisfying* (5.1)*, and the order symbol* $o(\alpha^n)$ *has the usual sense. The multilinear forms* $\mathfrak{l}_{j_1 \ldots j_k}$ *are uniquely determined by* $\mathfrak{F}$.

This theorem permits the asymptotic approximation of a memory *functional*, for "slow" histories, by a polynomial *function* of the derivatives at $s=0$ of the argument function of the functional. It is worth noting that the approximating expression

$$\mathfrak{F}_n(\boldsymbol{g}) = \sum_{(j_1, \ldots, j_k)} \mathfrak{l}_{j_1 \ldots j_k}(\overset{(j_1)}{\boldsymbol{g}}, \ldots, \overset{(j_k)}{\boldsymbol{g}}), \tag{5.6}$$

regarded as a functional of $\boldsymbol{g}$, is not a memory functional in the sense of the previous section because it is not even continuous at the zero function $\boldsymbol{g} = \boldsymbol{0} \in \mathscr{H}$.

For $n=1$ and $n=2$, (5.6) reduces to

$$\mathfrak{F}_1(\boldsymbol{g}) = \mathfrak{l}_1(\overset{(1)}{\boldsymbol{g}}), \qquad \mathfrak{F}_2(\boldsymbol{g}) = \mathfrak{l}_1(\overset{(1)}{\boldsymbol{g}}) + \mathfrak{l}_{11}(\overset{(1)}{\boldsymbol{g}}, \overset{(1)}{\boldsymbol{g}}) + \mathfrak{l}_2(\overset{(2)}{\boldsymbol{g}}) \tag{5.7}$$

respectively, where $\mathfrak{l}_1$ and $\mathfrak{l}_2$ are linear and $\mathfrak{l}_{11}$ is bilinear.

Proof of Theorem 2. For $n=0$, the theorem is a trivial consequence of (3.21). For $n \geqq 1$, the proof is based on a combination of the Fréchet-differentiability assumption (4.3) and equation (3.10) of Theorem 1 which, in the notation of (5.3), may be written as

$$\boldsymbol{g}_\alpha = \Pi_n \boldsymbol{g}_\alpha + o(\alpha^n). \tag{5.8}$$

Since $\overset{(0)}{\boldsymbol{g}} = \boldsymbol{0}$ by (4.6), the definition (3.6) gives

$$(\Pi_n \boldsymbol{g}_\alpha)(s) = \sum_{j=1}^{n} \alpha^j \frac{s^j}{j!} \overset{(j)}{\boldsymbol{g}}. \tag{5.9}$$

The result (3.22) applies to both $\boldsymbol{g} \in \mathscr{H}$ and $\Pi_n \boldsymbol{g} \in \mathscr{H}$:

$$\Pi_n \boldsymbol{g}_\alpha = O(\alpha), \tag{5.10}$$

$$\boldsymbol{g}_\alpha = O(\alpha). \tag{5.11}$$

Combining (5.11) with (4.3), (4.4), and (4.7), we obtain

$$\mathfrak{F}(\boldsymbol{g}_\alpha) = \sum_{k=1}^{n} \frac{1}{k!}\, \delta^k \mathfrak{F}(\boldsymbol{g}_\alpha) + o(\alpha^n). \tag{5.12}$$

Consider now a bounded homogeneous polynomial $\mathfrak{P}(\boldsymbol{g})$ of degree k, $1 \leqq k \leqq n$, in the sense of (2) of §4. The differentials $\delta^k\mathfrak{F}$, $k>1$ are such polynomials. The polar form of $\mathfrak{P}$ will also be denoted by $\mathfrak{P}$. Using the multilinearity of this polar form $\mathfrak{P}$, we obtain by (4.2), (5.8), and (5.10)

$$\begin{aligned}\mathfrak{P}(\boldsymbol{g}_\alpha) &= \mathfrak{P}(\Pi_n \boldsymbol{g}_\alpha + o(\alpha^n), \dots, \Pi_n \boldsymbol{g}_\alpha + o(\alpha^n)) \\ &= \mathfrak{P}(\Pi_n \boldsymbol{g}_\alpha, \dots, \Pi_n \boldsymbol{g}_\alpha) + \sum \mathfrak{P}(O(\alpha), \dots, o(\alpha^n), \dots, O(\alpha)),\end{aligned}$$

where each term of the sum contains at least one variable $o(\alpha^n)$. It follows from the boundedness (4.1) of $\mathfrak{P}$ that the terms in the sum are all $o(\alpha^{n+k-1})$. Since $k \geqq 1$, we have

$$\mathfrak{P}(\boldsymbol{g}_\alpha) = \mathfrak{P}(\Pi_n \boldsymbol{g}_\alpha) + o(\alpha^n). \tag{5.13}$$

We now investigate $\mathfrak{P}(\Pi_n \boldsymbol{g}_\alpha)$. Using (5.9) and the multilinearity of the polar form $\mathfrak{P}$, we obtain

$$\mathfrak{P}(\Pi_n \boldsymbol{g}_\alpha) = \sum_{j_1=1}^{n} \cdots \sum_{j_k=1}^{n} \frac{\alpha^{j_1+\cdots+j_k}}{j_1! \dots j_k!}\, \mathfrak{P}(s^{j_1} \overset{(j_1)}{\boldsymbol{g}}, \dots, s^{j_k} \overset{(j_k)}{\boldsymbol{g}}). \tag{5.14}$$

Due to the symmetry of $\mathfrak{P}$, all terms of (5.14) which differ only in the order of the indices $j_1, \dots, j_k$ are the same. Collecting these terms and separating all terms of order higher than n in α, we get

$$\mathfrak{P}(\Pi_n \boldsymbol{g}_\alpha) = \sum_{\substack{(j_1,\dots,j_k)\\ k \text{ fixed}}} \alpha^{j_1+\cdots+j_k} m_{j_1\dots j_k}\, \mathfrak{P}(s^{j_1} \overset{(j_1)}{\boldsymbol{g}}, \dots, s^{j_k} \overset{(j_k)}{\boldsymbol{g}}) + O(\alpha^{n+1}) \tag{5.15}$$

where the sum is to be extended over all k-tuples $(j_1, \dots, j_k)$ satisfying (5.1) and where the $m_{j_1\dots j_k}$ are positive rational numbers.

Now, the function

$$\mathfrak{p}_{j_1\dots j_k}(\boldsymbol{a}_1, \dots, \boldsymbol{a}_k) = \mathfrak{P}(s^{j_1} \boldsymbol{a}_1, \dots, s^{j_k} \boldsymbol{a}_k), \tag{5.16}$$

with variables $\boldsymbol{a}_i \in \mathscr{S}$ and values in $\mathscr{T}$ is clearly multilinear, because $\mathfrak{P}$ is. Also, $\mathfrak{p}_{j_1\dots j_k}$ is bounded. Indeed, application of (4.1) to (5.16) yields

$$|\mathfrak{p}_{j_1\dots j_k}(\boldsymbol{a}_1, \dots, \boldsymbol{a}_k)| \leqq M \|s^{j_1} \boldsymbol{a}_1\| \dots \|s^{j_k} \boldsymbol{a}_k\|,$$

and (3.11) shows

$$\|s^{j_l} \boldsymbol{a}_l\|_h = |\boldsymbol{a}_l| \sqrt[p]{L_{j_l}},$$

hence

$$|\mathfrak{p}_{j_1\dots j_k}(\boldsymbol{a}_1, \dots, \boldsymbol{a}_k)| \leqq \overline{M} |\boldsymbol{a}_1| \dots |\boldsymbol{a}_k|. \tag{5.17}$$

Substituting (5.16) into (5.15) and using (5.4) yields

$$\mathfrak{P}(\Pi_n \boldsymbol{g}_\alpha) = \sum_{\substack{(j_1,\dots,j_k)\\ k \text{ fixed}}} \mathfrak{l}_{j_1\dots j_k}(\overset{(j_1)}{\boldsymbol{g}}_\alpha, \dots, \overset{(j_k)}{\boldsymbol{g}}_\alpha) + O(\alpha^{n+1}) \tag{5.18}$$

where the $\mathfrak{l}_{j_1\dots j_k}$ are bounded k-linear forms.

Applying the results (5.13) and (5.18) to $\mathfrak{P}=\frac{1}{k!}\delta^k\mathfrak{F}$ in (5.12), we obtain formula (5.2).

The uniqueness of the $\mathfrak{l}_{j_1\ldots j_k}$ follows easily from the observation that they are linearly independent and of order $O(\alpha^n)$ or lower.

We remark that the multilinear forms $\mathfrak{l}_{j_1\ldots j_k}$ of (5.2) are not necessarily symmetric.

II. Applications

6. The Concept of a Simple Fluid

The notion of a simple fluid has been given a definition within the framework of a general theory of the mechanical behavior of materials*. This definition is based on the following two physical assumptions**:

(a) The present stress depends on the past history of the first spatial gradient of the displacement function.

(b) A fluid has no preferred configurations.

Using the principle of material objectivity ([1], § 11), it was shown in reference [1] that the constitutive equation of a simple fluid can be written in the form (22.12) of [1]. This functional relation, in a slightly different notation, reads

$$\boldsymbol{S}(t)=\underset{s=0}{\overset{\infty}{\mathfrak{H}}}\left(\boldsymbol{C}_t(t-s);\,\varrho(t)\right). \tag{6.1}$$

Here $\boldsymbol{S}(t)$ is the stress and $\varrho(t)$ the density at time t. $\boldsymbol{C}_t(\tau)$ is called the right Cauchy-Green tensor at time τ relative to the configuration at time t. This tensor is defined by

$$\boldsymbol{C}_t(\tau)=\boldsymbol{F}_t^T(\tau)\,\boldsymbol{F}_t(\tau) \tag{6.2}$$

where

$$\boldsymbol{F}_t(\tau)=\nabla_{\boldsymbol{x}}\boldsymbol{\chi}_t(\boldsymbol{x},\tau) \tag{6.3}$$

is the gradient the displacement function $\boldsymbol{\xi}=\boldsymbol{\chi}_t(\boldsymbol{x},\tau)$ which gives the position at time τ of the material point having the position $\boldsymbol{x}$ at time t. The stress tensor $\boldsymbol{S}(t)$ is symmetric. The Cauchy-Green tensor $\boldsymbol{C}_t(\tau)$ is positive definite and symmetric; for $\tau=t$ it reduces to the unit tensor $\boldsymbol{I}$:

$$\boldsymbol{C}_t(t)=\boldsymbol{I}. \tag{6.4}$$

The functional $\mathfrak{H}$ in (6.1) is isotropic; *i.e.*, $\mathfrak{H}$ obeys

$$\boldsymbol{Q}\underset{s=0}{\overset{\infty}{\mathfrak{H}}}\left(\boldsymbol{C}(s);\varrho\right)\boldsymbol{Q}^T=\underset{s=0}{\overset{\infty}{\mathfrak{H}}}\left(\boldsymbol{Q}\,\boldsymbol{C}(s)\,\boldsymbol{Q}^T;\varrho\right) \tag{6.5}$$

identically in the history $\boldsymbol{C}(s)=\boldsymbol{C}_t(t-s)$ and the orthogonal tensor $\boldsymbol{Q}$; here $\boldsymbol{Q}^T$ is the transpose of $\boldsymbol{Q}$. It follows from (6.5) that the value of $\mathfrak{H}$ for the "rest

* [1], § 21.

** In [3] we give a survey of the theory of simple fluids with emphasis on physical applications. In that paper we anticipate some of the results rigorously derived here. Although the language and the definitions of the present paper are slightly different from those used in §§ 6 and 7 of [3], the arguments presented here prove also the theorems stated there.

history" $\boldsymbol{C}(s) \equiv \boldsymbol{I}$ is a scalar multiple $-p(\varrho)$ of the unit tensor $\boldsymbol{I}$. Defining

$$\overset{\infty}{\underset{s=0}{\mathfrak{F}}}\left(\boldsymbol{G}(s), \varrho\right) = p(\varrho)\,\boldsymbol{I} + \overset{\infty}{\underset{s=0}{\mathfrak{H}}}\left(\boldsymbol{I} + \boldsymbol{G}(s), \varrho\right), \tag{6.6}$$

we may rewrite (6.1) in the form

$$\boldsymbol{S}(t) = -\,p\left(\varrho(t)\right)\boldsymbol{I} + \overset{\infty}{\underset{s=0}{\mathfrak{F}}}\left(\boldsymbol{C}_t(t-s) - \boldsymbol{I}, \varrho(t)\right). \tag{6.7}$$

The functional $\overset{\infty}{\underset{s=0}{\mathfrak{F}}}\left(\boldsymbol{G}(s);\varrho\right)$ is defined for functions $\boldsymbol{G}(s)$ with the property

$$\boldsymbol{G}(0) = \boldsymbol{0}, \tag{6.8}$$

and it has the value $\boldsymbol{0}$ for the zero function $\boldsymbol{G}(s) \equiv \boldsymbol{0}$; *i.e.*,

$$\overset{\infty}{\underset{s=0}{\mathfrak{F}}}(\boldsymbol{0}, \varrho) = \boldsymbol{0}. \tag{6.9}$$

It is also isotropic in the sense of (6.5).

We now assume that for each simple fluid defined by a constitutive equation (6.7) there exists an influence function h of an order r such that the functional $\mathfrak{F}$ of (6.7) is a memory functional of type (h, n) in the sense of §4. The domain of $\mathfrak{F}$ is a class $\mathscr{H}$ of histories $\boldsymbol{G}$ whose values $\boldsymbol{G}(s)$ are in the space $\mathscr{S}$ of all symmetric tensors. For the norm $|\ |$ in $\mathscr{S}$ we use

$$|\boldsymbol{A}| = \sqrt{\operatorname{tr} \boldsymbol{A}^2} \quad \text{for} \quad \boldsymbol{A} \in \mathscr{S}. \tag{6.10}$$

The range space $\mathscr{T}$ of $\mathfrak{F}$ is the same as the range space of the histories $\boldsymbol{G}$, *i.e.* the space $\mathscr{T} = \mathscr{S}$ of all symmetric tensors with norm (6.10). Equation (6.9) insures that $\mathfrak{F}$ has the normalization (4.7) required for a memory functional. The density ϱ enters into (6.7) only as a real parameter. The assumption that $\mathfrak{F}$ is a memory functional implies that its domain of definition contains a neighborhood of zero in a function space $\mathscr{H}$ which is defined by the conditions (α)–(δ) of §4. The condition (γ) expresses the assumption that all histories $\boldsymbol{G} \in \mathscr{H}$ correspond to motions which are continuous at the present instant $s = 0$ (*cf.* (6.8)).

If the simple fluid under consideration is *incompressible*, we must make some alterations in our starting assumptions. For every possible motion in such a fluid the density ϱ is constant and the tensor $\boldsymbol{C}_t(\tau)$ is unimodular. In addition, the stress is determined by the history of the motion only up to a hydrostatic pressure p. Consequently, the equation (6.7) must be replaced by

$$\boldsymbol{S}(t) = -\,p\,\boldsymbol{I} + \overset{\infty}{\underset{s=0}{\mathfrak{F}}}\left(\boldsymbol{C}_t(t-s) - \boldsymbol{I}\right) \tag{6.11}$$

in which the indeterminate pressure p and the functional $\mathfrak{F}$ may be normalized by

$$0 = \operatorname{tr} \boldsymbol{S}(t) + 3p = \operatorname{tr} \overset{\infty}{\underset{s=0}{\mathfrak{F}}}\left(\boldsymbol{C}_t(t-s) - \boldsymbol{I}\right). \tag{6.12}$$

If we were to limit the domain of the functional $\mathfrak{F}$ in (6.11) to kinematically possible histories, this domain would not contain a neighborhood of the zero function in an appropriate function space, $\mathscr{H}$; therefore, $\mathfrak{F}$ could not then be a memory functional. We assume, however, that $\mathfrak{F}$ becomes a memory functional when its domain is extended by putting

$$\mathfrak{F}_{s=0}^{\infty}\left(\boldsymbol{C}(s)-\boldsymbol{I}\right)=\mathfrak{F}_{s=0}^{\infty}\left((\det \boldsymbol{C}(s))^{-1}\boldsymbol{C}(s)-\boldsymbol{I}\right) \tag{6.13}$$

when $\boldsymbol{C}(s)$ is not unimodular.

Aside from the added properties (6.12) and (6.13), the functional $\mathfrak{F}$ occurring in (6.11) is assumed to be of the same type as that in (6.7) with respect to both isotropy and memory.

7. *Approximations of Order n for Simple Fluids*

We now apply Theorem 2, §5, to the memory functional occurring in the constitutive equations for simple fluids (6.7) or (6.11).

Suppose a motion with Cauchy-Green tensor $\boldsymbol{C}_t(\tau)$ is given. The corresponding history $\boldsymbol{G}$ is defined by

$$\boldsymbol{G}(s)=\boldsymbol{C}_t(t-s)-\boldsymbol{I}. \tag{7.1}$$

If $\boldsymbol{C}_t(\tau)$ is n times differentiable with respect to τ at $\tau=t$, the kth Rivlin-Ericksen tensor $\boldsymbol{A}_k$, $k=1, 2, \ldots, n$ is defined as follows:

$$\boldsymbol{A}_k=\frac{d^k}{d\tau^k}\boldsymbol{C}_t(\tau)\Big|_{\tau=t}=(-1)^k\overset{(k)}{\boldsymbol{G}}, \tag{7.2}$$

where $\overset{(k)}{\boldsymbol{G}}$ is the kth derivative of $\boldsymbol{G}(s)$ at $s=0$, as in §2.

We now consider histories $\boldsymbol{G}_\alpha$ obtained from $\boldsymbol{G}\in\mathscr{H}$ by retardation as in (5.3). The corresponding Rivlin-Ericksen tensors are

$$\boldsymbol{A}_k^\alpha=\alpha^k\boldsymbol{A}_k=(-1)^k\overset{(k)}{\boldsymbol{G}}_\alpha. \tag{7.3}$$

They differ only by the inessential factor $(-1)^k$ from the tensors $\overset{(k)}{\boldsymbol{G}}_\alpha$ to be substituted for $\overset{(k)}{\boldsymbol{g}}_\alpha$ in the approximation formula (5.2). This formula, applied to the constitutive equations (6.7) or (6.11), yields the following expression for the stress tensor $\boldsymbol{S}_\alpha$ corresponding to the retarded history $\boldsymbol{G}_\alpha$:

$$\boldsymbol{S}_\alpha=-p\boldsymbol{I}+\sum_{(j_1,\ldots j_k)}\mathfrak{m}_{j_1\ldots j_k}(\boldsymbol{A}_{j_1}^\alpha,\ldots,\boldsymbol{A}_{j_k}^\alpha)+o(\alpha^n), \tag{7.4}$$

where the summation is to be extended over all sets of indices $(j_1,\ldots,j_k)$ obeying (5.1). The terms $\mathfrak{m}_{j_1\ldots j_k}(\boldsymbol{A}_{j_1}^\alpha,\ldots,\boldsymbol{A}_{j_k}^\alpha)$ are linear in each of the variables. For compressible fluids, it is understood that p and $\mathfrak{m}_{j_1\ldots j_k}$ depend on the density ϱ.

The equation (7.4) remains valid even when the derivatives shown in (7.2) exist only in the generalized sense of (3.5).

The multilinear forms $\mathfrak{m}_{j_1\ldots j_k}$ in (7.4) are isotropic functions, which means that they obey the identities

$$\boldsymbol{Q}\,\mathfrak{m}_{j_1\ldots j_k}(\boldsymbol{U}_1,\ldots,\boldsymbol{U}_k)\,\boldsymbol{Q}^T=\mathfrak{m}_{j_1\ldots j_k}(\boldsymbol{Q}\,\boldsymbol{U}_1\,\boldsymbol{Q}^T,\ldots,\boldsymbol{Q}\,\boldsymbol{U}_k\,\boldsymbol{Q}^T) \tag{7.5}$$

for all orthogonal $\boldsymbol{Q}$ and all symmetric tensors $\boldsymbol{U}_1,\ldots,\boldsymbol{U}_k$. This proposition follows easily from the fact that the memory functional $\mathfrak{F}$ occurring in (6.7) or (6.11) is isotropic in the sense of (6.5), from the observation that the conjugation $\boldsymbol{G}(s)\to\boldsymbol{Q}\,\boldsymbol{G}(s)\,\boldsymbol{Q}^T$ leaves the norm $\|\boldsymbol{G}\|$ unchanged, and from the uniqueness of the multilinear forms $\mathfrak{m}_{j_1\ldots j_k}$ asserted in Theorem 2, §5.

It follows from known theorems on isotropic functions that each form $\mathfrak{m}_{j_1\ldots j_k}(\boldsymbol{A}_{j_1}, \ldots, \boldsymbol{A}_{j_k})$, because it is isotropic and multilinear, may be expressed as a sum

$$\mathfrak{m}_{j_1\ldots j_k}(\boldsymbol{A}_{j_1}, \ldots, \boldsymbol{A}_{j_k}) = \sum_i \boldsymbol{\zeta}_i \tag{7.6}$$

of products of the form

$$\boldsymbol{\zeta}_i = \mu_i \varphi_1 \varphi_2 \ldots \varphi_{m_i} (\boldsymbol{A}_{l_1} \boldsymbol{A}_{l_2} \ldots \boldsymbol{A}_{l_{q_i}} + \boldsymbol{A}_{l_{q_i}} \ldots \boldsymbol{A}_{l_2} \boldsymbol{A}_{l_1}) \tag{7.7}$$

where the φ_j's are traces of products of some of the tensors $\boldsymbol{A}_{j_r}$ and are such that each $\boldsymbol{A}_{j_r}$, $r=1, \ldots, k$, occurs precisely once in each term $\boldsymbol{\zeta}_i$. In the case of an incompressible simple fluid the coefficients μ_i are constants, whereas for simple fluids in general the μ_i are functions of the density. Thus, for simple fluids, a finite number of scalar material functions $p(\varrho)$, $\mu_i(\varrho)$ suffices to determine the stress $\boldsymbol{S}_\alpha$ to within terms of order n in α.

The case $n=1$ of (7.4) is of particular interest. With use of (7.6), (7.7), we obtain

$$\boldsymbol{S}_\alpha = -p\boldsymbol{I} + \eta \boldsymbol{A}_1^\alpha + \frac{\lambda}{2}(\operatorname{tr} \boldsymbol{A}_1^\alpha)\boldsymbol{I} + o(\alpha). \tag{7.8}$$

Now, the first Rivlin-Ericksen tensor $\boldsymbol{A}_1 = 2\boldsymbol{D}$ differs from the rate of deformation tensor $\boldsymbol{D}$ only by the factor 2*. It follows that (7.8) is, to within terms of order $o(\alpha)$, simply the constitutive equation (1.4) of a Newtonian fluid.

When the fluid is incompressible, the case $n=2$ of (7.4) takes a remarkably simple form. The observation that $\operatorname{tr} \boldsymbol{A}_1 = 0$ for isochoric motions and use of (7.6), (7.7) yield

$$\boldsymbol{S}_\alpha = -\tilde{p}\boldsymbol{I} + \eta \boldsymbol{A}_1^\alpha + \beta(\boldsymbol{A}_1^\alpha)^2 + \gamma \boldsymbol{A}_2^\alpha + o(\alpha^2), \tag{7.9}$$

where η, β, and γ are material constants and where $\tilde{p}$ is an indeterminate pressure. This pressure $\tilde{p}$ differs, in general, from the mean pressure p defined by the normalization (6.12), because it is obtained from p by absorbing all scalar multiples of $\boldsymbol{I}$ arising from $\mathfrak{m}_{11}$ and $\mathfrak{m}_2$ through use of (7.6) and (7.7).

Motivated by (7.9), we can define an *incompressible second-order fluid* by the constitutive equation

$$\boldsymbol{S} = -\tilde{p}\boldsymbol{I} + \eta \boldsymbol{A}_1 + \beta \boldsymbol{A}_1^2 + \gamma \boldsymbol{A}_2. \tag{7.10}$$

Incompressible Newtonian fluids correspond to the special case $\beta = \gamma = 0$ of (7.10).

In some dynamical situations equation (7.10) leads to a linear partial differential equation for the velocity, just as in the Newtonian case. For example, consider a rectilinear shearing flow which, in Cartesian coordinates x, y, z, has a velocity field with components

$$\{v^i\} = \{0, v(x,t), 0\}. \tag{7.11}$$

If the body forces are conservative, substitution of (7.10) and (7.11) into CAUCHY's dynamical equations leads to the following third-order partial differential equation for $v(x,t)$:

$$\frac{\partial v}{\partial t} = a\frac{\partial^2 v}{\partial x^2} + b\frac{\partial^3 v}{\partial x^2\,\partial t}, \qquad a = \frac{\eta}{\varrho}, \qquad b = \frac{\gamma}{\varrho}. \tag{7.12}$$

* *Cf.* [1], (9.7).

A non-steady flow of the Couette type, in cylindrical coordinates r, ϑ, z, has a velocity field with contravariant components

$$\{v^i\} = \{0, \omega(r, t), 0\}. \tag{7.13}$$

For such a flow, instead of (7.12), we get

$$r^3 \frac{\partial \omega}{\partial t} = \frac{\partial}{\partial r}\left[r^3\left(a \frac{\partial \omega}{\partial r} + b \frac{\partial^2 \omega}{\partial r\, \partial t}\right)\right]. \tag{7.14}$$

In the Newtonian case, since $b=0$, equations (7.12) and (7.14) reduce to diffusion equations. As for the diffusion equations, many physically interesting solutions of (7.12) and (7.14) may be obtained by separation of variables*. It would be desirable to develop a mathematical theory of third-order partial differential equations of the type (7.12), (7.14).

8. Simple Materials in General

The considerations of the previous two sections may easily be extended to the general simple materials defined in Reference [*1*], Part III.

The form of the constitutive equation of an *isotropic simple material* may be obtained from (6.1) by replacing the scalar density $\varrho(t)$ by the left Cauchy-Green tensor $\boldsymbol{B}(t)$, taken relative to an undistorted reference state (*cf.* [*1*] (22.10)). A consideration analogous to the one which led to (6.7) shows that the constitutive equation of an isotropic simple material may be written in the form

$$\boldsymbol{S}(t) = \mathfrak{h}\left(\boldsymbol{B}(t)\right) + \underset{s=0}{\overset{\infty}{\mathfrak{F}}}\left(\boldsymbol{C}_t(t-s) - \boldsymbol{I};\, \boldsymbol{B}(t)\right). \tag{8.1}$$

Here, the functional $\mathfrak{F}$ depends on a tensor parameter $\boldsymbol{B}$, instead of on a scalar parameter ϱ as in (6.7). We assume again that there is an influence function h of order n such that the functional $\mathfrak{F}$ in (8.1) is a memory functional of type (h, n). In place of (7.4), we then obtain the following approximation formula for the stress $\boldsymbol{S}_\alpha$ produced by a slow motion of an isotropic simple material:

$$\boldsymbol{S}_\alpha = \mathfrak{h}(\boldsymbol{B}) + \sum_{(j_1, \ldots, j_k)} \mathfrak{m}_{j_1 \ldots j_k}(\boldsymbol{A}^\alpha_{j_1}, \ldots, \boldsymbol{A}^\alpha_{j_k};\, \boldsymbol{B}) + o(\alpha^n), \tag{8.2}$$

where $\mathfrak{m}_{j_1 \ldots j_k}$ is linear in each of its first k tensor variables but not necessarily in the last tensor variable $\boldsymbol{B}$.

The formulae corresponding to (8.1) and (8.2) in the case of *anisotropic simple materials* are obtained from (8.1) and (8.2) simply by replacing all tensors $\boldsymbol{T}$ occurring in these formulae by their conjugates $\boldsymbol{R}^T \boldsymbol{T} \boldsymbol{R}$ with the rotation tensor $\boldsymbol{R} = \boldsymbol{R}(t)$ of the displacement from the reference state (*cf.* [*1*], (22.8)).

In the case of isotropic materials, the function $\mathfrak{h}$ and the functional $\mathfrak{F}$ in (8.1) are isotropic in the sense that they obey the identities

$$\boldsymbol{Q}\, \mathfrak{h}(\boldsymbol{B})\, \boldsymbol{Q}^T = \mathfrak{h}(\boldsymbol{Q}\boldsymbol{B}\boldsymbol{Q}^T), \tag{8.3}$$

$$\underset{s=0}{\overset{\infty}{\mathfrak{F}}}\left(\boldsymbol{Q}\,\boldsymbol{G}(s)\,\boldsymbol{Q}^T, \boldsymbol{Q}\boldsymbol{B}\boldsymbol{Q}^T\right) = \boldsymbol{Q} \underset{s=0}{\overset{\infty}{\mathfrak{F}}}\left(\boldsymbol{G}(s), \boldsymbol{B}\right) \boldsymbol{Q}^T \tag{8.4}$$

---

* In particular, the sinusoidal vibration problems discussed for Newtonian fluids in §§ 345—346 of LAMB's treatise [*5*] are easily solved for second-order fluids. Also, special solutions of (7.14) corresponding to sinusoidal vibrations of a fluid between coaxial cylinders can readily be found.

for all orthogonal tensors $\boldsymbol{Q}$. As in §7, it follows that the $\mathfrak{m}_{j_1\ldots j_k}$ in (8.2) are isotropic functions of all their variables; *i.e.*, they obey the identities

$$\boldsymbol{Q}\,\mathfrak{m}_{j_1\ldots j_k}(\boldsymbol{U}_1,\ldots,\boldsymbol{U}_k;\boldsymbol{B})\,\boldsymbol{Q}^T=\mathfrak{m}_{j_1\ldots j_k}(\boldsymbol{Q}\,\boldsymbol{U}_1\,\boldsymbol{Q}^T,\ldots,\boldsymbol{Q}\,\boldsymbol{U}_k\,\boldsymbol{Q}^T;\boldsymbol{Q}\,\boldsymbol{B}\,\boldsymbol{Q}^T) \quad (8.5)$$

for all orthogonal $\boldsymbol{Q}$ and all symmetric tensors $\boldsymbol{U}_1,\ldots,\boldsymbol{U}_k$, $\boldsymbol{B}$ for which $\mathfrak{m}_{j_1\ldots j_k}$ is defined. The methods developed by SPENCER & RIVLIN ([6], [7] and [8]) may be used to derive explicit representations for the $\mathfrak{m}_{j_1\ldots j_k}$ of a type analogous to but more complicated than (7.6), (7.7). Using such a representation, one can show that, in the case $n=1$, the approximation formula (8.2) reduces to

$$\boldsymbol{S}_\alpha=\mathfrak{h}(\boldsymbol{B})+\boldsymbol{A}_1^\alpha\,\mathfrak{t}_1(\boldsymbol{B})+\mathfrak{t}_1(\boldsymbol{B})\,\boldsymbol{A}_1^\alpha+\operatorname{tr}\left(\boldsymbol{A}_1^\alpha\,\mathfrak{t}_2(\boldsymbol{B})\right)\mathfrak{t}_3(\boldsymbol{B}),+o(\alpha), \quad (8.6)$$

where $\mathfrak{h}$, $\mathfrak{t}_1$, $\mathfrak{t}_2$ and $\mathfrak{t}_3$ are isotropic functions of the one variable $\boldsymbol{B}$ and hence have representations of the form

$$\mathfrak{h}(\boldsymbol{B})=\beta_0\,\boldsymbol{I}+\beta_1\,\boldsymbol{B}+\beta_2\,\boldsymbol{B}^2, \quad (8.7)$$

in which β_0, β_1 and β_2 are scalar functions of the three principal invariants of $\boldsymbol{B}$.

The first term $\mathfrak{h}(\boldsymbol{B})$ in the expression (8.2) for the stress $\boldsymbol{S}_\alpha$ corresponds to purely elastic response. The sum in (8.2) may be interpreted as representing the internal friction for slow motions.

Acknowledgement. The research leading to this paper was supported by the Air Force Office of Scientific Research under Contract AF 49(638)541 with Mellon Institute and by the National Science Foundation under Grant NSF-G 5250 to Carnegie Institute of Technology.

References

[1] NOLL, W.: A mathematical theory of the mechanical behavior of continuous media. Arch. Rational Mech. Anal. 2, 197–266 (1958).

[2] COLEMAN, B. D., & W. NOLL: On certain steady flows of general fluids. Arch. Rational Mech. Anal. 4, 289–303 (1959).

[3] COLEMAN, B. D., & W. NOLL: Recent results in the continuum theory of visco-elastic fluids. To appear in Ann. New York Acad. Sci.

[4] HILLE, E., & R. S. PHILLIPS: Functional Analysis and Semi-Groups. American Mathematical Society Colloquium Publications, Vol. XXXI. New York 1957.

[5] LAMB, H.: Hydrodynamics, Chap. XI, 6th Edit. Cambridge: Univ. Press 1932.

[6] SPENCER, A. J. M., & R. S. RIVLIN: The theory of matrix polynomials and its application to the mechanics of isotropic continua. Arch. Rational Mech. Anal. 3, 309–336 (1959).

[7] SPENCER, A. J. M., & R. S. RIVLIN: Finite integrity bases for five or fewer symmetric 3×3 matrices. Arch. Rational Mech. Anal. 3, 435–446 (1959).

[8] SPENCER, A. J. M., & R. S. RIVLIN: Further results in the theory of matrix polynomials. Arch. Rational Mech. Anal. 4, 214–230 (1960).

Mellon Institute
Pittsburgh, Pennsylvania
and
Mathematics Department
Carnegie Institute of Technology
Pittsburgh, Pennsylvania

(Received August 20, 1960)

Foundations of Linear Viscoelasticity*

BERNARD D. COLEMAN

AND

WALTER NOLL

1. INTRODUCTION

THE classical linear theory of viscoelasticity was apparently first formulated by Boltzmann[1] in 1874. His original presentation covered the three-dimensional case, but was restricted to isotropic materials. The extension of the theory to anisotropic materials is, however, almost immediately evident on reading Boltzmann's paper, and the basic hypotheses of the theory have not changed since 1874. Since that date, much work has been done on the following aspects of linear viscoelasticity: solution of special boundary value problems,[2a] reformulation[3,4] of the one-dimensional version of the theory in terms of new material functions (such as "creep functions" and frequency-dependent complex "impedances") which appear to be directly accessible to measurement, experimental determination[2b] of the material functions for those materials for which the theory appears useful, prediction of the form of the material functions from molecular models, and, recently, axiomatization[5,6] of the theory. In this article, instead of being concerned with these matters, we reexamine the fundamental hypotheses of linear viscoelasticity in the light of recent advances in nonlinear continuum mechanics.

The basic assumption of the classical linear theory of viscosity is a constitutive equation relating the stress tensor $T(t)$ at time t to the history of the infinitesimal strain tensor $E(t-s)$, $0 \leq s < \infty$. This assumption asserts that if $E(t-s)$, taken relative to a natural reference configuration corresponding to zero equilibrium stress, is *small* in magnitude for *all* s, then

$$T(t)=\mathbf{\Omega}\{E(t)\}+\mathbf{\Phi}(0)\{E(t)\}+\int_0^\infty \dot{\mathbf{\Phi}}(s)\{E(t-s)\}ds, \tag{1.1}$$

where

$$\dot{\mathbf{\Phi}}(s)=(d/ds)\mathbf{\Phi}(s), \tag{1.2}$$

and $\mathbf{\Phi}$ is such that

$$\lim_{s\to\infty} \mathbf{\Phi}(s)=0. \tag{1.3}$$

Here $\mathbf{\Phi}(s)\{\ \}$ (for each s) and $\mathbf{\Omega}\{\ \}$ are linear transformations of the space of symmetric tensors into itself. As a function of time, $\mathbf{\Phi}$ has a simple physical significance and is called the "stress relaxation function." For if we consider a deformation history such that the material is kept in its natural reference configuration ($E=0$) for all times $t<0$ and has the strain E^* for all times $t\geq 0$, then for such a history Eq. (1.1) yields

$$T(t)=\begin{cases}0 & \text{if } t<0 \\ \mathbf{\Phi}(t)\{E^*\}+\mathbf{\Omega}\{E^*\} & \text{if } t\geq 0.\end{cases} \tag{1.4}$$

In the familiar special case of an isotropic material, $\mathbf{\Phi}(t)$ is completely determined by two scalar-valued functions of time: the stress relaxation functions for shear and dilatation. The linear transformation $\mathbf{\Omega}$ characterizes the "linear equilibrium stress-strain law" of infinitesimal elasticity theory; i.e., Eqs. (1.3) and (1.4) yield

$$\lim_{t\to\infty} T(t)=\mathbf{\Omega}\{E^*\}. \tag{1.5}$$

For an isotropic solid, $\mathbf{\Omega}$ is determined by the two Lamé constants.

We refer to the classical linear theory based on Eqs. (1.1)–(1.3) as *infinitesimal viscoelasticity* because, roughly speaking, it can be applicable only to those situations in which the strain is small at all times.

In Sec. 4 we show how Eq. (1.1) must be modified when the reference configuration is arbitrary and not necessarily one in which the equilibrium stress is zero. In particular, in the case of a fluid, $T(t)$ should be replaced by $T(t)+p_r I$, where p_r is the equilibrium hydrostatic pressure corresponding to the reference configuration, and I denotes the unit (or identity) tensor. For a fluid $\mathbf{\Omega}$ is determined by the equilibrium compressibility.

It is often claimed that the theory of infinitesimal viscoelasticity can be derived from an assumption that on a microscopic level matter can be regarded as composed of "linear viscous elements" (also called

* This work was supported by the Air Force Office of Scientific Research under contract and by the National Science Foundation under Grant NSF-G5250.

[1] S. Boltzmann, Sitzber. Akad. Wiss. Wien, Math. naturw. Kl. **70**, 275 (1874); Pogg. Ann. Phys. **7**, 624 (1876).

[2] (a) E. H. Lee, in *Viscoelasticity*, edited by J. T. Bergen (Academic Press, Inc., New York, 1960), p. 1; (b) J. D. Ferry and K. Ninomiya, *ibid*., p. 55.

[3] B. Gross, *Mathematical Structure of the Theories of Viscoelasticity* (Hermann & Cie., Paris, 1953). A summary of relationships between those material functions which occur in the one-dimensional formulation of the theory is given in this reference and in reference 4.

[4] H. Leaderman, Trans. Soc. Rheol. **1**, 213 (1957).

[5] E. R. Love, Australian J. Phys. **9**, 1 (1956).

[6] H. Konig and J. Meixner, Math. Nachr. **19**, 265 (1958).

"dashpots") and "linear elastic elements" (called "springs") connected together in intricate "networks."[7] The motivation behind some of the recent work on spring and dashpot networks appears to be the hope that the consideration of such readily visualized models will suggest a formalism for immersing viscoelasticity in a general thermodynamical theory of irreversible processes.

We feel that the physicist's confidence in the usefulness of the theory of infinitesimal viscoelasticity does not stem from a belief that the materials to which the theory is applied are really composed of microscopic networks of springs and dashpots, but comes rather from other considerations. First, there is the observation that the theory works for many real materials. But second, and perhaps more important to theoreticians, is the fact that the theory looks plausible because it seems to be a mathematization of little more than certain intuitive prejudices about smoothness in macroscopic phenomena. It is natural to assume that the dependence of the stress on the history of the deformation should be, in some sense, a *smooth dependence.* (Smoothness assumptions are usually so "natural" to physicists that they are seldom made explicit.) Since we know that in small neighborhoods smooth dependences are approximately *linear,* it is felt that if only small deformations are considered, the stress should be given by a linear functional of the deformation history, and that this functional should yield the form exhibited in Eqs. (1.1)–(1.3).

This article tries to make precise these observations about smoothness, and in so doing seeks to obtain a mathematical derivation of infinitesimal viscoelasticity from plausible macroscopic assumptions. To do this one must first presume a nonlinear theory of the mechanical behavior of materials with memory, and, if the undertaking is to be at all worthwhile, the presumed nonlinear theory must rest on constitutive equations based only on very general physical principles. Our development starts with the recently formulated general theory[8] of "simple materials" (i.e., materials for which the stress depends in an arbitrary way on the history of the first spatial gradient of the displacement). The theory of simple materials is outlined in Sec. 3.

To make precise the notion of smoothness we must introduce a topology into the space of functions characterizing the history of the deformation; i.e., we must have a way of knowing when two histories are close to each other. We do this by defining a *norm.* The particular norm used here is one of those considered in our paper on memory functionals.[9] This norm has two important properties: first, it makes our space of histories a Hilbert space; second, it places greater emphasis on the deformations which occurred in the recent past than on those which occurred in the distant past. We believe that this second property is essential if one is to formulate a smoothness assumption for macroscopic phenomena that is compatible with the everyday observation that memories are imperfect. The memory of a macroscopic object for its past deformations fades in the sense that deformations which occurred in the distant past have a smaller effect on the present forces than have more recent deformations.

We mathematize the notion of smoothness by assuming that the constitutive functionals which give the stress in a simple material are *Fréchet differentiable* in our Hilbert space of histories.

In considering finite deformations in simple materials, it is often convenient to take the present configuration as the reference configuration for describing the history of the deformation. Indeed, when dealing with a fluid, this is the natural thing to do, because a fluid has no preferred configurations. However, we can do this even for solids, provided we maintain in the constitutive equations a tensor parameter which tells how the present configuration is related to a preferred configuration.

The function space norm which we use has the property that the norm of a history is small if the deformations have been small at all times in the past; indeed, our derivation of infinitesimal viscoelasticity is a combination of this fact with our differentiability assumption. However, when one takes the present state as a reference, the deformation at the present time is zero, and if one further notes that the distant past is of little importance, it becomes clear that there are several ways in which a history can be small in norm. In particular, any history for which the motion has been *slow* in the recent past has a small norm. This observation has suggested to us the consideration of a new linear approximation for the general constitutive functionals of simple materials. We call the theory based on this new approximation *finite linear viscoelasticity*; it includes the classical infinitesimal theory as a special case, but has the advantage of being meaningful in situations involving finite deformations. The arguments presented in Secs. 3 and 5 show that finite linear viscoelasticity furnishes a complete first-order approximation to the theory of simple materials in the limit in which the history of the deformation, taken relative to the present configuration, is small in norm.

The smoothness considerations presented can be extended to obtain higher order approximations to the general constitutive equations of simple materials. In Sec. 6 we discuss a second-order theory of viscoelasticity for incompressible simple fluids.

2. KINEMATICS

We present a brief outline of the kinematics required for a discussion of simple materials. For a more complete

[7] D. R. Bland, *The Theory of Viscoelasticity* (Pergamon Press, New York, 1960), Chap. 2.
[8] W. Noll, Arch. Ratl. Mech. Anal. 2, 197 (1958).
[9] B. D. Coleman and W. Noll, Arch. Ratl. Mech. Anal. 6, 356 (1960).

presentation which goes back to first principles, see Noll.[8]

Consider a particular material point X of a body $\mathcal{B}$. Suppose that X occupies the position $\mathbf{X}$ in Euclidean space $\mathcal{E}$ when $\mathcal{B}$ is in a reference configuration. Let $\boldsymbol{\xi}$ be the position of X in $\mathcal{E}$ at time τ. For the dependence of $\boldsymbol{\xi}$ on $\mathbf{X}$ and τ, we write

$$\boldsymbol{\xi}=\chi(\mathbf{X},\tau). \tag{2.1}$$

The gradient $F(\tau)$ of $\chi(\mathbf{X},\tau)$ with respect to $\mathbf{X}$,

$$F(\tau)=\nabla\chi(\mathbf{X},\tau), \tag{2.2}$$

is called the *deformation gradient* at the material point X at time τ. It is a tensor which possesses an inverse $F(\tau)^{-1}$. (Here the term "tensor" is used a synonym for "linear transformation of the three-dimensional Euclidean vector space into itself.") The value of $F(\tau)$ at each point of $\mathcal{B}$ is affected not only by the configuration of $\mathcal{B}$ at time τ but also by our choice of a reference configuration for $\mathcal{B}$. This reference configuration may be chosen for convenience and need not necessarily be a configuration actually occupied by the body during its motion.

It is often useful to employ the configuration at the present time t, rather than a fixed configuration, as the reference. The corresponding deformation gradient is denoted by $F_t(\tau)$ and called the *relative deformation gradient*. The deformation gradients enjoy the following important property, which is a direct consequence of the chain rule for the differentiation of composite vector-valued functions:

$$F(\tau)=F_t(\tau)F(t), \tag{2.3}$$

where the indicated multiplication is the usual composition of linear transformations (matrix product).

An immediate consequence of the definition of $F_t(t)$ is that

$$F_t(t)=I, \tag{2.4}$$

where I is the unit (or identity) tensor. From Eq. (2.3) we obtain the relation

$$F_t(\tau)=F(\tau)F(t)^{-1}. \tag{2.5}$$

Let $\rho(\tau)$ give the mass density at X as a function of τ; it follows from a theorem of kinematics that

$$\det F_t(\tau)=\rho(t)/\rho(\tau). \tag{2.6}$$

If $F(\tau)$ is independent of X, we say that the configuration of $\mathcal{B}$ at time τ and the reference configuration of $\mathcal{B}$ are related by a *homogeneous deformation*. If $F=F(\tau)$ is orthogonal, i.e., if

$$F^TF=FF^T=I, \tag{2.7}$$

in which F^T denotes the transpose of F, then this "homogeneous deformation" represents a rigid rotation of the body. If F is symmetric positive-definite, then the body has been subjected to a *pure stretch*; in this case the proper vectors of F give the principal directions of stretch and the proper numbers of F are the principal stretch ratios.

A theorem of algebra, called the *polar decomposition theorem*, states that any invertible tensor F can be written in two ways as the product of a symmetric positive-definite tensor and an orthogonal tensor:

$$F=RU, \tag{2.8}$$

$$F=VR. \tag{2.9}$$

Furthermore, the orthogonal tensor R and the symmetric positive-definite tensors U and V in these decompositions are uniquely determined by F and obey the following relations:

$$U^2=F^TF\equiv C, \tag{2.10}$$

$$V^2=FF^T\equiv B, \tag{2.11}$$

$$U=R^TVR. \tag{2.12}$$

Equations (2.8) and (2.9) have the following significance in kinematics: Any homogeneous deformation with deformation gradient F may be regarded as being the result of a pure stretch U followed by a rigid rotation R, or a rigid rotation R followed by a pure stretch V. These interpretations uniquely determine the pairs R, U and R, V. The rigid rotations entering these two interpretations are the same; however, the pure stretches U and V can be different. It follows from Eq. (2.12) that although these stretches may have different principal directions, they must yield the same stretch ratios. We call the tensor R the *rotation tensor* and the tensors U and V, respectively, the *right* and *left stretch tensors*. The symmetric positive-definite tensors C and B, defined by Eqs. (2.10) and (2.11), are called, respectively, the *right* and *left Cauchy-Green tensors*; they obviously contain the same information as the corresponding stretch tensors, and their components are often easier to compute.

The rotation tensor, the stretch tensors, and the Cauchy-Green tensors computed from the relative deformation gradient F_t are denoted by R_t, U_t, V_t, C_t, and B_t. The modifier *relative* is used to indicate that the present configuration (time t) is used as the reference. For example, $C_t(\tau)$, is called the *relative right Cauchy-Green tensor*.

The following formulas are consequences of Eq. (2.4):

$$U_t(t)=V_t(t)=C_t(t)=B_t(t)=R_t(t)=I. \tag{2.13}$$

For simplicity we have emphasized the interpretation for *homogeneous deformations* of the tensors defined by Eqs. (2.8)–(2.11). These definitions obviously apply also to nonhomogeneous deformations, and similar interpretations can be given to them in the nonhomogeneous case if one merely first observes that the deformations considered in continuum mechanics are sufficiently smooth to be approximately homogeneous in small regions of $\mathcal{B}$.

We note that there is no unique way to measure "the strain" corresponding to an arbitrary finite deformation.

We now establish the connection between the kinematics of *finite deformations* sketched in the foregoing and the more familiar kinematics of *infinitesimal deformations*.

The *magnitude* $|A|$ of a tensor A is defined by

$$|A|^2=\mathrm{Tr}(AA^T), \tag{2.14}$$

where Tr denotes the trace of a tensor. If Cartesian coordinates are used, then $|A|^2$ is the sum of the squares of the elements of the 3×3 matrix corresponding to A. We also use the definition (2.14) or magnitude when A is replaced by a linear transformation $\mathbf{\Gamma}$ of the six-dimensional space of symmetric tensors. In this case, the square of the magnitude $|\mathbf{\Gamma}|$ of $\mathbf{\Gamma}$ is the sum of the squares of the 6×6 matrix corresponding to $\mathbf{\Gamma}$.

Let a motion with deformation gradient $F=F(\tau)$ be given. We put

$$H=F-I \tag{2.15}$$

and

$$\epsilon=\sup_{\tau}|H(\tau)|. \tag{2.16}$$

H is the gradient of the displacement vector field. We say that the deformation corresponding to $F(\tau)$ is *infinitesimal at all times* τ if

$$\epsilon\ll 1. \tag{2.17}$$

The *infinitesimal strain tensor* $E=E(\tau)$ is defined by

$$E=\tfrac{1}{2}(H+H^T). \tag{2.18}$$

In the following we consider functions of τ which are determined by $H(\tau)$ and which have the property that for each τ their magnitude is less than $K\epsilon^n$, where K is a number independent of τ, the function $H(\tau)$, and ϵ. Any such function is denoted by the order symbol $O(\epsilon^n)$; i.e.,

$$|O(\epsilon^n)|<K\epsilon^n. \tag{2.19}$$

It is easy to show that

$$F=I+H=I+O(\epsilon), \tag{2.20}$$

$$F^{-1}=I-H+O(\epsilon^2)=I+O(\epsilon). \tag{2.21}$$

Also, it is not difficult to establish the following relations between the stretch tensors U, V and Cauchy-Green tensors C, B, on the one hand, and the infinitesimal strain tensor E, on the other hand:

$$U-I=E+O(\epsilon^2)=O(\epsilon), \tag{2.22}$$

$$V-I=E+O(\epsilon^2)=O(\epsilon), \tag{2.23}$$

$$C-I=2E+O(\epsilon^2)=O(\epsilon), \tag{2.24}$$

$$B-I=2E+O(\epsilon^2)=O(\epsilon). \tag{2.25}$$

Thus, if terms of order $O(\epsilon^2)$ can be neglected, the stretch tensors U, V and Cauchy-Green tensors C, B can be expressed in terms of E. For finite deformations, however, the infinitesimal strain tensor E is devoid of kinematical significance.

Finally, we note the following relations between the *infinitesimal rotation tensor* W, defined by

$$W=\tfrac{1}{2}(H-H^T), \tag{2.26}$$

and the finite rotation tensor R:

$$R=I+W+O(\epsilon^2)=I+O(\epsilon),$$
$$R^T=R^{-1}=I-W+O(\epsilon^2)=I+O(\epsilon). \tag{2.27}$$

In order to find an expression for the relative Cauchy-Green tensor $C_t(\tau)$, we first substitute Eqs. (2.20) and (2.21) into Eq. (2.5) and obtain

$$F_t(\tau)=I+H(\tau)-H(t)+O(\epsilon^2). \tag{2.28}$$

Equation (2.10), written for the relative tensors F_t and C_t, reads

$$C_t(\tau)=F_t(\tau)^TF_t(\tau). \tag{2.29}$$

Substitution of Eq. (2.28) into Eq. (2.29) and use of Eq. (2.18) yield

$$C_t(\tau)=I+2[E(\tau)-E(t)]+O(\epsilon^2)=I+O(\epsilon). \tag{2.30}$$

For finite deformations there is no simple relation between $C_t(\tau)$, $C(\tau)$, and $C(t)$.

3. FADING MEMORY

The theory of simple materials is based on the following physical assumption: *The present stress is given by a functional of the past history of the deformation gradient.*

Suppose the deformation gradient $F(\tau)$ is given (for all $\tau\leq t$) computed relative to a fixed reference configuration. The right Cauchy-Green tensor $C(t)$ and rotation tensor $R(t)$ corresponding to $F(t)$ are determined by Eqs. (2.8) and (2.10). On using Eqs. (2.5) and (2.29) we can compute the relative Cauchy-Green tensors $C_t(\tau)$ for all $\tau\leq t$. We now put

$$\bar{C}_t(\tau)=R^T(t)C_t(\tau)R(t). \tag{3.1}$$

If the material has always been at rest, we have, by Eqs. (2.13) and (3.1),

$$\bar{C}_t(\tau)\equiv I \quad \text{for } \tau\leq t. \tag{3.2}$$

The principle of material objectivity, which states that the properties of a material should appear the same to all observers, can be used to show that the general constitutive equation for simple materials reduces to the form

$$\bar{T}(t)\equiv R^T(t)T(t)R(t)=\underset{s=0}{\overset{\infty}{\mathfrak{R}}}(\bar{C}_t(t-s);C(t)), \tag{3.3}$$

where $T(t)$ is the stress tensor at time t and the symbol $\mathfrak{R}$ denotes a functional. [This may be compared with reference 8, Eq. (22.8). Here we use a somewhat more suggestive notation.]

It is useful to put Eq. (3.3) into a slightly different form by writing the right-hand side as the sum of an "equilibrium term" $\mathfrak{h}(C(t))$ and a term which vanishes when the material has always been at rest, i.e., when Eq. (3.2) holds:

$$\bar{T}(t)=\mathfrak{h}(C(t))+\underset{s=0}{\overset{\infty}{\mathfrak{F}}}(\bar{C}_t(t-s)-I;C(t)), \quad (3.4)$$

$$\underset{s=0}{\overset{\infty}{\mathfrak{F}}}(0;C(t))=0. \quad (3.5)$$

For present purposes it is sufficient to regard the constitutive equation (3.4) as the definition of a simple material.

We now add a new physical assumption: *The memory of a simple material fades in time.*

There is no unique way to give this statement a precise meaning. We consider a particular mathematical interpretation of it. For this purpose we first introduce the concept of an influence function which is used to characterize the rate at which the memory fades. (This definition of an influence function is slightly different and somewhat less technical than the one we gave in reference 9.) A function h is called an *influence function* of order $r>0$ if it satisfies the following conditions:

(a) $h(s)$ is defined for $0\leq s<\infty$ and has positive real values: $h(s)>0$.

(b) $h(s)$ decays to zero according to

$$\lim_{s\to\infty} s^r h(s)=0 \quad (3.6)$$

monotonically for large s. For example,

$$h(s)=(s+1)^{-p}$$

is an influence function of order r if $r<p$. An exponential

$$h(s)=e^{-\beta s}, \quad \beta>0$$

is an influence function of any order.

Any function $G(s)$, defined for $s\geq 0$ and with values which are symmetric tensors, is called a *history*. The argument function $G(s)=\bar{C}_t(t-s)-I$ of the functional $\mathfrak{F}$ of Eq. (3.4) is a history. The tensor $C(t)$ in Eq. (3.4) plays the role of a parameter.

Let an influence function $h(s)$ be given. We then define the norm $\|G(s)\|$ of a history $G(s)$ by

$$\|G(s)\|^2=\int_0^\infty |G(s)|^2 h(s)^2 ds, \quad (3.7)$$

where $|G(s)|$ is the magnitude of the tensor $G(s)$ defined by Eq. (2.14). The influence function $h(s)$ determines the influence assigned to the values of $G(s)$ in computing the norm $\|G(s)\|$. Since $h(s)\to 0$ as $s\to\infty$, the values of $G(s)$ for small s (recent past) have a greater weight than the values for large s (distant past).

The collection of all histories with finite norm (3.7) forms a Hilbert space $\mathcal{H}$. A history $G(s)$ belongs to the space $\mathcal{H}$ if it does not grow too fast as $s\to\infty$.

Consider now an influence function h and a functional

$$\underset{s=0}{\overset{\infty}{\mathfrak{F}}}(G(s))$$

which is defined on a neighborhood of the zero history in the Hilbert space $\mathcal{H}$ corresponding to h and whose values are symmetric tensors. Assume that the value of $\mathfrak{F}$ for the zero history is zero, i.e., that

$$\underset{s=0}{\overset{\infty}{\mathfrak{F}}}(0)=0. \quad (3.8)$$

We say that $\mathfrak{F}$ is *Fréchet-differentiable* at the zero history if there is a continuous linear functional $\delta\mathfrak{F}$ such that

$$\underset{s=0}{\overset{\infty}{\mathfrak{F}}}(G(s))=\delta\underset{s=0}{\overset{\infty}{\mathfrak{F}}}(G(s))+\underset{s=0}{\overset{\infty}{\mathfrak{R}}}(G(s)). \quad (3.9)$$

where the "remainder" $\mathfrak{R}$ is of order $o(\|G(s)\|)$ in the sense that

$$\lim_{\|G(s)\|\to 0} \|G(s)\|^{-1}\underset{s=0}{\overset{\infty}{\mathfrak{R}}}(G(s))=0. \quad (3.10)$$

The linear functional $\delta\mathfrak{F}$ is called the *first variation* or *Fréchet differential* of $\mathfrak{F}$ at the zero history.

We now translate our physical assumption of fading memory into the following mathematical requirement:

(F) *There exists an influence function $h(s)$ of an order $r>\frac{1}{2}$ such that, for each value of the tensor parameter C, the functional $\mathfrak{F}$ of the constitutive equation* (3.4) *is Fréchet-differentiable at the zero history in the Hilbert space $\mathcal{H}$ corresponding to $h(s)$.*

If we indicate the dependence on the tensor parameter C, Eq. (3.9) becomes

$$\underset{s=0}{\overset{\infty}{\mathfrak{F}}}(G(s);C)=\delta\underset{s=0}{\overset{\infty}{\mathfrak{F}}}(G(s);C)+\underset{s=0}{\overset{\infty}{\mathfrak{R}}}(G(s);C). \quad (3.11)$$

We now invoke the theorem of the theory of Hilbert spaces which states that every continuous linear functional may be written as an inner product. It follows from this theorem that the first variation $\delta\mathfrak{F}$ has an integral representation of the form

$$\delta\underset{s=0}{\overset{\infty}{\mathfrak{F}}}(G(s);C)=\int_0^\infty \mathbf{\Gamma}(s;C)\{G(s)\}ds. \quad (3.12)$$

Here $\mathbf{\Gamma}(s;C)\{\ \}$, for each s and each C, is a linear transformation of the space of symmetric tensors into itself with the property that

$$\int_0^\infty |\mathbf{\Gamma}(s;C)|^2 h(s)^{-2}ds<\infty, \quad (3.13)$$

where $|\Gamma(s;C)|$ is the magnitude of $\Gamma(s;C)$ as defined by Eq. (2.14). The property (3.13) shows that $\Gamma(s;C)$ must approach zero at a faster rate than the influence function $h(s)$ as $s\to\infty$. Substitution of Eqs. (3.12) and (3.11) into Eq. (3.4) yields

$$\bar{T}=\mathfrak{h}(C)+\int_0^\infty \Gamma(s;C)\{G(s)\}ds+\underset{s=0}{\overset{\infty}{\mathfrak{R}}}(G(s);C), \quad (3.14)$$

where

$$G(s)=\bar{C}_t(t-s)-I. \quad (3.15)$$

It is understood that the variables $\bar{T}$, C, and $G(s)$ depend on the present time t.

It seems natural to add to the requirement (F) the following two assumptions:

(F′) *The Fréchet-differentiability of* $\mathfrak{F}$ *postulated in* (F) *is uniform in the tensor parameter* C.

(D) *The tensor function* $\mathfrak{h}(C)$ *of* (3.14) *is continuously differentiable.*

By the assumption (F′) we mean that the first variation

$$\delta\underset{s=0}{\overset{\infty}{\mathfrak{F}}}(G(s);C)$$

depends continuously on C in the strong sense and that the convergence in Eq. (3.10) is uniform in C.

4. INFINITESIMAL VISCOELASTICITY

We first remark that any function of order $O(\epsilon^n)$ in the sense of Eq. (2.19) is also a function of order $O(\epsilon^n)$ with respect to the Hilbert-space norm (3.7); i.e., there is a constant $\bar{K}$, independent of ϵ, such that

$$\|O(\epsilon^n)\|<\bar{K}\epsilon^n. \quad (4.1)$$

In order to prove this inequality we substitute $O(\epsilon^n)$ for $G(s)$ in the definition (3.7) of the norm and use the inequality (2.19):

$$\|O(\epsilon^n)\|^2=\int_0^\infty |O(\epsilon^n)|^2h(s)^2ds<(K\epsilon^n)^2\int_0^\infty h(s)^2ds. \quad (4.2)$$

The requirement (F) of Sec. 3 ensures that the number r of Eq. (3.6) is greater than $\frac{1}{2}$. It follows that the integral $\int_0^\infty h^2(s)ds$ is finite and hence that the inequality (4.1) holds with

$$\bar{K}=K\left(\int_0^\infty h(s)^2ds\right)^{\frac{1}{2}}. \quad (4.3)$$

This remark shows that the order symbols in Eqs. (2.20)–(2.30) may be interpreted in terms of the convergence in the Hilbert space of histories defined in Sec. 3. This interpretation must be used to justify most of the subsequent considerations.

By combining Eqs. (3.1), (2.30), and (2.27), we find the following expression for the history

$$G(s)=\bar{C}_t(t-s)-I$$

which enters the constitutive equation (3.4) of a simple material:

$$G(s)=2[E(t-s)-E(t)]+O(\epsilon^2)=O(\epsilon). \quad (4.4)$$

On substituting Eq. (4.4) into Eq. (3.11) and using Eq. (3.10) and the linearity and continuity of the first variation $\delta\mathfrak{F}$, we obtain

$$\underset{s=0}{\overset{\infty}{\mathfrak{F}}}(G(s);C)=2\delta\underset{s=0}{\overset{\infty}{\mathfrak{F}}}(E(t-s)-E(t);C)+o(\epsilon). \quad (4.5)$$

where the order symbol $o(\epsilon)$ is used in the sense that

$$\lim_{\epsilon\to 0}\epsilon^{-1}|o(\epsilon)|=0. \quad (4.6)$$

It is not difficult to prove that the uniformity assumption (F′) of Sec. 3 implies that Eq. (4.5) remains valid if, on the right-hand side, the tensor $C=I+O(\epsilon)$ is replaced by the unit tensor I:

$$\underset{s=0}{\overset{\infty}{\mathfrak{F}}}(G(s);C)=2\delta\underset{s=0}{\overset{\infty}{\mathfrak{F}}}(E(t-s)-E(t);I)+o(\epsilon). \quad (4.7)$$

We now substitute the integral representation (3.12), for $C=I$, into Eq. (4.7), and obtain

$$\underset{s=0}{\overset{\infty}{\mathfrak{F}}}(G(s);C)=\int_0^\infty 2\Gamma(s)\{E(t-s)\}ds-\int_0^\infty 2\Gamma(s)ds\{E(t)\}+o(\epsilon). \quad (4.8)$$

On defining $\Phi(s)$ by

$$\Phi(s)=-2\int_s^\infty \Gamma(\sigma)d\sigma,\quad \dot{\Phi}(s)=\frac{d}{ds}\Phi(s)=2\Gamma(s), \quad (4.9)$$

we may rewrite Eq. (4.8) in the form

$$\underset{s=0}{\overset{\infty}{\mathfrak{F}}}(G(s);C)=\Phi(0)\{E(t)\}+\int_0^\infty \dot{\Phi}(s)\{E(t-s)\}ds+o(\epsilon), \quad (4.10)$$

where

$$\lim_{s\to\infty}\Phi(s)=0. \quad (4.11)$$

Assumption (D) of Sec. 3 and Eq. (2.24) imply that the equilibrium term $\mathfrak{h}(C)$ of Eq. (3.4) has the form

$$\mathfrak{h}(C(t))=T_r+\Omega\{E(t)\}+o(\epsilon). \quad (4.12)$$

Here, the linear transformation $\Omega\{\ \}$ of the space of symmetric tensors is the gradient of the tensor function $\mathfrak{h}(C)$ at $C=I$. The tensor

$$T_r=\mathfrak{h}(I) \quad (4.13)$$

is the *residual stress*, i.e., the stress the material would sustain if it had been held in the reference configuration at all times in the past.

Substitution of Eqs. (4.10) and (4.12) into the constitutive equation (3.4) yields

$$\bar{T}(t)=T_r+[\mathbf{\Omega}+\mathbf{\Phi}(0)]\{E(t)\}+\int_0^\infty \dot{\mathbf{\Phi}}(s)\{E(t-s)\}ds+o(\epsilon). \quad (4.14)$$

Finally, going back to the definition (3.3) of $\bar{T}$ and using Eqs. (2.27), we obtain the following expression for the stress tensor $T(t)$:

$$T(t)-T_r=W(t)T_r-T_rW(t)+[\mathbf{\Omega}+\mathbf{\Phi}(0)]\{E(t)\}+\int_0^\infty \dot{\mathbf{\Phi}}(s)\{E(t-s)\}ds+o(\epsilon). \quad (4.15)$$

When ϵ, given by Eq. (2.16), is small enough, the remainder term $o(\epsilon)$ can be neglected in comparison with the other terms on the right-hand side of Eq. (4.15), which are of order $O(\epsilon)$. Thus, the *constitutive equation of infinitesimal viscoelasticity* reads

$$T(t)-T_r=W(t)T_r-T_rW(t)+[\mathbf{\Omega}+\mathbf{\Phi}(0)]\{E(t)\}+\int_0^\infty \dot{\mathbf{\Phi}}(s)\{E(t-s)\}ds. \quad (4.16)$$

When the reference configuration is a natural state, we have $T_r=0$, and Eq. (4.16) reduces to the classical equation (1.1). Equation (4.16), with $T_r\neq 0$, applies to infinitesimal deformations superposed on a large deformation from an unstressed natural state. In this case, the reference configuration is not the natural state but the deformed state with equilibrium stress T_r. If T_r is a hydrostatic pressure $T_r=-pI$, the terms involving $W(t)$ in Eq. (4.16), cancel. The stress relaxation function $\mathbf{\Phi}(s)$ depends not only on the material but also on the configuration which has been taken as the reference.

We remark that the special case $\mathbf{\Phi}(s)\equiv 0$ of Eq. (4.16) corresponds to the theory of infinitesimal elastic deformations superposed on large deformations. The special case $\mathbf{\Phi}(s)\equiv 0$ and $T_r=0$ corresponds to the classical theory of infinitesimal elasticity.

5. FINITE LINEAR VISCOELASTICITY

Motivation

Let us return to Eq. (3.14), which, under our hypothesis (F), is equivalent to the fundamental constitutive equation (3.4). It follows from Eq. (3.10) that the remainder term of Eq. (3.14) is small compared to the term involving the integral, provided the history $G(s)=\bar{C}_t(t-s)-I$ has a small Hilbert-space norm. Thus, the equation

$$\bar{T}=\mathfrak{h}(C)+\int_0^\infty \mathbf{\Gamma}(s;C)\{G(s)\}ds \quad (5.1)$$

approximates the general constitutive equation of a simple material in the limit

$$\|G(s)\|\to 0, \quad (5.2)$$

and the error approaches zero faster than $\|G(s)\|$. We call the theory based on Eq. (5.1) *finite linear viscoelasticity.*

One way of achieving the limit (5.2) is to let ϵ, defined by Eq. (2.16), go to zero. The discussion of Sec. 4 shows that, in this case, Eq. (5.1) reduces to the constitutive equation (4.16) of infinitesimal viscoelasticity.

When we consider, however, the definition (3.7) of the norm $\|G(s)\|$, we see that the limit (5.2) may be achieved even when ϵ does not approach zero. In order for $\|G(s)\|$ to be small, it is not necessary that the deformation (relative to the configuration at the present time t) be small at *all* past times $\tau<t$, but only that the deformation be small in the *recent* past. In particular, $\|G(s)\|$ is small for "slow" motions. To make this remark precise we consider a history $G(s)$ which has finite norm and corresponds to a deformation which makes no jump at the present, so that

$$\lim_{s\to 0} G(s)=0. \quad (5.3)$$

We then construct for each α, $0<\alpha\leq 1$, a "retarded" history

$$G_\alpha(s)=G(\alpha s). \quad (5.4)$$

It follows from Eq. (3.21) of reference 9 that

$$\lim_{\alpha\to 0}\|G_\alpha(s)\|=0, \quad (5.5)$$

i.e., that the limit (5.2) may be achieved by retardation of a given process.

Aside from the fact that the finite theory based on Eq. (5.1) applies to a much larger class of problems than the infinitesimal theory, there is a fundamental difference between the two theories. The infinitesimal theory is physically meaningless for finite deformations because it does not have the invariance properties required by the principle of material objectivity. The finite linear theory, on the other hand, enjoys the correct invariance. Thus, it is conceivable that there exists *some* material which obeys Eq. (5.1) for *arbitrary* finite deformations. The infinitesimal theory cannot possibly apply to *any* material when finite deformations are considered.

Finally, we remark that in the derivation of Eq. (5.1) no assumption has been made about the magnitude of the tensor parameter C. Hence, the finite theory based on Eq. (5.1) is applicable when the present and the reference configuration are related by an arbitrary large deformation.

Isotropic Materials

When dealing with isotropic materials it is convenient to take the reference configuration to be *undis-*

torted. (A precise definition of this term is given in reference 8.) Then the equilibrium stress is hydrostatic. Furthermore, the results in Sec. 22 of reference 8 show that the constitutive equation (5.1) reduces to

$$T=\mathfrak{h}(B)+\int_0^\infty \mathbf{\Gamma}(s;B)\{J(s)\}ds, \qquad (5.6)$$

where $B=B(t)$ is the left Cauchy-Green tensor, defined by Eq. (2.11), and the history $J(s)$ is given by

$$J(s)=C_t(t-s)-I. \qquad (5.7)$$

Furthermore, the tensor function $\mathfrak{h}$ and the linear functional given by the integral in Eq. (5.6) are *isotropic* in the sense that they obey the identities

$$Q\mathfrak{h}(B)Q^T=\mathfrak{h}(QBQ^T), \qquad (5.8)$$

$$Q\int_0^\infty \mathbf{\Gamma}(s;B)\{J(s)\}ds\, Q^T = \int_0^\infty \mathbf{\Gamma}(s;QBQ^T)\{QJ(s)Q^T\}ds \qquad (5.9)$$

for all orthogonal tensors Q. A fundamental theorem of the theory of isotropic tensor functions (for an elegant recent proof see reference 10, Sec. 59) states that $\mathfrak{h}$ has a representation

$$\mathfrak{h}(B)=h_0I+h_1B+h_2B^2, \qquad (5.10)$$

where h_0, h_1, and h_2 are scalar invariants of B. Also, it can be shown that the identity (5.9) implies the following representation for $\mathbf{\Gamma}$:

$$\Gamma(s;B)\{J(s)\}=\mathfrak{k}_1(s;B)J(s)+J(s)\mathfrak{k}_1(s;B) + \mathrm{Tr}[J(s)\mathfrak{k}_2(s;B)]I+\mathrm{Tr}[J(s)\mathfrak{k}_3(s;B)]B + \mathrm{Tr}[J(s)\mathfrak{k}_4(s;B)]B^2. \qquad (5.11)$$

Here, for each s, the tensor functions $\mathfrak{k}_i(s;B)$ are isotropic in the sense of Eq. (5.8) and hence have representations of the form (5.10). The proof of this result is too technical to be included here. Equations (5.10) and (5.11) and the representations for the $\mathfrak{k}_i$ may be used to render the constitutive equation (5.6) explicit. The resulting formula shows that, in the finite theory of linear viscoelasticity, the behavior of an isotropic material is determined by 11 independent scalar material functions; three of these depend on three variables and the remaining eight on four variables. The assumption of isotropy alone yields no further simplification. The special case $\mathbf{\Gamma}\equiv 0$ of Eq. (5.6) corresponds to the theory of finite (nonlinear) isotropic elasticity.

Fluids

We now consider materials which not only obey a constitutive equation of the form (5.1) but which are also *simple fluids* in the sense of the definition given in reference 8. (Coleman and Noll[11] give a summary of the general theory of simple fluids with emphasis on physical applications.) Such materials are isotropic, and hence Eqs. (5.6)–(5.9) apply. Moreover, the functions $\mathfrak{h}(B)$ and $\mathbf{\Gamma}(s;B)$ in Eq. (5.6) depend on B only through the determinant of B or, equivalently, only through the present density $\rho=\rho(t)$. Thus, for a fluid, Eq. (5.6) becomes

$$T=\mathfrak{h}(\rho)+\int_0^\infty \mathbf{\Gamma}(s;\rho)\{J(s)\}ds. \qquad (5.12)$$

The isotropy identities (5.8) and (5.9) may be written in the form

$$Q\mathfrak{h}(\rho)Q^T=\mathfrak{h}(\rho), \qquad (5.13)$$

$$\int_0^\infty Q[\mathbf{\Gamma}(s;\rho)\{J(s)\}]Q^T - \mathbf{\Gamma}(s;\rho)\{QJ(s)Q^T\}ds=0. \qquad (5.14)$$

Since Eq. (5.13) is valid for all orthogonal tensors Q, it follows that $\mathfrak{h}(\rho)$ must reduce to a scalar multiple of the unit tensor:

$$\mathfrak{h}(\rho)=-p(\rho)I. \qquad (5.15)$$

We call $p(\rho)$ the ***equilibrium pressure***; it is the pressure the fluid would be supporting if it had remained at rest in its present configuration at all times in the past.

Equation (5.14) is valid for all orthogonal Q and for all possible histories $J(s)$ belonging to the Hilbert space $\mathfrak{H}$. The only element of a Hilbert space which is orthogonal to all elements of the space is the zero element. This fact implies that the integrand in Eq. (5.14) must be identically zero. Hence, the transformation $\mathbf{\Gamma}(s;\rho)\{\ \}$ satisfies the identity

$$Q[\Gamma(s;\rho)\{J\}]Q^T=\Gamma(s;\rho)\{QJQ^T\}. \qquad (5.16)$$

for all orthogonal tensors Q and all symmetric tensors J. In other words, for each s and ρ, $\mathbf{\Gamma}(\rho;s)\{\ \}$ is an *isotropic* linear transformation of the space of symmetric tensors. The representation theorem for such isotropic transformations [special case of the theorem embodied in Eq. (5.10) (see reference 10, Sec. 59)] states that $\mathbf{\Gamma}(s;\rho)\{J(s)\}$ must be of the form

$$\mathbf{\Gamma}(s;\rho)\{J(s)\}=\mu(s;\rho)J(s)+\lambda(s;\rho)(\mathrm{Tr}\,J(s))I, \qquad (5.17)$$

where $\mu(s;\rho)$ and $\lambda(s;\rho)$ are scalar functions of the time lapse s and the present density ρ. On substituting Eqs. (5.15) and (5.17) into Eq. (5.12), we obtain the following *constitutive equation of a simple fluid in the theory of finite linear viscoelasticity*:

$$T=-p(\rho)I+\int \mu(s;\rho)J(s)ds + \left[\int_0^\infty \lambda(s;\rho)\,\mathrm{Tr}\,J(s)ds\right]I. \qquad (5.18)$$

[10] J. Serrin, "Mathematical principles of classical fluid mechanics," in *Encyclopedia of Physics*, edited by S. Flügge (Springer-Verlag, Berlin, 1959), Vol. VIII/1.

[11] B. D. Coleman and W. Noll, Ann. N. Y. Acad. Sci. **89**, 672 (1961).

ı this theory, the mechanical behavior of a fluid is ·termined by the three scalar material functions $p(\rho)$, $s;\rho)$, and $\lambda(s;\rho)$.

If the fluid under consideration is *incompressible*, rtain modifications must be made in this analysis. In compressible materials, the motion determines the ress only up to a hydrostatic pressure. In other words, e constitutive equation gives only the *extra stress*

$$T_e = T + pI, \tag{5.19}$$

ıere p is an indeterminate pressure. In the incomessible case, the two terms in Eq. (5.18) which are alar multiples of the unit tensor I may be absorbed to the indeterminate pressure term pI. From these marks we see that in finite linear viscoelasticity the ress in an incompressible fluid is given by the remark-ly simple equation

$$T_e = T + pI = \int_0^\infty \mu(s) J(s) ds, \tag{5.20}$$

ıere, since the density is constant, $\mu(s)$ is a function of ly the time lapse s.

The "relaxation function" $\phi(s)$ determined by eologists from measurements of the decay of shearing ıctions for simple (infinitesimal) shear in incomessible fluids is related to the material function $\mu(s)$ follows:

$$\phi(s) = -2\int_s^\infty \mu(\sigma) d\sigma, \quad \mu(s) = \tfrac{1}{2}(d/ds)\phi(s). \tag{5.21}$$

ıus, the relaxation function $\phi(s)$ is sufficient to termine the mechanical behavior of incompressible ids in the theory of finite linear viscoelasticity.

For simple fluids, the property (3.13) is equivalent the conditions

$$\int_0^\infty |\mu(s;\rho)|^2 h(s)^{-2} ds < \infty,$$
$$\int_0^\infty |\lambda(s;\rho)|^2 h(s)^{-2} ds < \infty. \tag{5.22}$$

ese conditions relate the rate of decay of the influence ıction to the rate of decay of the material functions $;\rho)$ and $\lambda(s;\rho)$ as $s \to \infty$.

6. SECOND-ORDER VISCOELASTICITY

[n Sec. 3 we showed, on the basis of our assumption), that the (nonlinear) functional $\mathfrak{F}$ giving the stress a simple material may be approximated by a *linear* ıctional. The error in this approximation approaches o faster than the Hilbert-space norm $\|G(s)\|$ of the tory (3.15). The analysis of Sec. 3 may be generalized he assumption (F) is replaced by a stronger assumption which requires that the functional $\mathfrak{F}$ be not just once but n times Fréchet differentiable at the zero history. It is then possible to approximate $\mathfrak{F}$ by a *polynomial* functional of degree n with an error that approaches zero faster than the nth power of the norm $\|G(s)\|$. For example, when $n=2$, we find the following generalization of Eq. (3.14):

$$\bar{T} = \mathfrak{h}(C) + \int_0^\infty \Gamma(s;C)\{G(s)\} ds + \underset{s=0}{\overset{\infty}{\mathfrak{Q}}}(G(s);C) + \underset{s=0}{\overset{\infty}{\mathfrak{R}'}}(G(s);C). \tag{6.1}$$

Here, $\mathfrak{Q}$ is a continuous *quadratic* functional depending on the tensor parameter C; the remainder $\mathfrak{R}'$ is of order $o(\|G(s)\|^2)$, i.e.,

$$\lim_{\|G(s)\| \to 0} \|G(s)\|^{-2} \mathfrak{R}'(G(s);C) = 0. \tag{6.2}$$

Relation (6.1) shows that the equation

$$\bar{T} = \mathfrak{h}(C) + \int_0^\infty \Gamma(s;C)\{G(s)\} ds + \underset{s=0}{\overset{\infty}{\mathfrak{Q}}}(G(s);C) \tag{6.3}$$

approximates the general constitutive equation of a simple material in the limit $\|G(s)\| \to 0$, and the error approaches zero faster than $\|G(s)\|^2$. We call the theory based on Eq. (6.3) *second-order viscoelasticity*.

The quadratic functional $\mathfrak{Q}$ of Eq. (6.3) may be expressed in terms of a bounded symmetric operator on the Hilbert space of histories. It is not possible, in general, to represent $\mathfrak{Q}$ by integrals. However, an integral representation does exist if the operator corresponding to $\mathfrak{Q}$ is completely continuous. We consider only this special case.

Explicit forms of the constitutive equations for isotropic materials and for simple fluids in second-order viscoelasticity may be obtained by an analysis similar to the one given in Sec. 5 in finite linear viscoelasticity. The resulting formulas are too complicated to be included here in full. Without giving the details of the derivation, we state the *constitutive equation of an incompressible fluid in the second-order theory of viscoelasticity*:

$$T + pI = \int_0^\infty \mu(s)\, J(s)\, ds + \int_0^\infty \int_0^\infty [\alpha(s_1, s_2)\, J(s_1)\, J(s_2) + \beta(s_1, s_2)\{\mathrm{Tr}\, J(s_1)\}\, J(s_2)]\, ds_1\, ds_2. \tag{6.4}$$

Here, p is an indeterminate pressure, $J(s)$ is the history given by Eq. (5.7), and $\mu(s)$, $\alpha(s_1,s_2)$ and $\beta(s_1,s_2)$ are scalar material functions. The function $\mu(s)$ is the same as in Eq. (5.23). The functions α and β are symmetric, i.e.,

$$\alpha(s_1,s_2)=\alpha(s_2,s_1), \quad \beta(s_1,s_2)=\beta(s_2,s_1). \tag{6.5}$$

In order to illustrate the behavior predicted by Eq. (6.4), we consider a class of motions called *simple shearing motions*. These motions are defined by the property that the velocity field $\mathbf{v}(\mathbf{x})=\{v_x,v_y,v_z\}$, in some Cartesian coordinate system x, y, z, has the components

$$v_x=0, \quad v_y=v(x,t), \quad v_z=0. \tag{6.6}$$

It follows from Eqs. (5.6), (5.8), and (5.10) of reference 11 that the matrix function corresponding to the history $J(s)$ defined by Eq. (5.7) has the form

$$[J(s)]=\lambda_t(s)\begin{Vmatrix}0&1&0\\1&0&0\\0&0&0\end{Vmatrix}+\lambda_t(s)^2\begin{Vmatrix}1&0&0\\0&0&0\\0&0&0\end{Vmatrix}, \tag{6.7}$$

where

$$\lambda_t(s)=\int_0^s \frac{d}{dx}v(x,t-\sigma)d\sigma. \tag{6.8}$$

In order to obtain the components T_{xx}, T_{xy}, etc., of the stress tensor T, we substitute Eq. (6.7) into Eq. (6.4). After a simple calculation, we find

$$T_{xy}=\int_0^\infty \mu(s)\lambda_t(s)ds+\int_0^\infty\int_0^\infty \gamma(s_1,s_2) \times\lambda_t(s_1)^2\lambda_t(s_2)ds_1ds_2, \tag{6.9}$$

$$T_{xx}-T_{yy}=\int_0^\infty \mu(s)\lambda_t^2(s)ds+\int_0^\infty\int_0^\infty \gamma(s_1,s_2) \times\lambda_t(s_1)^2\lambda_t(s_2)^2ds_1ds_2, \tag{6.10}$$

$$T_{yy}-T_{zz}=\int_0^\infty\int_0^\infty \alpha(s_1,s_2)\lambda_t(s_1)\lambda_t(s_2)ds_1ds_2, \tag{6.11}$$

where

$$\gamma(s_1,s_2)=\alpha(s_1,s_2)+\beta(s_1,s_2). \tag{6.12}$$

Equations (6.9)–(6.11), together with Cauchy's equations of motion, lead to a rather complicated system of integro-differential equations.

We now consider the special case when

$$\epsilon=\sup_{s>0}|\lambda_t(s)| \tag{6.13}$$

is small. Physically, this case corresponds to shearing motions with the property that the configuration of the fluid at *all* past times differs from the present configuration only by a small deformation. Shearing vibrations of small amplitude have this property. It is clear from Eqs. (6.13) and (6.7) that the Hilbert space norm $\|J(s)\|$ is of order $O(\epsilon^2)$. But the terms involving double integrals in Eqs. (6.9) and (6.10) are of order $O(\epsilon^3)$ and $O(\epsilon^4)$, respectively. Therefore, for small ϵ, Eqs. (6.9) and (6.10) reduce to

$$T_{xy}=\int_0^\infty \mu(s)\lambda_t(s)ds, \tag{6.14}$$

$$T_{xx}-T_{yy}=\int_0^\infty \mu(s)\lambda_t^2(s)ds. \tag{6.15}$$

Equation (6.14) for the shearing stress T_{xy} is the same as the corresponding equation in the theory of infinitesimal viscoelasticity. The normal stress differences given by Eqs. (6.15) and (6.11), zero in the infinitesimal theory, do not vanish in the second-order theory. Equations (6.11) and (6.15) may be used, for example, for the interpretation of data on normal stresses obtained in experiments involving shearing vibrations of small amplitude. It is remarkable that the normal stress difference (6.15) depends only on the material function $\mu(s)$ or, equivalently, the shear relaxation function $\phi(s)$ given by Eq. (5.24).

These results on simple shearing motions can easily be generalized to motions that have a form similar to (6.6) in an appropriate curvilinear orthogonal coordinate system. (The method to be employed is analogous to the one used in Sec. 2 of reference 12.)

7. FINAL REMARKS

In our considerations in Secs. 3–6 we have used the relative right Cauchy-Green tensor C_t as a measure of strain. As we remarked at the end of Sec. 2, there is no unique "strain tensor" when finite deformations are considered. Instead of C_t we could also have used the relative right stretch tensor $U_t=(C_t)^{\frac{1}{2}}$, the inverse C_t^{-1}, $\log C_t$, or any other tensor related to C_t by a smooth one-to-one transformation. To different choices of the measure of strain correspond different theories of finite linear viscoelasticity. However, the difference of the stresses computed using two different such theories is of order $o(\|G(s)\|)$. Hence, since any finite linear theory can be expected to be accurate only when terms of order $o(\|G(s)\|)$ can be neglected, we can say that the various theories corresponding to the various measures of strain are equivalent.

To different choices of the measure of strain also correspond different theories of second-order viscoelasticity. These different theories are equivalent in the sense that the corresponding stresses differ only by terms of order $o(\|G(s)\|^2)$.

On the basis of a molecular model for certain incompressible fluids, Lodge[13] has derived a constitutive equation corresponding to Eq. (5.23) when $J(s)$ is

[12] B. D. Coleman and W. Noll, Arch. Ratl. Mech. Anal. 4, 289 (1959).
[13] A. S. Lodge, Trans. Faraday Soc. 52, 120 (1956).

nputed using C_t^{-1} rather than C_t as a measure of ain. Our analysis shows that any other molecular odel must give the same or an equivalent result, ovided only that terms of order $o(\|G(s)\|)$ may be glected.

As we have remarked in Sec. 5, the norm $\|G(s)\|$ is all in particular for "slow" motions, and hence the ite linear theory applies in this case. For slow flows simple fluids, the finite linear theory is actually uivalent to the classical theory of Newtonian fluids, ovided that the influence function h satisfies the ation (6.3) with $r>\frac{3}{2}$. This fact and analogous results fluids of higher order are proved in reference 9.

Rivlin and his co-workers[14] in recent years have developed memory theories involving multiple integrals similar to the second-order theory proposed in Sec. 6. The emphasis in their work has been on the representation theorems following from material objectivity and symmetry. In particular, the representations mentioned here in Secs. 5 and 6 can be derived using their results. An investigation of higher order theories of viscoelasticity based on the existence and complete continuity of Fréchet differentials of order >2 would make much more use of such representation theorems.

[14] A. J. M. Spencer and R. S. Rivlin, Arch. Ratl. Mech. Anal. 4, 214 (1960).

Those items in the **Erratum** published in
REVIEW OF MODERN PHYSICS, Vol. 36, No. 4, 1103, October 1964,
which have not been inserted in the text above

Page 239. In the second line of the second paragraph, place of "viscosity" read "viscoelasticity."

Page 239. Reference 1 should read:

[1] L. Boltzmann, Sitzber. Kaiserlich, Akad. Wiss. Vien), Math.-Naturwiss. Kl. **70**, Sect. II, 275–306 874).

Page 246. The sentence starting seven lines below q. (5.11) and reading "The resulting formula shows at . . ." should be deleted and replaced by:

The resulting formula shows that in finite linear viscoelasticity the behavior of an isotropic material is determined by 15 independent scalar-valued material functions.

Pages 247 and 248. The sentence containing Eq. (6.5) should be replaced by:

The function α is uniquely determined if and only if it is chosen to be symmetric, i.e.,

$$\alpha(s_1, s_2)=\alpha(s_2, s_1); \tag{6.5}$$

the function β need not be symmetric.

Motions with Constant Stretch History

WALTER NOLL

Introduction

The class of motions considered in the present paper has recently been introduced by COLEMAN* [1], who noticed that it includes almost all steady flows that have ever been studied in connection with non-classical fluids. Examples are: Simple shearing flow, laminar flow through pipes (circular or not), Couette flow, helical flow, torsional flow, and circular flow between coaxial cones or concentric spheres. All these examples are actually special cases of what is called here steady curvilineal flow (see Section 3). The curvilineal flows belong to a class of motions which have been called viscometric flows (see Section 3 of the present paper or Section 3 of [1]).

The main result of this paper is contained in Section 2, where a representation theorem is proved which characterizes all motions with constant stretch history. It turns out that the nature of such a motion is governed by one constant tensor $\boldsymbol{M}$.

Section 3 deals mainly with conditions which insure that $\boldsymbol{M}$ is nilpotent. In particular, the viscometric flows are shown to correspond to the case when $\boldsymbol{M}^2=\boldsymbol{0}$. For curvilineal flows it is shown how $\boldsymbol{M}$ can be computed explicitly. Some examples of motions with constant stretch history that do not reduce to viscometric flows are mentioned.

In Section 4 the representation theorem is used to obtain information about the behavior of incompressible simple fluids[1] in motions with constant stretch history. It is shown that this behavior is described by an isotropic tensor function of $\boldsymbol{M}$. This description leads, in particular, to a very efficient general treatment of viscometric flows of incompressible simple fluids.

The kinematical results of COLEMAN ([1], Section 6) are corollaries of the Theorem 3 presented here.[2]

1. Preliminaries

We collect for easy reference some definitions and results from the kinematics of continuous media. For more details we refer to [4] or [2].

Consider a continuous medium in smooth motion. Let $\boldsymbol{\xi}$ be the location at time τ of that material point which, at time t, is located at $\boldsymbol{x}$. Denote the

* He used the term "substantially stagnant motion".

[1] The concept of a simple fluid was introduced by the author in [2]. For details on the theory of simple fluids see [3] or [1] and the literature quoted there.

[2] The result derived here shows, moreover, that the condition (β) of COLEMAN'S Theorems 1 or 3 may be dropped and the condition (γ) may be weakened by replacing $\boldsymbol{A}_3$ by $\boldsymbol{A}_4$.

dependence of $\boldsymbol{\xi}$ on t, $\boldsymbol{x}$, and τ by

$$\boldsymbol{\xi} = \boldsymbol{\chi}_t(\boldsymbol{x}, \tau). \tag{1.1}$$

The gradient

$$\boldsymbol{F}_t(\tau) = \operatorname{grad}_{\boldsymbol{x}} \boldsymbol{\chi}_t(\boldsymbol{x}, \tau) \tag{1.2}$$

of $\boldsymbol{\xi}$ with respect to $\boldsymbol{x}$ is called the *relative deformation gradient*. It depends, of course, not only on t and τ but also on the material point. We shall fix our attention on only one particular material point throughout.

In addition to the variable times t and τ, we consider also a fixed time, taken to be 0. It is often convenient to think of t as the variable present time and of τ as a variable past time. The various values of the relative deformation gradient are related by

$$\boldsymbol{F}_0(\tau) = \boldsymbol{F}_t(\tau)\, \boldsymbol{F}_0(t), \qquad \boldsymbol{F}_t(t) = \boldsymbol{I}, \tag{1.3}$$

where $\boldsymbol{I}$ denotes the unit tensor.[3]

The *relative right Cauchy-Green tensor* is defined by

$$\boldsymbol{C}_t(\tau) = \boldsymbol{F}_t^T(\tau)\, \boldsymbol{F}_t(\tau), \tag{1.4}$$

where the superscript T denotes transposition. The tensor function $\boldsymbol{C}_t^t(s)$, $s \geqq 0$, defined by

$$\boldsymbol{C}_t^t(s) = \boldsymbol{C}_t(t-s) = \boldsymbol{F}_t^T(t-s)\, \boldsymbol{F}_t(t-s) \tag{1.5}$$

is a measure of the deformation history taking t as the present time. The proper numbers of $\boldsymbol{C}_t^t(s)$ are the squares of the principal stretches of the deformation carrying the present configuration (time t) into the past configuration s time units ago. The proper vectors of $\boldsymbol{C}_t^t(s)$ determine the directions of these principal stretches.

The following relations follow from (1.3) and (1.4):

$$\boldsymbol{C}_0(\tau) = \boldsymbol{F}_0^T(t)\, \boldsymbol{C}_t(\tau)\, \boldsymbol{F}_0(t), \qquad \boldsymbol{C}_t(t) = \boldsymbol{C}_t^t(0) = \boldsymbol{I}. \tag{1.6}$$

Differentiations, denoted by superimposed dots, are assumed not to affect parameters indicated by indices. For example:

$$\dot{\boldsymbol{F}}_t(\tau) = \frac{d}{d\tau} \boldsymbol{F}_t(\tau), \qquad \dot{\boldsymbol{F}}_t(t) = \frac{d}{d\tau} \boldsymbol{F}_t(\tau)\Big|_{\tau=t}. \tag{1.7}$$

The *velocity gradient tensor* $\boldsymbol{L}(t)$ is given by

$$\boldsymbol{L}(t) = \dot{\boldsymbol{F}}_t(t) = \boldsymbol{D}(t) + \boldsymbol{W}(t); \tag{1.8}$$

its symmetric part $\boldsymbol{D}(t) = \frac{1}{2}\left(\boldsymbol{L}(t) + \boldsymbol{L}^T(t)\right)$ is called the *stretching tensor*, and its skew part $\boldsymbol{W}(t)$ is called the *spin tensor*. The nth *Rivlin-Ericksen tensor* $\boldsymbol{A}_n(t)$ is defined by

$$\boldsymbol{A}_n(t) = \overset{(n)}{\boldsymbol{C}}_t(t) = (-1)^n \overset{(n)}{\boldsymbol{C}_t^t}(0), \qquad \boldsymbol{A}_1(t) = 2\boldsymbol{D}(t). \tag{1.9}$$

If $\boldsymbol{A}_n(t) = \boldsymbol{0}$ for all t, then $\boldsymbol{C}_t^t(s)$ is a polynomial of degree $< n$ in s.

The following results on exponentials of tensors will be needed. The exponential $e^{\boldsymbol{M}}$ of an arbitrary tensor $\boldsymbol{M}$ is defined by the convergent power series

$$e^{\boldsymbol{M}} = \sum_{n=0}^{\infty} \frac{1}{n!} \boldsymbol{M}^n. \tag{1.10}$$

[3] The term "tensor" is used as a synonym for linear transformation.

The following rules are valid:

$$(e^{\boldsymbol{M}})^T = e^{\boldsymbol{M}^T}, \qquad (e^{\boldsymbol{M}})^{-1} = e^{-\boldsymbol{M}}, \qquad e^{\boldsymbol{0}} = \boldsymbol{I}, \tag{1.11}$$

$$e^{\boldsymbol{M}+\boldsymbol{N}} = e^{\boldsymbol{M}} e^{\boldsymbol{N}} \quad \text{if} \quad \boldsymbol{M}\boldsymbol{N} = \boldsymbol{N}\boldsymbol{M}, \text{[4]} \tag{1.12}$$

$$\boldsymbol{Q} e^{\boldsymbol{M}} \boldsymbol{Q}^T = e^{\boldsymbol{Q}\boldsymbol{M}\boldsymbol{Q}^T} \quad \text{if} \quad \boldsymbol{Q} \text{ is orthogonal,} \tag{1.13}$$

$$\frac{d}{d\tau} e^{\tau \boldsymbol{M}} = \boldsymbol{M} e^{\tau \boldsymbol{M}} = e^{\tau \boldsymbol{M}} \boldsymbol{M}. \tag{1.14}$$

A tensor $\boldsymbol{M}$ is called *nilpotent* if $\boldsymbol{M}^n = \boldsymbol{0}$ for some integer $n \geqq 0$. In three dimensions, if $\boldsymbol{M}$ is nilpotent, then $\boldsymbol{M}^3 = \boldsymbol{0}$. Moreover, the matrix $[\boldsymbol{M}]$ of $\boldsymbol{M}$, relative to an appropriate orthonormal basis, has the form

$$[\boldsymbol{M}] = \begin{Vmatrix} 0 & 0 & 0 \\ \varkappa & 0 & 0 \\ \lambda & \nu & 0 \end{Vmatrix}. \tag{1.15}$$

If $\boldsymbol{M}^2 = \boldsymbol{0}$, then the basis may be chosen such that

$$[\boldsymbol{M}] = \varkappa \begin{Vmatrix} 0 & 0 & 0 \\ 1 & 0 & 0 \\ 0 & 0 & 0 \end{Vmatrix}. \tag{1.16}$$

The exponential $e^{\tau \boldsymbol{M}}$ is a polynomial in τ if and only if $\boldsymbol{M}$ is nilpotent. In this case,

$$e^{\tau \boldsymbol{M}} = \boldsymbol{I} + \tau \boldsymbol{M} + \frac{1}{2} \tau^2 \boldsymbol{M}^2. \tag{1.17}$$

2. The representation theorem

Definition. We say that a motion, along the path of a material point, has *constant stretch history* if there is an orthogonal tensor function $\boldsymbol{P}(t)$ such that the histories $\boldsymbol{C}_t^t(s)$ and $\boldsymbol{C}_0^0(s)$, for $s \geqq 0$, are related by

$$\boldsymbol{C}_t^t(s) = \boldsymbol{P}^T(t)\, \boldsymbol{C}_0^0(s)\, \boldsymbol{P}(t), \qquad s \geqq 0. \tag{2.1}$$

The condition (2.1) means physically that for an observer moving with the material point both the magnitudes of the principal stretches and the changes of direction of the principal axes of strain depend only on the time lapse s and not on the present time t.

Theorem 1. *A motion has constant stretch history if and only if the deformation gradient $\boldsymbol{F}_0(\tau)$ relative to some fixed time 0 has the representation*

$$\boldsymbol{F}_0(\tau) = \boldsymbol{Q}(\tau)\, e^{\tau \boldsymbol{M}}, \qquad \boldsymbol{Q}(0) = \boldsymbol{I}, \tag{2.2}$$

where $\boldsymbol{M}$ is a constant tensor and $\boldsymbol{Q}(\tau)$ is an orthogonal tensor function.

Before proving this theorem, we give a list of formulas which follow directly from (2.2) and the results given in the previous section. We use the abbreviations

$$\boldsymbol{M}_t = \boldsymbol{Q}(t)\, \boldsymbol{M}\, \boldsymbol{Q}^T(t), \qquad \boldsymbol{Z}(t) = \dot{\boldsymbol{Q}}(t)\, \boldsymbol{Q}^T(t). \tag{2.3}$$

Clearly, $\boldsymbol{Z}(t)$ is skew, *i.e.*, $\boldsymbol{Z}(t)^T = -\boldsymbol{Z}(t)$.

---

[4] Note that $e^{\boldsymbol{M}+\boldsymbol{N}} \neq e^{\boldsymbol{M}} e^{\boldsymbol{N}}$ if $\boldsymbol{M}$ and $\boldsymbol{N}$ do not commute.

(a) Relative deformation gradient $\boldsymbol{F}_t(\tau)$:

$$\boldsymbol{F}_t(\tau)=\boldsymbol{Q}(\tau)\, e^{(\tau-t)\boldsymbol{M}}\boldsymbol{Q}^T(t). \tag{2.4}$$

(b) Relative right Cauchy-Green tensor $\boldsymbol{C}_t^t(s)=\boldsymbol{C}_t(t-s)$:

$$\boldsymbol{C}_t^t(s)=e^{-s\boldsymbol{M}_t^T}e^{-s\boldsymbol{M}_t}=\boldsymbol{Q}(t)\, e^{-s\boldsymbol{M}^T}e^{-s\boldsymbol{M}}\boldsymbol{Q}^T(t). \tag{2.5}$$

(c) Velocity gradient $\boldsymbol{L}(t)$:

$$\boldsymbol{L}(t)=\boldsymbol{M}_t+\boldsymbol{Z}(t), \tag{2.6}$$

$$\dot{\boldsymbol{L}}(t)=\dot{\boldsymbol{Z}}(t)+\boldsymbol{Z}(t)\boldsymbol{L}(t)-\boldsymbol{L}(t)\boldsymbol{Z}(t). \tag{2.7}$$

(d) Stretching $\boldsymbol{D}(t)$ and spin $\boldsymbol{W}(t)$:

$$\boldsymbol{D}(t)=\frac{1}{2}(\boldsymbol{M}_t+\boldsymbol{M}_t^T), \qquad \boldsymbol{W}(t)=\frac{1}{2}(\boldsymbol{M}_t-\boldsymbol{M}_t^T)+\boldsymbol{Z}(t). \tag{2.8}$$

(e) Rivlin-Ericksen tensors $\boldsymbol{A}_n(t)$:

$$\boldsymbol{A}_n(t)=\sum_{k=0}^{n}\binom{n}{k}(\boldsymbol{M}_t^k)^T\boldsymbol{M}_t^{n-k}=\boldsymbol{Q}(t)\boldsymbol{A}_n(0)\boldsymbol{Q}^T(t). \tag{2.9}$$

It is often useful to consider an orthonormal basis $\boldsymbol{b}_1(t)$, $\boldsymbol{b}_2(t)$, $\boldsymbol{b}_3(t)$ which is attached to the material point and which rotates, as the point moves, according to the law

$$\boldsymbol{b}_k(t)=\boldsymbol{Q}(t)\boldsymbol{b}_k(0). \tag{2.10}$$

The matrix $[\boldsymbol{M}]$ of $\boldsymbol{M}_t$, relative to this rotating basis, is then independent of t. It follows from (2.5), $(2.8)_1$, and (2.9), that the matrices of $\boldsymbol{C}_t^t(s)$, $\boldsymbol{D}(t)$ and $\boldsymbol{A}_n(t)$ are also independent of t. If we define the matrix $[\boldsymbol{F}_t(\tau)]$ of $\boldsymbol{F}_t(\tau)$ by

$$\|\boldsymbol{b}_k(\tau)\cdot\boldsymbol{F}_t(\tau)\boldsymbol{b}_m(t)\|=[\boldsymbol{F}_t(\tau)], \tag{2.11}$$

then (2.4) is equivalent to

$$[\boldsymbol{F}_t(t-s)]=e^{-s[\boldsymbol{M}]}. \tag{2.12}$$

Proof of Theorem 1. Assume that the motion has constant stretch history. The function $\boldsymbol{P}(t)$ is not uniquely determined by (2.1), but in view of the assumed smoothness of the motion $\boldsymbol{P}(t)$ may be chosen to be a smooth function of t which satisfies $\boldsymbol{P}(0)=\boldsymbol{I}$. We abbreviate

$$\boldsymbol{E}(t)=\boldsymbol{P}(t)\boldsymbol{F}_0(t), \qquad \boldsymbol{H}(\tau)=\boldsymbol{C}_0(-\tau)=\boldsymbol{C}_0^0(\tau). \tag{2.13}$$

Note that $\boldsymbol{E}(0)=\boldsymbol{I}$, $\boldsymbol{H}(0)=\boldsymbol{I}$. We infer from (2.1), (1.6), and (2.13) that

$$\boldsymbol{H}(s-t)=\boldsymbol{C}_0(t-s)=\boldsymbol{E}^T(t)\boldsymbol{H}(s)\boldsymbol{E}(t), \qquad s\geqq 0. \tag{2.14}$$

We differentiate (2.14) with respect to t and then put $t=0$, obtaining

$$-\dot{\boldsymbol{H}}(s)=\boldsymbol{M}^T\boldsymbol{H}(s)+\boldsymbol{H}(s)\boldsymbol{M}, \qquad s\geqq 0, \tag{2.15}$$

where $\boldsymbol{M}=\dot{\boldsymbol{E}}(0)$. Now, (2.15) is a differential equation for $\boldsymbol{H}(s)$ which must have a unique solution satisfying the initial condition $\boldsymbol{H}(0)=\boldsymbol{I}$. This solution is given by

$$\boldsymbol{H}(s)=e^{-s\boldsymbol{M}^T}e^{-s\boldsymbol{M}}, \qquad s\geqq 0, \tag{2.16}$$

as is easily verified with the help of (1.14). Substituting (2.16) back into (2.14) and putting $s-t=\tau$ yields

$$\boldsymbol{H}(\tau)=\boldsymbol{E}^T(t)\,e^{-t\boldsymbol{M}^T}e^{-\tau\boldsymbol{M}^T}e^{-t\boldsymbol{M}}e^{-\tau\boldsymbol{M}}\boldsymbol{E}(t), \qquad \tau\geqq -t, \tag{2.17}$$

which shows that $\boldsymbol{H}(\tau)$ depends analytically on τ for $\tau\geqq -t$. Since t is arbitrary, it follows that $\boldsymbol{H}(\tau)$ is an analytic function of τ for all values of τ. From (2.16) and the principle of analytic continuation it follows that

$$\boldsymbol{H}(-\tau)=e^{\tau\boldsymbol{M}^T}e^{\tau\boldsymbol{M}}=\boldsymbol{C}_0(\tau)=\boldsymbol{F}_0^T(\tau)\,\boldsymbol{F}_0(\tau) \tag{2.18}$$

holds for all τ. (2.18) is equivalent to

$$(\boldsymbol{F}_0(\tau)\,e^{-\tau\boldsymbol{M}})^T(\boldsymbol{F}_0(\tau)\,e^{-\tau\boldsymbol{M}})=\boldsymbol{I}, \tag{2.19}$$

which shows that $\boldsymbol{Q}(\tau)=\boldsymbol{F}_0(\tau)\,e^{-\tau\boldsymbol{M}}$ is orthogonal and hence that (2.2) holds.

If, conversely, (2.2) is valid, then (2.1) follows directly from (2.5), with the choice $\boldsymbol{P}(t)=\boldsymbol{Q}(t)^T$. Q.E.D.

Remark. In certain degenerate cases the tensor $\boldsymbol{M}$ of (2.2) is not uniquely determined by the motion. Consider, for example, the case when $\boldsymbol{H}(s)=\boldsymbol{C}_0^0(s)\equiv\boldsymbol{I}$ for all s, which corresponds to a motion that is locally rigid along the path line under consideration. In this case, $\boldsymbol{F}_0(\tau)$ is itself orthogonal, and hence (2.2) holds when $\boldsymbol{M}$ is an arbitrary skew tensor, for $e^{\tau\boldsymbol{M}}$ is orthogonal when $\boldsymbol{M}$ is skew.

If, for some value of τ, the proper numbers of $\boldsymbol{H}(\tau)=\boldsymbol{C}_0^0(\tau)$ are distinct, then $\boldsymbol{M}$ is uniquely determined by the motion, as may be seen from the following argument. If

$$\boldsymbol{H}(\tau)=e^{-\tau\boldsymbol{M}_1^T}e^{-\tau\boldsymbol{M}_1}=e^{-\tau\boldsymbol{M}_2^T}e^{-\tau\boldsymbol{M}_2},$$

then $\boldsymbol{H}(\tau)$ satisfies

$$-\dot{\boldsymbol{H}}(\tau)=\boldsymbol{M}_i^T\boldsymbol{H}(\tau)+\boldsymbol{H}(\tau)\,\boldsymbol{M}_i, \qquad i=1,2. \tag{2.20}$$

Taking the difference of the two equations (2.20) gives

$$\boldsymbol{A}^T\boldsymbol{H}(\tau)+\boldsymbol{H}(\tau)\,\boldsymbol{A}=\boldsymbol{0}, \qquad \boldsymbol{A}=\boldsymbol{M}_1-\boldsymbol{M}_2. \tag{2.21}$$

For $\tau=0$ we obtain $\boldsymbol{A}^T+\boldsymbol{A}=\boldsymbol{0}$, which means that $\boldsymbol{A}$ must be skew. Fix the value of τ, and denote the proper numbers of $\boldsymbol{H}(\tau)$ by h_1, h_2, h_3. Let $\boldsymbol{f}_1, \boldsymbol{f}_2, \boldsymbol{f}_3$ be a corresponding orthonormal basis of proper vectors, and let A_{km} be the components of $\boldsymbol{A}$ relative to the basis $\boldsymbol{f}_k$. Equation (2.21) is then equivalent to

$$0=A_{mk}h_m+h_kA_{km}=A_{km}(h_k-h_m). \tag{2.22}$$

If the proper numbers h_k are distinct, it follows from (2.22) that $A_{km}=0$ for $k\neq m$. Since $\boldsymbol{A}$ is skew, we then must have $\boldsymbol{A}=\boldsymbol{M}_1-\boldsymbol{M}_2=\boldsymbol{0}$, *i.e.*, $\boldsymbol{M}_1=\boldsymbol{M}_2$.

3. Special cases

Theorem 2. *Consider a motion with constant stretch history. Then the following conditions are equivalent:*

(a) *One of the Rivlin-Ericksen tensors vanishes for all t.*

(b) *In the representation* (2.2) *the tensor* $\boldsymbol{M}$ *is nilpotent and hence satisfies* $\boldsymbol{M}^3=\boldsymbol{0}$.

(c) *The history* $\boldsymbol{C}_t^t(s)$ *has the form*

$$\boldsymbol{C}_t^t(s) = (\boldsymbol{I} - s\boldsymbol{M}_t^T + \tfrac{1}{2}s^2\boldsymbol{M}_t^{2\,T})(\boldsymbol{I} - s\boldsymbol{M}_t + \tfrac{1}{2}s^2\boldsymbol{M}_t^2), \tag{3.1}$$

where $\boldsymbol{M}_t = \boldsymbol{Q}(t)\boldsymbol{M}\boldsymbol{Q}(t)^T$.

(d) *The motion is isochoric, and* $\boldsymbol{A}_m(t) \equiv \boldsymbol{0}$ *for* $m \geqq 5$ *and all* t.

Proof. $\boldsymbol{A}_n(t) \equiv \boldsymbol{0}$ for some n implies that $\boldsymbol{C}_0^0(s)$ must be a polynomial in s. By (2.5), $\boldsymbol{C}_0^0(s)$ is the product of the two entire functions $e^{-s\boldsymbol{M}^T}$ and $e^{-s\boldsymbol{M}}$. It follows that $e^{-s\boldsymbol{M}}$ must be a polynomial in s. But this can be the case only when $\boldsymbol{M}$ is nilpotent. (3.1) follows then from (1.17) and (2.5). It is clear from (2.9) that $\boldsymbol{A}_m(t) \equiv \boldsymbol{0}$ for $m \geqq 5$. Since $\boldsymbol{M}$ is nilpotent, its trace is zero. Hence, by (2.8), $\operatorname{tr}\boldsymbol{D}(t) \equiv 0$, which shows that the motion must be isochoric. Q.E.D.

The following theorem is a corollary of Theorem 2:

Theorem 3. *Consider a motion with constant stretch history. Then the following conditions are equivalent:*

(a) $\boldsymbol{A}_4(t) = \boldsymbol{0}$ *for all* t.

(b) *The tensor* $\boldsymbol{M}$ *of the representation* (2.2) *satisfies* $\boldsymbol{M}^2 = \boldsymbol{0}$.

(c) *The history* $\boldsymbol{C}_t^t(s)$ *is of the form*

$$\boldsymbol{C}_t^t(s) = (\boldsymbol{I} - s\boldsymbol{M}_t^T)(\boldsymbol{I} - s\boldsymbol{M}_t) = \boldsymbol{I} - s(\boldsymbol{M}_t + \boldsymbol{M}_t^T) + s^2\boldsymbol{M}_t^T\boldsymbol{M}_t, \tag{3.2}$$

where $\boldsymbol{M}_t = \boldsymbol{Q}(t)\boldsymbol{M}\boldsymbol{Q}(t)^T$.

(d) *The motion is isochoric, and* $\boldsymbol{A}_m(t) \equiv \boldsymbol{0}$ *for* $n \geqq 3$ *and all* t.

The motions of the type described by Theorem 3 are called *locally viscometric motions*. A *viscometric flow* is a flow which is locally viscometric at *every* material point of the flowing medium.

Definition. We call *steady curvilineal flow* a motion whose velocity field $\boldsymbol{v}(\boldsymbol{x})$ has the contravariant components

$$v^1 = 0, \qquad v^2 = u(x^1), \qquad v^3 = w(x^1) \tag{3.3}$$

in an orthogonal curvilinear coordinate system x^k whose metric components g_{kk} are constant along the curves $\boldsymbol{\xi} = \boldsymbol{\xi}(s)$ defined by

$$\xi^1 = x^1, \qquad \xi^2 = x^2 + s\,u(x^1), \qquad \xi^3 = x^3 + s\,w(x^1). \tag{3.4}$$

The condition on the g_{kk} is satisfied in particular when the g_{kk} depend only on x^1 or when $w(x^1) \equiv 0$ and the g_{kk} depend only on x^1 and x^3.

Theorem 4. *Every steady curvilineal flow is viscometric. Let* $\tilde{\boldsymbol{e}}_k = \tilde{\boldsymbol{e}}_k(\boldsymbol{x})$ *be the orthonormal basis of the unit vectors in the direction of the coordinate lines, and let the orthonormal basis* $\boldsymbol{b}_k = \boldsymbol{b}_k(\boldsymbol{x})$ *be defined by*

$$\begin{aligned} \boldsymbol{b}_1 &= \tilde{\boldsymbol{e}}_1, \\ \boldsymbol{b}_2 &= \alpha\,\tilde{}_2 + \beta\tilde{\boldsymbol{e}}_3, \\ \boldsymbol{b}_3 &= -\beta\tilde{\boldsymbol{e}}_2 + \alpha\tilde{\boldsymbol{e}}_3, \end{aligned} \tag{3.5}$$

where

$$\alpha = \frac{u'}{\varkappa}\sqrt{\frac{g_{22}}{g_{11}}}, \qquad \beta = \frac{w'}{\varkappa}\sqrt{\frac{g_{33}}{g_{11}}}, \qquad \varkappa = \sqrt{\frac{u'^2 g_{22} + w'^2 g_{33}}{g_{11}}} \tag{3.6}$$

in which u' *and* w' *denote the derivatives of* $u(x^1)$ *and* $w(x^1)$ *with respect to* x^1.

The matrix $[\boldsymbol{M}]$ *of the tensor* $\boldsymbol{M}_t$ *of* (2.5) *relative to the basis* (3.5) *is then given by*

$$[\boldsymbol{M}]=\varkappa\begin{Vmatrix}0&0&0\\1&0&0\\0&0&0\end{Vmatrix}. \tag{3.7}$$

Proof. The deformation function (1.1) is obtained by solving the differential equations

$$\dot{\xi}^1=0,\qquad \dot{\xi}^2=u(\xi^1),\qquad \dot{\xi}^3=w(\xi^1) \tag{3.8}$$

with the initial conditions $\xi^k|_{\tau=t}=x^k$. The solution is given by

$$\xi^1=x^1,\qquad \xi^2=x^2+(\tau-t)\,u(x^1),\qquad \xi^3=x^3+(\tau-t)\,w(x^1). \tag{3.9}$$

Let $\boldsymbol{e}_k=\boldsymbol{e}_k(\boldsymbol{x})$ be the natural basis of the coordinate system used, and let $\boldsymbol{e}^k=\boldsymbol{e}^k(\boldsymbol{x})$ be the dual of $\boldsymbol{e}_k$ ($\boldsymbol{e}^k$ is characterized by $\boldsymbol{e}^k\cdot\boldsymbol{e}_m=\delta^k_m$). By (3.9) and (1.2) we then have

$$\|\boldsymbol{e}^k(\boldsymbol{\xi})\cdot\boldsymbol{F}_t(\tau)\,\boldsymbol{e}_m(\boldsymbol{x})\|=\left\|\frac{\partial\xi^k}{\partial x^m}\right\|=\begin{Vmatrix}1&0&0\\(\tau-t)\,u'&1&0\\(\tau-t)\,w'&0&1\end{Vmatrix}. \tag{3.10}$$

The unit vectors $\tilde{\boldsymbol{e}}_k$ in the direction of the coordinate lines are related to the $\boldsymbol{e}_k$ and $\boldsymbol{e}^k$ by

$$\boldsymbol{e}_k=\sqrt{g_{kk}}\,\tilde{\boldsymbol{e}}_k,\qquad \boldsymbol{e}^k=\sqrt{g_{kk}^{-1}}\,\tilde{\boldsymbol{e}}_k. \tag{3.11}$$

It follows from (3.9) and our hypothesis on the g_{kk} that the g_{kk} are constant along the path lines of the motion and hence that $g_{kk}(\boldsymbol{\xi})=g_{kk}(\boldsymbol{x})=g_{kk}$. Therefore, (3.10) can be rewritten, with use of (3.11), in the form

$$\|\tilde{\boldsymbol{e}}_k(\boldsymbol{\xi})\cdot\boldsymbol{F}_t(t-s)\,\tilde{\boldsymbol{e}}_m(\boldsymbol{x})\|=[\boldsymbol{I}]-s\begin{Vmatrix}0&0&0\\\sqrt{g_{22}\,g_{11}^{-1}}\,u'&0&0\\\sqrt{g_{33}\,g_{11}^{-1}}\,w'&0&0\end{Vmatrix} \tag{3.12}$$

where $[\boldsymbol{I}]$ is the unit matrix. Finally, the matrix $[\boldsymbol{F}_t(t-s)]$ in the sense of (2.11), for the basis defined by (3.5) and (3.6), is easily computed to be

$$[\boldsymbol{F}_t(t-s)]=[\boldsymbol{I}]-s\varkappa\begin{Vmatrix}0&0&0\\1&0&0\\0&0&0\end{Vmatrix}. \tag{3.13}$$

It follows that (2.12) is satisfied when $[\boldsymbol{M}]$ is given by (3.7). Since $\boldsymbol{M}^2=\boldsymbol{0}$, the motion is a viscometric flow. Q.E.D.

Consider a flow whose velocity field $\boldsymbol{v}(\boldsymbol{x})$ in Cartesian coordinates x, y, z has the components

$$u=0,\qquad v=\varkappa x,\qquad w=\lambda x+\nu y, \tag{3.14}$$

where $\varkappa$, λ, and ν are constants $\neq 0$. The components ξ,η,ζ of the deformation function (1.1) corresponding to (3.14) are easily computed:

$$\xi=x,\qquad \eta=(\tau-t)\,\varkappa x+y,\qquad \zeta=(\tau-t)\,(\lambda x+\nu y)+\tfrac{1}{2}(\tau-t)^2\nu\varkappa x+z. \tag{3.15}$$

The derivatives of ξ, η, ζ with respect to x, y, z are the components of the relative deformation gradient $\boldsymbol{F}_t(\tau)$. Thus, the matrix $[\boldsymbol{F}_t(\tau)]$ of $\boldsymbol{F}_t(\tau)$ is given by

$$[\boldsymbol{F}_t(\tau)] = [\boldsymbol{I}] + (\tau - t) \begin{Vmatrix} 0 & 0 & 0 \\ \varkappa & 0 & 0 \\ \lambda & \nu & 0 \end{Vmatrix} + \tfrac{1}{2}(\tau - t)^2 \begin{Vmatrix} 0 & 0 & 0 \\ 0 & 0 & 0 \\ \varkappa\nu & 0 & 0 \end{Vmatrix}, \tag{3.16}$$

which shows that $\boldsymbol{F}_t(\tau) = e^{(\tau - t)\boldsymbol{M}}$, where the matrix $[\boldsymbol{M}]$ of $\boldsymbol{M}$ is given by (1.15). We see that (3.14) describes a flow with constant stretch history such that $\boldsymbol{M}^3 = \boldsymbol{0}$ but $\boldsymbol{M}^2 \neq \boldsymbol{0}$. Hence, this flow is of the type described by Theorem 2 at all its material points, but it is not viscometric.

An example of a motion which corresponds to the case when the tensor $\boldsymbol{M}$ in the representation (2.2) is not nilpotent but symmetric is *steady extension*, which is treated in a separate paper [5].

4. Simple Fluids

Flows with constant stretch history are very important for the theory of incompressible simple fluids. Such fluids obey the constitutive equation

$$\boldsymbol{T} + p\boldsymbol{I} = \underset{s=0}{\overset{\infty}{\mathfrak{F}}} (\boldsymbol{C}_t^t(s)) \tag{4.1}$$

where $\boldsymbol{T}$ is the stress tensor and where p is an indeterminate pressure, which may be normalized so that $p = -\frac{1}{3}\operatorname{tr}\boldsymbol{T}$. The response functional $\mathfrak{F}$ satisfies the isotropy relation

$$\boldsymbol{Q}^T \underset{s=0}{\overset{\infty}{\mathfrak{F}}} (\boldsymbol{H}(s))\, \boldsymbol{Q} = \underset{s=0}{\overset{\infty}{\mathfrak{F}}} (\boldsymbol{Q}^T \boldsymbol{H}(s)\, \boldsymbol{Q}) \tag{4.2}$$

for all functions $\boldsymbol{H}(s)$ in a suitable class and for all orthogonal tensors $\boldsymbol{Q}$.

Consider a flow with constant stretch history. The formula (2.5) then applies and, in conjunction with (4.1) and (4.2), gives

$$\boldsymbol{Q}(t)\, \boldsymbol{T}(t)\, \boldsymbol{Q}^T(t) + p\boldsymbol{I} = \mathfrak{g}(\boldsymbol{M}), \tag{4.3}$$

where

$$\mathfrak{g}(\boldsymbol{M}) = \underset{s=0}{\overset{\infty}{\mathfrak{F}}} (e^{-\boldsymbol{M}^T s} e^{-\boldsymbol{M} s}). \tag{4.4}$$

The values $\mathfrak{g}(\boldsymbol{M})$ are symmetric tensors. The argument $\boldsymbol{M}$ need not be symmetric, but it must satisfy $\operatorname{tr}\boldsymbol{M} = 0$ in order that the motion be isochoric. (4.2) implies that $\mathfrak{g}(\boldsymbol{M})$ must be isotropic in the sense that

$$\boldsymbol{Q}^T \mathfrak{g}(\boldsymbol{M})\, \boldsymbol{Q} = \mathfrak{g}(\boldsymbol{Q}^T \boldsymbol{M} \boldsymbol{Q}) \tag{4.5}$$

holds for all $\boldsymbol{M}$ and all orthogonal $\boldsymbol{Q}$. Consider a rotating basis $\boldsymbol{b}_k(t)$ which satisfies (2.10), and let $[\boldsymbol{T}] = \|T\langle k\, m\rangle\|$ be the matrix of the stress tensor $\boldsymbol{T}(t)$ relative to that basis. Then (4.3) is equivalent to

$$\|T\langle k\, m\rangle\| = [\boldsymbol{T}] = -p[\boldsymbol{I}] + \hat{\mathfrak{g}}([\boldsymbol{M}]), \tag{4.6}$$

where $\hat{\mathfrak{g}}$ is the matrix function which corresponds to the tensor function $\mathfrak{g}$ relative to any orthonormal basis. It follows that the behavior of an incompressible simple fluid for flows with constant stretch history is completely determined by the matrix function $\hat{\mathfrak{g}}([\boldsymbol{M}])$.

For a viscometric flow, which is characterized by $\boldsymbol{M}^2=\boldsymbol{0}$, we may assume that $[\boldsymbol{M}]$ has the form (1.16), *i.e.*

$$[\boldsymbol{M}]=\varkappa\begin{Vmatrix}0 & 0 & 0\\ 1 & 0 & 0\\ 0 & 0 & 0\end{Vmatrix}. \tag{4.7}$$

In this case, the components of $[\boldsymbol{T}]+p[\boldsymbol{I}]$ are determined by $\varkappa$. This means that the off-diagonal components of $[\boldsymbol{T}]$ and the differences of the diagonal components of $[\boldsymbol{T}]$ are functions of $\varkappa$. We write

$$T\langle 12\rangle=\tau(\varkappa), \qquad T\langle 11\rangle-T\langle 33\rangle=\sigma_1(\varkappa), \qquad T\langle 22\rangle-T\langle 33\rangle=\sigma_2(\varkappa). \tag{4.8}$$

The other two off-diagonal terms $T\langle 13\rangle$ and $T\langle 23\rangle$ must be zero, as is shown by the following argument. The orthogonal matrix

$$[\boldsymbol{Q}]=\begin{Vmatrix}1 & 0 & 0\\ 0 & 1 & 0\\ 0 & 0 & -1\end{Vmatrix} \tag{4.9}$$

commutes with (4.7). It follows from (4.5) and (4.6) that $[\boldsymbol{T}]$ must commute with (4.9). It is easily seen that this can be the case only if $T\langle 13\rangle=T\langle 23\rangle=0$.

Making use of (4.5) with the choice

$$[\boldsymbol{Q}]=\begin{Vmatrix}1 & 0 & 0\\ 0 & -1 & 0\\ 0 & 0 & 1\end{Vmatrix}, \tag{4.10}$$

one can readily prove that the shear stress functions $\tau(\varkappa)$ must be odd while the normal stress functions $\sigma_1(\varkappa)$ and $\sigma_2(\varkappa)$ must be even in $\varkappa$.

The behavior of incompressible simple fluids in steady extension is discussed in [5].

Acknowledgement. This research was supported by the National Science Foundation under Grant NSF-G 16745 to Carnegie Institute of Technology.

References

[1] COLEMAN, B. D.: Kinematical concepts with applications in the mechanics and thermodynamics of incompressible viscoelastic fluids. Arch. Rational Mech. Anal. **9**, 273–300 (1961).

[2] NOLL, W.: A mathematical theory of the mechanical behavior of continuous media. Arch. Rational Mech. Anal. **2**, 197–226 (1958).

[3] COLEMAN, B. D., & W. NOLL: Simple fluids with fading memory. Symposium on Second Order Effects in Elasticity, Plasticity, and Fluid Dynamics. Haifa, Israel 1962.

[4] TRUESDELL, C., & R. A. TOUPIN: The Classical Field Theories. In: Encyclopedia of Physics, vol. III/1, edited by S. FLÜGGE. Berlin-Göttingen-Heidelberg: Springer 1960.

[5] COLEMAN, B. D., & W. NOLL: Steady extension of incompressible simple fluids. Phys. of Fluids **5**, 840–843 (1962).

The Johns Hopkins University
Baltimore, Maryland

(Received July 23, 1962)

LA MÉCANIQUE CLASSIQUE, BASÉE SUR UN AXIOME D'OBJECTIVITÉ [1]

par Walter NOLL

1. Introduction.

Il y a plusieurs points de vue sur l'axiomatisation d'une théorie physique. On peut les schématiser en parlant du point de vue du logicien, du point de vue du mathématicien et du point de vue du physicien.

Le logicien entreprend de donner des formalisations complètes en utilisant le symbolisme de la logique formelle. Il est intéressé plus dans la structure formelle d'une théorie que dans son contenu physique. L'accent du logicien est concentré sur les méta-théorèmes et non sur les théorèmes de la théorie.

Le physicien, au contraire, s'il fait l'axiomatique, entreprend d'ordonner les phénomènes d'expérience d'une façon assez logique sans exclure son intuition et sans être trop préoccupé des questions de rigueur. L'accent du physicien est concentré sur le contenu physique d'une théorie et non sur sa structure formelle.

Le mathématicien a une position intermédiaire. Il entreprend d'être rigoureux comme le logicien, mais son accent, comme celui du physicien, est concentré sur les théorèmes d'une théorie et non sur sa forme logique. Le but du mathématicien est de créer et d'analyser des structures complexes. Il emploie la logique et la théorie des ensembles d'un point de vue naïf ; mais, la théorie des ensembles admise, le reste doit être rigoureux. L'objectif de l'axiomatique en physique est d'achever la clarté et la précision des

(1) Ce rapport est basé sur des recherches supportées par la National Science Foundation, Grant NSF-G 5250 to Carnegie Institute of Technology.

concepts. C'est la théorie mathématique qui fournit la précision définitive des concepts de la physique.

Dans ce rapport je me propose de traiter la mécanique classique comme une structure mathématique. L'objectif de ce traitement est de donner des définitions mathématiques précises des concepts généraux de la mécanique classique en extrayant tout ce qui est commun à ses branches diverses : la mécanique des masses ponctuelles, la mécanique des corps rigides, la mécanique des fluides et des milieux élastiques, et la mécanique moderne des matériaux qui ne sont ni fluides ni élastiques dans le sens classique (voir [3] et les références données dans ce papier). Je crois que c'est une telle axiomatisation que Hilbert eut en vue quand il posa le sixième de ses célèbres problèmes en 1900 [1]. Je ne crois pas que la partie du sixième problème de Hilbert qui concerne la mécanique a trouvé une solution satisfaisante.

Dans une autre communication [2] j'ai présenté une tentative d'axiomatisation de la mécanique classique générale. Les axiomes fondamentaux étaient les lois de la balance des actions et des moments et ce que j'ai appelé le principe d'objectivité matérielle, qui demande l'invariance des équations constitutives par changement de repère arbitraire.

Récemment, j'ai observé qu'on peut achever une axiomatisation plus simple et plus naturelle en remplaçant les axiomes de la balance par un autre axiome d'objectivité qui demande l'invariance du travail par changement de repère arbitraire. Cet axiome est compatible avec les réalités physiques si l'on regarde les forces d'inertie comme des forces véritables qui sont les interactions entre les corps dans notre système solaire et la totalité des objets dans le reste de l'univers. Les repères inertiaux n'entrent plus dans la partie générale de la nouvelle axiomatisation. La loi d'inertie est regardée comme un postulat constitutif. Les repères privilégiés sont liés explicitement à la position du reste de l'univers.

Dans la suite j'esquisserai les idées essentielles de la nouvelle axiomatisation sans donner tous les détails techniques.

2. Univers matériel.

Soit $\mathcal{E}$ un espace euclidéen à trois dimensions et $\mathcal{V}$ l'espace vectoriel associé. Un *univers matériel* est un ensemble $\mathcal{U}$ muni d'une structure par un ensemble Φ d'applications biunivoques de $\mathcal{U}$

sur $\mathcal{E}$ et par une mesure positive finie m sur $\mathcal{U}$. On demande de Φ des propriétés convenables telles qu'on peut définir, à partir de Φ, sur $\mathcal{U}$ une structure de variété continûment différentiable isomorphe à $\mathcal{E}$.

Les éléments $X \in \mathcal{U}$ sont appelés les *points matériels* de l'univers, les applications $\chi \in \Phi$, les *configurations* de l'univers et la valeur $m(\mathcal{B})$ de la mesure m pour une partie $\mathcal{B}$ de $\mathcal{U}$ la *masse* de la partie $\mathcal{B}$.

Un *corps* $\mathcal{B}$ est défini comme un sous-ensemble fermé de $\mathcal{U}$ qui est assez régulier et qui n'a pas de sous-ensembles stricts fermés avec la même masse que $\mathcal{B}$. Deux corps $\mathcal{B}$ et $\mathcal{C}$ sont dits *séparés* si la masse de leur intersection est nulle, $m(\mathcal{B} \cap \mathcal{C}) = 0$. On demande que, pour tout corps $\mathcal{B}$, il existe un autre corps $\mathring{\mathcal{B}}$ séparé de $\mathcal{B}$ telle que $m(\mathcal{B} \cup \mathring{\mathcal{B}}) = m(\mathcal{U})$. Le corps $\mathcal{B}$ est appelé l'*extérieur* de $\mathcal{B}$.

3. Systèmes de forces.

Un *système de forces* pour un univers matériel $\mathcal{U}$ est défini comme une fonction f à valeurs vectorielles de deux corps séparés dans $\mathcal{U}$. La valeur $f(\mathcal{B}, \mathcal{C})$ est appelée la *force exercée sur le corps $\mathcal{B}$ par le corps $\mathcal{C}$*. On demande que f ait les propriétés suivantes :

a) Pour tout corps $\mathcal{B}$ il existe une fonction $f_{\mathcal{B}}$ vectorielle complètement additive définie sur des sous-ensembles de $\mathcal{B}$ et telle que :

$$f_{\mathcal{B}}(\mathcal{C}) = f(\mathcal{C}, \mathring{\mathcal{B}}) \tag{1}$$

si $\mathcal{C}$ est un corps contenu dans $\mathcal{B}$.

b) Pour tout corps $\mathcal{B}$ il existe une fonction $f^*_{\mathring{\mathcal{B}}}$ vectorielle complètement additive définie sur des sous-ensembles de l'extérieur $\mathring{\mathcal{B}}$ et telle que :

$$f^*_{\mathring{\mathcal{B}}}(\mathcal{C}) = f(\mathcal{B}, \mathcal{C}) \tag{2}$$

si $\mathcal{C}$ est un corps dans l'extérieur $\mathring{\mathcal{B}}$ de $\mathcal{B}$.

En vertu de ces propriétés on peut parler de l'intégrale de Stieltjès $\int_{\mathcal{B}} g \, df_{\mathcal{B}}$ d'une fonction g continue définie sur le corps $\mathcal{B}$.

On appelle :

$$f_{\mathcal{B}}(\mathcal{B}) = \int_{\mathcal{B}} df_{\mathcal{B}} = f(\mathcal{B}, \mathring{\mathcal{B}}) \tag{3}$$

la *force résultante* exercée sur $\mathcal{B}$.

Un système de forces f peut être une combinaison de *forces de contact* et de *forces de distance.* Pour un système de masses ponctuelles, c'est-à-dire si m est une mesure discrète, on ne peut avoir que des forces de distance. Les forces de contact sont essentielles dans les milieux continus.

4. Processus dynamiques.

Étant donné un univers matériel $\mathcal{U}$, on considère une paire $\{\chi_t, f_t\}$, $-\infty < t < +\infty$, d'une famille de configurations χ_t de $\mathcal{U}$ et d'une famille de systèmes de forces f_t pour $\mathcal{U}$. Le paramètre réel t de ces familles est appelé le *temps.* On emploie la notation $\chi_t(X) = \chi(X, t)$ et on demande que la *vitesse* :

$$v(X, t) = \frac{d}{dt}\chi(X, t) \tag{4}$$

existe et soit assez régulière. La famille χ_t ou la fonction $\chi(X, t)$ est appelée un *mouvement* de $\mathcal{U}$. Le point spatial $x = \chi(X, t)$ est la *position* du point matériel X au temps t.

Le *travail* par unité de temps des forces exercées sur un corps $\mathcal{B}$ est défini par

$$w(\mathcal{B}, t) = \int_{\mathcal{B}} v(X, t) \cdot df_{t,\mathcal{B}}, \tag{5}$$

où $f_{t,\mathcal{B}}$ est la mesure vectorielle associée à f_t par (1).

On dit que la paire $\{\chi_t, f_t\}$ définit un *processus dynamique* si l'axiome fondamental du § 6 ci-dessous est satisfait. Pour énoncer cet axiome il est d'abord nécessaire d'analyser la notion de changement de repère.

5. Changements de repère.

Un *événement* est une paire (x, t) d'un point x de l'espace $\mathcal{E}$, la *position* de l'événement, et d'un nombre réel t, le *temps* de l'événement. La totalité des événements est appelée l'*espace-temps.* On définit un *changement de repère* comme un automorphisme $(x, t) \leftrightarrow (x', t')$ de l'espace-temps qui préserve l'intervalle temporel de toute paire d'événements et la distance spatiale de toute paire d'événements simultanés. La géométrie analytique nous apprend que tout changement de repère a une représentation analytique de la forme :

$$\begin{aligned} x' &= c(t) + Q(t)x, \\ t' &= t + a, \end{aligned} \tag{6}$$

où les points $\boldsymbol{x}$, $\boldsymbol{x}'$, $\boldsymbol{c}(t)$ sont confondus avec leurs vecteurs de position par rapport à une origine arbitraire. $\boldsymbol{c}(t)$ est une fonction ponctuelle de t, $\mathbf{Q}(t)$ est une fonction de t à valeurs transformations orthogonales, et a est une constante réelle. On demande que $\boldsymbol{c}(t)$ et $\mathbf{Q}(t)$ soient continûment différentiables. Un changement de repère est déterminé par les données $\boldsymbol{c}(t)$, $\mathbf{Q}(t)$, a.

Si l'on demande que les relations entre les points, les scalaires, les vecteurs et les tenseurs soient préservées par changement de repère on obtient des lois définitives de transformation pour ces quantités. Par exemple, les scalaires ne changent pas par changement de repère et les vecteurs $\boldsymbol{u}$ subissent une transformation de la forme :

$$\boldsymbol{u}' = \mathbf{Q}(t)\boldsymbol{u}. \tag{7}$$

Deux processus dynamiques $\{ \boldsymbol{\chi}_t, \boldsymbol{f}_t \}$ et $\{ \boldsymbol{\chi}'_{t'}, \boldsymbol{f}'_{t'} \}$ sont dits *équivalents* s'ils sont liés par un changement de repère de la forme :

$$\boldsymbol{\chi}'_{t'} = \boldsymbol{c}(t) + \mathbf{Q}(t)\boldsymbol{\chi}_t, \tag{8}$$

$$\boldsymbol{f}'_{t'} = \mathbf{Q}(t)\boldsymbol{f}_t, \tag{9}$$

où $t' = t + a$.

L'interprétation physique d'un changement de repère c'est d'un changement de l'observateur seulement. Des processus équivalents ne rendent que des descriptions différentes d'une même réalité physique.

Une fonction définie à partir d'un processus dynamique est dite *objective* si elle transforme en accord avec les lois de changement de repère. Autrement elle est dite *relative*. Les systèmes de forces sont objectifs par définition (9). Par contre, les vitesses et les accélérations sont relatives parce qu'elles ne transforment pas en accord avec la loi (7). Il est aisé à voir que la partie symétrique du tenseur gradient de la vitesse est objective. Ce tenseur est une mesure du changement de déformation.

6. L'axiome fondamental.

En général, le travail (5) n'est pas invariant si l'on transforme $\boldsymbol{\chi}_t$ et $\boldsymbol{f}_t$ par (8) et (9). Autrement dit, le travail n'est pas objectif pour un pair $\{ \boldsymbol{\chi}_t, \boldsymbol{f}_t \}$ arbitraire. L'axiome fondamental c'est la restriction des considérations de la mécanique aux paires $\{ \boldsymbol{\chi}_t, \boldsymbol{f}_t \}$ pour lesquelles le travail est invariant.

Axiome fondamental. — *Pour un processus dynamique le travail est objectif. C'est-à-dire, pour tout corps* $\mathcal{B}$ *le travail* $w(\mathcal{B}, t)$ *défini par* (5) *est invariant par changement de repère arbitraire* (8) (9).

On peut regarder cet axiome comme la définition d'un processus dynamique.

7. Les lois fondamentales.

C'est une conséquence simple de (8) et (4) que la loi de transformation de la vitesse par changement de repère a la forme [1] :

$$v' = \mathbf{Q}(\boldsymbol{v} + \boldsymbol{v}_0 + \mathbf{W}\boldsymbol{x}) \tag{10}$$

où :

$$\boldsymbol{v}_0 = \mathbf{Q}^T\dot{\boldsymbol{c}}, \qquad \boldsymbol{x} = \chi(X, t), \qquad \mathbf{W} = \mathbf{Q}^T\dot{\mathbf{Q}}. \tag{11}$$

Ici, $\mathbf{Q}^T$ dénote la transposée de la transformation orthogonale $\mathbf{Q}$ et le point superposé dénote la dérivée par rapport au temps t. La transformation linéaire $\mathbf{W}$, produit des transformations $\dot{\mathbf{Q}}$ et $\mathbf{Q}^T$, est antisymétrique.

Si l'on substitue (9) et (10) dans la formule (5) pour le travail, on obtient :

$$w'(\mathcal{B}) = \int_{\mathcal{B}} \boldsymbol{v}' \cdot d\boldsymbol{f}'_{\mathcal{B}} = \int_{\mathcal{B}} \mathbf{Q}(\boldsymbol{v} + \boldsymbol{v}_0 + \mathbf{W}\boldsymbol{x}) \cdot \mathbf{Q} d\boldsymbol{f}_{\mathcal{B}}.$$

L'invariance du produit intérieur de deux vecteurs par transformation orthogonale donne :

$$w'(\mathcal{B}) = \int_{\mathcal{B}} (\boldsymbol{v} + \boldsymbol{v}_0 + \mathbf{W}\boldsymbol{x}) \cdot d\boldsymbol{f}_{\mathcal{B}},$$

d'où :

$$w'(\mathcal{B}) = w(\mathcal{B}) + \boldsymbol{v}_0 \cdot \int_{\mathcal{B}} d\boldsymbol{f}_{\mathcal{B}} + tr\left[\mathbf{W}\left(\int_{\mathcal{B}} \boldsymbol{x} \otimes d\boldsymbol{f}_{\mathcal{B}}\right)\right], \tag{12}$$

où tr dénote la trace d'une transformation linéaire et $\otimes$ le produit tensoriel de deux vecteurs [2].

L'axiome fondamental affirme que $w'(\mathcal{B}) = w(\mathcal{B})$ pour tout $\mathcal{B}$ et tout changement de repère. En vertu de (12) il résulte que :

$$\boldsymbol{v}_0 \cdot \int_{\mathcal{B}} d\boldsymbol{f}_{\mathcal{B}} + tr\left[\mathbf{W}\left(\int_{\mathcal{B}} \boldsymbol{x} \otimes d\boldsymbol{f}_{\mathcal{B}}\right)\right] = 0$$

---

[1] Dans ce qui suit, la dépendance des variables d'un processus dynamique du temps t et du point matériel X est entendue.

[2] Le produit tensoriel $\boldsymbol{a} \otimes \boldsymbol{b}$ est confondu avec la transformation linéaire définie par $(\boldsymbol{a} \otimes \boldsymbol{b})\boldsymbol{u} = \boldsymbol{a}(\boldsymbol{b} \cdot \boldsymbol{u})$ pour tout vecteur $\boldsymbol{u}$.

pour tout vecteur $\boldsymbol{v}_0$ et toute transformation antisymétrique $\mathbf{W}$. Il s'ensuit que :

$$\int_{\mathcal{B}} d\boldsymbol{f}_{\mathcal{B}} = \boldsymbol{f}_{\mathcal{B}}(\mathcal{B}) = \boldsymbol{f}(\mathcal{B}, \mathring{\mathcal{B}}) = 0 \tag{13}$$

et :

$$\int_{\mathcal{B}} \boldsymbol{x} \cdot \wedge \, d\boldsymbol{f}_{\mathcal{B}} = 0, \tag{14}$$

où $\wedge$ dénote le produit extérieur de deux vecteurs.

Les équations (13) et (14) sont le contenu des deux lois fondamentales de la mécanique :

Loi de la balance des actions. — *La force résultante exercée sur tout corps est nulle.*

Loi de la balance des moments. — *Le moment résultant des forces exercées sur tout corps est nul.*

Il faut se rappeler que nous n'avons pas dissocié les forces d'inertie des forces ordinaires. On obtient les énoncés conventionnels des lois fondamentales après avoir introduit une telle dissociation.

Les deux lois fondamentales ne sont pas seulement nécessaires, mais aussi suffisantes pour la validité de l'axiome fondamental. Toutes les autres lois générales de la dynamique classique sont des conséquences des deux lois fondamentales. En particulier, on a le théorème de l'action et de la réaction : pour deux corps séparés $\mathcal{B}$ et $\mathcal{C}$ assez réguliers, la force exercée sur $\mathcal{B}$ par $\mathcal{C}$ est l'opposée de la force exercée sur $\mathcal{C}$ par $\mathcal{B}$:

$$\boldsymbol{f}(\mathcal{B}, \mathcal{C}) = -\boldsymbol{f}(\mathcal{C}, \mathcal{B}). \tag{15}$$

8. Les postulats constitutifs.

La structure mathématique développée jusqu'ici n'est suffisante pour la description d'aucune situation physique particulière. Elle n'introduit qu'une langue générale qui est très commode dans la plupart des situations de la mécanique classique. Pour délimiter la nature particulière d'une situation mécanique il faut introduire des *postulats constitutifs.*

D'abord, on ne peut jamais traiter l'univers total. On concentre l'attention sur une partie $\mathcal{S}$ de l'univers $\mathcal{U}$ et on ne considère que le mouvement de $\mathcal{S}$ et les forces exercées sur les corps $\mathcal{B}$

dans $\mathcal{S}$. On appelle *système matériel* une telle partie $\mathcal{S}$ de l'univers matériel.

Un *postulat constitutif de la première espèce* spécifie l'action de l'extérieur $\mathring{\mathcal{S}}$ du système $\mathcal{S}$ sur les corps $\mathcal{B}$ dans $\mathcal{S}$. On suppose que les forces $\boldsymbol{f}(\mathcal{B}, \mathring{\mathcal{S}}) = \boldsymbol{f}_{\mathcal{S}}(\mathcal{B})$ exercées sur les corps $\mathcal{B}$ dans $\mathcal{S}$ par le reste de l'univers $\mathring{\mathcal{S}}$ dépendent d'une manière simple du mouvement de $\mathcal{B}$ relatif à un repère défini à partir de $\mathring{\mathcal{S}}$. Ce repère est appelé le *repère privilégié.*

Un exemple de postulat constitutif de la première espèce est le *postulat d'inertie.* Pour tout corps $\mathcal{B}$ dans le système $\mathcal{S}$, la force exercée sur $\mathcal{B}$ par le reste de l'univers $\mathring{\mathcal{S}}$ est donnée, relatif au repère privilégié, par :

$$\boldsymbol{f}(\mathcal{B}, \mathring{\mathcal{S}}) = \boldsymbol{f}_{\mathcal{S}}(\mathcal{B}) = -\frac{d}{dt}\int_{\mathcal{B}} \boldsymbol{v}\,dm. \tag{16}$$

On adopte ce postulat, en particulier, dans les situations physiques où $\mathcal{S}$ est le système solaire. En ce cas, le repère privilégié est le repère dans lequel les étoiles fixes sont au repos en moyenne.

Des postulats constitutifs de la première espèce plus généraux résultent, si l'on augmente le côté droit de (16) par une fonction prescrite du temps, des points matériels de $\mathcal{B}$ et de leurs positions et vitesses.

Un *postulat constitutif de la deuxième espèce* spécifie l'action mutuelle des corps contenus dans le système matériel $\mathcal{S}$. La forme générale d'un tel postulat est une relation fonctionnelle entre les restrictions des fonctions $\boldsymbol{\chi}(X, t)$ et $\boldsymbol{f}(\mathcal{B}, \mathcal{C})$ à $X \in \mathcal{S}$ et $\mathcal{B} \subset \mathcal{S}$, $\mathcal{C} \subset \mathcal{S}$. On appelle une telle relation une *équation constitutive.*

Un exemple d'une équation constitutive dans la mécanique des masses ponctuelles est donné par :

$$\boldsymbol{f}_t(X, Y) = \boldsymbol{h}(\boldsymbol{\chi}(X, t), \boldsymbol{\chi}(Y, t)) \tag{17}$$

où $\boldsymbol{h}$ est une fonction vectorielle prescrite de deux variables ponctuelles. En ce cas, il n'est pas nécessaire de distinguer les corps à un seul point matériel et les points matériels X, Y de masses positives.

Pour une théorie générale des équations constitutives dans la mécanique des milieux continus voir [3].

9. Le principe de l'objectivité matérielle.

Les postulats constitutifs ne peuvent pas être complètement arbitraires parce qu'on ne change pas une situation physique si l'on ne change que l'observateur.

PRINCIPE DE L'OBJECTIVITÉ MATÉRIELLE. — *Un postulat constitutif doit être objectif, c'est-à-dire invariant par changement de repère.*

Le contenu précis de ce principe est le suivant : si un postulat constitutif est satisfait pour un processus dynamique $\{ \chi_t, f_t \}$ le même postulat constitutif doit être satisfait pour tout processus dynamique $\{ \chi'_{t'}, f'_{t'} \}$ équivalent à $\{ \chi_t, f_t \}$. Du point de vue formel, le principe est une sorte de méta-axiome. Il dissocie les postulats constitutifs admissibles des postulats non admissibles.

Pour les postulats constitutifs de la première espèce, le principe d'objectivité matérielle permet le calcul des forces extérieures dans les repères non privilégiés. Par exemple, le principe affirme qu'on peut obtenir la forme du postulat d'inertie dans un repère arbitraire de la manière suivante : supposant que f' et v' correspondent au repère privilégié on a, d'après (16) :

$$f'_{\mathcal{S}}(\mathcal{B}) = -\frac{d}{dt'}\int_{\mathcal{B}} v'\,dm.$$

Après y substituer (9) et (10) on obtient la *forme générale du postulat d'inertie* :

$$f_{\mathcal{S}}(\mathcal{B}) = -\mathbf{Q}^{\mathsf{T}}\frac{d}{dt}\left\{\mathbf{Q}\int_{\mathcal{B}}(v + v_0 + \mathbf{W}x)\,dm\right\}. \tag{18}$$

Ici, v_0 est la vitesse de l'origine, $\mathbf{Q}$ est l'orientation et $\mathbf{W}$ est la vitesse angulaire du repère considéré relatif au repère privilégié.

Il est aisé de dériver de (18) la formule classique pour les forces d'inertie dans un repère non inerte. Le principe d'objectivité matérielle est nécessaire pour justifier cette formule. Pour les physiciens, le principe est trop évident pour prendre la peine de l'énoncer.

Les conséquences du principe de l'objectivité matérielle sont moins évidentes pour les postulats constitutifs de la deuxième espèce. Le principe affirme que la forme d'une équation constitutive doit être la même dans tout repère, privilégié ou non. Par exemple, on montre aisément que l'équation constitutive (17) n'est admissible que si la fonction vectorielle $\boldsymbol{h}$ a la forme spéciale :

$$\boldsymbol{h}(\boldsymbol{x}, \boldsymbol{y}) = \overrightarrow{\boldsymbol{x}\boldsymbol{y}}\, h(r), \tag{19}$$

où $h(r)$ est une fonction scalaire de la distance r des points $\boldsymbol{x}$ et $\boldsymbol{y}$. Dans le cas particulier $h(r) = \mathrm{G}r^{-3}$ on a l'équation constitutive de la gravitation.

Pour des applications moins triviales du principe de l'objectivité matérielle dans la mécanique des milieux continus voir [3].

Références

[1] D. Hilbert. — *Mathematische Probleme.* Göttinger Nachr., 1900, 253-297.
[2] W. Noll. — *The foundations of classical mechanics in the light of recent advances in continuum mechanics. Proceedings of the Berkeley Symposium on the Axiomatic Method*, 1959, 266-281.
[3] W. Noll. — A mathematical theory of the mechanical behavior of continuous media. *Arch. Rat. Mech. Anal.*, 1958, 2, 197-226.

The Thermodynamics of Elastic Materials with Heat Conduction and Viscosity

BERNARD D. COLEMAN & WALTER NOLL

1. Introduction

The basic physical concepts of classical continuum mechanics are *body, configuration* of a body, and *force system* acting on a body. In a formal rational development of the subject, one first tries to state precisely what mathematical entities represent these physical concepts: a body is regarded to be a smooth manifold whose elements are the material points; a configuration is defined as a mapping of the body into a three-dimensional Euclidean space, and a force system is defined to be a vector-valued function defined for pairs of bodies[1]. Once these concepts are made precise one can proceed to the statement of *general principles*, such as the principle of objectivity or the law of balance of linear momentum, and to the statement of specific *constitutive assumptions*, such as the assertion that a force system can be resolved into body forces with a mass density and contact forces with a surface density, or the assertion that the contact forces at a material point depend on certain local properties of the configuration at the point. While the general principles are the same for all work in classical continuum mechanics, the constitutive assumptions vary with the application in mind and serve to define the *material* under consideration. When one has stated the mathematical nature of bodies, configurations and forces, and has laid down the ways in which these concepts occur in the general principles and the constitutive assumptions, then the properties of these concepts are fixed, and one can present rigorous arguments without recourse to "operational definitions" and other metaphysical paraphernalia, which may be of some use in deciding on the applicability of a theory to a specific physical situation but seem to have no place in its mathematical development.

To discuss the *thermodynamics* of continua, it appears that to the concepts of continuum mechanics one must add five new basic concepts: these are *temperature*, specific *internal energy*[2], specific *entropy*[3], *heat flux*, and *heat supply*[4] (due to radiation). Once mechanics is axiomatized, it is easy to give the mathematical entities representing the thermodynamic concepts: temperature, specific internal energy, specific entropy and heat supply are scalar fields defined over

[1] For more extensive discussions of the foundations of continuum mechanics see references [*1*]–[*4*].

[2] Sometimes called "internal energy density".

[3] Sometimes called "entropy density".

[4] Sometimes called "density of absorbed radiation".

the body, while heat flux is a vector field over the body. We believe that in presenting thermodynamics one should retain all the general principles of mechanics but add to them two new principles: the *first law* of thermodynamics, *i.e.* the law of balance of energy[1], and the *second law*, which for continua takes the form of the Clausius-Duhem inequality[2]. The constitutive assumptions of our present work are the following:

(1) That there exists a caloric equation of state relating the specific internal energy to the "strain"[3] and the specific entropy.

(2) That there exists a temperature equation relating the temperature to the strain and specific entropy.

(3) That the stress tensor of mechanics is the sum of two terms, one of which, the "elastic" term, depends on only the strain and the specific entropy, and the other, the "viscous" term, depends on both these variables and, in addition, has a linear dependence on the "rate of strain"[4].

(4) That the heat flux depends on only the strain, the specific entropy and the spatial gradient of the temperature. (We assume smoothness but not linearity for this dependence.)

We allow the heat supply to be assignable in any way compatible with the general principles, just as body forces are often left assignable in mechanics.

The constitutive assumptions considered here are not the most general imaginable[5], but they are sufficiently general to cover many applications; in particular, they include the starting assumptions of thermoelasticity theory and the classical theory of viscous fluids with heat conduction.

We feel that a statement of the role of the thermodynamic fields in the general laws of thermodynamics and in specific constitutive assumptions should serve to fix the mathematics of thermodynamics and permit a rational development of the subject without requiring the introduction of non-mathematical concepts, such as Heat Reservoirs, Perpetual Motion Machines, and Reversibly Added Heat[6].

Here we proceed as follows. We first give a precise meaning to the term *thermodynamic process*; such a process is defined to be a time-dependent set of configurations, force systems, and temperature, internal energy, entropy, heat supply and heat flux fields, compatible with the principles of mechanics and the law of conservation of energy. A thermodynamic process is said to be *admissible* if it is compatible with the constitutive assumptions under consideration. We then demand that the Clausius-Duhem inequality hold for *all*

[1] *Cf.* §§ 241 and 242 of [*4*].

[2] *Cf.* § 257 of [*4*].

[3] More precisely, the deformation gradient.

[4] More precisely, the velocity gradient.

[5] For example, they do not allow for all the long range memory effects covered in the purely mechanical theory of simple materials; *cf.* [*5*], [*2*].

[6] We do not believe it impossible to assign a mathematical meaning to Heat Reservoirs, Perpetual Motion Machines, and Reversibly Added Heat, but we feel that this has not yet been done, and we doubt its necessity. In several of the extant "axiomatizations" of thermodynamics, reference to these vague concepts appears to disguise the absence of, rather than to furnish, a mathematical justification for the conclusions drawn.

admissible processes. This requirement places restrictions on the constitutive assumptions.

Our main interest here is in the rigorous derivation of a set of necessary and sufficient restrictions on the assumptions (1)—(4) for the Clausius-Duhem inequality to hold for all admissible processes. Each of the restrictions we find is, by itself, a familiar formula or inequality. What is new here is the logical connection.

In a future article we shall apply the general principles of the present framework to more general constitutive assumptions and in so doing obtain a thermodynamic theory compatible with long range memory and non-linear viscoelastic effects, such as stress relaxation.

2. Thermodynamic processes

Consider a body consisting of material points X. A thermodynamic process for this body is described by eight functions of X and the time t, with physical interpretations as follows:

(1) The spatial *position* $\boldsymbol{x}=\boldsymbol{\chi}(X,t)$; here the function $\boldsymbol{\chi}$, called the *deformation function*, describes a *motion* of the body.

(2) The symmetric *stress* tensor $\boldsymbol{T}=\boldsymbol{T}(X,t)$.

(3) The *body force* $\boldsymbol{b}=\boldsymbol{b}(X,t)$ per unit mass (exerted on the body by the external world).

(4) The specific *internal energy* $\varepsilon=\varepsilon(X,t)$.

(5) The *heat flux* vector $\boldsymbol{q}=\boldsymbol{q}(X,t)$.

(6) The *heat supply* $r=r(X,t)$ per unit mass and unit time (absorbed by the material and furnished by radiation from the external world).

(7) The specific *entropy* $\eta=\eta(X,t)$.

(8) The local *temperature* $\vartheta=\vartheta(X,t)$, which is assumed to be always positive, $\vartheta>0$.

We say that such a set of eight functions is a *thermodynamic process* if the following two conservation laws[1] are satisfied not only for the body but also for each of its parts $\mathscr{B}$:

(A) *The law of balance of linear momentum:*

$$\int_{\mathscr{B}} \ddot{\boldsymbol{x}}\, dm = \int_{\mathscr{B}} \boldsymbol{b}\, dm + \int_{\partial\mathscr{B}} \boldsymbol{T}\boldsymbol{n}\, ds. \tag{2.1}$$

(B) *The law of balance of energy*

$$\frac{1}{2}\frac{d}{dt}\int_{\mathscr{B}} \dot{\boldsymbol{x}}\cdot\dot{\boldsymbol{x}}\, dm + \int_{\mathscr{B}} \dot{\varepsilon}\, dm = \int_{\mathscr{B}} (\dot{\boldsymbol{x}}\cdot\boldsymbol{b} + r)\, dm + \int_{\partial\mathscr{B}} (\dot{\boldsymbol{x}}\cdot\boldsymbol{T}\boldsymbol{n} - \boldsymbol{q}\cdot\boldsymbol{n})\, ds. \tag{2.2}$$

In (2.1) and (2.2), dm denotes the element of mass in the body, $\partial\mathscr{B}$ the surface of $\mathscr{B}$, ds the element of surface area in the configuration at time t, and $\boldsymbol{n}$ the exterior unit normal vector to $\partial\mathscr{B}$ in the configuration at time t; superimposed dots denote time-derivatives.

[1] A thorough discussion of these conservation laws is given in [4], §§ 196—205, 240, 241.

The assumed symmetry of the stress tensor $\boldsymbol{T}$ insures that the moment of momentum is automatically balanced. Couple stresses, body couples and other mechanical interactions not included in $\boldsymbol{T}$ or $\boldsymbol{b}$ are assumed to be absent.

Under suitable smoothness assumptions the balance equations (2.1) and (2.2) in integral form are equivalent to the following two balance equations in differential form[1]:

$$\operatorname{div}\boldsymbol{T} - \varrho\,\ddot{\boldsymbol{x}} = -\varrho\,\boldsymbol{b}, \tag{2.3}$$

$$\operatorname{tr}\{\boldsymbol{T}\boldsymbol{L}\} - \operatorname{div}\boldsymbol{q} - \varrho\,\dot{\varepsilon} = -\varrho\,r; \tag{2.4}$$

here, ϱ denotes the mass density and $\boldsymbol{L}$ the velocity gradient.

In order to define a thermodynamic process it is sufficient to prescribe the six functions $\boldsymbol{\chi}$, $\boldsymbol{T}$, ε, $\boldsymbol{q}$, η, and ϑ. The two remaining functions $\boldsymbol{b}$ and r are then uniquely determined by (2.3) and (2.4).

3. The constitutive equations

A *material* is defined by a *constitutive assumption*, which is a restriction on the processes that are admissible in a body consisiting of the material. An *elastic material with heat conduction and viscosity* is defined by five *response functions* $\hat{\varepsilon}$, $\hat{\vartheta}$, $\hat{\boldsymbol{T}}$, $\mathfrak{l}$ and $\hat{\boldsymbol{q}}$. A process is said to be *admissible* in a homogeneous body consisting of such a material if the following constitutive equations hold at each material point X and at all times t:

$$\varepsilon = \hat{\varepsilon}(\boldsymbol{F}, \eta), \tag{3.1}$$

$$\vartheta = \hat{\vartheta}(\boldsymbol{F}, \eta), \tag{3.2}$$

$$\boldsymbol{T} = \hat{\boldsymbol{T}}(\boldsymbol{F}, \eta) + \mathfrak{l}(\boldsymbol{F}, \eta)[\boldsymbol{L}], \tag{3.3}$$

$$\boldsymbol{q} = \hat{\boldsymbol{q}}(\boldsymbol{F}, \eta, \operatorname{grad}\vartheta). \tag{3.4}$$

Here $\boldsymbol{F}$ denotes the deformation of gradient, at X and t, computed relative to a fixed homogeneous reference configuration, and $\boldsymbol{L}$ denotes, as before, the velocity gradient; *i.e.*, if we identify the material point X with its position $\boldsymbol{X}$ in the reference configuration, we have

$$\boldsymbol{F} = \nabla_{\boldsymbol{X}}\boldsymbol{\chi}(\boldsymbol{X}, t) \tag{3.5 a}$$

and

$$\boldsymbol{L} = \dot{\boldsymbol{F}}\boldsymbol{F}^{-1}. \tag{3.5 b}$$

The value $\mathfrak{l}(\boldsymbol{F}, \eta)$ of the response function $\mathfrak{l}$ is a linear transformation over the nine-dimensional space of tensors, and the square brackets in (3.3) indicate that this transformation operates on the tensor $\boldsymbol{L}$.[2] The response functions $\hat{\varepsilon}$, $\hat{\vartheta}$, $\hat{\boldsymbol{T}}$, $\mathfrak{l}$ and $\hat{\boldsymbol{q}}$ depend on the choice of the reference configuration.

[1] See the sections of [4] cited above.

[2] The assumed linear dependence of the "viscous stress" $\mathfrak{l}(\boldsymbol{F}, \eta)[\boldsymbol{L}]$ on $\boldsymbol{L}$ is not essential to our present arguments; for example, for our derivation of (5.5) it suffices that $\mathfrak{l}(\boldsymbol{F}, \eta)[\boldsymbol{L}] \to 0$ as $\boldsymbol{L} \to 0$. Our work on materials with fading memory [6] suggests, however, that if the extra-stress depends non-linearly on the velocity gradient then it should depend also on acceleration gradients. Since we here do not allow for an effect of acceleration gradients on the viscous stress, we prefer to let $\mathfrak{l}(\boldsymbol{F}, \eta)[\boldsymbol{L}]$ be linear in $\boldsymbol{L}$ and thus stay within the theory of "linearly viscous materials", a theory which gives, in the sense of [6], a complete first-order approximation to the non-linear theory of materials with fading memory.

To specify an admissible process, it is sufficient to prescribe the deformation function $\boldsymbol{\chi}$ and the entropy distribution η as functions of time. The deformation gradient $\boldsymbol{F}$ and the velocity gradient $\boldsymbol{L}$ can be computed from $\boldsymbol{\chi}$, and then ε, ϑ, and $\boldsymbol{T}$ can be determined from (3.1)—(3.3). With the temperature distribution ϑ thus obtained, the heat flux $\boldsymbol{q}$ can be calculated from (3.4). Finally $\boldsymbol{b}$ and r can be chosen so that (2.3) and (2.4) hold[1]. In Sections 4 and 5 we shall make frequent use of this observation that $\boldsymbol{\chi}$ and η can be chosen *independently* and that to every such choice corresponds a unique admissible thermodynamic process.

The constitutive equations (3.1)—(3.4) must obey the *principle of material objectivity*[2], which states that an admissible process must remain admissible after a change of frame or observer. Such a change of frame is defined by a time-dependent orthogonal tensor $\boldsymbol{Q}$. The scalars ε, η, and ϑ remain unaffected by a change of frame, but $\boldsymbol{F}$, $\boldsymbol{T}$, $\boldsymbol{L}$, and grad ϑ transform as follows:

$$\begin{aligned} \boldsymbol{F} &\to \boldsymbol{Q}\boldsymbol{F}, \\ \boldsymbol{T} &\to \boldsymbol{Q}\boldsymbol{T}\boldsymbol{Q}^T, \\ \boldsymbol{L} &\to \boldsymbol{Q}\boldsymbol{L}\boldsymbol{Q}^T + \dot{\boldsymbol{Q}}\boldsymbol{Q}^T, \\ \boldsymbol{q} &\to \boldsymbol{Q}\boldsymbol{q}, \\ \operatorname{grad}\vartheta &\to \boldsymbol{Q}\operatorname{grad}\vartheta. \end{aligned} \tag{3.6}$$

Using the methods of reference [7], it is easy to show that the principle of material objectivity is satisfied if and only if the response functions obey the following identities:

$$\begin{aligned} \hat{\varepsilon}(\boldsymbol{F},\eta) &= \hat{\varepsilon}(\boldsymbol{Q}\boldsymbol{F},\eta), \\ \hat{\vartheta}(\boldsymbol{F},\eta) &= \hat{\vartheta}(\boldsymbol{Q}\boldsymbol{F},\eta), \\ \boldsymbol{Q}\hat{\boldsymbol{T}}(\boldsymbol{F},\eta)\boldsymbol{Q}^T &= \hat{\boldsymbol{T}}(\boldsymbol{Q}\boldsymbol{F},\eta), \\ \boldsymbol{Q}\,\mathfrak{l}(\boldsymbol{F},\eta)[\boldsymbol{L}]\,\boldsymbol{Q}^T &= \mathfrak{l}(\boldsymbol{Q}\boldsymbol{F},\eta)[\boldsymbol{Q}\boldsymbol{L}\boldsymbol{Q}^T + \boldsymbol{W}], \\ \boldsymbol{Q}\hat{\boldsymbol{q}}(\boldsymbol{F},\eta,\boldsymbol{v}) &= \hat{\boldsymbol{q}}(\boldsymbol{Q}\boldsymbol{F},\eta,\boldsymbol{Q}\,\boldsymbol{v}) \end{aligned} \tag{3.7}$$

for all scalars η, all vectors $\boldsymbol{v}$, all tensors $\boldsymbol{F}$ and $\boldsymbol{L}$, all orthogonal tensors $\boldsymbol{Q}$, and all skew tensors $\boldsymbol{W}$.

---

[1] In many treatments of continuum thermo-mechanics $\boldsymbol{b}$ and r are regarded as assigned *a priori*. We do not follow this practice here. Nor are we disturbed by the fact that it might be difficult to control $\boldsymbol{b}$ and r experimentally. Physical considerations suggest, however, that for each set of assigned values of $\boldsymbol{F}$, $\boldsymbol{L}$, and η the heat supply r should have a negative lower bound $r_0 < 0$ whose magnitude $|r_0|$ is the radiation the body would emit into an environment at zero absolute temperature.

To account explicitly for this lower bound it would suffice to restrict the admissible fields $\boldsymbol{\chi}$ and η to those which give rise to values of $\boldsymbol{T}$, $\boldsymbol{L}$, div $\boldsymbol{q}$, ϱ, and $\dot{\varepsilon}$ such that the left side of (2.4) is greater than $\varrho\, r_0$. Our arguments would not be affected much by this restriction.

[2] See [*1*]—[*3*].

The identities (3.7) can be used to derive reduced forms for the constitutive equations (3.1)—(3.4). One set of such reduced constitutive equations is

$$\begin{aligned} \varepsilon &= \hat{\varepsilon}(\boldsymbol{U}, \eta), \\ \vartheta &= \hat{\vartheta}(\boldsymbol{U}, \eta), \\ \boldsymbol{R}^T \boldsymbol{T} \boldsymbol{R} &= \hat{\boldsymbol{T}}(\boldsymbol{U}, \eta) + \mathfrak{l}(\boldsymbol{U}, \eta)[\boldsymbol{R}^T \boldsymbol{D} \boldsymbol{R}], \\ \boldsymbol{q} &= \boldsymbol{R}\, \hat{\boldsymbol{q}}(\boldsymbol{U}, \eta, \boldsymbol{R}^T \operatorname{grad} \vartheta). \end{aligned} \tag{3.8}$$

Here $\boldsymbol{U}$ is the (symmetric and positive-definite) right stretch tensor and $\boldsymbol{R}$ the (orthogonal) rotation tensor, determined by the polar decomposition $\boldsymbol{F} = \boldsymbol{R}\boldsymbol{U}$, and $\boldsymbol{D} = \frac{1}{2}(\boldsymbol{L} + \boldsymbol{L}^T)$ is the stretching tensor.

An alternative set of reduced constitutive equations is

$$\begin{aligned} \varepsilon &= \hat{\varepsilon}(\boldsymbol{C}, \eta), \\ \vartheta &= \hat{\vartheta}(\boldsymbol{C}, \eta), \\ \boldsymbol{F}^T \boldsymbol{T} \boldsymbol{F} &= \tilde{\boldsymbol{T}}(\boldsymbol{C}, \eta) + \tilde{\mathfrak{l}}(\boldsymbol{C}, \eta)[\boldsymbol{F}^T \boldsymbol{D} \boldsymbol{F}], \\ \boldsymbol{q} &= \boldsymbol{F}\, \hat{\boldsymbol{q}}(\boldsymbol{C}, \eta, \boldsymbol{F}^T \operatorname{grad} \vartheta), \end{aligned} \tag{3.9}$$

where $\boldsymbol{C} = \boldsymbol{U}^2 = \boldsymbol{F}^T \boldsymbol{F}$ is the right Cauchy-Green tensor. Noting that

$$\boldsymbol{F}^T \operatorname{grad} \vartheta = \nabla_{\boldsymbol{X}} \vartheta, \tag{3.10}$$

where $\nabla_{\boldsymbol{X}} \vartheta$ is the gradient of the temperature with respect to the position $\boldsymbol{X}$ in the reference configuration, we find that the last equation of (3.9) has the form[1]

$$\boldsymbol{q} = \boldsymbol{F}\, \hat{\boldsymbol{q}}(\boldsymbol{C}, \eta, \nabla_{\boldsymbol{X}} \vartheta). \tag{3.11}$$

In rough terms, the *isotropy group* of a material is the set of density-preserving changes of reference configuration which leave the response of the material unaltered. A more formal definition is the following[2]. The isotropy group $\mathscr{I}$ of an elastic material with heat conduction and viscosity is the set of all unimodular tensors $\boldsymbol{H}$ for which the following identities hold:

$$\begin{aligned} \hat{\varepsilon}(\boldsymbol{F}, \eta) &= \hat{\varepsilon}(\boldsymbol{F}\boldsymbol{H}, \eta), \\ \hat{\vartheta}(\boldsymbol{F}, \eta) &= \hat{\vartheta}(\boldsymbol{F}\boldsymbol{H}, \eta), \\ \hat{\boldsymbol{T}}(\boldsymbol{F}, \eta) &= \hat{\boldsymbol{T}}(\boldsymbol{F}\boldsymbol{H}, \eta), \\ \mathfrak{l}(\boldsymbol{F}, \eta)[\boldsymbol{L}] &= \mathfrak{l}(\boldsymbol{F}\boldsymbol{H}, \eta)[\boldsymbol{L}], \\ \hat{\boldsymbol{q}}(\boldsymbol{F}, \eta, \boldsymbol{v}) &= \hat{\boldsymbol{q}}(\boldsymbol{F}\boldsymbol{H}, \eta, \boldsymbol{v}), \end{aligned} \tag{3.12}$$

for all scalars η, all vectors $\boldsymbol{v}$ and all tensors $\boldsymbol{F}$ and $\boldsymbol{L}$. It is easy to show that the tensors $\boldsymbol{H}$ in $\mathscr{I}$ form a group. Of course, the group $\mathscr{I}$ depends not only on the material under consideration but also on the choice of reference configuration. It can be shown, however, that the isotropy groups corresponding

[1] This is the form derived by PIPKIN & RIVLIN ([*8*], equation (17.2)), and by GREEN & ADKINS. ([*9*], equation (8.5.17)).

[2] This definition is analogous to that given in [*2*] and discussed in detail for elastic materials (without heat conduction or viscosity) in [*10*].

to two different reference configurations of the same material are conjugate and hence isomorphic.

The unit tensor $\boldsymbol{I}$ always belongs to $\mathscr{I}$. Suppose now that the inversion $-\boldsymbol{I}$ is in $\mathscr{I}$. (We note that if $-\boldsymbol{I}$ is in the isotropy group corresponding to one reference configuration of a material, then $-\boldsymbol{I}$ is in the isotropy groups corresponding to all other reference configurations of the same material.) In this case $(3.12)_5$ for $\boldsymbol{H}=-\boldsymbol{I}$ yields the identity

$$\hat{\boldsymbol{q}}(\boldsymbol{F},\eta,\boldsymbol{v})=\hat{\boldsymbol{q}}(-\boldsymbol{F},\eta,\boldsymbol{v}). \tag{3.13}$$

On the other hand, $(3.7)_5$ with $\boldsymbol{Q}=-\boldsymbol{I}$ and $\boldsymbol{v}$ replaced by $-\boldsymbol{v}$, gives

$$-\hat{\boldsymbol{q}}(\boldsymbol{F},\eta,-\boldsymbol{v})=\hat{\boldsymbol{q}}(-\boldsymbol{F},\eta,\boldsymbol{v}). \tag{3.14}$$

Combining (3.13) and (3.14), we obtain the identity

$$\hat{\boldsymbol{q}}(\boldsymbol{F},\eta,-\boldsymbol{v})=-\hat{\boldsymbol{q}}(\boldsymbol{F},\eta,\boldsymbol{v}). \tag{3.15}$$

In words: if $-\boldsymbol{I}$ is in $\mathscr{I}$, then, the heat flux must be given by an *odd* function of the temperature gradient. If we put $\boldsymbol{v}=\boldsymbol{0}$ in (3.15), it follows that

$$\hat{\boldsymbol{q}}(\boldsymbol{F},\eta,\boldsymbol{0})=\boldsymbol{0}. \tag{3.16}$$

Of course, (3.16) holds whenever $\hat{\boldsymbol{q}}(\boldsymbol{F},\eta,\boldsymbol{v})$ is linear in $\boldsymbol{v}$, *i.e.*, when Fourier's law is assumed. We have shown here that even if $\hat{\boldsymbol{q}}(\boldsymbol{F},\eta,\boldsymbol{v})$ is not linear in $\boldsymbol{v}$, the presence of the inversion $-\boldsymbol{I}$ in the isotropy group implies that $\boldsymbol{q}$ vanishes when $\operatorname{grad}\vartheta$ vanishes[1].

Pipkin & Rivlin [*8*], § 18, have referred to the equation (3.16) as expressing the "non-existence of a piezo-caloric effect".

4. The Clausius-Duhem inequality

We regard $\boldsymbol{q}/\vartheta$ to be a vectorial *flux of entropy* and r/ϑ to be a scalar *supply of entropy*. In other words, for each process we define the *production of entropy* in the body $\mathscr{B}$ to be

$$\Gamma=\frac{d}{dt}\int\limits_{\mathscr{B}}\eta\,dm-\int\limits_{\mathscr{B}}\frac{1}{\vartheta}r\,dm+\int\limits_{\partial\mathscr{B}}\frac{1}{\vartheta}\boldsymbol{q}\cdot\boldsymbol{n}\,ds. \tag{4.1}$$

Under suitable smoothness assumptions one can write

$$\Gamma=\int\limits_{\mathscr{B}}\gamma\,dm, \tag{4.2}$$

where

$$\begin{aligned}\gamma&=\dot{\eta}-\frac{r}{\vartheta}+\varrho^{-1}\operatorname{div}(\boldsymbol{q}/\vartheta)\\&=\dot{\eta}-\frac{r}{\vartheta}+\frac{1}{\varrho\vartheta}\operatorname{div}\boldsymbol{q}-\frac{1}{\varrho\vartheta^2}\boldsymbol{q}\cdot\operatorname{grad}\vartheta\end{aligned} \tag{4.3}$$

is the *specific production of entropy*.

[1] There are real materials, such as crystals in the pedial class, whose material symmetry does not imply that $-\boldsymbol{I}$ be in $\mathscr{I}$.

We now lay down the following

Postulate. *For every process admissible in a body consisting of a given material and for every part* $\mathscr{B}$ *of this body the inequality*

$$\Gamma \geqq 0 \tag{4.4}$$

is valid.

The inequality (4.4) is called *the Clausius-Duhem inequality*, and the postulate above is a statement of the second law of thermodynamics within the framework presented here. The main purpose of the present paper is to derive logical consequences from the postulate.

In order that the inequality (4.4) hold for *all* parts $\mathscr{B}$ of the body, it is necessary and sufficient that the specific entropy production (4.3) be non-negative at all material points X and at all times t. Using (2.4) and $\vartheta > 0$, we see that this condition is equivalent to

$$\varrho\,\vartheta\,\gamma = \varrho\,(\vartheta\,\dot{\eta} - \dot{\varepsilon}) + \operatorname{tr}\{\boldsymbol{T}\boldsymbol{L}\} - \frac{1}{\vartheta}\,\boldsymbol{q}\cdot\operatorname{grad}\vartheta \geqq 0. \tag{4.5}$$

It follows from (3.1) and the chain rule for the differentiation of composite functions that

$$\dot{\varepsilon} = \operatorname{tr}\{\hat{\varepsilon}_{\boldsymbol{F}}(\boldsymbol{F},\eta)\,\dot{\boldsymbol{F}}\} + \hat{\varepsilon}_\eta(\boldsymbol{F},\eta)\,\dot{\eta}, \tag{4.6}$$

where $\hat{\varepsilon}_{\boldsymbol{F}}$ is the gradient[1] of $\hat{\varepsilon}$ with respect to the tensor $\boldsymbol{F}$, and $\hat{\varepsilon}_\eta$ the derivative of $\hat{\varepsilon}$ with respect to the scalar η. Substituting (4.6) into (4.5) and using the relation $\dot{\boldsymbol{F}} = \boldsymbol{L}\boldsymbol{F}$, we obtain

$$\varrho\,[\vartheta - \hat{\varepsilon}_\eta(\boldsymbol{F},\eta)]\,\dot{\eta} + \operatorname{tr}\{[\boldsymbol{T} - \varrho\,\boldsymbol{F}\,\hat{\varepsilon}_{\boldsymbol{F}}(\boldsymbol{F},\eta)]\,\boldsymbol{L}\} - \frac{1}{\vartheta}\,\boldsymbol{q}\cdot\operatorname{grad}\vartheta \geqq 0. \tag{4.7}$$

5. Temperature relation, stress relation, dissipation inequality

Our postulate states that the inequality (4.7) must hold for *every* admissible process. We consider now processes that are defined by *homogeneous deformations* and *homogeneous entropy distributions*. For such processes, $\boldsymbol{F}$ and η are independent of the material point and depend on only the time. By (3.2), the temperature distribution ϑ is also homogeneous, and hence $\operatorname{grad}\vartheta = \boldsymbol{0}$. The inequality (4.7) then becomes, after substitution of (3.2) and (3.3),

$$\begin{aligned}\varrho\,[\hat{\vartheta}(\boldsymbol{F},\eta) - \hat{\varepsilon}_\eta(\boldsymbol{F},\eta)]\,\dot{\eta} + \operatorname{tr}\{[\hat{\boldsymbol{T}}(\boldsymbol{F},\eta) - \varrho\,\boldsymbol{F}\,\hat{\varepsilon}_{\boldsymbol{F}}(\boldsymbol{F},\eta)]\,\boldsymbol{L}\} &+ \\ + \operatorname{tr}\{\mathfrak{l}(\boldsymbol{F},\eta)\,[\boldsymbol{L}]\,\boldsymbol{L}\} \geqq 0.&\end{aligned} \tag{5.1}$$

As functions of time, $\boldsymbol{F}$ and η can be chosen arbitrarily. It follows that for a *particular* time, the values of $\boldsymbol{F}, \eta, \boldsymbol{L} = \dot{\boldsymbol{F}}\boldsymbol{F}^{-1}$, and $\dot{\eta}$ can be assigned arbitrarily. Taking $\boldsymbol{L} = \boldsymbol{0}$ in (5.1) shows that $[\hat{\vartheta}(\boldsymbol{F},\eta) - \hat{\varepsilon}_\eta(\boldsymbol{F},\eta)]\,\dot{\eta} \geqq 0$ must hold for *every* value of $\dot{\eta}$; this can be the case only if the *temperature relation*

$$\vartheta = \hat{\vartheta}(\boldsymbol{F},\eta) = \hat{\varepsilon}_\eta(\boldsymbol{F},\eta) \tag{5.2}$$

holds.

Choosing $\dot{\eta} = 0$ and substituting $\alpha\,\boldsymbol{L}$ for $\boldsymbol{L}$ in (5.1), where α is an arbitrary number, we obtain

$$\alpha\,\operatorname{tr}\{[\hat{\boldsymbol{T}}(\boldsymbol{F},\eta) - \varrho\,\boldsymbol{F}\,\hat{\varepsilon}_{\boldsymbol{F}}(\boldsymbol{F},\eta)]\,\boldsymbol{L}\} + \alpha^2\operatorname{tr}\{\boldsymbol{L}\,\mathfrak{l}(\boldsymbol{F},\eta)\,[\boldsymbol{L}]\} \geqq 0. \tag{5.3}$$

---

[1] The definition of the gradient used here is that given in reference [7], equations (C. 11) and (C. 11c). In components: $[\hat{\varepsilon}_{\boldsymbol{F}}(\boldsymbol{F},\eta)]_{ij} = \dfrac{\partial\varepsilon}{\partial F_{ji}}$.

This polynomial inequality in α must be satisfied for all values of α when $\boldsymbol{F}, \eta$, and $\boldsymbol{L}$ are kept fixed. It is clear that this is the case if and only if the coefficient of α vanishes and the coefficient of α^2 is non-negative. Thus we obtain the *stress relation*

$$\hat{\boldsymbol{T}}(\boldsymbol{F}, \eta) = \varrho\, \boldsymbol{F}\, \hat{\varepsilon}_{\boldsymbol{F}}(\boldsymbol{F}, \eta) \tag{5.4}$$

and the *dissipation inequality*

$$\operatorname{tr}\{\boldsymbol{L}\, \mathfrak{l}(\boldsymbol{F}, \eta)\, [\boldsymbol{L}]\} \geqq 0 \tag{5.5}$$

which must hold for all tensors $\boldsymbol{L}$.

Let $\boldsymbol{F} = \boldsymbol{R}\boldsymbol{U}$, where $\boldsymbol{R}$ is the rotation tensor and $\boldsymbol{U}$ the right stretch tensor. Writing $(3.7)_4$ with the choice $\boldsymbol{Q} = \boldsymbol{R}^T$ and $\boldsymbol{W} = \frac{1}{2}\boldsymbol{R}^T(\boldsymbol{L}^T - \boldsymbol{L})\boldsymbol{R}$, multiplying the result by $\boldsymbol{R}^T\boldsymbol{L}\boldsymbol{R}$, and then taking the trace, we obtain

$$\operatorname{tr}\{\boldsymbol{L}\, \mathfrak{l}(\boldsymbol{F}, \eta)\, [\boldsymbol{L}]\} = \operatorname{tr}\{\boldsymbol{S}\, \mathfrak{l}(\boldsymbol{U}, \eta)\, [\boldsymbol{S}]\}, \tag{5.6}$$

where $\boldsymbol{S} = \frac{1}{2}\boldsymbol{R}^T(\boldsymbol{L} + \boldsymbol{L}^T)\boldsymbol{R}$. It is evident that $\boldsymbol{L}$ can always be chosen so that $\boldsymbol{S}$ is an arbitrarily prescribed symmetric tensor. Therefore, by (5.6), the dissipation inequality (5.5) is equivalent to the requirement that

$$\operatorname{tr}\{\boldsymbol{S}\, \mathfrak{l}(\boldsymbol{U}, \eta)\, [\boldsymbol{S}]\} \geqq 0 \tag{5.7}$$

hold for all symmetric tensors $\boldsymbol{S}$.

6. Heat conduction inequality. Summary

Let us assume now that the temperature relation (5.2) can be solved for the entropy, so that (5.2) is equivalent to

$$\eta = \bar{\eta}(\boldsymbol{F}, \vartheta). \tag{6.1}$$

The invertibility of (5.2) is assured, in particular, when for each fixed $\boldsymbol{F}$, ε is a convex function of η.

Continuing to keep the homogeneous reference configuration fixed, let us consider deformations $\boldsymbol{\chi}$ that are both homogeneous and constant in time. Each such deformation is characterised by a unique value of $\boldsymbol{F}$, and, of course, we have $\boldsymbol{L} = \boldsymbol{0}$. Let us also consider time-independent temperature distributions $\vartheta = \vartheta(X)$. By (6.1), for each pair $\vartheta = \vartheta(X)$, $\boldsymbol{F} = \text{const.}$, a corresponding time-independent entropy distribution $\eta = \eta(X)$ is uniquely determined. Thus, by the last paragraph of Section 2, a prescribed constant $\boldsymbol{F}$ and a prescribed function $\vartheta(X)$ uniquely determine an admissible process. For such a process, the inequality (4.7) reduces to

$$-\boldsymbol{q} \cdot \operatorname{grad} \vartheta \geqq 0. \tag{6.2}$$

Since $\vartheta = \vartheta(X)$ can be so chosen that, at a particular material point, both ϑ and $\operatorname{grad} \vartheta$ have arbitrarily assigned values, it follows from (6.2) and (3.4) that the *heat conduction inequality*

$$\boldsymbol{v} \cdot \hat{\boldsymbol{q}}(\boldsymbol{F}, \eta, \boldsymbol{v}) \leqq 0 \tag{6.3}$$

must hold for all vectors $\boldsymbol{v}$, all tensors $\boldsymbol{F}$ and all scalars η.

Keeping $\boldsymbol{F}$ and η fixed, let us define a scalar function of the vector $\boldsymbol{v}$ by

$$\varphi(\boldsymbol{v}) = \boldsymbol{v} \cdot \hat{\boldsymbol{q}}(\boldsymbol{F}, \eta, \boldsymbol{v}). \tag{6.4}$$

The heat conduction inequality (6.3) states that $\varphi(\boldsymbol{v})$ has the maximum 0 at $\boldsymbol{v}=\boldsymbol{0}$. A rule of calculus tells us then that the gradient of $\varphi(\boldsymbol{v})$ with respect to $\boldsymbol{v}$ must vanish for $\boldsymbol{v}=\boldsymbol{0}$. Using the rules for the differentiation of products, we find from (6.4) that

$$\{[\nabla_{\boldsymbol{v}} \hat{\boldsymbol{q}}(\boldsymbol{F}, \eta, \boldsymbol{v})]^T \boldsymbol{v} + \hat{\boldsymbol{q}}(\boldsymbol{F}, \eta, \boldsymbol{v})\}|_{\boldsymbol{v}=\boldsymbol{0}} = \hat{\boldsymbol{q}}(\boldsymbol{F}, \eta, \boldsymbol{0}) = \boldsymbol{0}. \tag{6.5}$$

Therefore, the condition (3.15), expressing the non-existence of a piezo-caloric effect, is also a consequence of the heat conduction inequality and hence can be expected to hold even if $-\boldsymbol{I}$ does not belong to the isotropy group of the material. This is, in fact, a result already obtained by PIPKIN & RIVLIN [8], § 18.

We summarize the results obtained as a consequence of our postulate: *An elastic material with heat conduction and viscosity is determined by the three response functions* $\hat{\varepsilon}$, $\mathfrak{l}$, *and* $\hat{\boldsymbol{q}}$. *The temperature and stress are given by the relations* (5.2) *and* (5.4). *The response function* $\mathfrak{l}$ *is restricted by the dissipation inequality* (5.5). *If the temperature relation is invertible, the response function* $\hat{\boldsymbol{q}}$ *is restricted by the heat conduction inequality* (6.3) *and hence also by* (3.15).

Using (5.2), (3.3), and (5.4), we can rewrite the inequality (4.5) as follows:

$$\varrho\,\vartheta\,\gamma = \operatorname{tr}\{\boldsymbol{L}\,\mathfrak{l}(\boldsymbol{F}, \eta)[\boldsymbol{L}]\} - \frac{1}{\vartheta}\,\boldsymbol{q} \cdot \operatorname{grad}\vartheta \geqq 0. \tag{6.6}$$

It is clear from (6.6) that the temperature relation (5.2), the stress relation (5.4), the dissipation inequality (5.5), and the heat conduction inequality (6.3) are *sufficient* that the Clausius-Duhem inequality hold for all admissible processes.

7. Linearly viscous fluids

We say that a material is a *fluid* if its isotropy group $\mathscr{I}$ is the group $\mathscr{U}$ of all unimodular tensors[1]. If $\mathscr{I}=\mathscr{U}$ for one reference configuration, then $\mathscr{I}=\mathscr{U}$ for all reference configurations of the same material; hence, the property of being a fluid is intrinsic to the material.

For an *elastic fluid with heat conduction and viscosity* we have the identities (3.12) for all unimodular tensors $\boldsymbol{H}$, and, therefore, $\hat{\varepsilon}(\boldsymbol{F}, \eta)$, $\mathfrak{l}(\boldsymbol{F}, \eta)$ and $\hat{q}(\boldsymbol{F}, \eta, \boldsymbol{v})$ can depend on $\boldsymbol{F}$ only through $|\det \boldsymbol{F}|$ or, equivalently, through the specific volume $v = \frac{1}{\varrho}$. The temperature and stress relations now reduce to

$$\vartheta = \hat{\varepsilon}_\eta(v, \eta), \qquad \hat{\boldsymbol{T}}(v, \eta) = \hat{\varepsilon}_v(v, \eta)\,\boldsymbol{I}. \tag{7.1}$$

The identity $(3.7)_4$ implies that

$$\boldsymbol{Q}\,\mathfrak{l}(v, \eta)[\boldsymbol{D}]\,\boldsymbol{Q}^T = \mathfrak{l}(v, \eta)[\boldsymbol{Q}\boldsymbol{D}\boldsymbol{Q}^T] \tag{7.2}$$

holds for all symmetric tensors $\boldsymbol{D}$ and all orthogonal tensors $\boldsymbol{Q}$. It is well known[2] that the identity (7.2) holds if and only if $\mathfrak{l}$ is of the form

$$\mathfrak{l}(v, \eta)[\boldsymbol{D}] = 2\mu\boldsymbol{D} + \lambda(\operatorname{tr}\boldsymbol{D})\,\boldsymbol{I}, \tag{7.3}$$

[1] *Cf.* [2].
[2] *Cf.* [4], § 298.

where μ and λ are scalar functions of v and η. Substituting $(7.1)_2$ and (7.3) into $(3.8)_3$, we find

$$\boldsymbol{T} = -p\boldsymbol{I} + 2\mu\boldsymbol{D} + \lambda(\operatorname{tr}\boldsymbol{D})\boldsymbol{I} \tag{7.4}$$

where

$$p = -\hat{\varepsilon}_v(v, \eta). \tag{7.5}$$

Of course, (7.4) is the familiar constitutive equation for linearly viscous fluids. It is significant, however, that the pressure p is given by the same formula (7.5) which holds for perfect (inviscid) fluids. The dissipation inequality (5.7) states that

$$2\mu \operatorname{tr}\boldsymbol{S}^2 + \lambda(\operatorname{tr}\boldsymbol{S})^2 \geqq 0 \tag{7.6}$$

for all symmetric tensors $\boldsymbol{S}$. It is known[2] that the inequalities

$$\mu \geqq 0, \qquad \lambda + \tfrac{2}{3}\mu \geqq 0 \tag{7.7}$$

are equivalent to (7.6). They state that the shear viscosity μ and the bulk viscosity $\lambda + \frac{2}{3}\mu$ must both be non-negative.

For fluids, the identity $(3.7)_5$ reduces to

$$\boldsymbol{Q}\,\hat{\boldsymbol{q}}(v, \eta, \boldsymbol{v}) = \hat{\boldsymbol{q}}(v, \eta, \boldsymbol{Q}\boldsymbol{v}). \tag{7.8}$$

This identity holds if and only if $\hat{\boldsymbol{q}}$ is of the form

$$\hat{\boldsymbol{q}}(v, \eta, \boldsymbol{v}) = -\varkappa\,\boldsymbol{v} \tag{7.9}$$

where $\varkappa$ is a scalar function of v, η, and $\boldsymbol{v}\cdot\boldsymbol{v}$. By (7.9), the heat conduction inequality (6.3) reduces to

$$\varkappa \geqq 0, \tag{7.10}$$

which states that the heat conductivity $\varkappa$ cannot be negative.

Acknowledgements. We are grateful to Drs. J. L. ERICKSEN, A. E. GREEN, R. A. TOUPIN and C. TRUESDELL for stimulating discussions.

This research was supported by the Air Force Office of Scientific Research under Contract AF 49(638)541 with Mellon Institute and by the National Science Foundation under Grant G 11316 to The Johns Hopkins University.

References

[1] NOLL, W.: The foundations of classical mechanics in the light of recent advances in continuum mechanics. In: The Axiomatic Method with Special Reference to Geometry and Physics. Pp. 266—281. Amsterdam: North Holland Co. 1959.

[2] NOLL, W.: A mathematical theory of the mechanical behavior of continuous media. Arch. Rational Mech. Anal. 2, 197—226 (1958).

[3] NOLL, W.: La mécanique classique, basée sur un axiome d'objectivité. A paper read at the Colloque Internationale sur la Méthode Axiomatique in Mécanique Classique et Moderne, Paris, 1959 (to be published by Gauthier-Villars, Paris).

[4] TRUESDELL, C., & R. A. TOUPIN: The Classical Field Theories. In: Encyclopedia of Physics, Vol. III/1, edited by S. FLÜGGE. Berlin-Göttingen-Heidelberg: Springer 1960.

[5] GREEN, A. E., & R. S. RIVLIN: The mechanics of non-linear materials with memory. Arch. Rational Mech. Anal. 1, 1—34 (1957).

[6] COLEMAN, B. D., & W. NOLL: An approximation theorem for functionals, with applications in continuum mechanics. Arch. Rational Mech. Anal. 6, 355—370 (1960).

[7] NOLL, W.: On the continuity of the solid and fluid states. J. Rational Mech. Anal. **4**, 3–81 (1955).
[8] PIPKIN, A. C., & R. S. RIVLIN: The formulation of constitutive equations in continuum physics. Technical Report DA 4531/4 to the Department of the Army, Ordnance Corps (1958).
[9] GREEN, A. E., & J. E. ADKINS: Large Elastic Deformations and Non-Linear Continuum Mechanics. Oxford: Clarendon Press 1960.
[10] COLEMAN, B. D., & W. NOLL,: Material symmetry and thermostatic inequalities in finite elastic deformations, forthcoming.

Mellon Institute
Pittsburgh, Pennsylvania
and
The Johns Hopkins University
Baltimore, Maryland

(Received February 11, 1963)

Material Symmetry and Thermostatic Inequalities in Finite Elastic Deformations

BERNARD D. COLEMAN & WALTER NOLL

Contents

Introduction

This paper is concerned with *elastic materials*; these are substances for which the present stress S on a material point depends on only the present local configuration[1] M of that point:

$$S = \boldsymbol{S}(M). \tag{1}$$

There are several highly developed branches of physics which rest on special cases of the constitutive assumption (1): Euler's hydrodynamics of perfect fluids, the classical theory of infinitesimal elastic deformations, and the modern theory of finite elastic deformations. Our present interest is not in the solution of special problems in these subjects but rather in the theory of the form of the function $\boldsymbol{S}$ itself; here we investigate the general limitations placed on $\boldsymbol{S}$ by material symmetry and thermodynamic considerations.

Applicability of the Theory

Let us consider the situations in which we expect a theory based on (1) to be useful. It is known that there is a large, albeit not all-inclusive, class of

[1] In Sections 1 and 2 we explain more precisely some of the mechanical and kinematical terms used in this Introduction with only abbreviated and heuristic descriptions.

substances, called *simple materials*, whose behavior in general mechanical processes is described by the assertion that the stress at a material point is determined by a function of the history of the local configuration of the point:

$$S = \underset{s=0}{\overset{\infty}{\mathfrak{S}}} (M^s). \tag{2}$$

Here M^s denotes the local configuration s seconds ago, and $\mathfrak{S}$ is a functional whose argument is the function M^s, $0 \leqq s < \infty$, and whose value is the symmetric tensor S. Most of the literature on continuum mechanics deals with simple materials. For examples we have the theory of linear viscoelasticity, the hydrodynamics of perfect, viscous and non-Newtonian fluids, and also recent general theories of materials with nonlinear memory.[2,3]

Now, when we consider a material point which has been in its present local configuration at all times in the past, the function M^s reduces to a constant,

$$M^s \equiv M = \text{const.}, \qquad 0 \leqq s < \infty, \tag{3}$$

and (2) yields the result that S depends only on M:

$$S = \boldsymbol{S}(M). \tag{4}$$

Thus, the theory of elastic materials describes simple materials which have always been at rest.

Suppose that the material described by the functional $\mathfrak{S}$ in (2) has a "fading memory", *i.e.*, that $\mathfrak{S}$ is continuous in such a way that S is "more sensitive" to changes in the local configuration M^s at small s (recent past) than to changes in the local configuration at large s (distant past).[4] Then, given any history M^s, we can construct *retarded histories* $M^{\alpha s}$, $0 < \alpha < 1$, and prove that[7]

$$\lim_{\alpha \to 0} \underset{s=0}{\overset{\infty}{\mathfrak{S}}} (M^{\alpha s}) = \boldsymbol{S}(M^0). \tag{5}$$

The function $\boldsymbol{S}$ in Eqs. (4) and (5) is the same function.

Equation (5) tells us that for every history M^s there exists a retarded history $M^{\alpha s}$, "essentially the same as M^s but slower", such that the present stress corresponding to $M^{\alpha s}$ is given, to as good an approximation as we wish, by $\boldsymbol{S}(M^0)$. This result justifies, within the framework of the theory of simple materials, the use of the static stress-strain function $\boldsymbol{S}$ in the discussion of slow processes; *i.e.*, it gives a motivation to the *dynamic* (as distinguished from the static) theories of perfect fluids and perfectly elastic solids.

Thermostatic Inequalities

In the theory of *infinitesimal* elastic deformations from a natural state (*i.e.* a state with zero stress), it is usually assumed that the stress-strain function $\boldsymbol{S}$ is determined by the gradient of a *positive-definite* strain-energy function σ.

[2] GREEN, A. E., & R. S. RIVLIN, Arch. Rational Mech. Anal. **1**, 1 (1957).
[3] NOLL, W., Arch. Rational Mech. Anal. **2**, 197 (1958).
[4] This concept of "fading memory" is made precise in reference 5 and exploited further in reference 6.
[5] COLEMAN, B. D., & W. NOLL, Arch. Rational Mech. Anal. **6**, 355 (1960).
[6] COLEMAN, B. D., & W. NOLL, Rev. Mod. Phys. **33**, 239 (1961).
[7] The crucial step in proving (5) is Eq. (3.21) of reference 5.

It has often been supposed that the positive-definiteness of σ should be justifiable in terms of fundamental principles in classical thermodynamics, but rigorous arguments to this effect have not come forth. Nor has classical thermodynamics told us how to extend this assumption on σ (and hence on $\boldsymbol{S}$) for infinitesimal elasticity to an assertion about $\boldsymbol{S}$ that is in accord with experience in *finite* elasticity.[8]

Yet, Eq. (3) tells us that the function $\boldsymbol{S}$ must be appropriate for those physical situations in which a simple material can be regarded as having been forever at rest, and $\boldsymbol{S}$ must therefore describe "equilibrium states." Furthermore, our thermodynamic intuition suggests that an equilibrium state is one which maximizes an entropy or minimizes a free energy with respect to an appropriate class of disturbances, and this, in turn, suggests that the function $\boldsymbol{S}$ should be subject to certain inequalities. The failure of classical thermodynamics to yield these inequalities is, in part, due to the difficulty in deciding what to include in the "class of disturbances".

In 1959[9,10] we proposed a theory of thermostatics which led to a general inequality on $\boldsymbol{S}$. We refer to that inequality as the TI (Thermostatic Inequality). In references 9 and 11 consequences of the TI were obtained for various types of materials. In the theory of infinitesimal deformations from a natural state, the TI reduces to the classical assumption of the positive-definiteness of σ. The consequences of the TI for finite elastic deformations seem to be in accord with all experience in solids.

For example, consider a homogeneous deformation of an undistorted isotropic elastic cube in the direction of its three axes. The TI then implies that the greater stretch occurs always in the direction of the greater applied force.[9]

For fluids, the TI is equivalent to the assertion that the pressure be positive and that the compression modulus be greater than two-thirds of the pressure. That the pressure is positive seems to be in accord with experience. However, for a fluid whose critical point occurs at a high pressure there can be a range of densities at which the compression modulus is less than two-thirds of the pressure.

Our TI is equivalent to a requirement of stability against homogeneous disturbances at fixed surface forces. Such a requirement should be appropriate for solids, because it is the surface forces that are controlled in most measurements on solids. For fluids, however, it is usually pressure, and not surface force, that is controlled. When the pressure is fixed during a deformation, the surface forces do not remain fixed but change their direction so as to stay normal to the surface and change their magnitude so as to compensate for the variation of the surface area. The familiar statement that the compression modulus is positive can be derived from a requirement of stability against variations in volume at fixed pressure. Such a requirement is suitable for fluids but does not yield adequate restrictions on the behavior of solids; it does not even yield the positivity of the shear modulus for isotropic infinitesimal elasticity.

[8] Truesdell, C., Z. angew. Math. u. Mech. **36**, 97 (1956).
[9] Coleman, B. D., & W. Noll, Arch. Rational Mech. Anal. **4**, 97 (1959).
[10] A different formulation of the 1959 theory is given in reference 11.
[11] Coleman, B. D., Arch. Rational Mech. Anal. **9**, 172 (1962).

It is still an open question whether it is possible to find a single inequality from which one can derive all thermodynamic restrictions on the static behavior of elastic materials, whether they be fluids, solids, or neither.

In this article we make a detailed study of implications of the TI, stating, without proof, some of the known theorems and presenting several new theorems. Proofs are supplied for all new theorems stated.

Throughout the present discussion we ignore the dependence of the stress on a thermodynamic parameter, such as the temperature or the entropy density. Furthermore, we follow the procedure of a recent article by TRUESDELL & TOUPIN[12], and, whenever possible, we work with a generalization of the TI, called the WTI (Weakened Thermostatic Inequality) which can be expressed directly in terms of the stress-strain function $\boldsymbol{S}$ without recourse to mention of internal energy or free-energy functions. This deliberate suppression of mention of thermodynamic variables is done to emphasize the mechanical significance of our thermodynamic considerations.

Before going into the theory of the TI, we discuss the characterization of material symmetry in terms of stress-strain functions, giving simplified proofs of some known propositions and presenting several new results.

On Notation

The term *tensor* is used here as a synonym for linear transformation of a vector space into itself. We denote tensors by Latin majuscules: F, U, Q, R, I. We reserve the symbol X, however, to represent material points of a physical body. For the trace of a tensor F we write $\operatorname{tr} F$, and for the determinant of F we write $\det F$. We say that F is invertible if F has an inverse F^{-1}, which is the case if and only if $\det F \neq 0$. The transpose of F is denoted by F^T; F is symmetric if $F^T = F$ and skew if $F^T = -F$. The "unit tensor" (*i.e.* the identity transformation) is denoted by I. If $Q Q^T = Q^T Q = I$, then Q is orthogonal. If Q is orthogonal and such that $\det Q = +1$, then Q is proper orthogonal. If the inner product $\boldsymbol{v} \cdot U \boldsymbol{v}$ is strictly greater than zero for all nonzero vectors $\boldsymbol{v}$, then we say that the tensor U is positive-definite. If U is symmetric, then a necessary and sufficient condition that U be positive-definite is that all the proper numbers of U be positive.

1. Mechanical Preliminaries

For convenience, we briefly summarize some definitions and general results of continuum mechanics which are prerequisite to the present paper. Since detailed axiomatic treatments of this material have been given elsewhere,[3,13,14] the results stated here are asserted without proof.

---

[12] TRUESDELL, C., & R. A. TOUPIN, Arch. Rational Mech. Anal. **12**, 1 (1963). The inequality which we here call the "WTI" is called the "GCN condition" by TRUESDELL & TOUPIN.

[13] NOLL, W., in: Proceedings of the Berkeley Symposium on the Axiomatic Method, pp. 266–281. Amsterdam: North Holland.

[14] Another brief summary is given in §§ 1 and 5 of reference 9. For a thorough exposition of the foundations of continuum mechanics see C. TRUESDELL & R. A. TOUPIN, in: Encyclopedia of Physics, Edited by S. FLÜGGE, Vol. III/1. Berlin-Göttingen-Heidelberg: Springer 1959.

Kinematics

A *body* $\mathscr{B}$ is a smooth manifold whose elements X are called *material points*. A *configuration* $\boldsymbol{f}$ of $\mathscr{B}$ is a one-to-one mapping of $\mathscr{B}$ onto a region $\mathscr{R}$ in a three-dimensional Euclidean point space $\mathscr{E}$. The point $\boldsymbol{x}=\boldsymbol{f}(X)$ is called the *position of* X in the configuration $\boldsymbol{f}$, and $\mathscr{R}=\boldsymbol{f}(\mathscr{B})$ is called the *region* occupied by $\mathscr{B}$ in $\boldsymbol{f}$. The *mass distribution* m of $\mathscr{B}$ is a measure defined on subsets of $\mathscr{B}$. To each configuration $\boldsymbol{f}$ of $\mathscr{B}$ corresponds a mass density ϱ.

If we have two configurations $\boldsymbol{f}_1$ and $\boldsymbol{f}_2$ of the same body $\mathscr{B}$, we can consider the composition $\boldsymbol{g}_{2,1}$ of $\boldsymbol{f}_2$ with the inverse $\boldsymbol{f}_1^{-1}$ of $\boldsymbol{f}_1$; $\boldsymbol{g}_{2,1}$ is defined by

$$\boldsymbol{g}_{2,1}(\boldsymbol{x})=\boldsymbol{f}_2(\boldsymbol{f}_1^{-1}(\boldsymbol{x})).$$

Clearly, $\boldsymbol{g}_{2,1}$ is a mapping of the region $\mathscr{R}_1$ occupied by $\mathscr{B}$ in the configuration $\boldsymbol{f}_1$ onto the region $\mathscr{R}_2$ occupied by $\mathscr{B}$ in the configuration $\boldsymbol{f}_2$. The class of admissible configurations $\boldsymbol{f}$ is assumed to be such that for any two configurations $\boldsymbol{f}_1$ and $\boldsymbol{f}_2$, $\boldsymbol{g}_{2,1}$ is a smooth homeomorphism[15] of $\mathscr{R}_1$ onto $\mathscr{R}_2$. Now, since $\boldsymbol{g}_{2,1}$ is a mapping of one region of ordinary Euclidean space onto another, there is no difficulty in computing the gradient of $\boldsymbol{g}_{2,1}$, grad $\boldsymbol{g}_{2,1}$. Because $\boldsymbol{f}_1$ is one-to-one, grad $\boldsymbol{g}_{2,1}$ can be regarded as either a field over $\mathscr{R}_1$ or as a field over the manifold $\mathscr{B}$. Here it is convenient for us to take the latter point of view. We denote the values of grad $\boldsymbol{g}_{2,1}$ by $F_{2,1}(X)$. Of course $F_{2,1}(X)$ is a tensor; it is called the *deformation gradient* (at the material point X) *of the configuration* $\boldsymbol{f}_2$ *computed taking the configuration* $\boldsymbol{f}_1$ *as reference*. Because the function $\boldsymbol{g}_{2,1}$ is one-to-one, the tensors $F_{2,1}(X)$ are invertible.

If we have three configurations $\boldsymbol{f}_1, \boldsymbol{f}_2, \boldsymbol{f}_3$ of the same body $\mathscr{B}$ and compute the three deformation gradients $F_{3,1}(X)$, $F_{3,2}(X)$ and $F_{3,1}(X)$, then it follows from the chain rule for differentials of vector functions that

$$F_{3,1}(X)=F_{3,2}(X)\,F_{2,1}(X). \tag{1.1}$$

If $F_{2,1}$ is independent of X, we say that $\boldsymbol{f}_1$ and $\boldsymbol{f}_2$ are related by a homogeneous deformation. If, in this case, $F_{2,1}$ is a proper orthogonal tensor, then $F_{2,1}$ is said to characterize a *rigid rotation*. If $F_{2,1}$ is orthogonal (with no restriction on the sign of det $F_{2,1}$), then $F_{2,1}$ is said to correspond to an *orthogonal transformation, i.e.* a combination of rigid rotations and reflections. If $F_{2,1}$ has the form αQ, where α is a scalar and Q is orthogonal, then $F_{2,1}$ describes a *similarity transformation, i.e.* a combination of an orthogonal transformation and a uniform expansion by the factor α. If $F_{2,1}$ is a positive definite symmetric tensor, then $\boldsymbol{f}_2$ is said to be obtained from $\boldsymbol{f}_1$ by a *pure stretch*; the proper vectors of $F_{2,1}$ then give the *principal directions of stretch*, and the proper numbers of $F_{2,1}$ are the *principal stretches*.

Let us return to deformations which are not necessarily homogeneous; such deformations can still be regarded, in a sense which we shall now make precise, as homogeneous on a "local level".

If, at a particular material point X, $F_{2,1}(X)$ is the unit tensor I, then we say that $\boldsymbol{f}_1$ and $\boldsymbol{f}_2$ give rise to the same local configuration at X. To state this

[15] A smooth homeomorphism is a one-to-one function $\boldsymbol{g}$ such that both $\boldsymbol{g}$ and its inverse $\boldsymbol{g}^{-1}$ are continuously differentiable.

more formally, we observe that for each fixed X we can define an equivalence relation "$\sim$" among all the configurations $\boldsymbol{f}$ of $\mathscr{B}$ by the condition that $\boldsymbol{f}_1 \sim \boldsymbol{f}_2$ if $F_{2,1}(X) = I$; the resulting equivalence classes of configurations are the *local configurations* M of X. If M_1 is the equivalence class of $\boldsymbol{f}_1$ at X and if M_2 is the equivalence class of $\boldsymbol{f}_2$ at X, then $F_{2,1}(X)$ depends only on M_1 and M_2; it is also true that M_2 is uniquely determined by $F_{2,1}(X)$ and M_1. Hence we can call $F_{2,1}(X)$ simply *the deformation gradient from* M_1 *to* M_2. Dropping the symbol X in $F_{2,1}(X)$, we write

$$M_2 = F_{2,1} M_1. \tag{1.2}$$

If we denote the mass densities at X corresponding to the local configurations M_1 and M_2 by ϱ_1 and ϱ_2, respectively, then

$$\varrho_2 = \frac{1}{|\det F_{2,1}|}\, \varrho_1. \tag{1.3}$$

It is often convenient to employ a fixed *local reference configuration* M_0 and to characterize all other local configurations

$$M = F M_0 \tag{1.4}$$

by their deformation gradients F from M_0. It follows from (1.1) that if

$$M_1 = F_1 M_0, \qquad M_2 = F_2 M_0, \tag{1.5}$$

then the deformation gradient H from M_1 to M_2 is given by

$$H = F_2 F_1^{-1}, \qquad F_2 = H F_1. \tag{1.6}$$

A theorem of algebra, called the *polar decomposition theorem*, states that any invertible tensor F can be written in two ways as the product of a symmetric positive-definite tensor and an orthogonal tensor:

$$F = R U, \tag{1.7}$$

$$F = V R. \tag{1.8}$$

Furthermore, the orthogonal tensor R and the symmetric positive-definite tensors U and V in these decompositions are uniquely determined by F and obey the following relations:

$$U^2 = F^T F, \tag{1.9}$$

$$V^2 = F F^T, \tag{1.10}$$

$$U = R^T V R. \tag{1.11}$$

Equations (1.7) and (1.8) have the following significance in kinematics: Any deformation gradient F may be regarded as being the result of a pure stretch U followed by an orthogonal transformation R, or an orthogonal transformation R followed by a pure stretch V. These interpretations uniquely determine the pairs R, U and R, V. The orthogonal transformations entering these two interpretations are the same; however, the pure stretches U and V can be different. It follows from (1.11) that although U and V may have different principal directions, they must have the same proper numbers u_i. We call the orthogonal tensor R the *rotation tensor*[16] and the tensors U and V, respectively, the *right* and *left stretch tensors.* If $M_2=FM_1$, then we call U and V, respectively, the *right* and *left stretch tensors from M_1 to M_2*, and the proper numbers u_i of U and V are called the *principal stretches from M_1 to M_2*.

Stress

One distinguishes between contact forces and body forces. On assuming the principle of linear momentum and the principle of angular momentum, one can show, after laying down some strong continuity assumptions, that the contact force $d\boldsymbol{c}$ across an oriented surface element with unit normal $\boldsymbol{n}$ and area dA is given by

$$d\boldsymbol{c}=S\boldsymbol{n}dA\,, \tag{1.12}$$

where S is a symmetric tensor called the *stress tensor.*[17]

A proper vector of S is said to determine a *principal axis of stress.* Proper numbers of S are called *principal stresses.*

Since we are here interested in only "local phenomena", we ignore body forces throughout our discussion.

Changes of Frame

A *change of frame* is the mathematical embodiment of the physical concept of a change of observer.

In this paper we assume that the stress tensor S at a material point X is determined by the present value of the local configuration M at that point. It follows that, for our present purposes, we can define a change of frame to be the following simultaneous transformation of the local configuration M_X and the stress S_X at each material point X of a body:

$$M_X\to QM_X, \tag{1.13a}$$

$$S_X\to QS_XQ^{-1}; \tag{1.13b}$$

here Q is an orthogonal tensor independent of X. The change of frame (1.13) is said to be characterized by Q.

In other words, a change of frame is a simultaneous orthogonal transformation of both the present configuration of $\mathscr{B}$ and the present force system acting on $\mathscr{B}$.

The *principle of material objectivity* is a mathematization, for classical mechanics, of the physical idea that the behavior of a material should be independent of the observer. In our present theory this principle reduces to the statement that the dependence of the stress on the local configuration must be invariant under changes of frame; *i.e.*, this dependence must be such that if M is changed from M to QM, where Q is any orthogonal tensor, then the stress S corresponding to M must change to QSQ^{-1}.

2. The Response Function and the Isotropy Group

We are now prepared to state formally our

Fundamental Constitutive Assumption. To each local configuration M there corresponds a unique value of the stress tensor S.

[16] In applications R is usually proper orthogonal.
[17] For details, see reference 13.

Such a correspondence between S and M is called an *elastic material* (or, simply, "a material"). Since the behavior of a material should be independent of the observer, we also lay down an

Assumption of Material Objectivity. The dependence of the stress S on the local configuration M is such that if $M_2 = Q M_1$ where Q is any orthogonal tensor, then $S_2 = Q S_1 Q^T$ where S_1 and S_2 are the stresses corresponding to M_1 and M_2, respectively.

We follow standard procedure and express the dependence of S on M by picking a local reference configuration M_0, characterizing all local configurations by their deformation gradients F from M_0 in accordance with (1.4), and then regarding S as a function $\mathfrak{g}$ of F:

$$S = \mathfrak{g}(F). \tag{2.1}$$

The function $\mathfrak{g}$ is called the *response function*, of our material, *taken relative to* M_0.

It follows from our Assumption of Material Objectivity that $\mathfrak{g}$ must obey the identity

$$\mathfrak{g}(QF) = Q\mathfrak{g}(F)Q^{-1} \tag{2.2}$$

for all tensors F and all orthogonal tensors Q. Using Eqs. (2.2) and (1.7), we infer that Eq. (2.1) is equivalent to

$$S = R\mathfrak{g}(U)R^{-1}. \tag{2.3}$$

The function $\mathfrak{g}$ depends on the choice of the reference configuration M_0. To exhibit the nature of this dependence of $\mathfrak{g}$ on M_0, we note that it follows from (1.5) and (1.6) that if F and F' are two deformation gradients characterizing the same local configuration M with respect to two different local reference configurations M_0 and M_0', *i.e.*

$$M = F M_0 = F' M_0',$$

and if

$$M_0' = G M_0,$$

then

$$F = F' G.$$

If the response functions relative to M_0 and M_0' are denoted by $\mathfrak{g}$ and $\mathfrak{g}'$, respectively, we infer

$$S = \mathfrak{g}(F) = \mathfrak{g}'(F') = \mathfrak{g}'(F G^{-1}).$$

This proves

Proposition 1. *If $\mathfrak{g}$ and $\mathfrak{g}'$ are the response functions for the same material but taken relative to different local configurations M_0 and M_0', with $M_0' = G M_0$, then*

$$\mathfrak{g}(F) = \mathfrak{g}'(F G^{-1}) \tag{2.4}$$

for all F.

For a given material it may turn out that $\mathfrak{g}$ remains the same function if the local reference configuration M_0 is changed to another local reference configuration $M_0' = H M_0$ with the same density, *i.e.*, that

$$\mathfrak{g}(F) = \mathfrak{g}'(F) \tag{2.5}$$

for all F. (The physical interpretation of the identity (2.5) is that the two configurations M_0 and M_0' must be equal in their response to equal deformations, *i.e.*, that the classes of reference configurations for which the identity (2.5) holds must characterize the isotropy of the material under consideration.) Now, it follows from (2.4) that if (2.5) holds for M_0 and $M_0' = HM_0$, then $\mathfrak{g}$ satisfies the identity

$$\mathfrak{g}(F) = \mathfrak{g}(FH^{-1}) \tag{2.6}$$

for all F. Since we are here assuming that M_0' and M_0 have the same density, it is clear that $|\det H| = 1$; *i.e.*, H is here a unimodular transformation. If the identity (2.6) holds for H^{-1}, it holds also for H; furthermore, the set of unimodular transformations H for which the identity (2.6) holds forms a group. Let us now state the following important formal definition whose physical motivation should, we feel, now be clear.

Definition. The group $\mathscr{I}$ of unimodular tensors H for which the identity

$$\mathfrak{g}(FH) = \mathfrak{g}(F) \tag{2.7}$$

holds for all F, where $\mathfrak{g}$ is the response function relative to the reference configuration M_0, is called the *isotropy group relative to* M_0.

$\mathscr{I}$ depends on M_0, but we shall see that for a given material the isotropy groups $\mathscr{I}$ and $\mathscr{I}'$ relative to two distinct local reference configurations M_0 and M_0' are conjugate and hence isomorphic. In fact, combination of (2.4) and (2.7) shows that

$$\mathfrak{g}'(FG^{-1}) = \mathfrak{g}'(FHG^{-1}) \tag{2.8}$$

for all F. If we put $H' = GHG^{-1}$ and $F' = FG^{-1}$, then $FHG^{-1} = F'H'$, and (2.8) becomes the assertion that

$$\mathfrak{g}'(F') = \mathfrak{g}'(F'H') \tag{2.9}$$

for all F', but this simply says that H' is in $\mathscr{I}'$. Hence we have

Proposition 2. *If G is the deformation gradient from M_0 to M_0', then H belongs to $\mathscr{I}$ if and only if GHG^{-1} belongs to $\mathscr{I}'$, where $\mathscr{I}$ and $\mathscr{I}'$ are, respectively, the isotropy groups relative to M_0 and M_0'.*

$\mathscr{I} = \mathscr{I}'$ would mean that the change of reference configuration G does not affect the isotropy group. By Proposition 2 this is the case if and only if $G^{-1}HG$ belongs to $\mathscr{I}$. In the language of group theory, this condition states that G must belong to the *normalizer* group of $\mathscr{I}$ within the full linear group $\mathscr{L}$. Thus we have

Proposition 3. *A change of reference configuration from M_0 to GM_0 leaves the isotropy group $\mathscr{I}$ unchanged if and only if G belongs to normalizer of $\mathscr{I}$ in $\mathscr{L}$.*

Suppose Q is orthogonal and belongs to $\mathscr{I}$; then we can combine the identity (2.6), with $H = Q$, and the identity (2.2), to obtain

Proposition 4. *An orthogonal tensor Q belongs to the isotropy group $\mathscr{I}$ if and only if*

$$Q\,\mathfrak{g}(F)\,Q^{-1} = \mathfrak{g}(QFQ^{-1}) \tag{2.10}$$

for all tensors F.

Definitions. If an elastic material has a local configuration M_0 such that the corresponding isotropy group is either the orthogonal group $\mathcal{O}$ or a group which contains $\mathcal{O}$ as a subgroup, then we say that the material is *isotropic*, and M_0 is called an *undistorted* local configuration. If a material is such that for some M_0, $\mathcal{I}$ is a subgroup (which need not be proper) of $\mathcal{O}$, then we say that the material is a *solid*, and we again call M_0 *undistorted*. Hence for the undistorted configurations M_0 of an *isotropic solid* $\mathcal{I}=\mathcal{O}$. In general, we say that a configuration M_0 is *undistorted* if $\mathcal{I}$ is comparable to $\mathcal{O}$; *i.e.* if $\mathcal{I}$ either is itself a subgroup of $\mathcal{O}$ or contains $\mathcal{O}$ as a subgroup. We say that a material is a *fluid* if $\mathcal{I}$ is the full unimodular group $\mathcal{U}$. Since $\mathcal{O}$ is a subgroup of $\mathcal{U}$, an elastic fluid is isotropic. It follows from Proposition 2 that if $\mathcal{I}=\mathcal{U}$ for one local configuration M_0, then $\mathcal{I}'=\mathcal{U}$ for every other local configuration M_0'; hence, every configuration of a fluid is undistorted.

Note. Continuum mechanics does not yet have a standard terminology for the various mathematical concepts behind the intuitive notion of "isotropy". The particular definitions we give here are specializations to elastic materials of the definitions given by NOLL[3] for general simple materials. The reader is cautioned that, whereas we use *isotropic* as a quality of a *material*, other writers regard *isotropic* as a quality of both a *material and a configuration*. When these writers say that "the material is isotropic in the configuration M", we say that "M is an undistorted configuration of the isotropic material".

The following proposition is an immediate consequence of Proposition 4.

Proposition 5. *The response function* $\mathfrak{g}$ *of an isotropic material, when taken relative to an undistorted state, obeys the following identity for all tensors* F *and all orthogonal tensors* Q:

$$\mathfrak{g}(QFQ^{-1})=Q\,\mathfrak{g}(F)\,Q^{-1}. \tag{2.11}$$

Tensor-valued functions obeying (2.11) are called *isotropic functions*.

The next three propositions illustrate some of the physical motivation behind our formal definitions.

Proposition 6. *The stress on an undistorted state of an isotropic material is always a hydrostatic pressure.*

This proposition states that if $\mathcal{I}$ contains $\mathcal{O}$ as a subgroup (properly or improperly), then $\mathfrak{g}(I)=-pI$.

Proof. By hypothesis and Proposition 5 we have

$$Q\,\mathfrak{g}(I)\,Q^{-1}=\mathfrak{g}(QIQ^{-1})=\mathfrak{g}(I)$$

for all orthogonal tensors Q; *i.e.*, $\mathfrak{g}(I)$ must commute with all orthogonal tensors. This is possible only if $\mathfrak{g}(I)$ is a scalar multiple $(-pI)$ of the unit tensor I, *q.e.d.*

Proposition 7. *The stress on an elastic fluid is always a hydrostatic pressure* $-pI$ *which depends on only the density* ϱ.

In other words, for a simple fluid,

$$\mathfrak{g}(F)=-p(\varrho)\,I \tag{2.12}$$

where

$$\varrho=\frac{1}{|\det F|}\,\varrho_0, \tag{2.13}$$

ϱ being the density of the present local configuration FM and ϱ_0 the density of the local reference configuration M_o.

Proof. For any given tensor F, the tensor $H = |\det F|\, F^{-1}$ is obviously unimodular and hence belongs to the isotropy group $\mathscr{I} = \mathscr{U}$ of the fluid. Therefore, by (2.7),

$$\mathfrak{g}(F) = \mathfrak{g}(FH) = \mathfrak{g}(|\det F|\, I)$$

depends only on $|\det F|$ or, equivalently, on the density ϱ. That $\mathfrak{g}(F)$ is a hydrostatic pressure for all M and F is a consequence of Proposition 6 and the fact that a simple fluid is an isotropic material all of whose configurations are undistorted.

Proposition 8. *Two undistorted local configurations of an isotropic solid can differ by only a similarity transformation.*

Proof. By definition, the isotropy group of an isotropic solid, relative to an undistorted reference configuration, is the orthogonal group $\mathscr{O}$. Hence, by Proposition 3, a change G from one undistorted reference configuration to another must belong to the normalizer of $\mathscr{O}$ in $\mathscr{L}$. But it is a known result in group theory that the normalizer of $\mathscr{O}$ is the group of all similarity transformations,[18] *q.e.d.*

3. Undistorted States of Aeolotropic Solids

Proposition 6 of Section 2 shows that the stress on an undistorted local configuration of an isotropic material (solid or not) must be a hydrostatic pressure. Here we investigate the limitations material symmetry imposes upon the stress on undistorted configurations of various types of aeolotropic solids. We also extend Proposition 8 of Section 2 to obtain limitations on the possible strains which can relate undistorted states of general solids.

As far as purely mechanical behavior is concerned, the symmetry of an elastic solid is characterized by a corresponding "type" of isotropy group $\mathscr{I}$. By our definition of the term *solid*, $\mathscr{I}$ will be a subgroup of the orthogonal group $\mathscr{O}$ if an undistorted local configuration M_o is used as a reference. If $M_o' = GM_o$ is another undistorted reference configuration, then the isotropy group $\mathscr{I}'$ corresponding to M_o' will also be a subgroup of the orthogonal group. Moreover, according to Proposition 3 of Section 2, $\mathscr{I}'$ will be the conjugate $\mathscr{I}' = G\mathscr{I}G^{-1}$ of $\mathscr{I}$ under G; *i.e.*, every transformation Q' in $\mathscr{I}'$ will be of the form

$$Q' = GQG^{-1} \tag{3.1}$$

where Q is a member of $\mathscr{I}$. Let

$$G = RU \tag{3.2}$$

be the polar decomposition of G, so that U is the right stretch tensor and R the rotation tensor of the deformation carrying M_o into M_o'. Combining Eqs. (3.1) and (3.2), we find

$$\begin{aligned} Q'G &= GQ, \\ Q'RU &= RUQ = (RQ)(Q^{-1}UQ). \end{aligned} \tag{3.3}$$

[18] *Cf.* H. Weyl, The Classical Groups, p. 22. Princeton 1946.

Since both $Q'R$ and RQ are orthogonal tensors while U and $Q^{-1}UQ$ are both positive-definite and symmetric, (3.3) gives two polar decompositions of the one tensor $Q'G$. Therefore, the uniqueness of a polar decomposition implies that

$$Q'R = RQ \quad \text{and} \quad U = Q^{-1}UQ. \tag{3.4}$$

The first of the Eqs. (3.4) shows that every member Q' of $\mathscr{I}'$ is of the form

$$Q' = RQR^{-1}, \tag{3.5}$$

i.e., that $\mathscr{I}'$ is actually a *conjugate* $\mathscr{I}' = R\mathscr{I}R^{-1}$ *of* $\mathscr{I}$ *within the orthogonal group*. We say that two subgroups of the orthogonal group are *of the same type* if they are conjugate in this sense. The isotropy group of a solid depends on the choice of reference configuration, even when this configuration is restricted to be undistorted. The type to which this group belongs, however, represents an intrinsic property of the material.

Crystalline solids are classified into thirty-two classes, each of which is characterized by a certain type of symmetry group. The symmetry groups are finite subgroups of the orthogonal group. On putting $Q = -I$ in (2.2) we see that

$$\mathfrak{g}(F(-I)) = (-I)\,\mathfrak{g}(F)\,(-I) = \mathfrak{g}(F), \tag{3.6}$$

which shows that the inversion $-I$ is always a member of the isotropy group. This inversion, however, does not belong to the symmetry group of some of the crystal classes.[19] The following assumption seems to be natural:

Hypothesis on Crystalline Solids. Consider a crystal which, relative to some undistorted state M_0, has the crystallographic symmetry group (*i.e.* point group) $\mathscr{C}$. The isotropy group $\mathscr{I}$ of this crystal, relative to M_0, is assumed to be the group generated by $\mathscr{C}$ and the inversion $-I$.

The 32 types of symmetry groups[20] give rise to only 11 types of isotropy groups. In describing a group $\mathscr{I}$ we need not list all the elements of $\mathscr{I}$ but only a set of *generators* of $\mathscr{I}$, *i.e.* a set of members of $\mathscr{I}$ which, when they and their inverses are multiplied among themselves in various combinations, yield all the elements of $\mathscr{I}$. For an isotropy group relative to an undistorted state of a solid it is always possible to find a list of generators $-I, Q_1, \ldots, Q_m$ such that each Q_i is *proper orthogonal*. We denote by $R_{\boldsymbol{n}}^{\varphi}$ the right-handed rotation through the angle φ, $0 < \varphi < 2\pi$, about an axis in the direction of the unit vector $\boldsymbol{n}$. Each proper orthogonal transformation $\neq I$ is a rotation of the form $R_{\boldsymbol{n}}^{\varphi}$. Table 1 gives a description of the 11 types of isotropy groups for crystals. In this table, $\boldsymbol{i}, \boldsymbol{j}, \boldsymbol{k}$ denotes a right handed orthonormal basis and $\boldsymbol{d} = \frac{1}{\sqrt{3}}(\boldsymbol{i} + \boldsymbol{j} + \boldsymbol{k})$. A change from one group to another of the same type corresponds merely to a change of the orthonormal basis $\boldsymbol{i}, \boldsymbol{j}, \boldsymbol{k}$.

[19] If we were dealing here with vector phenomena, as in electromagnetic theory, then we should not automatically have $-I$ in $\mathscr{I}$, and we should have instead $\mathscr{I} = \mathscr{C}$. In other words, it is because our present theory involves only tensors of order two that our isotropy groups are sometimes bigger than the crystallographic point groups.

[20] Complete lists of members for all these groups are given by G. F. SMITH & R. S. RIVLIN, Trans. Am. Math. Soc. **88**, 175 (1958). The names we employ for the various crystal systems and classes are those used by SMITH & RIVLIN, who, in turn, state that they come from the 1952 Edition of DANA's Textbook of Mineralogy, revised by C. S. HURLBUT (New York: John Wiley).

Table 1. *Isotropy Groups for the Various Crystal Classes*

Ref. No.	Crystal Class	Proper Orthogonal Generators of $\mathcal{I}$	Number of Elements in $\mathcal{I}$
	Triclinic System		
1.	all classes	I	2
	Monoclinic System		
2.	all classes	R_k^π	4
	Rhombic System		
3.	all classes	R_i^π, R_j^π	8
	Tetragonal System		
4.	tetragonal-disphenoidal tetragonal-pyramidal tetragonal-dipyramidal	$R_k^{\pi/2}$	8
5.	tetragonal-scalenohedral ditetragonal-pyramidal tetragonal-trapezohedral ditetragonal-dipyramidal	$R_k^{\pi/2}$, R_i^π	16
	Cubic System		
6.	tetratoidal diploidal	R_i^π, R_j^π, $R_d^{2\pi/3}$	24
7.	hextetrahedral gyroidal hexoctahedral	$R_i^{\pi/2}$, $R_j^{\pi/2}$	48
	Hexagonal System		
8.	trigonal-pyramidal rhombohedral	$R_k^{2\pi/3}$	6
9.	ditrigonal-pyramidal trigonal-trapezohedral hexagonal-scalenohedral	$R_k^{2\pi/3}$, R_i^π	12
10.	trigonal-dipyramidal hexagonal-pyramidal hexagonal-dipyramidal	$R_k^{\pi/3}$	12
11.	ditrigonal-dipyramidal dihexagonal-pyramidal hexagonal-trapezohedral dihexagonal-dipyramidal	$R_k^{\pi/3}$, R_i^π	24

With respect to their elastic behavior alone, crystals can show only the 11 types of symmetry characterized by the 11 types of isotropy groups.

A type of symmetry appropriate not to crystals but to materials with a bundled or laminated structure or to drawn fibers is *transverse isotropy*. It is defined by the assumption that the isotropy group consists of $\pm I$ and $\pm R_k^\varphi$, for some fixed unit vector $\boldsymbol{k}$, and all angles φ, $0<\varphi<2\pi$.

Consider an elastic solid and a particular undistorted reference configuration M_o. Denote the corresponding response function by $\mathfrak{g}$ and the corresponding isotropy group by $\mathcal{I}$. The stress on M_o is given by

$$S_o=\mathfrak{g}(I). \tag{3.7}$$

The following is an immediate consequence of Eq. (2.10) for $F=I$:

Proposition 1. *The stress S_0 on an undistorted local configuration M_0 of a solid commutes with every member of the isotropy group $\mathscr{I}$ for M_0;* i.e.

$$Q S_0 = S_0 Q \tag{3.8}$$

holds for all Q in $\mathscr{I}$.

It is easily seen that (3.8) holds for all Q in $\mathscr{I}$ if it holds for all rotations Q_i, $i=1, \ldots, m$ in a list $-I, Q_1, \ldots, Q_m$ of generators of $\mathscr{I}$. Therefore, the problem of finding the possible stresses S_0 on an undistorted state reduces to the problem of finding those symmetric tensors that commute with a certain set of rotations of the form $R_{\boldsymbol{n}}^{\varphi}$. For each of the 11 cases given in Table 1 and for the case of transverse isotropy, this problem can easily be solved with the help of the following

Commutation Theorem[21]. *The symmetric tensor S commutes with the orthogonal tensor Q if and only if Q leaves each of the characteristic spaces of S invariant,* i.e., *if it maps each characteristic space into itself.*

Here, a *characteristic space* of S is defined to be a maximal subspace (of the ordinary three-dimensional vector space) consisting of only proper vectors of S. If S has three distinct proper numbers, then it has three one-dimensional characteristic spaces. If S has only two distinct proper numbers, then it has two characteristic spaces, one of which is one-dimensional and the other two-dimensional. In this case, S must be of the form[22]

$$S = -pI + q\boldsymbol{n} \otimes \boldsymbol{n}, \tag{3.9}$$

where $-p$ and $q-p$ are the proper numbers of S and $\boldsymbol{n}$ is a unit vector which generates the one-dimensional characteristic space. If the three proper numbers of S coincide, then the entire three-dimensional vector space is the only characteristic space of S, and S is of the form

$$S = -pI. \tag{3.10}$$

Now if $\varphi \neq \pi$, the only spaces left invariant by the rotation $R_{\boldsymbol{n}}^{\varphi}$ are the one-dimensional space of all multiples of $\boldsymbol{n}$ the two-dimensional space of all vectors perpendicular to $\boldsymbol{n}$, and the entire three-dimensional vector space. The rotation $R_{\boldsymbol{n}}^{\pi}$ leaves invariant, in addition, each one-dimensional space generated by a vector perpendicular to $\boldsymbol{n}$.

Proposition 1, Table 1, and the results from geometrical linear algebra just described enable one to establish easily the results collected in Table 2. In this table, $\boldsymbol{i}, \boldsymbol{j}, \boldsymbol{k}$ is the same orthonormal basis as is used for the isotropy groups in Table 1. Recall that a proper vector of the stress S_0 determines a principal axis of stress.

Let us now return to Eqs. (3.4). It follows from the second of these equations that $QU = UQ$, which is the content of

[21] This theorem is a corollary to Theorem 2, p. 77, and Theorem 3, p. 157, of P. R. HALMOS, Finite-Dimensional Vector Spaces. Princeton: Van Nostrand, 2nd ed., 1958.

[22] The symbol $\otimes$ denotes a tensor product; *i.e.*, $\boldsymbol{n} \otimes \boldsymbol{n}$ is the tensor with the property that $(\boldsymbol{n} \otimes \boldsymbol{n})\,\boldsymbol{v} = \boldsymbol{n}\,(\boldsymbol{n} \cdot \boldsymbol{v})$ for all vectors $\boldsymbol{v}$.

Proposition 2. *Let M_0 and M_0' be two local reference configurations of a solid, and let $\mathscr{I}$ be the isotropy group relative to M_0. If M_0 is undistorted, then M_0' is undistorted if and only if the right stretch tensor U from M_0 to M_0' commutes with every member of $\mathscr{I}$.*

Table 2. *Restrictions on the Stress for Undistorted States of Aeolotropic Solids*

Type of Isotropy	Ref. No. in Table 1	Restriction on S_0
Triclinic system	1	no restriction
Monoclinic system	2	$\boldsymbol{k}$ is a proper vector of S_0
Rhombic system	3	$\boldsymbol{i}, \boldsymbol{j}, \boldsymbol{k}$ are proper vectors of S_0
{ Tetragonal system, Hexagonal system, Transverse isotropy }	4, 5, 8, 9, 10, 11	$S_0 = -pI + q\,\boldsymbol{k} \otimes \boldsymbol{k}$
Cubic system	6, 7	$S_0 = -pI$

It follows from this proposition that if M_0 is an undistorted configuration of a solid with the symmetry listed in the first column of Table 2, then a necessary and sufficient condition that M_0' be an undistorted configuration of that solid is that the right stretch tensor U relating M_0' to M_0 obey the restrictions listed for S_0 in the second column.

4. Thermostatic Inequalities

As we mentioned in the Introduction, there are reasons, whose origins lie outside of mechanics, for believing that stress-strain functions used in physical applications should be subjected to restrictions beyond those following from Material Objectivity. These restrictions should have the form of inequalities and, we believe, should follow from a properly formulated theory of the thermodynamics of continuous media.

The inequalities we wish to consider here can be most simply expressed through use of the first Piola-Kirchhoff stress tensor, T, defined by

$$T = |\det F|\,(F^{-1}S)^T = \frac{\varrho_0}{\varrho}\,S F^{T^{-1}}, \tag{4.1}$$

where S is the ordinary stress and F the deformation gradient. The definition of T depends on the choice of the reference configuration. T need not be symmetric. We denote the relation between T and F by

$$T = \mathfrak{h}(F), \tag{4.2}$$

where the response function $\mathfrak{h}$ is related to the response function $\mathfrak{g}$ of Eq. (2.1) by

$$\mathfrak{h}(F) = |\det F|\,\mathfrak{g}(F) F^{T^{-1}}. \tag{4.3}$$

When the equation of material objectivity (2.2) is expressed in terms of $\mathfrak{h}$, it reads

$$\mathfrak{h}(QF) = Q\,\mathfrak{h}(F). \tag{4.4}$$

Using Eqs. (4.4) and (1.7), we see that Eq. (4.2) is equivalent to

$$T = R\mathfrak{h}(U), \tag{4.5}$$

where R is the rotation tensor and U the right stretch tensor corresponding to F.

In a work on the foundations of thermostatics[9] we have laid down postulates which, in our present theory, are equivalent to the following

Thermostatic Inequality (TI). There exists a (scalar-valued) *energy function* σ such that

$$\sigma(F^*) - \sigma(F) - \operatorname{tr}\{(F^* - F)\mathfrak{h}(F)^T\} > 0 \tag{4.6}$$

for all pairs of deformation gradients F, F^* such that $F^* \neq F$ and $F^* F^{-1}$ is positive-definite and symmetric.

The energy function $\sigma(F)$ in (4.6) can be interpreted as either the internal energy density (per unit volume in the reference configuration) or the Helmholtz free energy density, depending on whether one regards the entropy density or the temperature as the thermodynamic quantity being held fixed, and then suppressed, in defining $\mathfrak{g}$.

In reference 9 it is shown that (4.6) implies

$$\mathfrak{h}(F)^T = \operatorname{grad}_F \sigma(F). \tag{4.7}$$

Once (4.7) is established, our assumption of Material Objectivity, *i.e.* (2.2), is equivalent to the assertion that σ in (4.6) obey the identity

$$\sigma(QF) = \sigma(F) \tag{4.8}$$

for all F and all orthogonal Q.[23]

An alternative formulation of our TI is the following assertion.[24] Consider the Class $\mathfrak{C}$ of continuous rectifiable curves F_t in the space of all invertible tensors F, and let the parameter t for these curves vary from 0 to 1. Let $\mathfrak{C}'$ be the set of all curves F_t in $\mathfrak{C}$ for which $F_1 \neq F_0$ and $F_1 F_0^{-1}$ is both symmetric and positive-definite. Then, the following inequality must hold for curves F_t in $\mathfrak{C}'$:

$$\int_{F_0}^{F_1} \operatorname{tr}\{\mathfrak{h}(F_t)^T dF_t\} > \operatorname{tr}\{(F_1 - F_0)\mathfrak{h}(F_0)^T\}. \tag{4.9}$$

The integral on the left in (4.9) is to be interpreted as a line integral along the curve F_t from $t=0$ to $t=1$.

The quantities appearing on each side of (4.9) represent work, per unit volume in the reference configuration, done against contact forces at a material point X as the local configuration of X is deformed along path $F_t M_o$, $0 \leq t \leq 1$. The quantity on the left in (4.9) gives the "true" work done, *i.e.* the work done assuming that at each t the contact forces on each material surface at X are those which one calculates using the stress tensor $S = \mathfrak{g}(F_t)$ and the actual

[23] In essence, the proof of (4.8) is given in Theorem I of W. NOLL, J. Rational Mech. Anal. **4**, 3 (1955); see also § 1 of reference 11.

[24] The proof that (4.9) is equivalent to the assertion that σ exists and obeys (4.5) is given in reference 11. In writing (4.9) we have made use of the fact that $\mathfrak{g}(F)$ is symmetric. This enables us to simplify Eqs. (1.4) and the inequality (2.3) of reference 11.

configuration of the surface at time t; the quantity on the right in (4.9), however, gives the work which would be done along the path $F_t M_0$ if the contact forces were to remain fixed at their initial values.[25] Thus, (4.9) states that $\mathfrak{h}$ must be such that contact forces always change in a process which results in a pure stretch, and, furthermore, they always change in such a way that the work done against them is greater than that which would have been done had they remained fixed.

Let us now return to (4.6). Of course, when $F^* F^{-1}$ is symmetric and positive definite, so is $F F^{*-1}$, and we can interchange F and F^* in (4.6). If we do this and add the resulting inequality to (4.6), the terms involving σ cancel, and we get, as do TRUESDELL & TOUPIN[12],

$$\operatorname{tr}\{(F^* - F)[\mathfrak{h}(F^*) - \mathfrak{h}(F)]^T\} > 0. \tag{4.10}$$

We now state the

Weakened Thermostatic Inequality WTI. The response function $\mathfrak{h}$ is such that, for all pairs of tensors F^*, F for which $F^* \neq F$ and $F^* F^{-1}$ is positive-definite and symmetric, we have the inequality (4.10).

The WTI is equivalent to demanding that

$$\operatorname{tr}\{(U - I) F[\mathfrak{h}(UF) - \mathfrak{h}(F)]^T\} > 0 \tag{4.11}$$

for all F and all positive-definite symmetric $U \neq I$. Since the reference configuration can always be chosen such that $F = I$, the WTI is also equivalent to the requirement that

$$\operatorname{tr}\{(U - I)[\mathfrak{h}(U) - \mathfrak{h}(I)]\} > 0 \tag{4.12}$$

for all positive-definite symmetric U, no matter what reference configuration is used in defining the response function $\mathfrak{h}$.

Of course the TI implies the WTI; the WTI does not imply TI. It is possible to find (theoretical) materials obeying our WTI but yet such that $\mathfrak{h}$ is not derivable from an energy function σ through (4.7).

Remark. To say that a material obeys the TI is equivalent to the assertion that the response function $\mathfrak{h}$ is such that (4.9) holds for *all* curves F_t in $\mathfrak{C}$. To say that a material obeys the WTI, however, is to assert only that $\mathfrak{h}$ is such that (4.9) holds for those curves F_t in $\mathfrak{C}'$ which are straight lines.[26]

It follows from these observations that any special result derived from the WTI inequality (4.10) will also hold for materials obeying the TI inequality (4.6). Since most of the known implications of the TI, and also the implications of the TI which we wish to report here, can hold also under the weaker (*i.e.* more general) assumption of the WTI, we shall base our present discussion of thermostatics upon the WTI. Throughout the rest of this paper, if the WTI is not mentioned in the statement of a proposition, it is to be understood to be present as an axiom.

[25] Note that, when the configuration of a surface is changing, keeping contact forces fixed is not equivalent to keeping the stress tensor S fixed; rather, it is equivalent to keeping the Piola-Kirchhoff tensor T fixed.

[26] BRAGG, L. E., & B. D. COLEMAN, J. Math. Phys. **4**, 1074 (1963).

5. States of Pressure

Here we obtain some results which show that the WTI places limitations on the properties of local configurations which give rise to a hydrostatic pressure. Our main new result is Theorem 3, which tells us, among other things, that the stress tensor can vanish only in an undistorted configuration of a solid.

Lemma. *Consider two local configurations M_1 and M_2 of the same material, and suppose that they both correspond to hydrostatic pressures:*

$$S_1 = -p_1 I, \tag{5.1a}$$

$$S_2 = -p_2 I. \tag{5.1b}$$

Here p_1 may or may not equal p_2. Consider the right stretch U from M_1 to M_2; i.e., *put*

$$M_2 = F M_1, \qquad F = R U, \tag{5.2}$$

where R is orthogonal and U positive-definite and symmetric. Then, the following inequality must hold whenever $U \neq I$:

$$p_1(\operatorname{tr} U - 3) + p_2(\det U)(\operatorname{tr} U^{-1} - 3) > 0. \tag{5.3}$$

Proof. We use $M_1 = M_0$ as the reference configuration. By Eqs. (4.1), the Piola-Kirchhoff tensors corresponding to M_1 and M_2 are

$$T_1 = S_1 = -p_1 I, \tag{5.4a}$$

$$T_2 = |\det F| \, (F^{-1} S_2)^T = -p_2(\det U) R U^{-1}. \tag{5.4b}$$

Hence, by (4.5), we obtain

$$\mathfrak{h}(I) = -p_1 I, \qquad \mathfrak{h}(U) = -p_2(\det U) U^{-1}. \tag{5.5}$$

Substitution of Eqs. (5.5) into the form (4.12) of the WTI gives

$$\operatorname{tr}\{(U - I)[(-p_2)(\det U) U^{-1} + p_1 I]\} > 0, \tag{5.6}$$

whenever $U \neq I$; (5.6) is equivalent to (5.3), *q.e.d.*

Theorem 1. *Suppose that two local configurations M_1 and M_2 of a material correspond to the same hydrostatic stress $-pI$ with $p > 0$. Suppose further that M_1 and M_2 differ by more than an orthogonal transformation,* i.e., *that the principal stretches u_i, $i = 1, 2, 3$ from M_1 to M_2 are not all 1. Then the following two situations are both impossible:*

$$u_i \geqq 1 \quad \textit{for all } i, \tag{5.7a}$$

$$u_i \leqq 1 \quad \textit{for all } i. \tag{5.7b}$$

Proof. We use the Lemma, putting $p_1 = p_2 = p$ in (5.1). Of course, the u_i mentioned here are just the proper numbers of U defined in (5.2). Now the inequality (5.3) states that if $U \neq I$, *i.e.* if the u_i are not all 1, then

$$p g > 0, \tag{5.8}$$

where

$$\begin{aligned} g &= (\operatorname{tr} U - 3) + (\det U)(\operatorname{tr} U^{-1} - 3) \\ &= (u_1 + u_2 + u_3 - 3) + (u_2 u_3 + u_1 u_3 + u_2 u_3 - 3 u_1 u_2 u_3). \end{aligned} \tag{5.9}$$

Since, by hypothesis, $p>0$, the inequality (5.8) implies

$$g>0. \tag{5.10}$$

Differentiation of Eq. (5.9) yields equations of the type

$$\frac{\partial g}{\partial u_j}=u_k+u_l+1-3u_k u_l, \tag{5.11}$$

where j, k, and l are all distinct. Suppose now that the inequalities (5.7a) hold. We then also have inequalities of the form

$$u_k u_l \geqq \frac{u_k+u_l+1}{3}, \qquad k\neq l, \tag{5.12}$$

and hence, by Eq. (5.11),

$$\frac{\partial g}{\partial u_j}\leqq 0, \qquad j=1,2,3. \tag{5.13}$$

Since g reduces to 0 when $u_1=u_2=u_3=1$, it follows from (5.13) that $g\leqq 0$ when (5.7a) holds, which contradicts (5.10). Hence the inequalities (5.7a) are impossible. To show that the inequalities (5.7b) are also impossible, we need only interchange the roles of the configurations M_1 and M_2, *q.e.d.*

Corollary. *Suppose two local configurations M_1 and M_2 give rise to the same positive hydrostatic pressure p and differ only by a similarity transformation $F=\alpha Q$, Q orthogonal, $\alpha>0$. Then $\alpha=1$;* i.e., *M_1 and M_2 must have the same density and can differ by only a rotation or reflection.*

Theorem 2. *Suppose that, for a given material, M_1 and M_2 are two local configurations which have the same density and give rise to hydrostatic pressures p_1 and p_2, respectively, so that Eqs.* (5.1) *hold. If $p_1\leqq 0$ and $p_2\leqq 0$, then M_1 and M_2 can differ by only an orthogonal transformation, and we must have $p_1=p_2$.*

Proof. Since here M_1 and M_2 have the same density, the right stretch tensor U of (5.2) satisfies

$$\det U=\det U^{-1}=1. \tag{5.14}$$

Thus, if $U\neq I$, the inequality (5.3) becomes

$$p_1(\operatorname{tr} U-3)+p_2(\operatorname{tr} U^{-1}-3)>0. \tag{5.15}$$

The arithmetic mean of a set of positive numbers is strictly greater than the geometric mean unless the numbers are all equal to one. Applied to the proper numbers of a positive-definite, symmetric tensor A, this observation states that

$$\sqrt[3]{\det A}<\frac{\operatorname{tr} A}{3} \quad \text{if} \quad A\neq I. \tag{5.16}$$

On putting $A=U$ and $A=U^{-1}$ in (5.16) and using (5.14), we obtain

$$\operatorname{tr} U-3>0, \qquad \operatorname{tr} U^{-1}-3>0, \tag{5.17}$$

whenever $U\neq I$. If $p_1\leqq 0$ and $p_2\leqq 0$, the inequalities (5.15) and (5.17) are inconsistent. Hence $U=I$, which means that M_1 and M_2 can differ only by an orthogonal transformation $F=R$ (see Eq. (5.2)). Moreover, using Eqs. (2.1) and (2.2) with M_1 as the reference configuration, $F=I$ and $Q=R$, we find

$$S_1 = -p_1 I = \mathfrak{g}(I), \tag{5.18a}$$

$$S_2 = -p_2 I = \mathfrak{g}(R) = R\mathfrak{g}(I)R^{-1} = R(-p_1 I)R^{-1} = -p_1 I; \tag{5.18b}$$

i.e. $p_1 = p_2$, q.e.d.

Corollary. *If* $\mathfrak{g}(H)=\mathfrak{g}(I)=-pI$ *where* H *is unimodular but non-orthogonal, then* $p>0$.

Theorem 3. *If there exists a local configuration* M_0 *such that the corresponding stress* S_0 *vanishes or is a negative hydrostatic pressure, then the material under consideration is a solid, and* M_0 *is an undistorted configuration.*

The theorem states that if $S_0=-pI$, with $p\leqq 0$, then the isotropy graup $\mathscr{I}$ relative to M_0 is a subgroup of $\mathscr{O}$.

Proof. By the definition of $\mathscr{I}$, if H is in $\mathscr{I}$, then $|\det H|=1$ and $\mathfrak{g}(H)=\mathfrak{g}(I)=S_0$. Since we here have $S_0=-pI$ with $p\leqq 0$, the Corollary to Theorem 2 shows that H must be orthogonal, *q.e.d.*

Let us now consider elastic fluids. On turning back to Proposition 7 of Section 2 and observing that the stress in a fluid is characterized by a scalar-valued function p of a scalar ϱ, we see that our WTI is equivalent to our TI for a fluid.[27] It follows from Theorem 3 that $p(\varrho)$ in (2.12) is always positive. Furthermore, it follows from Theorem 1 that $p(\varrho)$ is an invertible function. These observations are sharpened in the following theorem.[11]

Theorem 4. *For an elastic fluid, a necessary and sufficient condition for the WTI (and hence also for the TI) is that the function* $p(\varrho)$ *in* (2.12) *obey the following inequalities for all* ϱ:

$$p(\varrho) > 0, \tag{5.19}$$

$$\varrho \frac{dp(\varrho)}{d\varrho} \geqq \frac{2}{3} p(\varrho). \tag{5.20}$$

In (5.20) *it is to be understood that equality holds on only a nowhere dense set of values of* ϱ.

Theorem 4 has the same content as Theorem 6 of reference 9, as is easily seen by observing that (5.20) is equivalent to the statement that the internal energy density in a fluid is a convex function of the cube root of the specific volume.[28]

Various implications of the WTI for isotropic materials are derived by COLEMAN & NOLL[29], TRUESDELL & TOUPIN[12], BRAGG & COLEMAN[30], and NOLL & TRUESDELL[31, 32].

[27] This is not the case when thermodynamic variables such as the temperature or entropy density are considered.

[28] An extension of the inequality (5.20) to a statement about the mean pressure on arbitrary states of strain in general elastic materials is given in reference 26.

[29] Section 12 of reference 9.

[30] BRAGG, L. E., & B. D. COLEMAN, J. Math. Phys. 4, 424 (1963).

[31] NOLL, W., & C. TRUESDELL, in: Encyclopedia of Physics, Edited by S. FLÜGGE, Vol. III/3. Berlin-Göttingen-Heidelberg: Springer (forthcoming).

[32] For its discussion of inequalities in isotropic materials the article of TRUESDELL & TOUPIN[12] is the most exhaustive currently in print. It also contains some interesting remarks on uniqueness questions for boundary value problems in materials of arbitrary symmetry.

6. Differential Inequalities

We assume now that the response function $\mathfrak{g}$ of Eq. (2.1) or, equivalently, the response function $\mathfrak{h}$ of Eq. (4.2) is continuously differentiable and investigate the restriction that the WTI imposes on the gradients of $\mathfrak{g}$ and $\mathfrak{h}$.

Let D be an arbitrary non-zero symmetric tensor, and consider

$$U = I + \tau D, \tag{6.1}$$

where τ is a real parameter $\neq 0$. It is clear that when $|\tau|$ is sufficiently small the tensor U given by Eq. (6.1) is not only symmetric but also positive-definite and $\neq I$. Let us make the abbreviation

$$f(\tau) = \mathrm{tr}\{D\mathfrak{h}(I + \tau D)^T\}. \tag{6.2}$$

Starting with (4.12) and letting U be given by (6.1), we find that the WTI may be written

$$\tau[f(\tau) - f(0)] > 0. \tag{6.3}$$

Dividing (6.3) by τ^2 and taking the limit $\tau \to 0$, we obtain

$$\dot{f}(0) \geqq 0, \tag{6.4}$$

where the superimposed dot denotes the derivative. The directional derivative $\nabla\mathfrak{h}[D]$ of $\mathfrak{h}$ at I, in the direction of the tensor D, is defined by

$$\frac{d}{d\tau}\mathfrak{h}(I + \tau D)\Big|_{\tau=0} = \nabla\mathfrak{h}[D]. \tag{6.5}$$

Here $\nabla\mathfrak{h}$, the *gradient* of $\mathfrak{h}$ at I, is a linear transformation of the six-dimensional space of symmetric tensors D into the nine-dimensional space of arbitrary tensors. We now define a quadratic form $\mathbf{\Omega}$ on the six-dimensional space of all symmetric tensors D by

$$\mathbf{\Omega}(D) = \mathrm{tr}\{D(\nabla\mathfrak{h}[D])^T\}. \tag{6.6}$$

Equations (6.2), (6.5), and (6.6) show that the inequality (6.4) states that $\mathbf{\Omega}(D) \geqq 0$. Thus we have

Theorem 1. *In order for the WTI to hold it is necessary that the quadratic form* $\mathbf{\Omega}$ *defined by* (6.6) *be positive-semidefinite,* i.e. *that*

$$\mathbf{\Omega}(D) \geqq 0 \tag{6.7}$$

for all symmetric tensors D.

Of course, the response function $\mathfrak{h}$ and hence the quadratic form $\mathbf{\Omega}$ depend on the choice of the reference configuration; $\mathbf{\Omega}$ must be positive semi-definite for *every* such choice.

To cast (6.7) into a more transparent form we consider a smooth one-parameter family of deformation gradients $F(\tau)$ such that $F(0) = I$. The corresponding rotation tensors are denoted by $R(\tau)$ and the corresponding right stretch tensors by $U(\tau)$. We then have $R(0) = U(0) = I$. If we put

$$\dot{R}(0) = W, \qquad \dot{U}(0) = D, \tag{6.8}$$

we find, by differentiating Eq. (1.7) with respect to τ and then putting $\tau = 0$, that

$$\dot{F}(0) = W + D. \tag{6.9}$$

Moreover, it is clear that D is symmetric and it is easy to show that W is skew (by differentiating $R^T R = I$).

Let $S(\tau)$ and $T(\tau)$ be the stress and the Piola-Kirchhoff tensor, respectively, corresponding to $F(\tau)$. Equation (4.1) shows that $T(0) = S(0)$. We use the notation

$$S(0) = T(0) = S_o, \qquad \dot{S}(0) = \dot{S}_o, \qquad \dot{T}(0) = \dot{T}_o. \tag{6.10}$$

On differentiating (4.5) we find

$$\dot{T}_o = W\mathfrak{h}(I) + \nabla\mathfrak{h}[D], \tag{6.11}$$

where $\nabla\mathfrak{h}$ is the gradient of $\mathfrak{h}$ at I and where the notations (6.9) and (6.10) have been used. Since $S_o = \mathfrak{h}(I)$, Eq. (6.11) is equivalent to

$$\dot{T}_o - W S_o = \nabla\mathfrak{h}[D]. \tag{6.12}$$

Let us denote the density of the local configuration corresponding to $F(\tau)$ by $\varrho(\tau)$ and use the notation $\varrho_o = \varrho(0)$, $\dot{\varrho}_o = \dot{\varrho}(0)$. If we differentiate Eq. (4.1) with respect to τ and then put $\tau = 0$, we find, using (6.8)—(6.10),

$$\varrho_o \dot{S}_o = \dot{\varrho}_o S_o + \varrho_o S_o (D - W) + \varrho_o \dot{T}_o. \tag{6.13}$$

The equation of continuity[33] implies $\dot{\varrho}_o = -(\operatorname{tr} D)\varrho_o$, hence (6.13) is equivalent to

$$\dot{S}_o = -S_o(\operatorname{tr} D) + S_o(D - W) + \dot{T}_o. \tag{6.14}$$

Combining Eqs. (6.6), (6.12), and (6.14), we find

$$\boldsymbol{\Omega}(D) = \operatorname{tr}\{(\dot{S}_o + S_o W - W S_o) D\} + \operatorname{tr} D \operatorname{tr}(S_o D) - \operatorname{tr}(S_o D^2). \tag{6.15}$$

If we differentiate Eq. (2.3) with respect to τ and then put $\tau = 0$, we find

$$\dot{S}_o + S_o W - W S_o = \nabla\mathfrak{g}[D], \tag{6.16}$$

where $\nabla\mathfrak{g}$ is the gradient of the response function $\mathfrak{g}$ at the tensor I. We may regard $\nabla\mathfrak{g}$ as a linear transformation of the six-dimensional space of symmetric tensors into itself. When components are used, the matrix of $\nabla\mathfrak{g}$ has 36 components, which are the elastic coefficients for an infinitesimal deformation from the reference configuration. Of course, these elastic coefficients depend on the choice of the reference configuration.

It follows from Eqs. (6.15) and (6.16) that the quadratic form $\boldsymbol{\Omega}(D)$ may be expressed in terms of $\nabla\mathfrak{g}$ by

$$\boldsymbol{\Omega}(D) = \operatorname{tr}\{D\,\nabla\mathfrak{g}[D]\} + (\operatorname{tr} D)\operatorname{tr}(S_o D) - \operatorname{tr}(S_o D^2). \tag{6.17}$$

We call a one-parameter family $M = M(t)$ of local configurations a *deformation path*. If we define $F(\tau)$ by $M(t+\tau) = F(\tau)M(t)$, the tensors D and W are called the *stretching* and the *spin* of the deformation path. In the notation of Ebs. (6.10)—(6.17) the stress corresponding to $M(t)$ is S_o. We now omit the subscript o and note that Eq. (6.15) and Theorem 1 yield

Theorem 2. *In order for the WTI to hold it is necessary that for every deformation path*

$$\operatorname{tr}(S^* D) \geqq 0, \tag{6.18}$$

[33] Or, equivalently, (1.3).

where D is the stretching, and S^* is the invariant stress rate defined by

$$S^* = \dot{S} + SW - WS + S\,\mathrm{tr}D - \tfrac{1}{2}(SD + DS). \tag{6.19}$$

In (6.19) *S is the stress and W the spin.*

In his "Recherches sur l'élasticité," P. Duhem[34] makes the following assertion.

"Considérons d'abord un système défini par un certain nombre de variables normales $\alpha, \beta, \ldots, \lambda$, hors la température absolue T. Supposons qu'à une certaine température T, le système prenne un état d'équilibre lorsqu'on le soumet aux actions extérieures $A, B, \ldots, L$, et que cet état d'équilibre varie d'une manière continue lorsque, sans faire varier la température T, on fait varier les valeurs $A, B, \ldots, L$ des actions extérieures. Si les actions $A, B, \ldots, L$ éprouvent des variations infiniment petites $dA, dB, \ldots, dL$, que nous nommerons des *actions perturbatrices*, les valeurs des variables $\alpha, \beta, \ldots, \lambda$ qui conviennent à l'équilibre éprouvent des variations $d\alpha, d\beta, \ldots, d\lambda$, que nous nommerons des *perturbations*; l'expression

$$dA\,d\alpha + dB\,d\beta + \cdots + dL\,d\lambda$$

sera nommée le *travail perturbateur isothermique*.

"Dire qu'un travail perturbateur est positif, c'est dire, sous une forme mathématique précise, que la perturbation se produit dans le sens vers lequel tendent les actions perturbatrices. Il est clair que *les systèmes que la nature nous offre seront tels, en général, que tout travail perturbateur isothermique, accompli à partir d'un état d'équilibre, soit positif*. C'est ce que nous experimerons en disant qu'ils sont soumis à la *loi du déplacement isothermique de l'équilibre*."

Our Theorem 2 above furnishes a precise mathematization, for elastic materials, of Duhem's "law of isothermal displacement of equilibrium".

Isotropy

We now seek the limitations which the WTI places on the elastic coefficients characterizing the response of an isotropic material to an infinitesimal deformation from an undistorted state.

Taking the gradient of Eq. (2.11) with respect to F at $F = I$, we find that

$$\nabla\mathfrak{g}[QDQ^{-1}] = Q\nabla\mathfrak{g}[D]Q^{-1} \tag{6.20}$$

holds for all orthogonal tensors Q in the isotropy group and all symmetric tensors D. Here we assume that $\mathfrak{g}$ is the response function for an undistorted configuration of an isotropic material; it then follows that $\nabla\mathfrak{g}[D]$ is a linear isotropic tensor function of D. It is a consequence of this remark that $\nabla\mathfrak{g}$ has the representation[35]

$$\nabla\mathfrak{g}[D] = \lambda(\mathrm{tr}D)I + 2\mu D, \tag{6.21}$$

where λ and μ (the Lamé coefficients) are elastic coefficients which depend on the reference configuration. By Proposition 6 of Section 2, the stress S_0 on the undistorted reference configuration must be of the form

$$S_0 = -pI. \tag{6.22}$$

Substitution of Eqs. (6.21) and (6.22) into (6.17) gives

$$\boldsymbol{\Omega}(D) = (\lambda - p)(\mathrm{tr}D)^2 + (2\mu + p)\,\mathrm{tr}D^2. \tag{6.23}$$

[34] Duhem, P., Ann. École Norm. **22**, 143—217 (1905), p. 193.

[35] *Cf.* A. E. H. Love, A Treatise on the Mathematical Theory of Elasticity (Cambridge, 4th Edition, 1927), §§ 106—110. Love's assumption of the existence of a strain energy function does not affect the generality of his results for isotropic materials or cubic crystals.

Now, quadratic forms of the type

$$q=a(d_1+d_2+d_3)^2+b(d_1^2+d_2^2+d_3^2) \tag{6.24}$$

are well known. A necessary and sufficient condition that q be positive semi-definite in the triple (d_1, d_2, d_3) is that both

$$b\geqq 0 \quad \text{and} \quad 3a+2b\geqq 0. \tag{6.25}$$

If we take $a=\lambda-p$, $b=2\mu+p$, and d_i to be the proper numbers of D, the quadratic form (6.24) reduces to (6.23). Therefore, a necessary and sufficient condition that $\mathbf{\Omega}(D)$ be positive semi-definite is that

$$2\mu+p\geqq 0 \quad \text{and} \quad 3(\lambda-p)+2(2\mu+p)\geqq 0. \tag{6.26}$$

From Theorem 1 we now obtain the following result.

Theorem 3. *The WTI implies that for each undistorted configuration of an isotropic material the following inequalities hold:*

$$\mu\geqq -\tfrac{1}{2}p, \tag{6.27a}$$

$$\varkappa\geqq \tfrac{2}{3}p; \tag{6.27b}$$

here p is the pressure, μ the shear modulus, and $\varkappa=\lambda+\frac{2}{3}\mu$ the compression modulus; of course p, μ, and $\varkappa$ can all depend on the choice of the reference configuration.

If $p=0$, *i.e.*, if the reference configuration is a natural state of an isotropic solid, then Theorem 3 yields the familiar assertion that

$$\mu\geqq 0 \quad \text{and} \quad \varkappa\geqq 0. \tag{6.28}$$

For an elastic fluid we have $\mu=0$ and $\varkappa=\varrho\, dp(\varrho)/d\varrho$, and hence Theorem 3 yields

$$p\geqq 0, \qquad \varrho\frac{dp(\varrho)}{d\varrho}\geqq\frac{2}{3}p.$$

This condition is just a bit weaker than the necessary and sufficient condition for the WTI stated in Theorem 4 of Section 5, for our present Theorem 3 does not contain the precious result that $p=0$ is *impossible* in a fluid. Yet we think Theorem 3 to be not without interest. Through its conclusion (6.27a) it yields a condition on μ for a particular negative pressure p to be possible: If the configuration is varied among undistorted configurations in such a way that p decreases and becomes negative, μ must always remain greater than $-\frac{1}{2}p$, otherwise the material will not support the negative pressure.[36]

Cubic Symmetry

When the material has cubic symmetry and the reference configuration is undistorted, it follows from (6.20) that there exists an orthonormal basis $\boldsymbol{h}_i$ such that the components of $\nabla\mathfrak{g}[D]$ and D are related by[37]

$$\nabla\mathfrak{g}[D]_{ii}=\alpha D_{ii}+\beta(D_{jj}+D_{kk}), \tag{6.29a}$$

$$\nabla\mathfrak{g}[D]_{ij}=2\gamma D_{ij}. \tag{6.29b}$$

[36] We take this to mean that cavitation will occur, as seems to be the case with fluids.

[37] The result is essentially given in LOVE's treatise.[35] The three coefficients α, β, γ, are related to those on p. 160 of reference 35 by $C_{11}=\alpha$, $C_{12}=\beta$, $C_{44}=\gamma$.

No summation convention is used here, and i, j, k is any permutation of 1, 2, 3. It follows from the last entry in Table 2 that the stress S_o in the reference configuration must be a hydrostatic pressure:

$$S_o = -pI. \tag{6.30}$$

Substitution of (6.29) and (6.30) into (6.17) yields

$$\boldsymbol{\Omega}(D) = \alpha \sum_{i=1}^{n} D_{ii}^2 + 2(\beta - p) \sum_{\substack{i,j=1\\ j>i}}^{3} D_{ii} D_{jj} + (4\gamma + 2p) \sum_{\substack{i,j=1\\ j>i}}^{3} D_{ij}^2. \tag{6.31}$$

Thus $\boldsymbol{\Omega}(D)$ is the sum of the two quadratic forms

$$\boldsymbol{\Omega}_1(D) = \alpha \sum_{i=1}^{3} D_{ii}^2 + (2\beta - p) \sum_{\substack{i,j=1\\ j>i}}^{3} D_{ii} D_{jj}, \tag{6.32a}$$

and

$$\boldsymbol{\Omega}_2(D) = (4\gamma + 2p) \sum_{\substack{i,j=1\\ j>i}}^{3} D_{ij}^2. \tag{6.32b}$$

The form $\boldsymbol{\Omega}_1(D)$ involves the three diagonal components of D, which can be chosen at will, while the form $\boldsymbol{\Omega}_2(D)$ involves only the three off-diagonal components of D, which also can be chosen at will. Hence, in order for $\boldsymbol{\Omega}(D)$ to be positive-semidefinite it is necessary and sufficient that both of the forms (6.32) be positive-semidefinite.

It is clear that the form $\boldsymbol{\Omega}_2(D)$ of (6.32b) is positive-semidefinite if and only if $4\gamma + 2p \geqq 0$. The form $\boldsymbol{\Omega}_1(D)$ of (6.32a) is of the type (6.24) when we put $a + b = \alpha$, $2a = 2\beta - p$, and $d_i = D_{ii}$, $i = 1, 2, 3$. Using (6.25), it follows that the form $\boldsymbol{\Omega}(D)$ of (6.31) is positive-semidefinite if and only if the three inequalities

$$2\gamma + p \geqq 0, \qquad \alpha - \beta + \tfrac{1}{2}p \geqq 0, \qquad \beta + 2\alpha - \tfrac{1}{2}p \geqq 0 \tag{6.33}$$

hold. We summarize:

Theorem 4. *The WTI implies that for each undistorted configuration of a material with cubic symmetry, the elastic coefficients α, β, γ and the pressure p obey the inequalities*

$$\gamma \geqq -\tfrac{1}{2}p, \qquad \alpha - \beta \geqq -\tfrac{1}{2}p, \qquad \beta + 2\alpha \geqq \tfrac{1}{2}p. \tag{6.34}$$

Acknowledgement. This research was supported by the Air Force Office of Scientific Research under Contract AF 49(638)541 with Mellon Institute and by the National Science Foundation under Grant NSF-G 16745 to Carnegie Institute of Technology.

Mellon Institute
Pittsburgh
and
Department of Mathematics
Carnegie Institute of Technology
Pittsburgh

(Received August 16, 1963)

EUCLIDEAN GEOMETRY AND MINKOWSKIAN CHRONOMETRY

WALTER NOLL

1. Introduction. The term "Minkowskian Chronometry" is used here for the study of the structure of space-time appropriate to Einstein's special relativity. This term suggests a parallel with Euclidean geometry. Originally, geometry was an empirical science which dealt with measuring of distances on earth. The Greeks transformed this science into a beautiful mathematical discipline. Chronometry is the science of the measurement of time intervals. Einstein's first paper was a critical study of time measurements. It was Minkowski, however, who made a geometrical discipline out of Einstein's chronometry. More recently, Synge ([2], [3]) has been an able advocate of the geometrical point of view in relativity.

The present paper is an attempt to axiomatize Minkowskian chronometry using direct coordinate-free methods. I believe that the coordinate-free approach fosters the cultivation of intuition, a scarce commodity in relativity because the phenomena this theory is intended to describe are as yet rather remote from our daily experience. I hope, moreover, that Minkowskian chronometry will become, as Euclidean geometry did, a branch of mathematics that is of interest purely for its esthetic value.

Section 2 contains a collection of definitions and results on vector spaces with an inner product that is not necessarily positive definite. We give a new proof of the inertia theorem of Sylvester, a proof that remains valid for infinite dimensional spaces.

In Section 3 we present certain inequalities valid in inner product spaces of index one. The most notable of these are the "reversed Schwarz inequality" and the "reversed triangle inequality."

Section 4 consists of an axiomatic introduction to pseudo-Euclidean geometry, based on the fact that the structure of a pseudo-Euclidean space is determined by its "separation function," which in the Euclidean case is identical to the square of the metric. Associated with each pseudo-Euclidean space is a unique "translation space," which is a vector space with inner product.

"Inner product" is just another term for "nondegenerate bilinear form." Usually the term is used, however, only when the form is positive definite. But indefinite inner products can be treated in the same geometric spirit as definite inner products usually are, and we adopt here the terminology that has become standard so far only for the case of definite inner products (see, e.g. Halmos [5]). The theory of bilinear forms is extensive, much of it going back to the nineteenth century. In most textbooks, however, the subject is treated in analytical language with components. A notable exception is the treatise of Bourbaki [1], to which we refer for further information on the topics treated in Sections 2 and 4.

When the index of the translation space is zero, pseudo-Euclidean geometry reduces to Euclidean geometry, and when the index is one, it reduces to Minkowskian chronometry. From a physical point of view it is reasonable to use distance as a primitive notion in Euclidean geometry, because there are yardsticks to measure distances. In this case, the separation of two points is just the square of their distance. There are no "separation meters," however, to measure the separation of two arbitrary events in relativistic space-time. In Section 5 we give an axiomatic introduction to Minkowskian chronometry based on primitive notions that have a more direct physical meaning: observers, clock-readings, signals. We show—and this is the least trivial part of the paper—that these primitive data determine the separation function uniquely. The fact that the index of the translation space is one is not an assumption of the theory but comes out as a theorem.

Section 6 deals with temporal order, i.e. with the possibility of distinguishing future from past. It turns out that such order, in Minkowskian chronometry, is related to the distinction between emission and reception of signals. We note that in classical space-time, such distinction is not sufficient to determine temporal order. The axioms given here may answer a problem proposed by Suppes ([4], Sect. 4).

We do not impose anywhere a restriction on the dimension, which may even be infinite. Of course, in the presently known physical applications of Minkowskian chronometry the dimension is 4.

2. Inner product spaces. (For details see [1].) Let $\mathcal{V}$ be a real vector space. A nondegenerate symmetric bilinear form on $\mathcal{V}$ will be called an inner product. The inner product of $\mathbf{u}, \mathbf{v} \in \mathcal{V}$ will be denoted by $\mathbf{u}\cdot\mathbf{v}$, and we will abbreviate $\mathbf{u}\cdot\mathbf{u}=\mathbf{u}^2$. Nondegeneracy means that

$$(2.1) \qquad \mathbf{u}\cdot\mathbf{v} = 0 \text{ for all } \mathbf{v} \in \mathcal{V} \text{ implies } \mathbf{u} = \mathbf{0}.$$

A *quadratic form* Φ on $\mathcal{V}$ is a function $\Phi\colon \mathcal{V}\to\mathcal{R}$ ($\mathcal{R}$ = set of real numbers) such that $\Phi(\mathbf{v}) = \mathbf{v}\cdot\mathbf{v}$ for some inner product. The quadratic form determines the inner product uniquely.

An *inner product space* $\mathcal{V}$ is a vector space with an additional structure defined by an inner product. If $\mathcal{U}$ is a subspace of $\mathcal{V}$, then $\mathcal{U}^{\perp} = \{\mathbf{v} \mid \mathbf{v}\cdot\mathbf{u} = 0$ for all

$\mathbf{u}\in\mathfrak{U}\}$ is called the *orthogonal complement* of $\mathfrak{U}$. If dim $\mathfrak{U}$ is finite, then $\mathfrak{U}^{\perp\perp}=\mathfrak{U}$ and dim $\mathfrak{U}+$dim $\mathfrak{U}^{\perp}=$dim $\mathfrak{V}$. A subspace $\mathfrak{U}$ is called *regular* if $\mathfrak{U}\cap\mathfrak{U}^{\perp}=\{\mathbf{0}\}$ and *singular* if $\mathfrak{U}\cap\mathfrak{U}^{\perp}\neq\{\mathbf{0}\}$. If dim $\mathfrak{U}$ is finite and $\mathfrak{U}$ is regular, then $\mathfrak{V}$ has the direct decomposition $\mathfrak{V}=\mathfrak{U}\oplus\mathfrak{U}^{\perp}$.

A linear transformation Q: $\mathfrak{V}\rightarrow\mathfrak{V}$ is called *orthogonal* if it preserves the inner product. In view of (2.1), Q is then also one-to-one and hence an automorphism of the inner product space $\mathfrak{V}$.

The following terminology is suggested by the physical applications. The sets of vectors

$$\mathfrak{V}_{+} = \{\mathbf{v}\mid \mathbf{v}^2 > 0 \text{ or } \mathbf{v} = \mathbf{0}\},$$
$$\mathfrak{V}_{-} = \{\mathbf{v}\mid \mathbf{v}^2 < 0 \text{ or } \mathbf{v} = \mathbf{0}\}, \qquad (2.2)$$
$$\mathfrak{V}_{0} = \{\mathbf{v}\mid \mathbf{v}^2 = 0\}$$

will be called the *space-cone*, the *time-cone*, and the *signal-cone*, respectively. These cones have only the zero vector $\mathbf{0}$ in common. A vector $\mathbf{v}$ is said to be *space-like*, *time-like*, or a *signal vector* depending on whether $\mathbf{v}\in\mathfrak{V}_{+}$, $\mathbf{v}\in\mathfrak{V}_{-}$, or $\mathbf{v}\in\mathfrak{V}_{0}$. The maximal dimension of the time-like subspaces of $\mathfrak{V}$, (i.e. subspaces contained in $\mathfrak{V}_{-}$) will be called the *index* of $\mathfrak{V}$; it will be denoted by $i=$ind $\mathfrak{V}$. (The customary definition is ind $\mathfrak{V}=$maximal dimension of the signal subspaces of $\mathfrak{V}$. Our definition will give the same value if the sign of the inner product is properly adjusted, as is shown in Theorem 2 below.)

THEOREM 1. *If $\mathfrak{U}$ is a time-like subspace of maximal dimension $i=$ind $\mathfrak{V}$ and if $i<\infty$, then the complement $\mathfrak{U}^{\perp}$ is space-like and $\mathfrak{V}$ has the direct decomposition*

$$\mathfrak{V} = \mathfrak{U} \oplus \mathfrak{U}^{\perp}, \qquad \mathfrak{U} \subset \mathfrak{V}_{-}, \ \mathfrak{U}^{\perp} \subset \mathfrak{V}_{+}. \qquad (2.3)$$

In addition, for any decomposition of the type (2.3), *we have* dim $\mathfrak{U}=i$.

Proof. Let $\mathfrak{U}\subset\mathfrak{V}_{-}$, dim $\mathfrak{U}=i$. $\mathfrak{U}$ is then regular and $\mathfrak{V}$ has the direct decomposition $\mathfrak{V}=\mathfrak{U}\oplus\mathfrak{U}^{\perp}$. Assume that $\mathfrak{U}^{\perp}$ is not space-like. Then there is a vector $\mathbf{w}\in\mathfrak{U}^{\perp}$ such that $\mathbf{w}^2\leqq 0$, $\mathbf{w}\neq\mathbf{0}$. By (2.1), there is a vector $\mathbf{v}\in\mathfrak{V}$ such that $\mathbf{w}\cdot\mathbf{v}=\alpha\neq 0$. Let

$$\mathbf{v} = \mathbf{u} + \mathbf{z}, \qquad \mathbf{u} \in \mathfrak{U},\ \mathbf{z} \in \mathfrak{U}^{\perp}$$

be the decomposition of $\mathbf{v}$. We have $\alpha=\mathbf{w}\cdot\mathbf{v}=\mathbf{w}\cdot\mathbf{u}+\mathbf{w}\cdot\mathbf{z}=\mathbf{w}\cdot\mathbf{z}$. Since $\mathbf{w}$, $\mathbf{z}\in\mathfrak{U}^{\perp}$, the vector $\mathbf{p}=\mathbf{z}+\beta\mathbf{w}$ also belongs to $\mathfrak{U}^{\perp}$ for any choice of β. Now,

$$\mathbf{p}^2 = \mathbf{z}^2 + 2\beta\mathbf{w}\cdot\mathbf{z} + \beta^2\mathbf{w}^2 \leqq \mathbf{z}^2 + 2\beta\alpha.$$

Since $\alpha\neq 0$, we can adjust β such that $\mathbf{p}^2<0$. If $\mathbf{x}\in\mathfrak{U}$ and $\mathbf{y}=\mathbf{x}+\lambda\mathbf{p}$, then $\mathbf{y}^2=\mathbf{x}^2+\lambda^2\mathbf{p}^2<0$, unless $\mathbf{y}=\mathbf{0}$. Therefore, the space $\mathfrak{U}'$ spanned by $\mathfrak{U}$ and $\mathbf{p}$ is time-like and of dimension $i+1$, which contradicts the assumption that $\mathfrak{U}$ is of maximal dimension.

Consider now an arbitrary decomposition of the form (2.3) and assume that $\hat{\mathfrak{U}}$ is a time-like subspace of maximal dimension i. Every $\hat{\mathbf{u}}\in\hat{\mathfrak{U}}$ has a unique decomposition $\hat{\mathbf{u}} = \mathbf{u} + \mathbf{z}$, $\mathbf{u} \in \mathcal{U}$, $\mathbf{z} \in \mathcal{U}^\perp$. This decomposition defines a linear transformation $\hat{\mathbf{u}} \to \mathbf{u} = L\hat{\mathbf{u}}$ of $\hat{\mathcal{U}}$ into $\mathcal{U}$. If $\mathbf{u} = \mathbf{0}$, then $\hat{\mathbf{u}} = \mathbf{0} + \mathbf{z} \in \mathcal{U}^\perp \subset \mathcal{V}_+$, and hence $\hat{\mathbf{u}} = \mathbf{0}$. Thus L is one-to-one, which implies $i = \dim \hat{\mathcal{U}} \leqq \dim \mathcal{U}$. On the other hand, $\dim \mathcal{U} \leqq \dim \hat{\mathcal{U}} = i$ by definition of i. Q.E.D.

In the case when $\dim \mathcal{V}$ is finite, Theorem 1 reduces to a version of the classical inertia theorem of Sylvester (cf. [1], Chapter 7, no. 2).

Lincoln E. Bragg has disclosed to me a counterexample which shows that the conclusion of Theorem 1 may be false when $i = \infty$. From now on we assume that the index i of $\mathcal{V}$ is finite ($\dim \mathcal{V}$ need not be finite).

COROLLARY. *Let a decomposition of the type* (2.3) *be given. Then any other decomposition of the same type has the form*

$$\mathcal{V} = Q(\mathcal{U}) \oplus Q(\mathcal{U}^\perp), \tag{2.4}$$

where Q is orthogonal.

THEOREM 2. *If* $\dim \mathcal{V} - i \geqq i$, *then the maximal dimension of the signal subspaces of $\mathcal{V}$ is also given by i.*

Proof. Let $\mathcal{S}$ be a signal subspace of maximal dimension i', and consider a decomposition of the form (2.3). By Theorem 1 we have $\dim \mathcal{U} = i$, $\dim \mathcal{U}^\perp = \dim \mathcal{V} - i \geqq i$. The unique decomposition $\mathbf{s} = \mathbf{u} + \mathbf{z}$, $\mathbf{u} \in \mathcal{U}$, $\mathbf{z} \in \mathcal{U}^\perp$ defines two linear transformations $\mathbf{s} \to \mathbf{u} = L\mathbf{s}$ and $\mathbf{s} \to \mathbf{z} = N\mathbf{s}$ of $\mathcal{S}$ into $\mathcal{U}$ and $\mathcal{U}^\perp$, respectively. It is easily seen that both L and N are one-to-one. It follows that $i' = \dim \mathcal{S} \leqq \dim \mathcal{U} = i$. If $i' < i$, $L(\mathcal{S})$ and $N(\mathcal{S})$ must be proper subspaces of $\mathcal{U}$ and $\mathcal{U}^\perp$, respectively. It is then possible to find $\mathbf{x} \in \mathcal{U}$ and $\mathbf{y} \in \mathcal{U}^\perp$ such that $\mathbf{x}$ is orthogonal to $L(\mathcal{S})$ and $\mathbf{y}$ is orthogonal to $N(\mathcal{S})$. If $\mathbf{x}$ and $\mathbf{y}$ are normalized such that $\mathbf{x}^2 = -1$, $\mathbf{y}^2 = +1$, the space spanned by $\mathcal{S}$ and $\mathbf{x} + \mathbf{y}$ is easily seen to be a signal subspace of dimension $i' + 1$, which contradicts the definition of i'. Q.E.D.

3. Spaces of index one. From now on we assume that $\mathcal{V}$ is an inner product space of index one. The following two theorems are corollaries of the results of the previous section.

THEOREM 1. *Let $\mathbf{l}$ be a time-like unit vector ($\mathbf{l}^2 = -1$). Then every vector $\mathbf{v} \in \mathcal{V}$ has unique decomposition of the form*

$$\mathbf{v} = \xi \mathbf{l} + \mathbf{w}, \qquad \mathbf{w} \cdot \mathbf{l} = 0, \qquad \mathbf{w} \in \mathcal{V}_+. \tag{3.1}$$

THEOREM 2. *If a vector is orthogonal to a nonzero time-like vector then it must be space-like.*

In an indefinite inner product space (ind $\mathcal{V} \neq 0$), Schwarz's inequality is of course not valid. When ind $\mathcal{V} = 1$ the following two theorems partially replace the Schwarz inequality:

THEOREM 3 (reversed Schwarz inequality). *If* **u** *and* **v** *are both time-like, then*

$$(\mathbf{u}\cdot\mathbf{v})^2 \geqq \mathbf{u}^2\mathbf{v}^2, \tag{3.2}$$

and equality holds only when **u** *and* **v** *are linearly dependent.*

THEOREM 4. *For any three nonzero time-like vectors* **u**, **v**, *and* **w**,

$$(\mathbf{u}\cdot\mathbf{v})(\mathbf{v}\cdot\mathbf{w})(\mathbf{w}\cdot\mathbf{u}) < 0. \tag{3.3}$$

Proof. Let $\mathbf{u}, \mathbf{v}, \mathbf{w} \in \mathcal{V}_-$, all $\neq \mathbf{0}$. Consider

$$\mathbf{z} = \alpha\mathbf{u} - \beta\mathbf{v}, \qquad \alpha = \mathbf{v}\cdot\mathbf{w}, \qquad \beta = \mathbf{u}\cdot\mathbf{w}.$$

We have $\mathbf{z}\cdot\mathbf{w} = \alpha\beta - \beta\alpha = 0$. It follows from Theorem 2 that

$$0 \leqq \mathbf{z}^2 = \alpha^2\mathbf{u}^2 + \beta^2\mathbf{v}^2 - 2\alpha\beta\mathbf{u}\cdot\mathbf{v},$$

i.e.

$$2(\mathbf{v}\cdot\mathbf{w})(\mathbf{u}\cdot\mathbf{w})(\mathbf{u}\cdot\mathbf{v}) \leqq \alpha^2\mathbf{u}^2 + \beta^2\mathbf{v}^2. \tag{3.4}$$

Here equality can hold only when $\mathbf{z} = \mathbf{0}$, i.e. when **u** and **v** are linearly dependent. Since **w** is time-like and $\neq \mathbf{0}$, Theorem 2 shows that $\alpha = \mathbf{v}\cdot\mathbf{w}$ and $\beta = \mathbf{u}\cdot\mathbf{w}$ cannot be zero. Therefore, the right-hand side of (3.4) is negative, which proves Theorem 4. Theorem 3 is trivial when $\mathbf{u} = \mathbf{0}$ or $\mathbf{v} = \mathbf{0}$. Otherwise it follows from (3.4) when we put $\mathbf{w} = \mathbf{u}$ and observe that $\mathbf{u}^2 < 0$. Q.E.D.

Consider the relation $\mathbf{u}\cdot\mathbf{v} < 0$ within the set $\hat{\mathcal{V}}_-$ of all nonzero time-like vectors. Theorem 4 shows that this relation is transitive. Clearly, it is also reflexive and symmetric. Hence $\mathbf{u}\cdot\mathbf{v} < 0$ is an equivalence relation in $\hat{\mathcal{V}}_-$. The vectors $\mathbf{u} \in \hat{\mathcal{V}}_-$ and $-\mathbf{u} \in \hat{\mathcal{V}}_-$ belong to different equivalence classes, hence there are at least two such classes. But there are no more than two classes, because if **u**, **v**, **w** belonged to three different classes, then the inner products on the left side of (3.3) would be all positive, which contradicts (3.3). We adjoin the zero vector **0** to each of the two equivalence classes and denote the resulting sets by $\mathcal{V}_-^1$ and $\mathcal{V}_-^2$, respectively. It is easily verified that $\mathcal{V}_-^1$ and $\mathcal{V}_-^2$ are in fact convex cones. We summarize:

THEOREM 5. *The time cone* $\mathcal{V}_-$ *is the union of two convex cones* $\mathcal{V}_-^1$ *and* $\mathcal{V}_-^2$ *which have only the zero vector in common. Two time-like vectors* **u**, **v** *belong to the same cone if and only if* $\mathbf{u}\cdot\mathbf{v} \leqq 0$. *If* $\mathbf{u} \in \mathcal{V}_-$ *belongs to one of the cones, then* $-\mathbf{u}$ *belongs to the other, i.e.,* $\mathcal{V}_-^1 = -\mathcal{V}_-^2$.

We call $\mathcal{V}_-^1$ and $\mathcal{V}_-^2$ the two *directed time cones.*

If **v** is time-like, we call $\tau(\mathbf{v}) = \sqrt{-\mathbf{v}^2}$ the *duration* of **v**. We have $\tau(\mathbf{v}) \geqq 0$, and $\tau(\mathbf{v}) = 0$ only if $\mathbf{v} = \mathbf{0}$. The following "reversed triangle inequality" is a consequence of Theorems 3 and 5. The proof is analogous to that of the ordinary triangle inequality.

THEOREM 6 (reversed triangle inequality). *If the two vectors* $\mathbf{u}$ *and* $\mathbf{v}$ *both belong to one of the directed time cones* $(\mathbf{u}\cdot\mathbf{v}\leqq 0)$, *then*

$$\tau(\mathbf{u}+\mathbf{v}) \geqq \tau(\mathbf{u}) + \tau(\mathbf{v}), \tag{3.5}$$

and equality holds only when $\mathbf{u}$ *and* $\mathbf{v}$ *are linearly dependent.*

Remark. The reversed triangle inequality expresses what is often misleadingly called "the relativistic clock paradox."

4. Pseudo-Euclidean Geometry. Let a set $\mathcal{E}$ of points $x, y, \cdots$ and a function

$$\sigma: \mathcal{E}\times\mathcal{E}\to\mathcal{R} \tag{4.1}$$

be given, where $\mathcal{R}$ denotes the real line. The function σ defines a certain structure on $\mathcal{E}$. The automorphisms of this structure are the one-to-one mappings $\mathfrak{a}$ of $\mathcal{E}$ onto itself which satisfy

$$\sigma(\mathfrak{a}(x), \mathfrak{a}(y)) = \sigma(x, y) \tag{4.2}$$

for all $x, y\in\mathcal{E}$. We denote by $\mathcal{A}$ the group of all these automorphisms. We impose restrictive conditions on σ by assuming that $\mathcal{A}$ contains a subgroup $\mathcal{V}$ which satisfies the axioms (E_1)–(E_4) stated below.

(E_1) $\mathcal{V}$ *is commutative.*

(E_2) $\mathcal{V}$ *is transitive.*

(E_3) *If* $\mathbf{v}\in\mathcal{V}$ *maps some point* $x\in\mathcal{E}$ *onto itself, then* $\mathbf{v}$ *is the identity mapping.*

We write the group operation in $\mathcal{V}$ additively and denote the identity mapping by $\mathbf{0}$. We write $x+\mathbf{v}\in\mathcal{E}$ for the image $\mathbf{v}(x)$ of $x\in\mathcal{E}$ under $\mathbf{v}\in\mathcal{V}$. It follows from (E_2) and (E_3) that for any two points $x, y\in\mathcal{E}$ there is a *unique* $\mathbf{v}\in\mathcal{V}$ which maps x onto y. The mapping $\mathbf{v}$ determined in this way will be denoted by $\mathbf{v}=y-x$. The "sum" $x+\mathbf{v}$ and "point-difference" $y-x$ thus defined obey the rules suggested by the notation.

The following proposition is easily established: The value $\sigma(x, y)$ depends only on the point difference $y-x$; i.e., there is a function $\Phi: \mathcal{V}\to\mathcal{R}$ such that

$$\Phi(y - x) = \sigma(x, y) \tag{4.3}$$

for all $x, y\in\mathcal{E}$.

The last condition required of $\mathcal{V}$ is the following.

(E_4) $\mathcal{V}$ *is the underlying additive group of a real vector space and* Φ *is a nondegenerate quadratic form on* $\mathcal{V}$; *i.e.,* $\mathcal{V}$ *can be given the structure of an inner product space such that*

$$(y - x)^2 = (y - x)\cdot(y - x) = \sigma(x, y). \tag{4.4}$$

In view of (E_4) we refer to the automorphism $\mathbf{v}$ in $\mathcal{V}$ as *vectors.*

UNIQUENESS THEOREM. *There is at most one subgroup* $\mathcal{V}$ *of the automorphism group* $\mathcal{A}$ *such that* $\mathcal{V}$ *satisfies the axioms* (E_1)–(E_4). *If such a subgroup exists then its structure as an inner product space, as required for* (E_4), *is unique.*

Proof. Assume that $\mathfrak{V}$ and $\hat{\mathfrak{V}}$ are subgroups of $\mathfrak{A}$ which both satisfy (E_1)–(E_4). If two points $x, y \in \mathcal{E}$ are given, we denote by $y-x$ the unique vector in $\mathfrak{V}$ which maps x onto y and by $y \hat{-} x$ the unique vector in $\hat{\mathfrak{V}}$ which maps x onto y.

We choose a fixed point $q \in \mathcal{E}$ and define a mapping $\mathfrak{f}$ of $\mathfrak{V}$ into $\hat{\mathfrak{V}}$ by

$$\hat{\mathbf{v}} = \mathfrak{f}(\mathbf{v}) = (q + \mathbf{v}) \hat{-} q, \tag{4.5}$$

so that

$$q + \mathbf{v} = q + \mathfrak{f}(\mathbf{v}) = q + \hat{\mathbf{v}} \tag{4.6}$$

for all $\mathbf{v} \in \mathfrak{V}$. It is clear that $\mathfrak{f}$ is one-to-one and onto, and that $\mathfrak{f}(\mathbf{0}) = \hat{\mathbf{0}} = \mathbf{0}$.

By (4.4) we have

$$\sigma(x, y) = (y - x)^2 = (y \hat{-} x)^2. \tag{4.7}$$

Substituting $x = q + \mathbf{u}$ and $y = q + \mathbf{v}$ into (4.7) we find that

$$(\mathbf{v} - \mathbf{u})^2 = (\hat{\mathbf{v}} - \hat{\mathbf{u}})^2 \tag{4.8}$$

holds for all $\mathbf{u}, \mathbf{v} \in \mathfrak{V}$. In particular, when $\mathbf{u} = \mathbf{0} = \hat{\mathbf{0}} = \hat{\mathbf{u}}$, (4.8) shows that $\mathbf{v}^2 = \hat{\mathbf{v}}^2$ for all $\mathbf{v} \in \mathfrak{V}$. Hence, if we expand (4.8),

$$\mathbf{v}^2 - 2\mathbf{v}\cdot\mathbf{u} + \mathbf{u}^2 = \hat{\mathbf{v}}^2 - 2\hat{\mathbf{v}}\cdot\hat{\mathbf{u}} + \hat{\mathbf{u}}^2, \tag{4.9}$$

the square terms cancel and we obtain

$$\mathbf{v}\cdot\mathbf{u} = \hat{\mathbf{v}}\cdot\hat{\mathbf{u}} = \mathfrak{f}(\mathbf{v})\cdot\hat{\mathbf{u}}. \tag{4.10}$$

Repeated use of (4.10) shows that

$$\begin{aligned}\mathfrak{f}(\alpha\mathbf{v} + \beta\mathbf{w})\cdot\hat{\mathbf{u}} &= (\alpha\mathbf{v} + \beta\mathbf{w})\cdot\mathbf{u} = \alpha(\mathbf{v}\cdot\mathbf{u}) + \beta(\mathbf{w}\cdot\mathbf{u}) \\ &= [\alpha\mathfrak{f}(\mathbf{v}) + \beta\mathfrak{f}(\mathbf{w})]\cdot\hat{\mathbf{u}}.\end{aligned}$$

Hence

$$[\alpha\mathfrak{f}(\mathbf{v}) + \beta\mathfrak{f}(\mathbf{w}) - \mathfrak{f}(\alpha\mathbf{v} + \beta\mathbf{w})]\cdot\hat{\mathbf{u}} = 0,$$

which holds for all $\hat{\mathbf{u}} \in \hat{\mathfrak{V}}$. Since ($E_4$) requires that the inner product is non-degenerate it follows from (2.1) that

$$\alpha\mathfrak{f}(\mathbf{v}) + \beta\mathfrak{f}(\mathbf{w}) = \mathfrak{f}(\alpha\mathbf{v} + \beta\mathbf{w}) \tag{4.11}$$

is valid for all $\mathbf{v}, \mathbf{w} \in \mathfrak{V}$ and all real α, β. It is the content of (4.10), (4.11) and the remark after (4.6) that $\mathfrak{f}$ is an inner product space isomorphism of $\mathfrak{V}$ onto $\hat{\mathfrak{V}}$.

Let $x \in \mathcal{E}$ and $\mathbf{v} \in \mathfrak{V}$ be given. Put $\mathbf{u} = x - q$ so that $x = q + \mathbf{u}$. Using (4.6) twice and (4.11) once we derive

$$x + \mathbf{v} = q + \mathbf{u} + \mathbf{v} = q + \mathfrak{f}(\mathbf{u} + \mathbf{v}) = q + \mathfrak{f}(\mathbf{u}) + \mathfrak{f}(\mathbf{v}) = q + \mathbf{u} + \mathfrak{f}(\mathbf{v}) = x + \mathfrak{f}(\mathbf{v}).$$

Thus, $x + \mathbf{v} = x + \mathfrak{f}(\mathbf{v})$ is valid for all $x \in \mathcal{E}$, which shows that the mappings $\mathbf{v}$ and $\mathfrak{f}(\mathbf{v})$ of $\mathcal{E}$ onto $\mathcal{E}$ are the same. Therefore, $\mathfrak{f}$ is the identity mapping of $\mathfrak{V}$ onto itself, which completes the proof that $\mathfrak{V}$ and $\hat{\mathfrak{V}}$ coincide as inner product spaces.

DEFINITION. *A set $\mathcal{E}$ which is endowed with a structure defined by a real valued function σ on $\mathcal{E}\times\mathcal{E}$ is called a pseudo-Euclidean space if the axioms* (E_1)–(E_4) *are satisfied. The function σ will be called the separation function of $\mathcal{E}$. The inner product space $\mathcal{V}$ determined by σ is called the translation space of $\mathcal{E}$.* (The definition given here differs from that of Bourbaki ([1] Chapter 6, no. 6) in that the translation space $\mathcal{V}$ is *not* regarded as part of the defining structure.)

The uniqueness theorem insures that the translation space is well defined.

If the separation function σ is nonnegative and if the translation space $\mathcal{V}$ is finite-dimensional, then $\mathcal{E}$ may be regarded as a Euclidean space in the ordinary sense.

The following is a corollary of the uniqueness theorem.

REPRESENTATION THEOREM. *Let $\mathcal{E}$ be a pseudo-Euclidean space and q a point in $\mathcal{E}$. Then every automorphism $\mathfrak{a}$ of $\mathcal{E}$ has a unique representation of the form*

$$\mathfrak{a}(x) = \mathfrak{a}(q) + Q(x - q), \tag{4.12}$$

where Q is an orthogonal transformation of the translation space $\mathcal{V}$.

Proof. The mapping Q defined by

$$Q(\mathbf{v}) = \mathfrak{a} \circ \mathbf{v} \circ \mathfrak{a}^{-1} \tag{4.13}$$

is an isomorphism of the subgroup $\mathcal{V}$ of $\mathfrak{A}$ onto its conjugate $\mathcal{V}^* = \mathfrak{a} \circ \mathcal{V} \circ \mathfrak{a}^{-1}$ in $\mathfrak{A}$. The mapping Q may be used to transport the inner product space structure of $\mathcal{V}$ to $\mathcal{V}^*$. It is clear that $\mathcal{V}^*$ then satisfies the conditions (E_1)–(E_4). Hence, by the uniqueness theorem, it must coincide with $\mathcal{V}$, and Q must be an automorphism of the inner product space $\mathcal{V}$. Since we use the notation $\mathbf{u}(q) = q + \mathbf{u}$ when $\mathbf{u} \in \mathcal{V}$, the image of $q \in \mathcal{E}$ under the mapping $Q\mathbf{v} \circ \mathfrak{a} = \mathfrak{a} \circ \mathbf{v}$ (see (4.13)) is given by

$$\mathfrak{a}(q) + Q\mathbf{v} = \mathfrak{a}(q + \mathbf{v}). \tag{4.14}$$

We obtain (4.12) by substituting $\mathbf{v} = x - q$ into (4.14).

Remark I. In the Euclidean case (4.12) is the well-known formula for rigid displacements. Many textbooks, however, derive this formula under unnecessary a priori assumptions of smoothness or even linearity of $\mathfrak{a}$; a theorem that does not require such assumptions appears as Exercise 21a, chapter 6, in Bourbaki [1]. The uniqueness theorem and the representation theorem, including proofs, remain valid when the field of real numbers is replaced by an arbitrary commutative ring $\mathfrak{R}$ of characteristic $\neq 2$. In this case, the translation space is a module over $\mathfrak{R}$.

Remark II. One may be tempted to define a subspace $\mathfrak{F}$ of a pseudo-Euclidean space $\mathcal{E}$ by the condition that the restriction $\sigma_{\mathfrak{F}}$ to $\mathfrak{F}$ of the separation function σ of $\mathcal{E}$ satisfy the axioms (E_1)–(E_4) and hence give $\mathfrak{F}$ the structure of a pseudo-Euclidean space. Unfortunately, it may happen that such a "subspace" $\mathfrak{F}$ is not a subspace in the customary sense, i.e., a set of the form

$$\mathfrak{F} = \{x \mid x = p + \mathbf{u}, \mathbf{u} \in \mathcal{U}\}, \tag{4.15}$$

where $\mathcal{U}$ is a subspace of the translation space of $\mathcal{E}$. For example, let $\mathcal{E} = \mathcal{R} \times \mathcal{R} \times \mathcal{R}$ with σ defined by

$$\sigma(x, y) = (x_1 - y_1)^2 - (x_2 - y_2)^2 - (x_3 - y_3)^2. \tag{4.16}$$

The set $\mathfrak{F}$ of all triples of the form $(\phi(\xi), \phi(\xi), \xi)$, where ϕ is an arbitrary function $\phi: \mathcal{R} \to \mathcal{R}$, is a "subspace," but not of the type (4.15) when ϕ is not linear. But for this fact Theorem 1 of the following section would be trivial.

Remark III. The uniqueness theorem given in this Section shows that the complete structure of a pseudo-Euclidean space, including the translation space, is determined by a knowledge of the separation function σ alone. One may ask whether σ itself may not be uniquely determined by the prescription of even less information. A result in this direction was found by Suppes [2]. Under the assumption that the translation space has index 1 and dimension 4, he showed that σ is uniquely determined by its values $\sigma(x, y)$ for all pairs (x, y) such that $\sigma(x, y) < 0$. This result is easily obtained by an adaptation of the reasoning leading from (5.10) to (5.14) given in the following section. The uniqueness theorem of the following section is a different result of the same type.

5. Minkowskian Chronometry. We assume that the following data are given:

(a) a set $\mathcal{E}$, whose elements $x, y, \cdots$ will be called *events*;

(b) a family Ω of subsets of $\mathcal{E}$ which covers $\mathcal{E}$. The members $\mathcal{L}$ of Ω will be called *observers*;

(c) for each observer $\mathcal{L} \in \Omega$, a nonpositive separation function $\sigma_{\mathcal{L}}$ on $\mathcal{L}$ which gives $\mathcal{L}$ the structure of a one-dimensional pseudo-Euclidean space;

(d) a symmetric binary relation $\sim$ on $\mathcal{E}$ with the following property: Given any observer $\mathcal{L}$ and any event $x \notin \mathcal{L}$, there are at least two events y_1 and y_2 in $\mathcal{L}$ that are related to x. The relation $\sim$ will be called the *signal relation* and a pair (x, y) of events related by $\sim$ will be called a *signal*.

Remarks on physical interpretation: $x \in \mathcal{L}$ means physically that x is an event which is experienced by the observer $\mathcal{L}$. We imagine that each observer is equipped with a clock. If $\tau_{\mathcal{L}}(x, y)$ is the time-difference of the two clock readings at the events x and y of $\mathcal{L}$, then the separation $\sigma_{\mathcal{L}}(x, y)$ of $x, y \in \mathcal{L}$ is assumed to be given by $\sigma_{\mathcal{L}}(x, y) = -(\tau_{\mathcal{L}}(x, y))^2$. The two events of a signal (x, y) are interpreted to be the emission and reception of a light, radio, or other electromagnetic signal. If $x \notin \mathcal{L}$, we may imagine x to be the event of reflection of a signal which is sent out by $\mathcal{L}$ at y_1 and returns to $\mathcal{L}$ at y_2.

The data described under (a)–(d) define a certain structure on $\mathcal{E}$. We impose restrictions on this structure by assuming that there exists a separation function σ on all of $\mathcal{E}$ which endows $\mathcal{E}$ with the structure of a pseudo-Euclidean space and which satisfies the following two axioms:

(M_1) *σ is an extension of the separation function $\sigma_{\mathcal{L}}$ for each observer $\mathcal{L}\in\Omega$. In other words,*

$$\sigma(x, y) = \sigma_{\mathcal{L}}(x, y) \tag{5.1}$$

holds whenever $x, y\in\mathcal{L}$.

(M_2) *The pair (x, y) is a signal if and only if the separation of x and y is zero, i.e. $\sigma(x, y)=0$ if and only if $y\sim x$.*

UNIQUENESS THEOREM. *There is at most one separation function σ which satisfies the axioms* (M_1) *and* (M_2).

We first prove a number of preliminary theorems, assuming that some separation function σ satisfying (M_1) and (M_2) on $\mathcal{E}$ is given. The corresponding translation space is denoted by $\mathcal{V}$, as in Section 4.

THEOREM 1. *Every observer $\mathcal{L}$ is a time-like straight line in $\mathcal{E}$. More precisely: There is a vector $\mathbf{l}\in\mathcal{V}$ with the following properties*
(i) *$\mathbf{l}$ is a time-like unit vector,* i.e. $\mathbf{l}^2=-1$.
(ii) *if $q\in\mathcal{L}$ is given, then $x\in\mathcal{L}$ if and only if $x=q+\xi\mathbf{l}$, and $\xi\in\mathcal{R}$.*

A vector $\mathbf{l}$ with the properties (i) and (ii) will be called a *direction vector* of the observer $\mathcal{L}$. It is clear that if $\mathbf{l}$ is a direction vector, then $-\mathbf{l}$ is also one and there can be no others.

Proof. Let $\mathcal{V}'$ be the one-dimensional translation space corresponding to $\sigma_{\mathcal{L}}$. The event-difference in $\mathcal{V}'$ of the two events $x, y\in\mathcal{L}$ will be denoted by $y\stackrel{\prime}{-}x$. This difference must be carefully distinguished from the event difference $y-x$ in $\mathcal{V}$, which corresponds to the separation function σ. It follows from axiom (M_1) and from (4.3) that

$$(y-x)^2 = (y\stackrel{\prime}{-}x)^2 = \sigma(x, y) = \sigma_{\mathcal{L}}\cdot(x, y) \tag{5.2}$$

for $x, y\in\mathcal{L}$. Since $\mathcal{V}'$ is assumed to be one-dimensional and $\sigma_{\mathcal{L}}$ nonpositive, $\mathcal{L}$ can be represented in the form

$$\mathcal{L} = \{x \mid x = q + \xi\mathbf{l}', \xi\in\mathcal{R}\},$$

where q is a fixed event in $\mathcal{L}$ and $\mathbf{l}'\in\mathcal{V}'$ is such that $\mathbf{l}'^2=-1$.

We now define a mapping $\mathbf{f}$ of $\mathcal{R}$ into $\mathcal{V}$ by $\mathbf{f}(\xi)=(q+\xi\mathbf{l}')-q$. $\mathcal{L}$ is then the set of events x that are of the form

$$x = q+\xi\mathbf{l}' = q+\mathbf{f}(\xi), \qquad \xi\in\mathcal{R}. \tag{5.3}$$

The same argument as the one that led from (4.5) to (4.10) shows that for all $\xi, \eta\in\mathcal{R}$ we must have

$$-\xi\eta = \mathbf{f}(\xi)\cdot\mathbf{f}(\eta). \tag{5.4}$$

We fix ξ and η and consider the vector $\mathbf{s}\in\mathcal{V}$ given by

$$\mathbf{s} = \eta\mathbf{f}(\xi) - \xi\mathbf{f}(\eta). \tag{5.5}$$

It follows from (5.4) that

$$\mathbf{s}^2 = \eta^2(-\xi^2) - 2\xi\eta(-\xi\eta) + \xi^2(-\eta^2) = 0,$$

i.e. that $\mathbf{s}$ is a signal vector. Consider the event $z=q+\mathbf{s}$. By (5.2) we have

$$\mathbf{s}^2 = (z-q)^2 = \sigma(q, z) = 0. \tag{5.6}$$

If $z\in\mathcal{L}$ then $\mathbf{s}=z-q$ must be of the form $\mathbf{s}=\mathbf{f}(\zeta)$. By (5.4) it then follows that $0=\mathbf{s}^2=-\zeta^2$ and hence that $\mathbf{s}=\mathbf{f}(0)=\mathbf{0}$.

Assume now that $\mathbf{s}\neq\mathbf{0}$, in which case $z\notin\mathcal{L}$. Axiom (M_2) and (5.6) imply that $q\in\mathcal{L}$ and $z\notin\mathcal{L}$ must be related by a signal. The signal relation has the property that there must be another event $p\in\mathcal{L}$, $p\neq q$, which is also related to z. Using axiom (M_2) again, we find

$$0 = \sigma(p, z) = (z-p)^2 = (\mathbf{s}+\mathbf{v})^2 = 2\mathbf{s}\cdot\mathbf{v} + \mathbf{v}^2, \tag{5.7}$$

where $\mathbf{v}=q-p$. Since $p\in\mathcal{L}$, $\mathbf{v}$ must be of the form $\mathbf{v}=\mathbf{f}(\lambda)$. By (5.4) and (5.5), we obtain

$$\mathbf{s}\cdot\mathbf{v} = \eta(-\xi\lambda) - \xi(-\eta\lambda) = 0$$

and hence, by (5.7), $0=\mathbf{v}^2=-\lambda^2$. Consequently, $\mathbf{v}=\mathbf{f}(0)=\mathbf{0}=q-p$, i.e. $p=q$, which contradicts $p\neq q$.

We conclude that always $\mathbf{s}=\mathbf{0}$ and hence, by (5.5), that

$$\eta\,\mathbf{f}(\xi) = \xi\,\mathbf{f}(\eta) \tag{5.8}$$

holds for all real ξ and η. Putting $\eta=1$ and $\mathbf{f}(1)=\boldsymbol{l}\in\mathcal{V}$, we see that (5.8) gives $\mathbf{f}(\xi)=\xi\boldsymbol{l}$. It follows from (5.3) that $\mathcal{L}$ is the set of events of the form $x=q+\xi\boldsymbol{l}$. The relation $\boldsymbol{l}^2=-1$ is a consequence of (5.4). Hence $\boldsymbol{l}$ has the properties (i) and (ii). Q.E.D.

THEOREM 2. *Let $\mathcal{L}$ be an observer with direction vector $\boldsymbol{l}$ and let $q\in\mathcal{L}$, $x\in\mathcal{E}$. An event $y=q+\eta\boldsymbol{l}\in\mathcal{L}$ is then related to x by a signal if and only if η is a root of the equation*

$$\eta^2+2\boldsymbol{l}\cdot(x-q)\eta + (x-q)^2 = 0. \tag{5.9}$$

When $x\notin\mathcal{L}$, there are exactly two events

$$y_1 = q+\eta_1\boldsymbol{l}, \qquad y_2 = q+\eta_2\boldsymbol{l} \tag{5.10}$$

in $\mathcal{L}$ that are related to x by a signal and we have

$$\sigma(q, x) = (x-q)^2 = -\eta_1\eta_2, \qquad (x-q)\cdot\boldsymbol{l} = -(\eta_1+\eta_2), \tag{5.11}$$

$$[\boldsymbol{l}\cdot(x-q)]^2 + 4(x-q)^2 > 0. \tag{5.12}$$

Proof. By axiom (M_2), $y=q+\eta\boldsymbol{l}$ is related to x by a signal if and only if

$$\sigma(x, y) = (y-x)^2 = (q-x+\eta\boldsymbol{l})^2 = -\eta^2-2\boldsymbol{l}\cdot(x-q)\eta + (x-q)^2 = 0,$$

which proves the first part of the theorem.

When $x \notin \mathcal{L}$, the property of the signal relation described in (d) requires that the roots of (5.9) must be real and distinct. (5.12) is a necessary and sufficient condition for this requirement. The roots of (5.9) always satisfy (5.11).

Proof of the uniqueness theorem. Let q and x be any two events in $\mathcal{E}$. Since the family of all observers covers $\mathcal{E}$, there is at least one observer such that $q \in \mathcal{L}$. If $x \in \mathcal{L}$ also, then by axiom (M_1)

$$\sigma(q, x) = \sigma_{\mathcal{L}}(q, x). \tag{5.13}$$

If $x \notin \mathcal{L}$, we consider the events $y_1, y_2 \in \mathcal{L}$, determined as in Theorem 2. With the notation of Theorem 2, we have

$$\begin{aligned} \sigma(q, y_i) &= (y_i - q)^2 = \eta_i^2 l^2 = -\eta_i^2, & i = 1, 2 \\ \sigma(y_1, y_2) &= (y_2 - y_1)^2 = (\eta_2 - \eta_1)^2 l^2 = -\eta_2^2 + 2\eta_1\eta_2 - \eta_1^2. \end{aligned}$$

It follows that

$$\eta_1\eta_2 = \tfrac{1}{2}[\sigma(y_1, y_2) - \sigma(q, y_1) - \sigma(q, y_2)].$$

Noting that $y_1, y_2, q \in \mathcal{L}$, we infer from axiom ($M_1$) and $(5.11)_1$ that

$$\sigma(q, x) = -\tfrac{1}{2}[\sigma_{\mathcal{L}}(y_1, y_2) - \sigma_{\mathcal{L}}(q, y_1) - \sigma_{\mathcal{L}}(y, y_2)]. \tag{5.14}$$

Equations (5.13) and (5.14) show that $\sigma(q, x)$ is uniquely determined when $\sigma_{\mathcal{L}}$ is given. Q.E.D.

DEFINITION. *A set $\mathcal{E}$ which is endowed with a structure defined by Ω, $\{\sigma_{\mathcal{L}} \mid \mathcal{L} \in \Omega\}$, and $\sim$ as described under* (a)–(d) *is called a Minkowskian domain if the axioms* (M_1) *and* (M_2) *are satisfied.*

The uniqueness theorem shows that a Minkowskian domain is also endowed, in a canonical way, with the structure of a pseudo-Euclidean space.

THEOREM 3. *The translation space $\mathcal{V}$ of a Minkowskian domain has index* 1.

Proof. Let $\mathcal{L}$ be an observer with direction vector l, and let $\mathcal{U} = \{\mathbf{u} \mid \mathbf{u} = \xi l, \xi \in \mathcal{R}\}$ be the one-dimensional subspace of $\mathcal{V}$ generated by l. Consider a nonzero vector $\mathbf{v} \in \mathcal{U}^\perp$, which then satisfies $l \cdot \mathbf{v} = 0$. Choose an event $q \in \mathcal{L}$ and put $x = q + \mathbf{v}$. It is clear that $x \notin \mathcal{L}$. Since $l \cdot \mathbf{v} = l \cdot (x - q) = 0$, (5.12) states that $(x - q)^2 = \mathbf{v}^2 > 0$, which shows that $\mathbf{v}$ is space-like. It follows that $\mathcal{V} = \mathcal{U} \oplus \mathcal{U}^\perp$ is a direct decomposition with the property that $\mathcal{U} \subset \mathcal{V}_-$, $\mathcal{U}^\perp \subset \mathcal{V}_+$, i.e., it is a decomposition of the type (2.3). Theorem 1 of Section 2 implies that ind $\mathcal{V} = \dim \mathcal{U} = 1$.

Remark. As is shown by Theorem 1, the family Ω of observers is a congruence of straight lines in $\mathcal{E}$. Since Ω covers $\mathcal{E}$, there must be at least one line through each event. For example, Ω could be the set of all straight lines parallel to a given time-like direction. Unless Ω contains "very many" observers, a knowledge of $\sigma(x, y)$ for all pairs (x, y) such that x, y belong to the same observer is *not* sufficient to determine σ. It is the interconnection of σ with the signal relation, as provided by axiom (M_2), which renders σ unique.

6. Temporal order. As in the previous section we assume that data as described under (a), (b), and (c) are given. Instead of (d), we suppose that we have

(d′) a binary relation $\rightharpoonup$ on $\mathcal{E}$ with the following property: Given any observer $\mathcal{L}$ and any event $x \in \mathcal{E}$, there is an event $y_1 \in \mathcal{L}$ such that $y_1 \rightharpoonup x$ and there is an event $y_2 \in \mathcal{L}$ such that $x \rightharpoonup y_2$. Moreover, $x \rightharpoonup y$ and $y \rightharpoonup x$ can both be valid only if $x = y$. The relation $\rightharpoonup$ will be called the *directed signal relation* and the relation defined by

$$x \sim y \text{ if and only if } x \rightharpoonup y \text{ or } y \rightharpoonup x \tag{6.1}$$

the (undirected) signal relation associated with $\rightharpoonup$.

It is clear that $\sim$ has the property described under (d) of the previous section.

Remark on physical interpretation. The change from (d) to (d′) corresponds to introducing the possibility of distinguishing the emission x from the reception y of a signal $x \rightharpoonup y$.

We require here the existence of a separation function σ on $\mathcal{E}$ which satisfies not only the axioms (M_1) and (M_2) but also

(M_3) *If $\mathcal{L}$ is an observer and if y_1, y_2 and z_1, z_2 are events in $\mathcal{L}$ such that*

$$y_1 \rightharpoonup p \rightharpoonup y_2, \qquad z_1 \rightharpoonup q \rightharpoonup z_2 \tag{6.2}$$

for some p, $q \in \mathcal{E}$, then

$$\sigma_{\mathcal{L}}(y_1, z_1) + \sigma_{\mathcal{L}}(y_2, z_2) - \sigma_{\mathcal{L}}(y_1, z_2) - \sigma_{\mathcal{L}}(y_2, z_1) \geqq 0. \tag{6.3}$$

DEFINITION. *A set $\mathcal{E}$ which is endowed with a structure defined by Ω, $\{\sigma_{\mathcal{L}} \mid \mathcal{L} \in \Omega\}$, and $\rightharpoonup$ as described under* (a)–(c) *and* (d′) *is called a directed Minkowskian domain if the axioms* (M_1), (M_2), *and* (M_3) *are satisfied.*

From now on we shall exclude the trivial case when the domain $\mathcal{E}$ coincides with one of the observers. The dimension of the translation space $\mathcal{V}$ of $\mathcal{E}$ is then at least two.

DEFINITION. *We say that the event $x \in \mathcal{E}$ is earlier than the event $y \in \mathcal{E}$, and we write $x \prec y$, if there is an event p such that $x \rightharpoonup p \rightharpoonup y$.*

THEOREM 1. *Every observer $\mathcal{L}$ has a unique direction vector $\boldsymbol{l}$ with the following property: For any two events x, $y \in \mathcal{E}$ the relation $x \prec y$ holds if and only if*

$$(y - x)^2 \leqq 0, \qquad (y - x) \cdot \boldsymbol{l} \leqq 0. \tag{6.4}$$

The direction vector $\boldsymbol{l}$ determined by the condition (6.4) will be called the *proper direction vector* of $\mathcal{L}$.

LEMMA. *Let $\mathcal{L}$ be an observer with direction vector $\boldsymbol{l}$ and let x_1, $x_2 \in \mathcal{E}$. Assume that $q \in \mathcal{L}$ and $z_i = q + \xi_i \boldsymbol{l}$ are the two events related to x_i by a signal* ($i = 1, 2$). *Then*

$$(x_2 - x_1) \cdot \boldsymbol{l} = (\xi_1 - \xi_2), \qquad \xi_1 \xi_2 (x_2 - x_1)^2 \geqq 0, \tag{6.5}$$

and $(6.5)_2$ *can reduce to equality only if* $(x_2 - x_1)^2 = 0$.

Proof. Theorem 2 of Section 5 applies when we replace there q, x, y_1, y_2, η_1, η_2 by q, x_i, q, z_i, 0, ξ_i, respectively. $(5.11)_2$ then yields

$$\mathbf{s}_i \cdot \boldsymbol{l} = -\xi_i, \tag{6.6}$$

where $\mathbf{s}_i = x_i - q$. Hence $(x_2 - x_1)\cdot \boldsymbol{l} = (\mathbf{s}_2 - \mathbf{s}_1)\cdot \boldsymbol{l} = -(\xi_2 - \xi_1)$, which proves $(6.5)_1$.

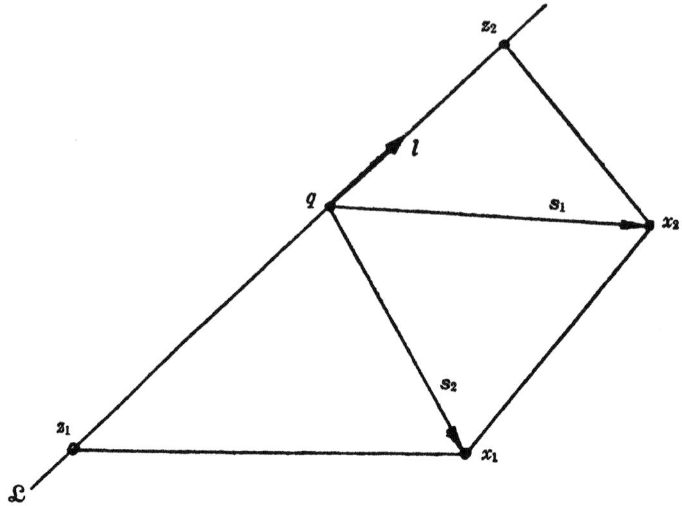

Consider the vector $\mathbf{u} = \xi_2 \mathbf{s}_1 - \xi_1 \mathbf{s}_2$. It follows from (6.6) that $\mathbf{u}\cdot \boldsymbol{l} = 0$. Therefore, by Theorem 3 of Section 5 and Theorem 2 of Section 3, $\mathbf{u}$ must be space-like. Since $\mathbf{s}_i^2 = 0$, we obtain

$$\mathbf{u}^2 = -2\xi_1\xi_2 \mathbf{s}_1\cdot \mathbf{s}_2 \geqq 0. \tag{6.7}$$

On the other hand we have

$$(x_2 - x_1)^2 = (\mathbf{s}_2 - \mathbf{s}_1)^2 = -2\mathbf{s}_1\cdot \mathbf{s}_2,$$

which shows that (6.7) is equivalent to $(6.5)_2$. Equality can hold in (6.7) only if $\mathbf{u} = \xi_2\mathbf{s}_1 - \xi_1\mathbf{s}_2 = \mathbf{0}$, in which case $x_2 - x_1 = \mathbf{s}_2 - \mathbf{s}_1$ must be a signal vector. Q.E.D.

Proof of Theorem 1. The proof will be carried out in four steps.

I. Assume first that $x, y \in \mathcal{L}$. Let $\boldsymbol{l}'$ be one of the two direction vectors of $\mathcal{L}$ (see Theorem 1 of the previous section) and let $y - x = \alpha \boldsymbol{l}'$. Since the translation space $\mathcal{V}$ of $\mathcal{E}$ is such that ind $\mathcal{V} = 1$, dim $\mathcal{V} \geqq 2$, there is a space-like unit vector orthogonal to $\boldsymbol{l}'$, i.e. a vector $\mathbf{u}$ such that $\mathbf{u}\cdot \boldsymbol{l}' = 0$, $\mathbf{u}^2 = 1$. Put $q = x + \frac{1}{2}\alpha\, \boldsymbol{l}' \in \mathcal{L}$ and $z = q + \frac{1}{2}\alpha \mathbf{u}$, so that

$$x = q - \frac{\alpha}{2}\boldsymbol{l}', \qquad y = q + \frac{\alpha}{2}\boldsymbol{l}', \qquad z - q = \frac{\alpha}{2}\mathbf{u}.$$

The two roots of the equation

$$\eta^2 + \boldsymbol{l}'\cdot(z - q)\eta - (z - q)^2 = \eta^2 - \left(\frac{\alpha}{2}\right)^2$$

are $\eta = \pm\frac{1}{2}\alpha$. Hence, by Theorem 2 of Section 5, x and y are the events in $\mathcal{L}$ that are related to z by a signal. Therefore x and y, in some order, must be the two events in $\mathcal{L}$ that correspond to z as described in (d′), and we must have either $x \to z \to y$, i.e. $x \prec y$, or $y \to z \to x$, i.e. $y \prec x$. Both $x \prec y$ and $y \prec x$ can hold only when $x = y = z$.

II. We choose two events $z_1, z_2 \in \mathcal{L}$ such that $(z_2 - z_1)^2 = -1$. By the result of I we may assume that z_1, z_2 are arranged such that $z_1 \prec z_2$. We put $l = z_2 - z_1$. Let $x, y \in \mathcal{L}$ and assume $x \prec y$. An easy calculation, starting from (6.3) with y_1, y_2 replaced by x, y, shows that

$$\sigma_{\mathcal{L}}(x, z_1) + \sigma_{\mathcal{L}}(y, z_2) - \sigma_{\mathcal{L}}(x, z_2) - \sigma_{\mathcal{L}}(y, z_1) = -2(y - x)\cdot l \leqq 0.$$

Under the restrictive hypothesis $x, y \in \mathcal{L}$ the conclusion of the theorem now follows. We note that if $x, y \in \mathcal{L}$, then

$$(6.8) \qquad y - x = \xi\, l, \quad \text{where} \quad \begin{matrix} \xi \geqq 0 \text{ if } x \prec y \\ \xi \leqq 0 \text{ if } y \prec x. \end{matrix}$$

(6.8) characterizes the proper direction vector of $\mathcal{L}$.

III. Assume $x \prec y$ holds for two events $x, y \in \mathcal{E}$, i.e. $x \to q \to y$ for some $q \in \mathcal{E}$. Let $\mathcal{L}'$ be an observer passing through q and let l' be the proper direction vector of $\mathcal{L}'$. According to the requirements described in (d′), there are events $z_1, z_2 \in \mathcal{L}'$ such that

$$z_1 \to x \to q \quad \text{and} \quad q \to y \to z_2,$$

i.e. such that $z_1 \prec q \prec z_2$. By II, (6.8) applies with x, y replaced by z_1, q or q, z_2. Hence

$$(6.9) \qquad z_i = q + \xi_i l', \qquad \xi_1 \leqq 0, \qquad \xi_2 \geqq 0.$$

The lemma and (6.9) show that

$$(6.10) \qquad (y - x)\cdot l' \leqq 0, \qquad (y - x)^2 \leqq 0.$$

Therefore the conclusion (6.4) holds for the particular observer $\mathcal{L}'$.

Let $\mathcal{L}$ be another observer and let $q' \in \mathcal{L}'$ be chosen arbitrarily. It is clear that there are events $x_1, x_2 \in \mathcal{L}$ and $z_1', z_2' \in \mathcal{L}'$ such that $z_1' \to x_1 \to q' \to x_2 \to z_2'$. If $\mathcal{L}' \neq \mathcal{L}$ we must have $x_1 \neq x_2$ and hence

$$(6.11) \qquad x_2 - x_1 = \alpha\, l, \qquad \alpha > 0,$$

where l is the proper direction vector of $\mathcal{L}$. Also, (6.9) remains valid when q, z_i, ξ_i are replaced by q', z_i', ξ_i'. Thus, the lemma applies again and we find, using (6.11), that

$$(6.12) \qquad (x_2 - x_1)\cdot l' = \alpha\, l\cdot l' \leqq 0, \qquad l\cdot l' \leqq 0.$$

If $y \neq x$, it follows from (6.10), (6.12) and Theorem 2 of Section 3 that actually

$$(y - x)^2 \leqq 0, \qquad (y - x)\cdot l' < 0, \qquad l\cdot l' < 0.$$

Theorem 4 of Section 3 shows that we must have $(y-x)\cdot l<0$. If $y=x$, $(y-x)\cdot l=0$ is trivially true. Hence (6.4) holds for the proper direction vector l of an *arbitrary* observer.

IV. Assume that $(y-x)^2\leqq 0$. It is not hard to see that it is then possible to choose an observer $\mathcal{L}$ and events q, z_1, $z_2\in\mathcal{L}$ such that the hypotheses of the lemma are satisfied for $x_2=y$, $x_1=x$. If $(y-x)^2<0$, $(6.5)_2$ implies $\xi_1\xi_2<0$, which states that ξ_1 and ξ_2 have opposite sign. From part II we conclude that either $z_1\rightharpoonup x\rightharpoonup q\rightharpoonup y\rightharpoonup z_2$ or $z_2\rightharpoonup y\rightharpoonup q\rightharpoonup x\rightharpoonup z_1$. Hence, we must have either $x\prec y$ or $y\prec x$. If $(y-x)^2=0$, axiom (M_2) shows that x and y are related by a signal and hence that $x\prec y$ or $y\prec x$, trivially. Thus, x and y are comparable. This observation, together with the result of III, completes the proof of Theorem 1. Q.E.D.

Remark. The number $-(y-x)\cdot l$ is the "time-difference" of the events x and y relative to the observer $\mathcal{L}$. Theorem 1 states, therefore, that x is earlier than y if and only if it is "earlier" for *every* observer.

An argument based on Theorem 1 and Theorem 4 of Section 3 will prove

THEOREM 2. *The relation $\prec$ is a partial order on $\mathcal{E}$, which has the property that x and y are comparable if and only if $\sigma(x, y)\leqq 0$.*

One of the two directed time-cones described in Theorem 5 of Section 3, say $\mathcal{V}^1_-$, is singled out by the property that $y-x\in\mathcal{V}^1_-$ implies $x\prec y$. We may call $\mathcal{V}^1_-$ the *future time-cone* and $\mathcal{V}^2_-=-\mathcal{V}^1_-$ the *past time-cone.*

The research leading to this paper was done under Grant NSF-G16745 by the National Science Foundation to Carnegie Institute of Technology.

References

1. N. Bourbaki, Algèbre, Chap. 9, Hermann, Paris, 1959.
2. J. L. Synge, Relativity, The Special Theory, North Holland, Amsterdam, 1956.
3. ———, Relativity, The General Theory, North Holland, Amsterdam, 1960.
4. P. Suppes, Axioms for Relativistic Kinematics with or without Parity. Symposium on the Axiomatic Method, with special reference to geometry and physics, North Holland, Amsterdam, 1959, p. 291–307.
5. P. R. Halmos, Finite Dimensional Vector Spaces, Van Nostrand, Princeton, 1958.

Proof of the Maximality of the Orthogonal Group in the Unimodular Group

WALTER NOLL

Proof of the Maximality of the Orthogonal Group in the Unimodular Group

WALTER NOLL

Let $\mathscr{D}$ be a real inner product space of dimension n. The inner product is assumed to be positive definite. The *unimodular group* $\mathscr{U}$ and the *orthogonal group* $\mathscr{O}$ of $\mathscr{D}$ are defined by

$$\mathscr{U}=\{\boldsymbol{H}\,|\,\boldsymbol{H}\in\mathscr{L},\ |\det\boldsymbol{H}|=1\} \tag{1}$$

and

$$\mathscr{O}=\{\boldsymbol{Q}\,|\,\boldsymbol{Q}\in\mathscr{L},\ \boldsymbol{Q}^T=\boldsymbol{Q}^{-1}\}, \tag{2}$$

where $\mathscr{L}$ is the algebra of all linear transformations of $\mathscr{D}$ into itself. It is the purpose of this note to give a simple proof of the fact that *$\mathscr{O}$ is a maximal subgroup of $\mathscr{U}$.*

This theorem has an important corollary in the theory of simple materials. In the terminology introduced in [*1*], the theorem implies that every isotropic simple material is either a solid or a fluid.

I convinced myself of the validity of the theorem about two years ago, on the basis of a tedious and ugly argument. Since I could not believe that the theorem was new, I asked a number of distinguished mathematicians for references, but I had no success[1]. I still believe that the theorem must have been discovered previously, but it is certainly not *well* known.

The polar decomposition theorem states that every invertible linear transformation is the product of an orthogonal transformation and one that is positive definite and symmetric. It follows that every group $\mathscr{G}$ containing $\mathscr{O}$ is generated by its positive definite and symmetric members. It is therefore sufficient to prove the

Theorem. *Assume that $\mathbf{S}\in\mathscr{U}$ is positive definite and symmetric and has at least two distinct proper numbers s and t, $s>t$. Then every positive definite and symmetric $\boldsymbol{H}\in\mathscr{U}$ belongs to the group $\mathscr{G}$ generated by $\mathscr{O}$ and $\mathbf{S}$.*

The proof is based on the following

Lemma. *For every ξ such that*

$$\left(\frac{t}{s}\right)^2\leqq\xi\leqq\left(\frac{s}{t}\right)^2 \tag{3}$$

[1] The proof of BRAUER [*2*], which came to my attention after this note was written, is being published in response to my inquiries.

there exists an $\boldsymbol{R}\in\mathcal{O}$ *such that the symmetric transformation*

$$\boldsymbol{T}=\boldsymbol{S}^{-1}\boldsymbol{R}\boldsymbol{S}^{2}\boldsymbol{R}^{-1}\boldsymbol{S}^{-1} \tag{4}$$

has the proper numbers $(\xi, \xi^{-1}, 1, \ldots, 1)$.

Proof. Let $\boldsymbol{g}$ and $\boldsymbol{f}$ be unit proper vectors of $\boldsymbol{S}$ corresponding to the proper numbers s and t, respectively. Let $\mathscr{W}$ be the $(n-2)$-dimensional subspace of $\mathscr{D}$ that is orthogonal to both $\boldsymbol{g}$ and $\boldsymbol{f}$. We define $\boldsymbol{R}_\vartheta\in\mathcal{O}$ by

$$\boldsymbol{R}_\vartheta\,\boldsymbol{w}=\boldsymbol{w} \quad \text{if} \quad \boldsymbol{w}\in\mathscr{W},$$

$$\boldsymbol{R}_\vartheta\,\boldsymbol{g}=\cos\vartheta\,\boldsymbol{g}+\sin\vartheta\,\boldsymbol{f},$$

$$\boldsymbol{R}_\vartheta\,\boldsymbol{f}=-\sin\vartheta\,\boldsymbol{g}+\cos\vartheta\,\boldsymbol{f},$$

and we put

$$\boldsymbol{T}_\vartheta=\boldsymbol{S}^{-1}\boldsymbol{R}_\vartheta\boldsymbol{S}^{2}\boldsymbol{R}_\vartheta^{-1}\boldsymbol{S}^{-1}.$$

It is easily seen that $\boldsymbol{T}_\vartheta\boldsymbol{w}=\boldsymbol{w}$ if $\boldsymbol{w}\in\mathscr{W}$ and

$$\boldsymbol{T}_{(\pi/2)}\,\boldsymbol{g}=\left(\frac{t}{s}\right)^{2}\boldsymbol{g}, \qquad \boldsymbol{T}_{(\pi/2)}\,\boldsymbol{f}=\left(\frac{s}{t}\right)^{2}\boldsymbol{f}.$$

Hence $\boldsymbol{T}_{(\pi/2)}$ has the proper numbers $\left(\frac{t}{s}\right)^{2}, \left(\frac{t}{s}\right)^{-2}, 1, \ldots, 1$. Since $\boldsymbol{T}_0=\boldsymbol{1}$ and since $\boldsymbol{T}_\vartheta$ depends continuously on ϑ, it follows that there exists a ϑ such that the assertion of the lemma holds for $\boldsymbol{T}=\boldsymbol{T}_\vartheta$, QED.

Proof of the theorem. Let $\boldsymbol{e}_1, \ldots, \boldsymbol{e}_n$ be an orthonormal basis of proper vectors of $\boldsymbol{H}$, with corresponding proper numbers $h_1, \ldots, h_n$. Let $\boldsymbol{H}_k$ be the symmetric tensor that has the same proper vectors $\boldsymbol{e}_1, \ldots, \boldsymbol{e}_n$ as $\boldsymbol{H}$, but with corresponding proper numbers, $1, \ldots, 1, \eta_k, \eta_k^{-1}, 1, \ldots, 1$, where η_k corresponds to $\boldsymbol{e}_k$ and η_k^{-1} to $\boldsymbol{e}_{k+1}$. If the η_k are chosen such that

$$\eta_1=h_1, \qquad \eta_k=\eta_{k-1}\,h_k, \qquad k=1, 2, \ldots, n-1,$$

then, observing that $h_1 h_2 \ldots h_n=1$, we infer

$$\boldsymbol{H}=\boldsymbol{H}_1\boldsymbol{H}_2\ldots\boldsymbol{H}_{n-1}. \tag{5}$$

We now fix k and choose m large enough that $\xi=\sqrt[m]{\eta_k}$ satisfies the inequality (3). We then determine $\boldsymbol{T}$ according to the lemma. Let $\bar{\boldsymbol{g}}$ and $\bar{\boldsymbol{f}}$ be unit proper vectors of $\boldsymbol{T}$ which correspond to the proper numbers ξ and ξ^{-1}, respectively. We can find an orthogonal transformation $\boldsymbol{Q}$ which maps the basis $\boldsymbol{e}_1, \ldots, \boldsymbol{e}_n$ onto an orthonormal basis of proper vectors of $\boldsymbol{T}$ in such a way that

$$\boldsymbol{Q}\,\boldsymbol{e}_k=\bar{\boldsymbol{g}}, \qquad \boldsymbol{Q}\,\boldsymbol{e}_{k+1}=\bar{\boldsymbol{f}}.$$

With ξ, $\boldsymbol{T}$, and $\boldsymbol{Q}\in\mathcal{O}$ chosen in this manner, we easily see that

$$\boldsymbol{H}_k=\boldsymbol{Q}^{-1}\boldsymbol{T}^{m}\boldsymbol{Q}. \tag{6}$$

Of course, $\boldsymbol{T}$, $\boldsymbol{Q}$, and m depend on k.

If we substitute (4) into (6) and the result into (5), and if we do so for each of the $\boldsymbol{H}_k$, we see that $\boldsymbol{H}$ is indeed generated by $\boldsymbol{S}$ and orthogonal transformations. QED.

Acknowledgement. This work was supported by the National Science Foundation under Grant NSF-G 16745 to Carnegie Institute of Technology.

References

[1] NOLL, W., A mathematical theory of the mechanical behavior of continuous media. Arch. Rational Mech. Anal. **2**, 197—226 (1958).
[2] BRAUER, R., On the relation between the orthogonal group and the unimodular group. Arch. Rational Mech. Anal. **18**, 97 (1965).

Carnegie Institute of Technology
Pittsburgh, Pennsylvania

(Received August 4, 1964)

Space-Time Structures in Classical Mechanics

Walter Noll

Space-Time Structures in Classical Mechanics

WALTER NOLL

1. Introduction

The English language contains many words that denote spatial or temporal concepts: 'now', 'later', 'soon', 'simultaneous', 'here', 'there', 'far', 'location', 'equidistant', etc. The grammar is in part organized in accordance with temporal categories: present, past, future. If we tried to remove all words, prefixes, and suffixes with a temporal or spatial meaning from the language we would surely all but destroy it. The system of temporal and spatial concepts of a natural language such as English constitutes a *verbal space-time structure*. It is not a very precise system, but it serves very well as a framework for the common experiences of human life.

The geometry of the ancient Greeks and the spatial and temporal concepts of the mechanics of GALILEO and NEWTON may be viewed as being refinements of the intuitive verbal space-time structure, refinements which resulted in a very precise mathematical system. This system, which I call the *classical space-time structure*, provides the basis for several very successful branches of physics, chief among them the mechanics of particle systems of NEWTON and the mechanics of rigid bodies of EULER. In Sect. 2 I shall give a brief outline of a modern version of the classical space-time structure.

Until the beginning of the 19-th century there were very few people, if there were any, who could even imagine a system that might replace classical space-time. KANT, for example, regarded the valid statements of classical geometry and mechanics as being "a priori" and "synthetic"; i.e., he considered them to be truths about reality not derived from experience and yet not mere tautologies. This judgement reflects the view that the classical space-time structure is not just an expedient framework for physical experience, but is indeed the only conceivable such framework. It is a very understandable view, because before the invention of non-Euclidean geometry it must have been impossible to imagine how another space-time structure might be formulated.

A non-classical space-time structure cannot easily be described entirely in words, because words have connotations that imply the classical structure. The term "space-time" itself has misleading connotations when applied to a nonclassical system. The natural language keeps us in the prison of classical space-time. The only language within which a non-classical structure can be unambiguously formulated is the language of axiomatic mathematics. The methods of modern mathematics make it possible to fabricate almost at will structures that can play the role of classical space-time.

One may ask why anybody would wish to consider a nonclassical space-time at all. The reason is, of course, that the classical structure, while adequate as a basis for the concepts of ordinary experience and of the older branches of physics, is inadequate for some of the newer physical disciplines, inadequate, in particular, for relativistic and quantum physics. These disciplines require space-time structures that radically deviate from the classical one. It is not the purpose of this lecture to elaborate on these structures. Rather, I here content that classical space-time is not well suited even for some venerable branches of mechanics, and I shall develop another structure which I believe to be more appropriate for these branches. This structure, which I call *neo-classical space-time*, will be presented in Sect. 3; and in Sects. 4—6 I shall indicate how it can be used for developing classical mechanics.

Many of the concepts of classical space-time reflect the fact that we humans live on this solid earth, which in daily life is always available as a frame of reference for specifying "locations". The first blow to the classical system was dealt by COPERNICUS, who deprived the earth of its once secure place as the universal frame of reference. The concept of a "location" in interstellar space is much more problematical than that of a location on earth. NEWTON'S *absolute space*, as the set of all possible locations, has since its inception been regarded with unease by most thinkers. Nonetheless, the "Newtonian" mechanics built upon this concept of absolute space has been very successful. Actually, use of the concept of absolute space is one among several ways of accounting for the phenomenon of inertia. Classical space-time with its absolute space has been most successful in those branches of mechanics in which inertia plays the central role. Such is not the case, however, in many of the branches of the mechanics of continuous media, where inertia is often of minor importance or sometimes even altogether neglected. In these branches of mechanics absolute space is an artificial and inappropriate concept. If it is used anyway, it is necessary to compensate for its arbitrariness by introducing a requirement of invariance, called the *principle of frame-indifference or objectivity* (c.f. [1], Sect. 19). The description of mechanics in terms of the neo-classical space-time, which has no room for an absolute space, shows clearly why this principle is needed.

2. The Event-World of Classical Space-Time

The *absolute space* is a set E consisting of *points (locations)* $x, y, \ldots$. The set E is endowed with a mathematical structure defined by a *distance function* δ which associates with each pair x, y of points a number $\delta(x, y)$, *the distance from* x *to* y. The distance function δ is subject to certain axioms which ensure that δ is a Euclidean metric that gives E the structure of a *Euclidean space*. This Euclidean structure makes it possible to define a unique *translation space* V_E of E, which is a vector space with inner product consisting of automorphisms of E. (For details, see [2], Sect. 4.)

The *event-world* of classical space-time is the set $W = E \times R$ of all pairs (x, t), where $x \in E$ and $t \in R$, the set of real numbers. The point x is called the *location*, the number t the *time* of the event (x, t).

3. The Event-World of Neo-Classical Space-Time

The *event-world* of neo-classical space-time is a set W consisting of *events* $e, f, \ldots$. The set W is endowed with a mathematical structure defined by a *time-lapse function* τ and a *distance function* δ, subject to the axioms $(T_1)-(T_4)$ and $(D_1)-(D_3)$ stated below.

(T_1) The time-lapse function τ assigns to each pair e, f of events a number $\tau(e, f)$, called the *time-lapse* between e and f.

(T_2) For any $e, f \in W$,

$$\tau(e, f) = -\tau(f, e). \tag{3.1}$$

(T_3) For any $e, f, g \in W$,

$$\tau(e, f) + \tau(f, g) = \tau(e, g). \tag{3.2}$$

(T_4) For any $e \in W$ and any $t \in R$ there is a $f \in W$ such that $\tau(e, f) = t$.

We say that e is *earlier* than, *later* than, or *simultaneous* with f according to whether $\tau(e, f) > 0$, < 0, or $= 0$.

The set of all pairs of simultaneous events

$$S = \{(e, f) \mid \tau(e, f) = 0\} \tag{3.4}$$

is an equivalence relation on W, as can easily be seen to follow from (T_2) and (T_3). This equivalence relation determines a partition Γ of W into classes T of simultaneous events such that

$$S = \bigcup_{T \in \Gamma} T \times T. \tag{3.5}$$

The equivalence classes T will be called ***instantaneous spaces*** or simply ***instants***. If $e \in T$ we say that *the event e happens at the instant T*.

The value of the time-lapse $\tau(e, f)$ depends only on the instants T and S at which e and f happen. Therefore, we can define unambiguously a time-lapse

$$\bar{\tau}(T, S) = \tau(e, f) \quad \text{if} \quad e \in T, f \in S \tag{3.6}$$

between two instants T and S.

(D_1) The distance function δ assigns to each pair $(e, f) \in S$, i.e., to each pair of simultaneous events, a number $\delta(e, f)$, called the *instantaneous distance* between e and f.

(D_2) For each instant T, the restriction δ_T of δ to $T \times T$ is a Euclidean metric on T.

To say that δ_T is a Euclidean metric means that it gives to the instant T the structure of a Euclidean space. The translation space of T will be denoted by V_T.

(D_3) For each instant T, the dimension of the translation space V_T is 3.

Physically, the values $\tau(e, f)$ of the time-lapse function τ are to be interpreted as the results of time-measurements with clocks. The axioms (T_2)—(T_4) reflect familiar experiences with such measurements. The values $\delta(e, f)$ of the distance function δ are to be interpreted as the results of distance measurements with measuring sticks. The value $\delta(e, f)$ is defined only when e and f are simultaneous because each distance measurement is made at a particular instant. The axioms (D_2) and (D_3) are the abstract of thousands of years of experience with distance measurements.

An *automorphism* α of the event-world W is a one-to-one mapping of W onto itself which preserves time-lapses and distances. Thus, an automorphism α satisfies

$$\tau(\alpha(e), \alpha(f)) = \tau(e, f) \quad \text{for all} \quad e, f \in W \tag{3.7}$$

and

$$\delta(\alpha(e), \alpha(f)) = \delta(e, f) \quad \text{for all} \quad (e, f) \in S. \tag{3.8}$$

If $U \subset W$ we write

$$U^\alpha = \{\alpha(e) \mid e \in U\} \tag{3.9}$$

for the set of all images under α of events in U. In this way, the mapping $T \to T^\alpha$ of Γ onto itself is an automorphism of Γ in the sense that it preserves the time-lapse function $\bar{\tau}$ defined by (3.6).

An automorphism α of W also induces isomorphisms $V_T \to V_{T^\alpha}$ between the translation spaces of the instantaneous spaces.

4. Material Universes, Motions

A *material universe*[1] is a set $\mathcal{U}$ consisting of *bodies* $B, C, \ldots$. It is assumed that $\mathcal{U}$ is partially ordered by a relation $<$, and we say that C *is a part of* B if $C < B$. The body B is said to be *separate* from the body C if B and C have no part in common, i.e., if there is no $D \in \mathcal{U}$ such that both $D < B$ and $D < C$.

The material universe appropriate to a system of discrete particles consists of the collection of all subsets of a finite set, whose members represent the particles. In this case, the relation $<$ is the set-inclusion. For continuum mechanics, however, more complicated material universes must be considered.

A *motion* of the material universe $\mathcal{U}$ is a function M which assigns to each body $B \in \mathcal{U}$ a subset $M(B)$ of W such that

$$M(B) \subset M(C) \quad \text{if} \quad B < C \tag{4.1}$$

and

$$M(B) \cap T \neq \emptyset \quad \text{for each} \quad T \in \Gamma. \tag{4.2}$$

The set $M(B)$ is called the *set of events experienced by* B during the motion M, or simply the *world-tube* of B.

The requirement (4.1) states that a part experiences fewer events than the whole, and (4.2) expresses the fact that bodies cannot appear out of nothing nor disappear into nothing.

An automorphism α of the event-world W induces a transformation $M \to M^\alpha$ on motions; it is defined by

$$M^\alpha(B) = \big(M(B)\big)^\alpha \quad \text{for all} \quad B \in \mathcal{U}. \tag{4.3}$$

The above is only the beginning of a kinematics based on the neoclassical space-time. A more detailed development will be presented in future publications.

5. Force Systems, Dynamical Processes

A *force system* for a material universe $\mathcal{U}$ is a function φ which assigns to every triple (B, C, T), where B and C are separate bodies in $\mathcal{U}$ and $T \in \Gamma$ is an instant, a vector $\varphi(B, C, T)$ in the translation space V_T of T. The value $\varphi(B, C, T)$ is called the *force exerted by the body* C *on the body* B *at the instant* T. Force systems are subject to restrictions which will not be stated here. (They are similar to the ones given in [3], Sect. 3, and [4], Sect. 4.)

[1] The concept of material universe used here differs from the one of references [3] and [4]. In these papers, bodies are assumed to be certain subsets of a universal set, while here they need not be sets at all.

A *dynamical process* for the material universe $\mathscr{U}$ is a pair $\Pi = (M, \varphi)$, where M is a motion of $\mathscr{U}$ and φ a force system for $\mathscr{U}$, such that the fundamental laws of balance of forces and moments are statisfied. A precise statement of these laws in the present framework and a detailed treatment of dynamics will be given in future publications. The totality of all dynamical processes for $\mathscr{U}$ will be denoted by $\mathscr{D}$.

An automorphism α of W induces a transformation $\varphi \to \varphi^\alpha$ on force systems; it is defined by

$$\varphi^\alpha(B, C, T^\alpha) = \alpha(\varphi(B, C, T)), \tag{5.1}$$

where the right hand side is the image under the isomorphism $V_T \to V_{T\alpha}$ induced by α. The automorphism α induces the transformation

$$\Pi = (M, \varphi) \to \Pi^\alpha = (M^\alpha, \varphi^\alpha) \tag{5.2}$$

on the set $\mathscr{D}$ of all dynamical processes, where M^α is defined by (4.3) and φ^α by (5.1). The fundamental laws of balance are invariant under transformations of the form (5.2), so that dynamical processes are transformed into dynamical processes.

In order to have consistency of the theory described here with the conventional approaches to classical mechanics one must include inertial forces in the force systems φ on an equal footing with other kinds of forces (cf. [3]).

6. Constitutive Classes

The nature of many problems in mechanics can roughly be described as follows: among all conceivable dynamical processes for a material universe, select the one that will actually occur. In order to make such a selection, one must know something about the particular material properties of the bodies which belong to the material universe. Conventionally such properties are described by "force laws", "stress-strain relations", or similar types of constitutive laws.

A way of making precise the concept of a material property within the present framework is that of using the notion of a constitutive class: A *constitutive class* for a pair B, C of separate bodies is a subset $\mathscr{C}(B, C)$ of the set $\mathscr{D}$ of all dynamical processes, subject to the following requirements:

(I) If $\Pi_1 = (M_1, \varphi_1)$ and $\Pi_2 = (M_2, \varphi_2)$ are two dynamical processes such that

$$M_1(D) = M_2(D) \tag{6.1}$$

for all parts $D < B$ or $D < C$ and

$$\varphi_1(B, C, T) = \varphi_2(B, C, T) \tag{6.2}$$

for all instants $T \in \Gamma$, then $\Pi_1 \in \mathscr{C}(B, C)$ if and only if $\Pi_2 \in \mathscr{C}(C, B)$.

(II) $\mathscr{C}(B, C)$ is stable under automorphisms; i.e., if $\Pi \in \mathscr{C}(B, C)$ and if α is an automorphism of W, then $\Pi^\alpha \in \mathscr{C}(B, C)$, where Π^α is defined by (5.2).

The first of these requirements is a principle of irrelevance, stating that the material properties of the bodies B and C concern only B and C, and not anything else in the universe. The second requirement is a principle of homogeneity for the event-world, expressing the condition that events may have no individuality beyond the one conferred to them by motions of material universes.

In conventional treatments of mechanics material properties are defined by means of *constitutive equations*. Such constitutive equations actually define classes of dynamical processes in the sense of Sect. 5 if a suitable concept of *frame of reference* is employed. One can show that these constitutive equations must satisfy the principle of material frame-indifference and the principle of local action (see [*1*], Sect. 23) if they are to define constitutive classes in the sense described above.

REFERENCES

[*1*] TRUESDELL, C., and W. NOLL: The non-linear field theories of mechanics. In: Handbuch der Physik (S. FLÜGGE, ed.), vol. III/3. Berlin-Göttingen-Heidelberg: Springer 1965.

[*2*] NOLL, W.: Euclidean geometry and Minkowskian chronometry. Am. Math. Monthly **71**, 129—144 (1964).

[*3*] — La Mécanique Classique Basée Sur Un Axiome D'Objectivité, Colloque internat. sur la méthode axiomatique dans les mécaniques classiques et nouvelles 1959, p. 47—56. Paris 1963.

[*4*] — The foundations of classical mechanics in the light of recent advances in continuum mechanics. Proc. of the Berkeley Symposium on the Axiomatic Method, p. 266—281, Amsterdam 1959.

Materially Uniform Simple Bodies with Inhomogeneities

WALTER NOLL

Contents

1. Introduction

The basic concepts of the theory of simple materials have been introduced in reference [*1*] (see also the exposition in [*2*], Chapter C III). Here I present a detailed study of the structure of bodies that consist of a uniform simple material yet are not necessarily homogeneous.

After assembly of the necessary mathematical tools in Sects. 2–4, the concept of a *simple body* is introduced in Sect. 5. This concept is more inclusive than the one described in [*1*] because it can be appropriate not only to mechanical material properties, but also to thermal, optical, electrical, magnetic, or any other type of material properties. A body may be simple with respect to any particular such material property or to any combination of them. The physical theory relevant to these properties need not be made explicit.

In Sect. 6 a precise definition of a *materially uniform simple body* is given. The nature of the coherence of a uniform body with respect to the local material properties under consideration can be described mathematically in terms of what I call a *material uniformity* or in terms of what I call a *uniform reference*. In general, neither of these is uniquely determined by the simple body structure, but the degree of non-uniqueness can be delimited precisely. There may or may not exist uniform references that are gradients of global configurations. If they do exist, the body is homogeneous, and the theory becomes trivial.

This paper supersedes an unpublished preliminary study written by the author in 1963. Section 34 of reference [*2*] is a summary of that study.

Sections 7–9 contain an exposition of the mathematical prerequisites necessary to describe the local behavior of material uniformities and uniform references that possess a degree of smoothness. In the remainder of the paper, such smoothness is always assumed. A material uniformity is then equivalent to an affine connection, which is defined in Sect. 10 and called a *material connection*. The Cartan torsion of this connection describes locally the deviation from homogeneity and is therefore called, in Sect. 11, the *inhomogeneity* of the given material uniformity.

Associated with each smooth uniform reference is also a Riemannian structure on the body, and the relation of this structure to the material connection is studied in Sect. 12. The difference between the Riemannian connection and the material connection determines what I call the *contortion* of the given uniform reference. Contortion and inhomogeneity determine one another.

Of particular interest is a special type of non-homogeneity called *contorted aeolotropy* in Section 13. It generalizes the more familiar *curvilinear aeolotropy*. In contorted aeolotropy, the deviation from homogeneity is given by a distribution of rotations on a suitable global configuration, and the contortion describes the local behavior of this distribution. The curvature of the Riemannian structure mentioned before describes locally the deviation from contorted aeolotropy.

Section 14 contains a number of results that apply when the response functions of the body have special properties, especially with respect to material symmetry.

The usual version of CAUCHY's equation of balance (*cf.* [*2*], (16.6)) is very useful only when applied to bodies that are homogeneous. For applications to materially uniform but inhomogeneous bodies, a new version of CAUCHY's equation, derived in Sect. 15, is much more suitable than the usual one. This new version gives rise, for example, to a definite differential equation for the theory of inhomogeneous but materially uniform elastic bodies.

Unfortunately, there is no easily accessible exposition of the coordinate-free type of modern differential geometry that is the most appropriate for the applications in this study. The monograph of LANG [*3*], although it explains some of the concepts used here, does not contain sufficient material and emphasizes matters not relevant in the present context. For this reason I develop in this paper all mathematical tools as they come to be needed, tailored to the requirements of the intended applications.

There is a large literature on theories of *continuous distributions of dislocations*, proposed in various forms by KONDO, NYE, BILBY, BULLOUGH, SMITH, SEEGER, KRÖNER, GÜNTHER, and others[1]. Motivated by heuristic considerations, mostly concerning lattice defects in crystals, these authors lay down *a priori* certain geometric structures to describe distributions of dislocations. These geometric structures are formally of the same type as some of those occurring in the present paper. The conceptual status of the theory presented here, however, is very different. I show that once a constitutive assumption defining a materially uniform simple body is laid down, the geometric structures of the body are determined. The geometry is thus the *natural outcome*, not the first assumption, of the theory. Since the underlying constitutive assumption is very general, the real materials to which the theory can be expected to apply need be neither crystalline, nor elastic, nor solid.

[1] For details and references I refer to the expository articles [*4*] and [*5*].

2. Deformations

We shall employ the concept of *absolute physical space*[2], as is customary in classical physics. This space $\mathscr{E}$, whose elements $\boldsymbol{x}, \boldsymbol{y}, \ldots$, we call *spatial points*, has the structure of a three-dimensional Euclidean point space[3]. The translation space of $\mathscr{E}$ is denoted by $\mathscr{V}$; it is a three-dimensional inner product space. The elements $\boldsymbol{u}, \boldsymbol{v}, \ldots$, of $\mathscr{V}$ are called *spatial vectors*. The translation which carries $\boldsymbol{x} \in \mathscr{E}$ to $\boldsymbol{y} \in \mathscr{E}$ is denoted by $\boldsymbol{y}-\boldsymbol{x} \in \mathscr{V}$, and $\boldsymbol{x}+\boldsymbol{u}$ denotes the point into which $\boldsymbol{x} \in \mathscr{E}$ is carried by the translation $\boldsymbol{u} \in \mathscr{V}$. The inner product of two spatial vectors $\boldsymbol{u}, \boldsymbol{v} \in \mathscr{V}$ is denoted by $\boldsymbol{u} \cdot \boldsymbol{v}$. Of course, $\boldsymbol{u} \cdot \boldsymbol{v} \in \mathscr{R}$, where $\mathscr{R}$ is the set of all real numbers.

The set of all linear transformations $\boldsymbol{L}$: $\mathscr{V} \to \mathscr{V}$ of $\mathscr{V}$ into itself is denoted by $\mathscr{L}$. The composition of $\boldsymbol{L} \in \mathscr{L}$ with $\boldsymbol{M} \in \mathscr{L}$ is denoted by $\boldsymbol{ML} \in \mathscr{L}$. The identity transformation on $\mathscr{V}$ is denoted by $\mathbf{1} \in \mathscr{L}$. The transpose of $\boldsymbol{L} \in \mathscr{L}$ is denoted by $\boldsymbol{L}^T$, so that $\boldsymbol{u} \cdot \boldsymbol{L}\boldsymbol{v} = \boldsymbol{L}^T \boldsymbol{u} \cdot \boldsymbol{v}$ holds for all $\boldsymbol{u}, \boldsymbol{v} \in \mathscr{V}$. The trace and determinant of $\boldsymbol{L} \in \mathscr{L}$ are denoted by $\operatorname{tr} \boldsymbol{L}$ and $\det \boldsymbol{L}$, respectively. The set $\mathscr{L}$ of all linear transformations has the natural structure of a nine-dimensional algebra. It is also endowed with a natural inner product, whose values are given by $\boldsymbol{L} \cdot \boldsymbol{M} = \operatorname{tr}(\boldsymbol{L}\boldsymbol{M}^T)$. A transformation $\boldsymbol{L} \in \mathscr{L}$ is said to be *invertible* if it is a bijection (*i.e.*, one-to-one and onto). In this case, there exists an *inverse* $\boldsymbol{L}^{-1} \in \mathscr{L}$ so that $\boldsymbol{L}\boldsymbol{L}^{-1} = \boldsymbol{L}^{-1}\boldsymbol{L} = \mathbf{1}$. The invertible members of $\mathscr{L}$ form a group $\ell \subset \mathscr{L}$ under composition; it is called the *linear group* of $\mathscr{V}$. Important subgroups of ℓ are the *unimodular group*

$$u = \{\boldsymbol{H} \in \ell \,|\, |\det \boldsymbol{H}| = 1\}$$

and the *orthogonal group*

$$o = \{\boldsymbol{Q} \in \ell \,|\, \boldsymbol{Q}\boldsymbol{Q}^T = \mathbf{1}\}.$$

Of course, o is a subgroup of u.

Consider a mapping $\boldsymbol{\varphi}$: $\mathscr{G} \to \mathscr{E}'$ of an open subset $\mathscr{G} \subset \mathscr{E}$ into a point-space or vector-space $\mathscr{E}'$. Let $\mathscr{V}'$ be the translation space of $\mathscr{E}'$ ($\mathscr{V}' = \mathscr{E}'$ if $\mathscr{E}'$ is already a vectorspace) and let $\mathscr{L}(\mathscr{V}, \mathscr{V}')$ be the space of all linear transformations of $\mathscr{V}$ into $\mathscr{V}'$. We say that $\boldsymbol{\varphi}$ is *of class* C^1 if there is a continuous mapping $\nabla \boldsymbol{\varphi}$: $\mathscr{G} \to \mathscr{L}(\mathscr{V}, \mathscr{V}')$ such that

$$\boldsymbol{\varphi}(\boldsymbol{x}+\boldsymbol{u}) = \boldsymbol{\varphi}(\boldsymbol{x}) + (\nabla \boldsymbol{\varphi}(\boldsymbol{x}))\,\boldsymbol{u} + \boldsymbol{\sigma}(\boldsymbol{x}, \boldsymbol{u}),$$

where

$$\lim_{|\boldsymbol{u}| \to 0} \frac{1}{|\boldsymbol{u}|}\, \boldsymbol{\sigma}(\boldsymbol{x}, \boldsymbol{u}) = \mathbf{0}$$

holds for all $\boldsymbol{x} \in \mathscr{G}$. The mapping $\nabla \boldsymbol{\varphi}$, if it exists, is uniquely determined by $\boldsymbol{\varphi}$ and is called the *gradient* of $\boldsymbol{\varphi}$. If $\nabla \boldsymbol{\varphi}$ exists and is itself of class C^1, we say that $\boldsymbol{\varphi}$ is *of class* C^2. The gradient of $\nabla \boldsymbol{\varphi}$ is denoted by $\nabla^{(2)} \boldsymbol{\varphi}$ and is called the *second gradient* of $\boldsymbol{\varphi}$. Continuing in this manner, we say that $\boldsymbol{\varphi}$ is *of class* C^r, if it is of class C^{r-1} and if its $(r-1)^{\text{st}}$ gradient $\nabla^{(r-1)} \boldsymbol{\varphi}$ is of class C^1. The gradient of $\nabla^{(r-1)} \boldsymbol{\varphi}$ is denoted by $\nabla^{(r)} \boldsymbol{\varphi}$. We say that $\boldsymbol{\varphi}$ is of class C^0 if it is merely continuous. If $\boldsymbol{\varphi}$ is of class C^2, its second gradient has the symmetry property $((\nabla^{(2)} \boldsymbol{\varphi})\, \boldsymbol{u})\, \boldsymbol{v} = ((\nabla^{(2)} \boldsymbol{\varphi})\, \boldsymbol{v})\, \boldsymbol{u}$, $\boldsymbol{u}, \boldsymbol{v} \in \mathscr{V}$.

[2] The considerations of this paper can be adapted to the neoclassical space-time explained in [6]. When this is done, absolute space must be replaced by suitably defined "instantaneous spaces".

[3] The exact meaning of this term is explained in [7], Sect. 4.

The modifier "of class C^r" may apply, in particular, to a *scalar field, i.e.* a mapping $f: \mathscr{G} \to \mathscr{R}$, a *vector field, i.e.* a mapping $\boldsymbol{h}: \mathscr{G} \to \mathscr{V}$, or a *tensor field, i.e.*, a mapping $\boldsymbol{T}: \mathscr{G} \to \mathscr{L}$. A one-to-one mapping $\boldsymbol{\lambda}: \mathscr{G} \to \mathscr{E}$ is called a *deformation of class* C^r ($r \geqq 1$) if it is not only of class C^r but if also the values of its gradient are invertible, *i.e.* if $\nabla \boldsymbol{\lambda}(\boldsymbol{x}) \in \ell$ for all $\boldsymbol{x} \in \mathscr{G}$.

The members of the linear group ℓ are also called *local deformations*, so that a (global) deformation has a gradient whose values are local deformations.

3. Continuous Bodies

A physical object can often be described mathematically by the concept of a *body* $\mathscr{B}$, which is a set whose members $X, Y, \ldots$, are called *material points*, and which is endowed with a structure defined by a class C of mappings $\boldsymbol{\kappa}: \mathscr{B} \to \mathscr{E}$. The mappings $\boldsymbol{\kappa} \in \mathsf{C}$ are called the *configurations* of $\mathscr{B}$ (in the space $\mathscr{E}$). The spatial point $\boldsymbol{\kappa}(X) \in \mathscr{E}$ is called the *place* of the material point $X \in \mathscr{B}$ in the configuration $\boldsymbol{\kappa}$.

We say that $\mathscr{B}$ is a *continuous body of class* C^p ($p \geqq 1$) if the class C of configurations satisfies the following axioms:

(C 1) Every $\boldsymbol{\kappa} \in \mathsf{C}$ is one-to-one and its range $\boldsymbol{\kappa}(\mathscr{B})$ is an open subset of $\mathscr{E}$, which is called the *region occupied by* $\mathscr{B}$ *in the configuration* $\boldsymbol{\kappa}$.

(C 2) If $\boldsymbol{\gamma}, \boldsymbol{\kappa} \in \mathsf{C}$ then the composite[4] $\boldsymbol{\lambda} = \boldsymbol{\gamma} \circ \overset{-1}{\boldsymbol{\kappa}}: \boldsymbol{\kappa}(\mathscr{B}) \to \boldsymbol{\gamma}(\mathscr{B})$ is a deformation of class C^p, which is called the *deformation of* $\mathscr{B}$ *from the configuration* $\boldsymbol{\kappa}$ *into the configuration* $\boldsymbol{\gamma}$.

(C 3) If $\boldsymbol{\kappa} \in \mathsf{C}$ and if $\boldsymbol{\lambda}: \boldsymbol{\kappa}(\mathscr{B}) \to \mathscr{E}$ is a deformation of class C^p, then $\boldsymbol{\lambda} \circ \boldsymbol{\kappa} \in \mathsf{C}$. The mapping $\boldsymbol{\lambda} \circ \boldsymbol{\kappa}$ is called the *configuration obtained from the configuration* $\boldsymbol{\kappa}$ *by the deformation* $\boldsymbol{\lambda}$.

In the remainder of this paper we shall always assume that $\mathscr{B}$ is a continuous body of class C^p, $p \geqq 1$.

The axioms (C 1)–(C 3) ensure that the class C endows the body $\mathscr{B}$ with the structure of a "C^p-manifold modelled on $\mathscr{E}$" in the sense of LANG ([3], Ch. II, § 1). Topologically, it is a very simple manifold because it can be mapped out with a single "chart" ("configuration" in our terminology).

Of central importance for the present paper is the concept of a *local configuration*[5] *at a material point* X. Two (global) configurations $\boldsymbol{\kappa}$ and $\boldsymbol{\gamma}$ are said to be *equivalent at* X, and we write[6]

$$\boldsymbol{\kappa} \sim_X \boldsymbol{\gamma} \quad \text{if} \quad \nabla(\boldsymbol{\kappa} \circ \overset{-1}{\boldsymbol{\gamma}})|_{\boldsymbol{\gamma}(X)} = \mathbf{1}. \tag{3.1}$$

It is an immediate consequence of the chain rule for gradients that $\sim_X$ is an equivalence relation on C. The resulting partition of C is denoted by $\mathscr{C}_X$, and its members $\boldsymbol{K}_X, \boldsymbol{G}_X, \ldots$, *i.e.* the equivalence classes, are called *local configurations* at X. Instead of writing $\boldsymbol{\kappa} \in \boldsymbol{K}_X$ when $\boldsymbol{\kappa}$ is a member of the class $\boldsymbol{K}_X$ we often write

$$\nabla \boldsymbol{\kappa}(X) = \boldsymbol{K}_X \tag{3.2}$$

[4] Composition of mappings other than linear mappings is denoted by $\circ$. The inverse of a one-to-one mapping $\boldsymbol{\kappa}$ is denoted by $\overset{-1}{\boldsymbol{\kappa}}$.

[5] The term "configuration gradient" was used and another meaning was assigned to the term "local configuration" in [1].

[6] For better reading, we sometimes write $f|_x$ instead of $f(x)$ for the value of f at x.

and say that the local configuration $\boldsymbol{K}_X$ is the *gradient at X* of the (global) configuration $\boldsymbol{\kappa}$.

Let $\boldsymbol{K}_X, \boldsymbol{G}_X \in \mathscr{C}_X$ be two local configurations, and let $\boldsymbol{\kappa} \in \boldsymbol{K}_X$, $\gamma \in \boldsymbol{G}_X$. It is easily seen that the local deformation $\nabla(\gamma \circ \overset{-1}{\boldsymbol{\kappa}})|_{\boldsymbol{\kappa}(X)} \in \ell$ depends only on $\boldsymbol{K}_X$ and $\boldsymbol{G}_X$, and not on the particular choices of $\boldsymbol{\kappa} \in \boldsymbol{K}_X$ and $\gamma \in \boldsymbol{G}_X$. We denote this local deformation by $\boldsymbol{G}_X \boldsymbol{K}_X^{-1}$ and call it the *local deformation from the local configuration* $\boldsymbol{K}_X$ *into the local configuration* $\boldsymbol{G}_X$. Using the notation (3.2), we then have

$$\nabla(\gamma \circ \overset{-1}{\boldsymbol{\kappa}})|_{\boldsymbol{\kappa}(X)} = \nabla\gamma(X)\,[\nabla\boldsymbol{\kappa}(X)]^{-1}. \tag{3.3}$$

If $\boldsymbol{K}_X \in \mathscr{C}_X$ is a local configuration and $\boldsymbol{L} \in \ell$ any local deformation, we can define a new local configuration $\boldsymbol{L}\boldsymbol{K}_X \in \mathscr{C}_X$ by

$$\boldsymbol{L}\boldsymbol{K}_X = \{\lambda \circ \boldsymbol{\kappa} \,|\, \nabla\lambda|_{\boldsymbol{\kappa}(X)} = \boldsymbol{L},\ \nabla\boldsymbol{\kappa}(X) = \boldsymbol{K}_X\}. \tag{3.4}$$

We call $\boldsymbol{L}\boldsymbol{K}_X$ the *local configuration obtained from the local configuration* $\boldsymbol{K}_X$ *by the local deformation* **L**. Clearly, we have the rules

$$(\boldsymbol{G}_X \boldsymbol{K}_X^{-1})\boldsymbol{K}_X = \boldsymbol{G}_X, \qquad (\boldsymbol{L}\boldsymbol{K}_X)\boldsymbol{K}_X^{-1} = \boldsymbol{L}. \tag{3.5}$$

4. Tangent Spaces

Consider pairs $(\boldsymbol{K}_X, \boldsymbol{u})$, where $\boldsymbol{K}_X \in \mathscr{C}_X$ is a local configuration at X and $\boldsymbol{u} \in \mathscr{V}$ a spatial vector. We say that two such pairs $(\boldsymbol{K}_X, \boldsymbol{u})$ and $(\boldsymbol{G}_X, \boldsymbol{v})$ are *equivalent* if

$$(\boldsymbol{K}_X \boldsymbol{G}_X^{-1})\,\boldsymbol{v} = \boldsymbol{u}. \tag{4.1}$$

It follows from the rules (3.5) that (4.1) does indeed define an equivalence relation. The resulting equivalence classes are called *tangent vectors* $\mathfrak{u}_X, \mathfrak{v}_X, \ldots$ at X. The totality of all these tangent vectors is denoted by $\mathscr{T}_X$ and is called the *tangent space* at $X \in \mathscr{B}$. Let $\mathfrak{u}_X \in \mathscr{T}_X$ and $\boldsymbol{K}_X \in \mathscr{C}_X$ be given and let $(\boldsymbol{G}_X, \boldsymbol{v})$ be any pair belonging to the class $\mathfrak{u}_X$. Now, if $(\boldsymbol{K}_X, \boldsymbol{u})$ is to belong to $\mathfrak{u}_X$ then (4.1) must hold. Therefore, we see that $\mathfrak{u}_X \in \mathscr{T}_X$ and $\boldsymbol{K}_X \in \mathscr{C}_X$ determine a unique spatial vector $\boldsymbol{u} \in \mathscr{V}$ such that $(\boldsymbol{K}_X, \boldsymbol{u}) \in \mathfrak{u}_X$. We can therefore use the notation

$$\boldsymbol{u} = \boldsymbol{K}_X \mathfrak{u}_X, \qquad \mathfrak{u}_X = \boldsymbol{K}_X^{-1}\boldsymbol{u} \quad \text{if} \quad (\boldsymbol{K}_X, \boldsymbol{u}) \in \mathfrak{u}_X, \tag{4.2}$$

and we see that $\boldsymbol{K}_X$ determines a one-to-one mapping of the tangent space $\mathscr{T}_X$ onto the space $\mathscr{V}$ of spatial vectors. The tangent space $\mathscr{T}_X$ has the natural structure of a three-dimensional vector space, with addition defined by

$$\mathfrak{u}_X + \mathfrak{v}_X = \boldsymbol{K}_X^{-1}(\boldsymbol{u} + \boldsymbol{v}) \quad \text{if} \quad \mathfrak{u}_X = \boldsymbol{K}_X^{-1}\boldsymbol{u}, \qquad \mathfrak{v}_X = \boldsymbol{K}_X^{-1}\boldsymbol{v} \tag{4.3}$$

and multiplication with scalars by

$$a\,\mathfrak{u}_X = \boldsymbol{K}_X^{-1}(a\,\boldsymbol{u}) \quad \text{if} \quad \mathfrak{u}_X = \boldsymbol{K}_X^{-1}\boldsymbol{u}, \qquad a \in \mathscr{R}. \tag{4.4}$$

It is immediately seen that these definitions of $\mathfrak{u}_X + \mathfrak{v}_X$ and $a\,\mathfrak{u}_X$ are legitimate because the results are independent of the choice of the local configuration $\boldsymbol{K}_X$ used to represent $\mathfrak{u}_X$ and $\mathfrak{v}_X$ in $\mathscr{V}$. The local configurations can be identified with the invertible linear transformations of $\mathscr{T}_X$ onto $\mathscr{V}$.

Given a local configuration $\boldsymbol{K}_X \in \mathscr{C}_X$, we can define an inner product $\mathfrak{u}_X * \mathfrak{v}_X$ of $\mathfrak{u}_X, \mathfrak{v}_X, \in \mathscr{T}_X$ by

$$\mathfrak{u}_X * \mathfrak{v}_X = (\boldsymbol{K}_X \mathfrak{u}_X) \cdot (\boldsymbol{K}_Y \mathfrak{v}_X). \tag{4.5}$$

However, we obtain different inner products on $\mathscr{T}_X$ for different choices of $\boldsymbol{K}_X$, and hence $\mathscr{T}_X$ is *not* naturally an inner product space.

5. Simple Bodies, Material Isomorphisms, Intrinsic Isotropy Groups

To describe mathematically the physical characteristics of a body $\mathscr{B}$ we must endow $\mathscr{B}$ with additional structure. Some of these characteristics, such as elasticity, viscosity, heat capacity, and electrical conductivity, are local, *i.e.*, they are attached to the individual material points $X \in \mathscr{B}$ rather than to the body as a whole. Other characteristics, such as mutual gravitation and internal radiative heat transfer, involve more than one material point. We deal here only with local characteristics. The physical response of the body $\mathscr{B}$ at a particular material point $X \in \mathscr{B}$ and a particular time will depend on the configuration $\boldsymbol{\kappa}$ of $\mathscr{B}$ at that time. It may happen that only the local configuration $\nabla \boldsymbol{\kappa}(X)$ at X determined by $\boldsymbol{\kappa}$, and no other properties of $\boldsymbol{\kappa}$, has an influence on the response. If this is the case, we say that the *material at X is simple*. We say that the whole of $\mathscr{B}$ is simple or that $\mathscr{B}$ is a *simple body* if the material at X is simple for all $X \in \mathscr{B}$.

We assume that a possible physical response at a material point is given mathematically by specifying an element from a set R of mathematical objects. The nature of R depends on the particular physical phenomena to be described. For example, in the theory of elasticity R consists of all possible 'stress tensors', *i.e.*, of all symmetric linear transformations of $\mathscr{V}$ into $\mathscr{V}$. In the mechanical theory of simple materials with fading memory, R consists of "memory functionals" that relate relative deformation histories to stresses and are subject to certain smoothness requirements. In theories that include non-mechanical effects R consists of functions or functionals whose independent and dependent variables have interpretations as local temperatures, energy or entropy densities, heat fluxes, electric or magnetic field strengths, polarizations, magnetizations, electric currents, *etc.* For the purpose of the present paper, no specific assumptions about the nature of R need be made.

We can now make our definition of a simple body precise:

Definition 1. *Let* R *be a set, whose elements we call* ***response descriptors****. A continuous body* $\mathscr{B}$ *of class* C^p *will be called a* ***simple body*** *with respect to* R *if it is endowed with a structure by a function* $\mathfrak{G}$ *which assigns to each material point* $X \in \mathscr{B}$ *a mapping*

$$\mathfrak{G}_X \colon \mathscr{C}_X \to \mathsf{R}. \tag{5.1}$$

The value $\mathfrak{G}_X(\boldsymbol{G}_X)$ *is the response descriptor of the material at X in any configuration* γ *of* $\mathscr{B}$ *such that* $\nabla \gamma(X) = \boldsymbol{G}_X$.

The mappings $\mathfrak{G}_X$ cannot be entirely arbitrary, for they are subject to restrictions imposed by general physical principles such as the principle of frame-indifference and the principle of dissipation. These restrictions need not be made explicit here.

We would like to give now a precise meaning to the statement that the material at one point $X \in \mathscr{B}$ is the same as the material at another point $Y \in \mathscr{B}$. We cannot construe this statement to mean that $\mathfrak{G}_X$ and $\mathfrak{G}_Y$ are the same, for they have different domains and hence cannot be directly compared. However, we can connect the domains $\mathscr{C}_X$ and $\mathscr{C}_Y$ if an isomorphism $\Phi_{XY}: \mathscr{T}_Y \to \mathscr{T}_X$ of the tangent space at Y onto the tangent space at X is given. Recalling that a local configuration $G_X \in \mathscr{C}_X$ can be regarded as a mapping $G_X: \mathscr{T}_X \to \mathscr{V}$, we can let $G_X \in \mathscr{C}_X$ correspond to the composition $G_X \Phi_{XY} \in \mathscr{C}_Y$. We are thus led to the following definition:

Definition 2. *An invertible linear transformation $\Phi_{XY}: \mathscr{T}_Y \to \mathscr{T}_X$ is called a* ***material isomorphism*** *from $\mathscr{T}_Y$ onto $\mathscr{T}_X$ if*

$$\mathfrak{G}_X(G_X) = \mathfrak{G}_Y(G_X \Phi_{XY}) \tag{5.2}$$

holds for all $G_X \in \mathscr{C}_X$.

To say that the material at X is the same as the material at Y means that there exists a material isomorphism from $\mathscr{T}_X$ onto $\mathscr{T}_Y$.

It follows immediately from Definition 2 that if $\Phi_{XY}: \mathscr{T}_Y \to \mathscr{T}_X$ and $\Phi_{YZ}: \mathscr{T}_Z \to \mathscr{T}_Y$ are material isomorphisms, so is their composition $\Phi_{XY} \Phi_{YZ}: \mathscr{T}_Z \to \mathscr{T}_X$. Also, if $\Phi_{XY}: \mathscr{T}_Y \to \mathscr{T}_X$ is a material isomorphisms, so is its inverse $\Phi_{XY}^{-1}: \mathscr{T}_X \to \mathscr{T}_Y$. If we denote the set of all material isomorphisms from $\mathscr{T}_X$ onto $\mathscr{T}_Y$ by g_{YX}, these facts can be expressed by[7]

$$g_{ZY} g_{YX} = g_{ZX}, \quad g_{YX} = g_{XY}^{-1}, \qquad \text{valid if these sets are not empty.} \tag{5.3}$$

It is clear that g_{XX}, the set of all material isomorphisms of $\mathscr{T}_X$ onto itself, is a subgroup of the linear group ℓ_X of $\mathscr{T}_X$, which consists of all invertible linear transformations of $\mathscr{T}_X$. We write

$$g_X = g_{XX} \tag{5.4}$$

and call g_X the *intrinsic isotropy group* of the material at X. For any $\Phi_{XY} \in g_{XY}$ one easily establishes the relations

$$g_{XY} = g_X \Phi_{XY} g_Y, \quad g_X = \Phi_{XY} g_Y \Phi_{XY}^{-1}. \tag{5.5}$$

It follows from $(5.5)_2$ that if a material isomorphism $\Phi_{XY}: \mathscr{T}_Y \to \mathscr{T}_X$ exists, *i.e.* if the material at X is the same as the material at Y, then the intrinsic isotropy groups g_X and g_Y are isomorphic.

6. Material Uniformity, Uniform References, Relative Isotropy Groups

A simple body $\mathscr{B}$ is said to be *materially uniform* if the material at any two of its points is the same, *i.e.* if g_{XY} is never empty. From now on we assume that $\mathscr{B}$ is a materially uniform simple body. We select a member $\Phi'(X, Y)$ from each g_{XY} and thereby define a function Φ' which assigns to each pair (X, Y) of material points of $\mathscr{B}$ a material isomorphism from $\mathscr{T}_Y$ onto $\mathscr{T}_X$. Choose $X_0 \in \mathscr{B}$ arbitrarily

[7] If g and $\hbar$ are sets of linear transformations of any kind such that the composition LM makes sense whenever $L \in g$, $M \in \hbar$, we write $g\hbar = \{LM | L \in g, M \in \hbar\}$. If the $L \in g$ are invertible, we write $g^{-1} = \{L^{-1} | L \in g\}$. Also, we write $Kg = \{KL | L \in g\}$ if KL makes sense for all $L \in g$.

and define Φ by

$$\Phi(X, Y) = \Phi'(X, X_0)\,\Phi'(Y, X_0)^{-1}. \tag{6.1}$$

It follows from (5.3) that $\Phi(X, Y) \in g_{XY}$. Moreover, we have

$$\Phi(Z, Y)\,\Phi(Y, X) = \Phi(Z, X), \qquad \Phi(X, X) = \mathbf{1}_X, \tag{6.2}$$

where $\mathbf{1}_X$ is the identity transformation of $\mathscr{T}_X$.

Definition 3. *A function Φ which assigns to each pair (X, Y) of material points of $\mathscr{B}$ a material isomorphism $\Phi(X, Y) \in g_{XY}$ is called a* **material uniformity** *if (6.2) holds.*

The construction (6.1) shows that the materially uniform bodies are those that admit material uniformities. It follows from (5.5) and (6.2) that any two material uniformities Φ and $\hat{\Phi}$ are related by

$$\hat{\Phi}(X, Y) = \mathfrak{P}(X)\,\Phi(X, Y)\,\mathfrak{P}(Y)^{-1}, \tag{6.3}$$

where $\mathfrak{P}$ is a function on $\mathscr{B}$ whose values $\mathfrak{P}(X)$ belong to the intrinsic isotropy groups g_X.

Definition 4. *A function $\boldsymbol{K}$ on $\mathscr{B}$ whose values $\boldsymbol{K}(X) \in \mathscr{C}_X$ are local configurations is called a* **reference** *for $\mathscr{B}$. If, moreover,*

$$\Phi(X, Y) = \boldsymbol{K}(X)^{-1}\,\boldsymbol{K}(Y) \tag{6.4}$$

is a material isomorphism of $\mathscr{T}_Y$ onto $\mathscr{T}_X$ for any $X, Y \in \mathscr{B}$, then $\boldsymbol{K}$ is called a **uniform reference** *for $\mathscr{B}$.*

Actually, (6.2) holds if Φ is defined by (6.4), so that Φ is a material uniformity if $\boldsymbol{K}$ is a uniform reference. Hence, every uniform reference $\boldsymbol{K}$ determines a material uniformity Φ through (6.4). Conversely, if a material uniformity Φ and a local configuration $\boldsymbol{K}_{X_0} \in \mathscr{C}_{X_0}$ for a particular material point $X_0 \in \mathscr{B}$ are given, then there exist a unique uniform reference $\boldsymbol{K}$ such that (6.4) and $\boldsymbol{K}(X_0) = \boldsymbol{K}_{X_0}$ hold. In fact, $\boldsymbol{K}$ is given by

$$\boldsymbol{K}(X) = \boldsymbol{K}_{X_0}\,\Phi(X_0, X). \tag{6.5}$$

Therefore, every material uniformity has representations (6.4) in terms of uniform references.

If $\boldsymbol{\kappa}$ is a (global) configuration, then $\nabla\boldsymbol{\kappa}$, which assigns to X the local configuration $\nabla\boldsymbol{\kappa}(X)$ at X, *i.e.*, the equivalence class to which $\boldsymbol{\kappa}$ belongs, is a reference, called the *gradient* of the configuration $\boldsymbol{\kappa}$. We say that a body is *homogeneous* if it admits a gradient as a uniform reference. Of course, not every reference is a gradient, and it may happen that none of the uniform references of a materially uniform body is a gradient.

Let $\boldsymbol{K}$ be a uniform reference. Every local configuration $\boldsymbol{G}_X \in \mathscr{C}_X$ can be characterized by the local deformation $\boldsymbol{F} = \boldsymbol{G}_X\,\boldsymbol{K}(X)^{-1} \in \ell$ from $\boldsymbol{K}(X)$ into $\boldsymbol{G}_X$, so that

$$\boldsymbol{G}_X = \boldsymbol{F}\,\boldsymbol{K}(X). \tag{6.6}$$

Substituting (6.6) and (6.4) into (5.2) with the choice $\Phi_{XY}=\Phi(X,Y)$, we see that

$$\mathfrak{G}_X(F\,K(X))=\mathfrak{G}_Y(F\,K(Y)) \tag{6.7}$$

must hold for all $F\in\ell$ and all $X, Y\in\mathscr{B}$. Conversely, if (6.7) holds for all $F\in\ell$ and all $X, Y\in\mathscr{B}$, then K is a uniform reference. This result may be formulated as follows:

Theorem 1. *A reference K for $\mathscr{B}$ is uniform if and only if there is a function $\mathfrak{H}_K\colon \ell\to \mathbf{R}$ which satisfies*

$$\mathfrak{H}_K(F)=\mathfrak{G}_X(F\,K(X)) \tag{6.8}$$

for all $X\in\mathscr{B}$ and all $F\in\ell$.

The function $\mathfrak{H}_K$, which assigns to each local deformation a response descriptor, will be called the *response function* of the body *relative to the uniform reference* K.

Let K be uniform reference. If we substitute (6.4) for Φ_{XY} in (5.5), we see that

$$K(X)\,g_X\,K(X)^{-1}=K(Y)\,g_Y\,K(Y)^{-1}, \tag{6.9}$$

i.e. that

$$g_K=K(X)\,g_X\,K(X)^{-1} \tag{6.10}$$

is independent of X. The group g_K is a subgroup of the linear group ℓ. We call g_K *the isotropy group of the body $\mathscr{B}$ relative to the uniform reference* K. In view of (6.10), all the intrinsic isotropy groups g_X, $X\in\mathscr{B}$, are isomorphic to the relative isotropy group g_K. It is easily seen that g_K is given in terms of the response function $\mathfrak{H}_K$ by

$$g_K=\{P\in\ell\mid \mathfrak{H}_K(F)=\mathfrak{H}_K(F\,P)\ \text{for all}\ F\in\ell\}. \tag{6.11}$$

The relation between two uniform references and the corresponding response functions and isotropy groups is described by the following theorem:

Theorem 2. *Any two uniform references K and $\hat{K}$ are related by*

$$\hat{K}(X)=L\,P(X)\,K(X), \tag{6.12}$$

where $L\in\ell$ and where P is a function on $\mathscr{B}$ with values in g_K.

The isotropy groups g_K and $g_{\hat{K}}$ relative to K and $\hat{K}$ are conjugate:

$$g_{\hat{K}}=L\,g_K\,L^{-1}. \tag{6.13}$$

The response functions $\mathfrak{H}_{\hat{K}}$ and $\mathfrak{H}_K$ are related by the identity

$$\mathfrak{H}_{\hat{K}}(F)=\mathfrak{H}_K(F\,L)\qquad \text{for all}\ F\in\ell. \tag{6.14}$$

Proof. The two material uniformities Φ and $\hat{\Phi}$ given by

$$\Phi(X,Y)=K(X)^{-1}K(Y),\qquad \hat{\Phi}(X,Y)=\hat{K}(X)^{-1}\hat{K}(Y)$$

must be related by (6.3). It follows that

$$\hat{K}(Y)\,\mathfrak{P}(Y)\,K(Y)^{-1}=\hat{K}(X)\,\mathfrak{P}(X)\,K(X)^{-1}=L\in\ell$$

is independent of $X\in\mathscr{B}$. Hence (6.12) holds with the choice

$$P(X)=K(X)\,\mathfrak{P}(X)^{-1}\,K(X)^{-1}. \tag{6.15}$$

It follows from (6.10) that $\boldsymbol{P}(X)\in g_{\boldsymbol{K}}$ for all $X\in\mathscr{B}$, which proves the first assertion of the theorem. If we write (6.10) with $\boldsymbol{K}$ replaced by $\hat{\boldsymbol{K}}$ and substitute (6.12), we obtain

$$\begin{aligned} g_{\hat{\boldsymbol{K}}} &= \boldsymbol{L}\,\boldsymbol{P}(X)\,\boldsymbol{K}(X)\,g_X\,\boldsymbol{K}(X)^{-1}\,\boldsymbol{P}(X)^{-1}\,\boldsymbol{L}^{-1} \\ &= \boldsymbol{L}\,\boldsymbol{P}(X)\,g_{\boldsymbol{K}}\,\boldsymbol{P}(X)^{-1}\,\boldsymbol{L}^{-1}. \end{aligned}$$

Since $\boldsymbol{P}(X)\in g_{\boldsymbol{K}}$ we have $\boldsymbol{P}(X)\,g_{\boldsymbol{K}}\,\boldsymbol{P}(X)^{-1}=g_{\boldsymbol{K}}$ and hence (6.13). The identity (6.14) is derived by writing (6.8) with $\boldsymbol{K}$ replaced by $\hat{\boldsymbol{K}}$, then substituting (6.12) and observing (6.11). Q.E.D.

The theory of isotropy groups relative to a local reference configuration at a single material point[8] extends without change to isotropy groups relative to a uniform reference $\boldsymbol{K}$ of a whole materially uniform body. In particular, we say that the uniform reference $\boldsymbol{K}$ is *undistorted* if $g_{\boldsymbol{K}}$ is comparable, with respect to inclusion, to the orthogonal group o, *i.e.*, if either $g_{\boldsymbol{K}}\subset o$ or $o\subset g_{\boldsymbol{K}}$. If there are uniform references $\boldsymbol{K}$ such that $g_{\boldsymbol{K}}\supset o$, we say that $\mathscr{B}$ is a *uniform isotropic body*; if there are uniform references $\boldsymbol{K}$ such that $g_{\boldsymbol{K}}\subset o$, we say that $\mathscr{B}$ is a *uniform solid body*. It is possible that a uniform simple body has no undistorted uniform references at all; such a body would be neither a solid nor isotropic.

7. Vector and Tensor Fields

As before, we assume that $\mathscr{B}$ is a continuous body of class C^p, $p\geqq 1$.

A mapping ψ: $\mathscr{B}\to\mathscr{E}'$ of $\mathscr{B}$ into some point-space or vector-space $\mathscr{E}'$ is said to be *of class* C^r, $0\leqq r\leqq p$, if for every configuration $\kappa\in\mathsf{C}$, the mapping $\psi\circ\overset{-1}{\kappa}$: $\kappa(\mathscr{B})\to\mathscr{E}'$ is of class C^r. In view of the axioms for $\mathscr{B}$ it is clear that $\psi\circ\overset{-1}{\kappa}$ is of class C^r for *every* $\kappa\in\mathsf{C}$ if it is of class C^r for *some* $\kappa\in\mathsf{C}$. These definitions apply, in particular, to *functions (scalar fields)* on $\mathscr{B}$, *i.e.* mappings f: $\mathscr{B}\to\mathscr{R}$, to *vector fields* on $\mathscr{B}$, *i.e.* mappings $\boldsymbol{h}$: $\mathscr{B}\to\mathscr{V}$, and to *tensor fields* on $\mathscr{B}$, *i.e.* mappings $\boldsymbol{T}$: $\mathscr{B}\to\mathscr{L}$.

A mapping $\mathfrak{h}$ which assigns to each material point $X\in\mathscr{B}$ a tangent vector $\mathfrak{h}(X)\in\mathscr{T}_X$ is called a *tangent vector field.* We say that such a tangent vector field $\mathfrak{h}$ is of class C^r, $0\leqq r\leqq p-1$, if the vector field $(\nabla\kappa)\,\mathfrak{h}$ on $\mathscr{B}$ defined by

$$(\nabla\kappa)\,\mathfrak{h}|_X=(\nabla\kappa(X))\,\mathfrak{h}(X) \tag{7.1}$$

is of class C^r for some — and hence every — configurafion $\kappa\in\mathsf{C}$.

The algebra of all linear transformations of the tangent space $\mathscr{T}_X$ into itself will be denoted by $\mathscr{I}_X$. A mapping $\mathfrak{T}$ which assigns to each material point $X\in\mathscr{B}$ a linear transformation $\mathfrak{T}(X)\in\mathscr{I}_X$ is called an *intrinsic tensor field.* We say that $\mathfrak{T}$ is of class C^r, $0\leqq r\leqq p-1$, if the tensor field $(\nabla\kappa)\,\mathfrak{T}(\nabla\kappa)^{-1}$ on $\mathscr{B}$ defined by

$$(\nabla\kappa)\,\mathfrak{T}(\nabla\kappa)^{-1}|_X=\nabla\kappa(X)\,\mathfrak{T}(X)\,(\nabla\kappa(X))^{-1} \tag{7.2}$$

is of class C^r for some — and hence every — configuration $\kappa\in\mathsf{C}$.

We shall use the term *field on* $\mathscr{B}$ for any mapping that assigns to every $X\in\mathscr{B}$ an element of some vector space (which may consist of linear or multilinear transformations).

[8] This theory was initiated in [*1*], §§19—21. An exposition is given in [*2*], §§ 31—33.

We shall employ the following scheme of notation:

$\mathscr{F}_{\mathscr{B}}^r$ = set of all functions (scalar fields) of class C^r on $\mathscr{B}$.

$\mathscr{V}_{\mathscr{B}}^r$ = set of all vector fields of class C^r on $\mathscr{B}$.

$\mathscr{T}_{\mathscr{B}}^r$ = set of all tangent vector fields of class C^r on $\mathscr{B}$.

$\mathscr{L}_{\mathscr{B}}^r$ = set of all tensor fields of class C^r on $\mathscr{B}$.

$\mathscr{I}_{\mathscr{B}}^r$ = set of all intrinsic tensor fields of class C^r on $\mathscr{B}$.

The set $\mathscr{F}_{\mathscr{B}}^r$ is a commutative algebra under pointwise addition and multiplication. The sets $\mathscr{V}_{\mathscr{B}}^s$, $\mathscr{T}_{\mathscr{B}}^s$, $\mathscr{L}_{\mathscr{B}}^s$, and $\mathscr{I}_{\mathscr{B}}^s$ can be made modules with respect to any of the algebras $\mathscr{F}_{\mathscr{B}}^r$, $s \leqq r \leqq p-1$, by defining addition and scalar multiplication with functions in $\mathscr{F}_{\mathscr{B}}^r$ pointwise. For example, if $\mathfrak{h}$, $\mathfrak{k} \in \mathscr{T}_{\mathscr{B}}^r$ and $f \in \mathscr{F}_{\mathscr{B}}^r$, we define $\mathfrak{h}+\mathfrak{k} \in \mathscr{T}_{\mathscr{B}}^r$ and $f\,\mathfrak{h} \in \mathscr{T}_{\mathscr{B}}^r$ by

$$(\mathfrak{h}+\mathfrak{k})|_X = \mathfrak{h}(X)+\mathfrak{k}(X), \qquad (f\,\mathfrak{h})|_X = f(X)\,\mathfrak{h}(X), \qquad X \in \mathscr{B}. \tag{7.3}$$

The sets $\mathscr{L}_{\mathscr{B}}^r$ and $\mathscr{I}_{\mathscr{B}}^r$ become associative (but not commutative) algebras over $\mathscr{F}_{\mathscr{B}}^r$ if multiplication is defined pointwise.

It is evident that we have $\mathscr{F}_{\mathscr{B}}^r \subset \mathscr{F}_{\mathscr{B}}^s$ if $s \leqq r$ and similar inclusions for the other sets in the list given above. Actually, $\mathscr{F}_{\mathscr{B}}^r$ is a subalgebra of $\mathscr{F}_{\mathscr{B}}^s$. Also, $\mathscr{V}_{\mathscr{B}}^r$ is not only a $\mathscr{F}_{\mathscr{B}}^r$-module, but also a submodule of $\mathscr{V}_{\mathscr{B}}^s$, regarded as a $\mathscr{F}_{\mathscr{B}}^r$-module. Analogous observations apply to the other modules and algebras of the list above.

If $\boldsymbol{T} \in \mathscr{L}_{\mathscr{B}}^s$ and $\boldsymbol{h} \in \mathscr{V}_{\mathscr{B}}^r$ or $\mathfrak{T} \in \mathscr{I}_{\mathscr{B}}^s$ and $\mathfrak{h} \in \mathscr{T}_{\mathscr{B}}^r$, we define $\boldsymbol{T}\boldsymbol{h}$ or $\mathfrak{T}\mathfrak{h}$ pointwise, *i.e.* by

$$\boldsymbol{T}\boldsymbol{h}|_X = \boldsymbol{T}(X)\,\boldsymbol{h}(X), \qquad \mathfrak{T}\mathfrak{h}|_X = \mathfrak{T}(X)\,\mathfrak{h}(X). \tag{7.4}$$

When $s \leqq r$, one can see that $\boldsymbol{T}\boldsymbol{h} \in \mathscr{V}_{\mathscr{B}}^s$, $\mathfrak{T}\mathfrak{h} \in \mathscr{T}_{\mathscr{B}}^s$. It is evident from $(7.4)_1$ that the rules

$$\boldsymbol{T}(\boldsymbol{h}+\boldsymbol{k}) = \boldsymbol{T}\boldsymbol{h} + \boldsymbol{T}\boldsymbol{k}, \qquad \boldsymbol{T}(f\,\boldsymbol{h}) = f\,\boldsymbol{T}\boldsymbol{h} \tag{7.5}$$

are valid. Hence every $\boldsymbol{T} \in \mathscr{L}_{\mathscr{B}}^s$ gives rise to a mapping

$$\boldsymbol{T}\colon \mathscr{V}_{\mathscr{B}}^r \to \mathscr{V}_{\mathscr{B}}^s \tag{7.6}$$

which satisfies the rules (7.5) for $\boldsymbol{h}, \boldsymbol{k} \in \mathscr{V}_{\mathscr{B}}^r$, $f \in \mathscr{F}_{\mathscr{B}}^r$. Mappings of the type (7.6) satisfying the rules (7.5) are homomorphism with respect to the $\mathscr{F}_{\mathscr{B}}^r$-module structures of $\mathscr{V}_{\mathscr{B}}^r$ and $\mathscr{V}_{\mathscr{B}}^s$. We also call them *$\mathscr{F}$-linear mappings*. Thus, every $\boldsymbol{T} \in \mathscr{L}_{\mathscr{B}}^s$ gives rise to an $\mathscr{F}$-linear mapping (7.6). It is remarkable that the converse is also true, *i.e.* that every $\mathscr{F}$-linear mapping of the type (7.6) arises from a tensor field of class C^s on $\mathscr{B}$:

Proposition 1. *If $\boldsymbol{T}\colon \mathscr{V}_{\mathscr{B}}^r \to \mathscr{V}_{\mathscr{B}}^s (s \leqq r)$ is $\mathscr{F}$-linear, then there exists a unique tensor field $\bar{\boldsymbol{T}} \in \mathscr{L}_{\mathscr{B}}^s$ such that $\boldsymbol{T}\boldsymbol{h} = \bar{\boldsymbol{T}}\boldsymbol{h}$ holds for all $\boldsymbol{h} \in \mathscr{V}_{\mathscr{B}}^r$.*

Proof. Let $(\boldsymbol{e}_1, \boldsymbol{e}_2, \boldsymbol{e}_3)$ be a basis of $\mathscr{V}$. The vectors $\boldsymbol{e}_i$ can be regarded as constant vector fields on $\mathscr{B}$, so that $\boldsymbol{e}_i \in \mathscr{V}_{\mathscr{B}}^{p-1} \subset \mathscr{V}_{\mathscr{B}}^r$. Every $\boldsymbol{h} \in \mathscr{V}_{\mathscr{B}}^r$ has a unique component representation

$$\boldsymbol{h} = \sum_i h^i \boldsymbol{e}_i, \qquad h^i \in \mathscr{F}_{\mathscr{B}}^r. \tag{7.7}$$

Applying the given $\mathscr{F}$-linear mapping $\boldsymbol{T}$ to (7.7), we obtain

$$\boldsymbol{T}\boldsymbol{h} = \sum_i h^i \boldsymbol{T}\boldsymbol{e}_i. \tag{7.8}$$

Now, if there is a tensor field $\bar{T}$ such that $\bar{T}\boldsymbol{h}=\boldsymbol{T}\boldsymbol{h}$ for all $\boldsymbol{h}\in\mathscr{V}_{\mathscr{B}}^r$, we must have, in particular, $\boldsymbol{T}\boldsymbol{e}_i=\bar{T}\boldsymbol{e}_i$, *i.e.*

$$\bar{T}(X)\,\boldsymbol{e}_i=(\boldsymbol{T}\boldsymbol{e}_i)|_X \tag{7.9}$$

for all $X\in\mathscr{B}$. But since $(\boldsymbol{e}_1,\boldsymbol{e}_2,\boldsymbol{e}_3)$ is a basis of $\mathscr{V}$, we can find, for each $X\in\mathscr{B}$, exactly one $\bar{T}(X)\in\mathscr{L}$ such that (7.9) holds. Since the vector fields $\boldsymbol{T}\boldsymbol{e}_i$ are of class C^s, it is easily seen that the tensor field $\bar{T}$ obtained in this way is also of class C^s. Moreover, in view of (7.8), (7.9) and the $\mathscr{F}$-linearity of $\bar{T}$ we have

$$\bar{T}\boldsymbol{h}=\sum_i h^i\,\bar{T}\boldsymbol{e}_i=\sum_i h^i\,\boldsymbol{T}\boldsymbol{e}_i=\boldsymbol{T}\big(\sum_i h^i\boldsymbol{e}_i\big)=\boldsymbol{T}\boldsymbol{h}$$

for all $\boldsymbol{h}$. Q.E.D.

Proposition 1 enables us to identify the set of all $\mathscr{F}$-linear mappings of the type (7.6) with the set $\mathscr{L}_{\mathscr{B}}^s$ of all tensor fields of class C^s on $\mathscr{B}$. Similarly, we can identify the set of all $\mathscr{F}$-linear mappings of the type

$$\mathfrak{T}\colon\ \mathscr{T}_{\mathscr{B}}^r\to\mathscr{T}_{\mathscr{B}}^s \qquad (s\leqq r)$$

with the set $\mathscr{I}_{\mathscr{B}}^s$ of all intrinsic tensor fields of class C^s on $\mathscr{B}$. The proof of this fact follows from Proposition 1 by choosing a configuration $\boldsymbol{\kappa}$ of $\mathscr{B}$ and letting $\mathfrak{T}$ correspond to $\bar{T}=\nabla\boldsymbol{\kappa}\,\mathfrak{T}(\nabla\boldsymbol{\kappa})^{-1}\colon\mathscr{V}_{\mathscr{B}}^r\to\mathscr{V}_{\mathscr{B}}^s$. The result just stated is a special case of a general proposition referring to $\mathscr{F}$-multilinear mappings. For later application we state another special case:

Proposition 2. *If*

$$\mathfrak{S}\colon\ \mathscr{T}_{\mathscr{B}}^r\times\mathscr{T}_{\mathscr{B}}^r\to\mathscr{T}_{\mathscr{B}}^s \qquad (or\ \mathscr{I}_{\mathscr{B}}^s) \tag{7.10}$$

is $\mathscr{F}$-bilinear (i.e. $\mathscr{F}$-linear in each of the two variables), then there exists a unique field $\bar{\mathfrak{S}}$ on $\mathscr{B}$ whose values $\bar{\mathfrak{S}}(X)$ are bilinear mappings

$$\bar{\mathfrak{S}}(X)\colon\ \mathscr{T}_X\times\mathscr{T}_X\to\mathscr{T}_X \qquad (or\ \mathscr{I}_X) \tag{7.11}$$

such that

$$\bar{\mathfrak{S}}(X)\,(\mathfrak{h}(X),\mathfrak{k}(X))=\mathfrak{S}(\mathfrak{h},\mathfrak{k})|_X \tag{7.12}$$

holds for all $\mathfrak{h},\mathfrak{k}\in\mathscr{T}_{\mathscr{B}}^r$ and all $X\in B$. The function $\bar{\mathfrak{S}}$ is of class C^s (in the obvious sense).

8. Relative Gradients, Brackets

From now on we assume that $\mathscr{B}$ is a continuous body of class C^p with $p\geqq 2$.

Let $\psi\colon\mathscr{B}\to\mathscr{E}'$ be a mapping of class $\mathscr{C}^r$, $1\leqq r\leqq p$, where $\mathscr{E}'$ is some point space or vector space. Given a configuration $\boldsymbol{\kappa}$ of $\mathscr{B}$, we can then define

$$\nabla_{\boldsymbol{\kappa}}\psi\colon\ \mathscr{B}\to\mathscr{L}(\mathscr{V},\mathscr{V}') \tag{8.1}$$

where $\mathscr{V}'$ is the translation space of $\mathscr{E}'$, by $\nabla_{\boldsymbol{\kappa}}\psi=\nabla(\psi\circ\overset{-1}{\boldsymbol{\kappa}})\circ\boldsymbol{\kappa}$, *i.e.*

$$\nabla_{\boldsymbol{\kappa}}\psi|_X=\nabla(\psi\circ\overset{-1}{\boldsymbol{\kappa}})|_{\boldsymbol{\kappa}(X)},\qquad X\in\mathscr{B}. \tag{8.2}$$

We call $\nabla_{\boldsymbol{\kappa}}\psi$ the *gradient of ψ relative to the configuration $\boldsymbol{\kappa}$*. It is clear that $\nabla_{\boldsymbol{\kappa}}\psi$ is of class C^{r-1}.

Let $\boldsymbol{\kappa},\gamma\in\mathrm{C}$ be two configurations. Taking the gradient of $\psi\circ\overset{-1}{\boldsymbol{\kappa}}=(\psi\circ\overset{-1}{\gamma})\circ(\gamma\circ\overset{-1}{\boldsymbol{\kappa}})$ and using the chain rule, we see with the help of (3.3) that the gradients of ψ relative to κ and γ are related by

$$\nabla_\kappa \psi|_X = \nabla_\gamma \psi|_X \circ [\nabla\gamma(X)(\nabla\kappa(X))^{-1}]. \tag{8.3}$$

Let $\boldsymbol{K}$ and $\boldsymbol{G}$ be two references for $\mathscr{B}$ (see Definition 4). We define $\boldsymbol{K}\boldsymbol{G}^{-1}$ pointwise, *i.e.* by

$$\boldsymbol{K}\boldsymbol{G}^{-1}|_X = \boldsymbol{K}(X)\boldsymbol{G}(X)^{-1}. \tag{8.4}$$

Recalling that the configuration gradients $\nabla\kappa$ and $\nabla\gamma$ are references, we see that (8.3) can then be written as

$$\nabla_\kappa \psi = \nabla_\gamma \psi \circ (\nabla\gamma(\nabla\kappa)^{-1}). \tag{8.5}$$

We say that a reference $\boldsymbol{K}$ is *of class* C^r, $r \leq p-1$, if for some – and hence every – configuration $\kappa \in \mathsf{C}$ the tensor field $(\nabla\kappa)\boldsymbol{K}^{-1}$ is of class C^r, *i.e.* belongs to $\mathscr{L}^r_{\mathscr{B}}$. It is clear that every gradient reference $\nabla\kappa$ is of class C^{p-1}.

Let a local configuration $\boldsymbol{K}_X \in \mathscr{C}_X$ be given. If κ and γ both belong to the equivalence class that defines $\boldsymbol{K}_X$, which means that $\nabla\kappa(X) = \nabla\gamma(X) = \boldsymbol{K}_X$, we have, by (8.3), $\nabla_\kappa\psi(X) = \nabla_\gamma\psi(X)$. Hence, $\nabla_\kappa\psi(X)$ depends on κ only through the equivalence class $\boldsymbol{K}_X \in \mathscr{C}_X$ to which κ belongs, and it is legitimate to define

$$\nabla_{\boldsymbol{K}_X}\psi(X) = \nabla_\kappa\psi(X) \quad \text{if} \quad \kappa \in \boldsymbol{K}_X. \tag{8.6}$$

If $\boldsymbol{K}$ is a reference, we define the *gradient of* ψ *relative to the reference* $\boldsymbol{K}$ by

$$\nabla_{\boldsymbol{K}}\psi|_X = \nabla_{\boldsymbol{K}(X)}\psi(X), \qquad X \in \mathscr{B}. \tag{8.7}$$

If $\boldsymbol{K}$ and $\boldsymbol{G}$ are any two references for $\mathscr{B}$, we see that (8.3) and the definitions (8.6) and (8.7) yield the formula

$$\nabla_{\boldsymbol{K}}\psi = \nabla_{\boldsymbol{G}}\psi \circ (\boldsymbol{G}\boldsymbol{K}^{-1}), \tag{8.8}$$

which generalizes (8.5). By writing (8.8) with $\boldsymbol{G} = \nabla\gamma$, where $\gamma \in \mathsf{C}$, we infer that $\nabla_{\boldsymbol{K}}\psi$ is of class C^{r-1} if ψ is of class C^r and $\boldsymbol{K}$ of class C^{r-1}.

When the range of ψ coincides with the set $\mathscr{R}$ of real numbers, in which case we write f instead of ψ, we can identify $\nabla_\kappa f$ with a vector field on $\mathscr{B}$. Thus, if $f \in \mathscr{F}^r_{\mathscr{B}}$ then $\nabla_\kappa f \in \mathscr{V}^{r-1}_{\mathscr{B}}$. The formula (8.5) becomes

$$\nabla_\kappa f = (\nabla\gamma(\nabla\kappa)^{-1})^T \nabla_\gamma f. \tag{8.9}$$

Let $\mathfrak{h} \in \mathscr{T}^{r-1}_{\mathscr{B}}$ and $f \in \mathscr{F}^r_{\mathscr{B}}$. The function $\mathfrak{h}(f)$ on $\mathscr{B}$ defined by

$$\mathfrak{h}(f) = \nabla_\kappa f \cdot (\nabla\kappa)\mathfrak{h}, \tag{8.10}$$

where the inner product is defined pointwise, does not depend on the choice of the configuration $\kappa \in \mathsf{C}$, as is easily seen with the help of (8.9). Moreover, $\mathfrak{h}(f)$ is of class C^{r-1}. Therefore, every $\mathfrak{h} \in \mathscr{T}^{r-1}_{\mathscr{B}}$ gives rise to a mapping

$$\mathfrak{h}\colon\ \mathscr{F}^r_{\mathscr{B}} \to \mathscr{F}^{r-1}_{\mathscr{B}}. \tag{8.11}$$

Actually, every $\mathfrak{h} \in \mathscr{T}^{r-1}_{\mathscr{B}}$ can be *identified* with a mapping of the type (8.11), because it is easily seen from (8.10) that $\mathfrak{h}_1(f) = \mathfrak{h}_2(f)$ cannot hold for all $f \in \mathscr{F}^r_{\mathscr{B}}$ unless $\mathfrak{h}_1 = \mathfrak{h}_2$. The mapping (8.11) defined by (8.10) has the following basic property, which follows immediately from the chain rule.

Proposition 3. *If* $\mathfrak{h}\in\mathcal{T}_{\mathscr{B}}^{r-1}$, *if* H *is a real-valued function of class* C^r *of any number* m *of real variables, and if* $f_1, f_2, \ldots, f_m \in \mathcal{F}_{\mathscr{B}}^r$, *then*

$$\mathfrak{h}(H(f_1, f_2, \ldots, f_m)) = \sum_{k=1}^{m} H_{,k}(f_1, f_2, \ldots, f_m)\,\mathfrak{h}(f_k), \tag{8.12}$$

where $H_{,k}$ *denotes the derivative of* H *with respect to its* k^{th} *variable.*

Actually, the property described in Proposition 3 characterizes the tangent vector fields of class C^{r-1} and hence could have been used for their definition, but we shall neither use nor prove this fact.

Applying (8.12) to the cases when $H(\xi_1, \xi_2)=\xi_1+\xi_2$ and $H(\xi_1, \xi_2)=\xi_1\xi_2$, we obtain

$$\mathfrak{h}(f+g)=\mathfrak{h}(f)+\mathfrak{h}(g), \qquad \mathfrak{h}(f\,g)=f\,\mathfrak{h}(g)+g\,\mathfrak{h}(f). \tag{8.13}$$

Let $\mathfrak{h}, \mathfrak{k}\in\mathcal{T}_{\mathscr{B}}^{r-1}$ with $r\geqq 2$. Since $\mathcal{T}_{\mathscr{B}}^{r-1}\subset\mathcal{T}_{\mathscr{B}}^{r-2}$ and hence also $\mathfrak{h}, \mathfrak{k}\in\mathcal{T}_{\mathscr{B}}^{r-2}$ we can identify $\mathfrak{h}$ and $\mathfrak{k}$ not only with mappings from $\mathcal{F}_{\mathscr{B}}^r$ into $\mathcal{F}_{\mathscr{B}}^{r-1}$, but also with mappings from $\mathcal{F}_{\mathscr{B}}^{r-1}$ into $\mathcal{F}_{\mathscr{B}}^{r-2}$. Therefore, we can form the compositions $\mathfrak{h}\circ\mathfrak{k}$ and $\mathfrak{k}\circ\mathfrak{h}$ as mappings from $\mathcal{F}_{\mathscr{B}}^r$ into $\mathcal{F}_{\mathscr{B}}^{r-2}$. By themselves, these compositions do not correspond to tangent vector fields, but it is remarkable that the difference

$$[\mathfrak{h}, \mathfrak{k}] = \mathfrak{h}\circ\mathfrak{k} - \mathfrak{k}\circ\mathfrak{h}\colon\ \mathcal{F}_{\mathscr{B}}^r \to \mathcal{F}_{\mathscr{B}}^{r-2}, \tag{8.14}$$

called the *bracket* of $\mathfrak{h}$ and $\mathfrak{k}$, has values that belong to $\mathcal{F}_{\mathscr{B}}^{r-1}(\subset\mathcal{F}_{\mathscr{B}}^{r-2})$ and does correspond to a tangent vector field:

Proposition 4. *The bracket of two tangent vector fields* $\mathfrak{h}, \mathfrak{k}\in\mathcal{T}_{\mathscr{B}}^{r-1}(r\geqq 2)$ *can be identified with the tangent vector field of class* C^{r-2} *given by*

$$[\mathfrak{h}, \mathfrak{k}] = (\nabla\kappa)^{-1}[(\nabla_\kappa \boldsymbol{k})\boldsymbol{h} - (\nabla_\kappa \boldsymbol{h})\boldsymbol{k}], \qquad \boldsymbol{h}=(\nabla\kappa)\mathfrak{h}, \qquad \boldsymbol{k}=(\nabla\kappa)\mathfrak{k}, \tag{8.15}$$

where κ *is an arbitrary configuration.*

Proof. We denote the tangent vector field of class C^{r-2} defined by the right-hand side of (8.15) by $\mathfrak{b}$, so that

$$(\nabla\kappa)\mathfrak{b} = [(\nabla_\kappa \boldsymbol{k})\boldsymbol{h} - (\nabla_\kappa \boldsymbol{h})\boldsymbol{k}]. \tag{8.16}$$

Now let $f\in\mathcal{F}_{\mathscr{B}}^r$. In view of (8.10), it follows from (8.16) that

$$\mathfrak{b}(f) = \nabla_\kappa f\cdot[(\nabla_\kappa \boldsymbol{k})\boldsymbol{h} - (\nabla_\kappa \boldsymbol{h})\boldsymbol{k}] \tag{8.17}$$

and from $(8.15)_{2,3}$ that

$$(\mathfrak{h}\circ\mathfrak{k})(f) = \mathfrak{h}(\mathfrak{k}(f)) = \nabla_\kappa(\nabla_\kappa f\cdot\boldsymbol{k})\cdot\boldsymbol{h}. \tag{8.18}$$

The rules of ordinary differential calculus yield $\nabla_\kappa(\nabla_\kappa f\cdot\boldsymbol{k})\cdot\boldsymbol{h} = \boldsymbol{h}\cdot(\nabla_\kappa^{(2)}f)\boldsymbol{k} + \nabla_\kappa f\cdot(\nabla_\kappa \boldsymbol{k})\boldsymbol{h}$. Hence, since $\nabla_\kappa^{(2)}f$ is symmetric, if we write (8.18) with $\mathfrak{h}$ and $\mathfrak{k}$ interchanged, take the difference, and then compare with (8.17), we obtain

$$\mathfrak{b}(f) = (\mathfrak{h}\circ\mathfrak{k})(f) - (\mathfrak{k}\circ\mathfrak{h})(f) = [\mathfrak{h}, \mathfrak{k}](f)$$

i.e. the desired result $\mathfrak{b}=[\mathfrak{h}, \mathfrak{k}]$. Q.E.D.

The bracket $[\mathfrak{h}, \mathfrak{k}]$ depends linearly (but not $\mathscr{F}$-linearly) on $\mathfrak{h}$ and $\mathfrak{k}$ and satisfies for $\mathfrak{h}, \mathfrak{k}, \mathfrak{l} \in \mathscr{T}_{\mathscr{B}}^{r-1}$, $f \in \mathscr{F}_{\mathscr{B}}^{r}$, $2 \leqq r$, the identities

$$[\mathfrak{h}, \mathfrak{k}] = -[\mathfrak{k}, \mathfrak{h}], \tag{8.19}$$

$$[\mathfrak{h}, f\mathfrak{k}] = f[\mathfrak{h}, \mathfrak{k}] + \mathfrak{h}(f)\,\mathfrak{k}, \tag{8.20}$$

and for $\mathfrak{h}, \mathfrak{k}, \mathfrak{l} \in \mathscr{T}_{\mathscr{B}}^{r}$, $2 \leqq r \leqq p-1$, the Jacobi-identity

$$\sum_{\text{cyclic}} [\mathfrak{h}, [\mathfrak{k}, \mathfrak{l}]] = 0, \tag{8.21}$$

where the sum is taken of all terms obtained from the one written by cyclic permutation of $\mathfrak{h}, \mathfrak{k}, \mathfrak{l}$. The identity (8.19) is obvious from (8.14), and (8.21) is the result of a trivial calculation. The identity (8.20) follows from (8.14) and (8.13).

It would have been possible to define the bracket $[\mathfrak{h}, \mathfrak{k}] \in \mathscr{T}_{\mathscr{B}}^{r-1}$ for $\mathfrak{h}, \mathfrak{k} \in \mathscr{T}_{\mathscr{B}}^{r}$, $1 \leqq r \leqq p-1$, directly by (8.15), for it is easy to see that the right-hand side of (8.15) does not depend on the choice of the configuration $\boldsymbol{\kappa}$.

9. Affine Connections, Torsion, Curvature

From now on we assume that $\mathscr{B}$ is a continuous body of class C^p, $p \geqq 3$.

A mapping

$$\Gamma\colon\ \mathscr{T}_{\mathscr{B}}^{r} \to \mathscr{I}_{\mathscr{B}}^{r-1} \tag{9.1}$$

is called an *affine connection* of class $C^{r-1}(1 \leqq r \leqq p-1)$ on $\mathscr{B}$ if

$$\Gamma(\mathfrak{h} + \mathfrak{k}) = \Gamma\mathfrak{h} + \Gamma\mathfrak{k} \tag{9.2}$$

holds for all $\mathfrak{h}, \mathfrak{k} \in \mathscr{T}_{\mathscr{B}}^{r}$ and

$$\Gamma(f\mathfrak{h})\,\mathfrak{k} = f(\Gamma\mathfrak{h})\,\mathfrak{k} + \mathfrak{k}(f)\,\mathfrak{h} \tag{9.3}$$

holds for all $\mathfrak{h} \in \mathscr{T}_{\mathscr{B}}^{r}$, $f \in \mathscr{F}_{\mathscr{B}}^{r}$ and all $\mathfrak{k} \in \mathscr{T}_{\mathscr{B}}^{r-1}$.

If a is a real constant, then $\mathfrak{k}(a) = 0$ by the definition (8.10). Hence (9.3) reduces to $\Gamma(a\mathfrak{h}) = a\Gamma\mathfrak{h}$ when $a \in \mathscr{R}$, $\mathfrak{h} \in \mathscr{T}_{\mathscr{B}}^{r}$. Thus, Γ is a linear mapping, but it is never $\mathscr{F}$-linear. The rule (9.3) resembles one of the product rules for gradient operators.

A triple $(\mathfrak{e}_1, \mathfrak{e}_2, \mathfrak{e}_3)$ of tangent vector fields of class C^r, $r \leqq p-1$, is called a *frame* of class C^r if the values $\mathfrak{e}_i(X)$ form a basis of the tangent space $\mathscr{T}_X$ for each $X \in \mathscr{B}$. Frames of class C^{p-1} (and hence of class C^r, $r \leqq p-1$) exist. For example, if $(\boldsymbol{e}_1, \boldsymbol{e}_2, \boldsymbol{e}_3)$ is a basis of $\mathscr{V}$ and $\boldsymbol{\kappa}$ a configuration, then $\mathfrak{e}_i = (\nabla\boldsymbol{\kappa})^{-1}\boldsymbol{e}_i$ defines a frame of class C^{p-1}. Every tangent vector field $\mathfrak{h} \in \mathscr{T}_{\mathscr{B}}^{r}$ has a component representation

$$\mathfrak{h} = \sum_i h^i\,\mathfrak{e}_i \tag{9.4}$$

with respect to a given frame $(\mathfrak{e}_1, \mathfrak{e}_2, \mathfrak{e}_3)$ of class C^r such that the component functions h^i belong to $\mathscr{F}_{\mathscr{B}}^{r}$.

Now let Γ be a connection of class C^{r-1}. Substituting (9.4) into $(\Gamma\mathfrak{h})\,\mathfrak{e}_j$ and using the rules (9.2) and (9.3), we obtain

$$(\Gamma\mathfrak{h})\,\mathfrak{e}_j = \sum_i \left[h^i(\Gamma\mathfrak{e}_i)\,\mathfrak{e}_j + \mathfrak{e}_j(h^i)\,\mathfrak{e}_i\right]. \tag{9.5}$$

The components Γ_{ij}^k of the three intrinsic tensor fields $\Gamma \mathfrak{e}_i$ with respect to the frame $(\mathfrak{e}_1, \mathfrak{e}_2, \mathfrak{e}_3)$ are defined by

$$(\Gamma \mathfrak{e}_i)\,\mathfrak{e}_j = \sum_k \Gamma_{ij}^k\, \mathfrak{e}_k\,. \tag{9.6}$$

These components Γ_{ij}^k belong to $\mathscr{F}_{\mathscr{B}}^{r-1}$ and are called the *components of the connection* Γ with respect to the frame $(\mathfrak{e}_1, \mathfrak{e}_2, \mathfrak{e}_3)$. If we prescribe a frame $(\mathfrak{e}_1, \mathfrak{e}_2, \mathfrak{e}_3)$ of class C^r and 27 functions $\Gamma_{ij}^k \in \mathscr{F}_{\mathscr{B}}^{r-1}$ on $\mathscr{B}$ arbitrarily, then (9.5) and (9.6) determine a unique affine connection of class C^{r-1}.

Let Γ be a connection of class C^{r-1} having components $\Gamma_{ij}^k \in \mathscr{F}_{\mathscr{B}}^{r-1}$ with respect to the frame $(\mathfrak{e}_1, \mathfrak{e}_2, \mathfrak{e}_3)$ of class C^r. When $1 \leqq s \leqq r$, then $(\mathfrak{e}_1, \mathfrak{e}_2, \mathfrak{e}_3)$ is also of class C^s and $\Gamma_{ij}^k \in \mathscr{F}_{\mathscr{B}}^{s-1} \supset \mathscr{F}_{\mathscr{B}}^{r-1}$. Hence, (9.5) and (9.6) define an affine connection of class C^{s-1}. Therefore, the mapping $\Gamma\colon \mathscr{T}_{\mathscr{B}}^r \to \mathscr{I}_{\mathscr{B}}^{r-1}$ has a unique extension to $\mathscr{T}_{\mathscr{B}}^s$ that is an affine connection of class C^{s-1}. We denote this extension by the same symbol Γ. With this convention, we can say that every affine connection of class C^{r-1} is also of class C^{s-1} when $1 \leqq s \leqq r \leqq p-1$.

Let Γ be a connection of class C^{r-1}, and hence also of class C^{s-1} when $1 \leqq s \leqq r$. Using the notation

$$\Gamma_{\mathfrak{h}}\,\mathfrak{k} = (\Gamma\,\mathfrak{k})\,\mathfrak{h}\,, \tag{9.7}$$

we can identify $\Gamma_{\mathfrak{h}}$ with a mapping

$$\Gamma_{\mathfrak{h}}\colon\ \mathscr{T}_{\mathscr{B}}^s \to \mathscr{T}_{\mathscr{B}}^{s-1} \tag{9.8}$$

for any choice of s, $1 \leqq s \leqq r$, and any choice of $\mathfrak{h} \in \mathscr{T}_{\mathscr{B}}^{s-1}$. In terms of $\Gamma_{\mathfrak{h}}$ the rule (9.3) reads

$$\Gamma_{\mathfrak{h}}(f\mathfrak{k}) = f(\Gamma_{\mathfrak{h}}\,\mathfrak{k}) + \mathfrak{h}(f)\,\mathfrak{k}\,. \tag{9.9}$$

Moreover, $\Gamma_{\mathfrak{h}}$ depends $\mathscr{F}$-linearly on $\mathfrak{h}$.

The *Cartan-torsion* (or simply *torsion*) of the connection Γ is the mapping

$$\mathfrak{S}\colon\ \mathscr{T}_{\mathscr{B}}^r \times \mathscr{T}_{\mathscr{B}}^r \to \mathscr{T}_{\mathscr{B}}^{r-1} \tag{9.10}$$

defined by

$$\mathfrak{S}(\mathfrak{h}, \mathfrak{k}) = \Gamma_{\mathfrak{h}}\,\mathfrak{k} - \Gamma_{\mathfrak{k}}\,\mathfrak{h} - [\mathfrak{h}, \mathfrak{k}]. \tag{9.11}$$

In view of (8.19) it is obvious that $\mathfrak{S}$ is skew in the sense that

$$\mathfrak{S}(\mathfrak{h}, \mathfrak{k}) = -\,\mathfrak{S}(\mathfrak{k}, \mathfrak{h})\,. \tag{9.12}$$

It is an almost immediate consequence of (9.9) and the rule (8.20) that $\mathfrak{S}$ is $\mathscr{F}$-bilinear. Hence, by Proposition 2 (Sect. 7), the torsion $\mathfrak{S}$ can be identified with a field on $\mathscr{B}$ of class C^{r-1} whose value $\mathfrak{S}(X)$ at $X \in \mathscr{B}$ is a bilinear mapping from $\mathscr{T}_X \times \mathscr{T}_X$ into $\mathscr{T}_X$. It follows that $\mathfrak{S}(\mathfrak{h}, \mathfrak{k})$ remains meaningful for any, even discontinuous, tangent vector fields $\mathfrak{h}, \mathfrak{k}$, and that $\mathfrak{S}(\mathfrak{h}, \mathfrak{k})|_X = \mathfrak{S}(X)(\mathfrak{h}(X), \mathfrak{k}(X))$ depends on $\mathfrak{h}$ and $\mathfrak{k}$ only through their values at X.

Let $\mathfrak{k}, \mathfrak{h} \in \mathscr{T}_{\mathscr{B}}^{r-1}$, $2 \leqq r \leqq p-1$. In view of (9.8) we can regard $\Gamma_{\mathfrak{h}}$ and $\Gamma_{\mathfrak{k}}$ as mappings from $\mathscr{T}_{\mathscr{B}}^r$ into $\mathscr{T}_{\mathscr{B}}^{r-1}$ and also as mappings from $\mathscr{T}_{\mathscr{B}}^{r-1}$ into $\mathscr{T}_{\mathscr{B}}^{r-2}$. Hence we can form the compositions $\Gamma_{\mathfrak{h}} \circ \Gamma_{\mathfrak{k}}$ and $\Gamma_{\mathfrak{k}} \circ \Gamma_{\mathfrak{h}}$ and the bracket

$$[\Gamma_{\mathfrak{h}}, \Gamma_{\mathfrak{k}}] = \Gamma_{\mathfrak{h}} \circ \Gamma_{\mathfrak{k}} - \Gamma_{\mathfrak{k}} \circ \Gamma_{\mathfrak{h}} \tag{9.13}$$

as mappings from $\mathscr{T}_{\mathscr{B}}^{r}$ into $\mathscr{T}_{\mathscr{B}}^{r-2}$. Since $[\mathfrak{h},\mathfrak{k}]\in\mathscr{T}_{\mathscr{B}}^{r-1}$, we can regard $\Gamma_{[\mathfrak{h},\mathfrak{k}]}$ as a mapping from $\mathscr{T}_{\mathscr{B}}^{r}(\subset\mathscr{T}_{\mathscr{B}}^{r-1})$ into $\mathscr{T}_{\mathscr{B}}^{r-2}$. Hence we can define

$$\mathfrak{R}(\mathfrak{h},\mathfrak{k})\colon\ \mathscr{T}_{\mathscr{B}}^{r}\to\mathscr{T}_{\mathscr{B}}^{r-2} \tag{9.14}$$

by

$$\mathfrak{R}(\mathfrak{h},\mathfrak{k})=[\Gamma_{\mathfrak{h}},\Gamma_{\mathfrak{k}}]-\Gamma_{[\mathfrak{h},\mathfrak{k}]}\,. \tag{9.15}$$

An easy calculation, based on the definition (9.13) and the rules (9.9) and (8.20), shows that the mapping (9.15) is $\mathscr{F}$-linear. Hence, by an analogue of Proposition 1 (Sect. 7) for intrinsic tensor fields, $\mathfrak{R}(\mathfrak{h},\mathfrak{k})$ can be identified with an element of $\mathscr{I}_{\mathscr{B}}^{r-2}$ and $\mathfrak{R}$ can be regarded as a mapping

$$\mathfrak{R}\colon\ \mathscr{T}_{\mathscr{B}}^{r-1}\times\mathscr{T}_{\mathscr{B}}^{r-1}\to\mathscr{I}_{\mathscr{B}}^{r-2}, \tag{9.16}$$

which is called the *Riemann-curvature* (or simply *curvature*) of the connection Γ. It is obvious that $\mathfrak{R}$ is skew in the sense that

$$\mathfrak{R}(\mathfrak{h},\mathfrak{k})=-\mathfrak{R}(\mathfrak{k},\mathfrak{h})\,. \tag{9.17}$$

A short calculation shows that the mapping (9.16) is $\mathscr{F}$-bilinear. Therefore, by Proposition 2 (Sect. 7), we can identify the curvature $\mathfrak{R}$ with a field on $\mathscr{B}$ of class C^{r-2} whose value $\mathfrak{R}(X)$ at $X\in\mathscr{B}$ is a bilinear transformation from $\mathscr{T}_X\times\mathscr{T}_X$ into $\mathscr{I}_X$. We have $\mathfrak{R}(\mathfrak{h},\mathfrak{k})|_X=\mathfrak{R}(X)(\mathfrak{h}(X),\mathfrak{k}(X))$, which shows that $\mathfrak{R}(\mathfrak{h},\mathfrak{k})$ is meaningful for any tangent vector fields $\mathfrak{h}$, $\mathfrak{k}$.

There is an important relation between the torsion and the curvature of an affine connection:

Proposition 5. *Let Γ be an affine connection of class C^{r-1} on $\mathscr{B}$, $2\leqq r\leqq p-1$. The torsion $\mathfrak{S}$ and the curvature $\mathfrak{R}$ of Γ satisfy*

$$\sum_{\text{cyclic}}\{\Gamma_{\mathfrak{l}}(\mathfrak{S}(\mathfrak{h},\mathfrak{k}))+\mathfrak{S}(\mathfrak{k},[\mathfrak{l},\mathfrak{h}])-\mathfrak{R}(\mathfrak{l},\mathfrak{h})\mathfrak{k}\}=0 \tag{9.18}$$

for all $\mathfrak{h},\mathfrak{k},\mathfrak{l}\in\mathscr{T}_{\mathscr{B}}^{r}$. The sum is to be taken over all terms obtained from the one written by cyclic permutation of $\mathfrak{h},\mathfrak{k},\mathfrak{l}$.

The identity (9.18) is often called the *First Bianchi Identity*.

Proof. Operating with $\Gamma_{\mathfrak{l}}$ on (9.11) gives

$$\Gamma_{\mathfrak{l}}(\mathfrak{S}(\mathfrak{h},\mathfrak{k}))-(\Gamma_{\mathfrak{l}}\circ\Gamma_{\mathfrak{h}})\mathfrak{k}+(\Gamma_{\mathfrak{l}}\circ\Gamma_{\mathfrak{k}})\mathfrak{h}+\Gamma_{\mathfrak{l}}[\mathfrak{h},\mathfrak{k}]=0\,.$$

The cyclic sum of the left side of this equation remains unchanged if the third term is changed by one and the fourth by two cyclic permutations of $\mathfrak{h},\mathfrak{k},\mathfrak{l}$. Hence we have

$$\begin{aligned}&\sum_{\text{cyclic}}\{\Gamma_{\mathfrak{l}}(\mathfrak{S}(\mathfrak{h},\mathfrak{k}))-(\Gamma_{\mathfrak{l}}\circ\Gamma_{\mathfrak{h}})\mathfrak{k}+(\Gamma_{\mathfrak{h}}\circ\Gamma_{\mathfrak{l}})\mathfrak{k}+\Gamma_{\mathfrak{k}}[\mathfrak{l},\mathfrak{h}]\}\\ &\qquad=\sum_{\text{cyclic}}\{\Gamma_{\mathfrak{l}}(\mathfrak{S}(\mathfrak{h},\mathfrak{k}))-[\Gamma_{\mathfrak{l}},\Gamma_{\mathfrak{h}}]\mathfrak{k}+\Gamma_{\mathfrak{k}}[\mathfrak{l},\mathfrak{h}]\}=0\,.\end{aligned}$$

Using the definitions (9.15) and (9.11), we obtain

$$\sum_{\text{cyclic}}\{\Gamma_{\mathfrak{l}}(\mathfrak{S}(\mathfrak{h},\mathfrak{k}))-\mathfrak{R}(\mathfrak{l},\mathfrak{h})\mathfrak{k}+\mathfrak{S}(\mathfrak{k},[\mathfrak{l},\mathfrak{h}])+[\mathfrak{k},[\mathfrak{l},\mathfrak{h}]]\}=0\,.$$

In view of the Jacobi identity (8.21) the last term gives no contribution and (9.18) results. Q.E.D.

Let Γ and $\overset{*}{\Gamma}$ be two connections of class C^{r-1} on $\mathscr{B}$. Using the notation (9.7) and observing the rule (9.9), we see that, for each s, $1 \leqq s \leqq r$, and each $\mathfrak{h} \in \mathscr{T}_{\mathscr{B}}^{s-1}$, the difference

$$\mathfrak{D}_{\mathfrak{h}} = \Gamma_{\mathfrak{h}} - \overset{*}{\Gamma}_{\mathfrak{h}}: \ \mathscr{T}_{\mathscr{B}}^{s} \to \mathscr{T}_{\mathscr{B}}^{s-1} \tag{9.19}$$

is actually $\mathscr{F}$-linear. Hence $\mathfrak{D}_{\mathfrak{h}}$ can be identified with an intrinsic tensor field of class C^{s-1}, *i.e.* $\mathfrak{D}_{\mathfrak{h}} \in \mathscr{I}_{\mathscr{B}}^{s-1}$. Since $\mathfrak{D}_{\mathfrak{h}}$ depends $\mathscr{F}$-linearly on $\mathfrak{h}$, $\mathfrak{D}$ can be regarded as an $\mathscr{F}$-linear mapping

$$\mathfrak{D}: \ \mathscr{T}_{\mathscr{B}}^{s-1} \to \mathscr{I}_{\mathscr{B}}^{s-1}, \qquad 1 \leqq s \leqq r, \tag{9.20}$$

and hence can be identified with a field of class C^{r-1} whose values $\mathfrak{D}(X)$ are linear transformations from $\mathscr{T}_X$ into $\mathscr{I}_X$. The possibility of identifying $\mathfrak{D}_{\mathfrak{h}}$ and $\mathfrak{D}$ with fields on $\mathscr{B}$ follows from analogues of Proposition 1 (Sect. 7).

Let $\mathfrak{S}$ and $\overset{*}{\mathfrak{S}}$ denote the torsions and $\mathfrak{R}$ and $\overset{*}{\mathfrak{R}}$ the curvatures of Γ and $\overset{*}{\Gamma}$, respectively. If we write the definition (9.11) of the torsion for both Γ and $\overset{*}{\Gamma}$ and take the difference, we obtain

$$\mathfrak{S}(\mathfrak{h}, \mathfrak{k}) - \overset{*}{\mathfrak{S}}(\mathfrak{h}, \mathfrak{k}) = \mathfrak{D}_{\mathfrak{h}} \mathfrak{k} - \mathfrak{D}_{\mathfrak{k}} \mathfrak{h}. \tag{9.21}$$

If we write the definition (9.15) of the curvature for $\overset{*}{\Gamma}$ and substitute $\overset{*}{\Gamma}_{\mathfrak{h}} = \Gamma_{\mathfrak{h}} - \mathfrak{D}_{\mathfrak{h}}$, we find

$$\overset{*}{\mathfrak{R}}(\mathfrak{h}, \mathfrak{k}) = \mathfrak{R}(\mathfrak{h}, \mathfrak{k}) - [\Gamma_{\mathfrak{h}}, \mathfrak{D}_{\mathfrak{k}}] - [\mathfrak{D}_{\mathfrak{h}}, \Gamma_{\mathfrak{k}}] + [\mathfrak{D}_{\mathfrak{h}}, \mathfrak{D}_{\mathfrak{k}}] + \mathfrak{D}_{[\mathfrak{h}, \mathfrak{k}]}. \tag{9.22}$$

10. Material Connections

Let Φ be a material uniformity for the simple body $\mathscr{B}$ of class C^p. (See Definition 3, Sect. 6.) We say that a tangent vector field $\mathfrak{c}$ is *materially constant* if

$$\mathfrak{c}(X) = \Phi(X, Y)\, \mathfrak{c}(Y) \tag{10.1}$$

holds for all $X, Y \in \mathscr{B}$. If $X_0 \in \mathscr{B}$ is fixed and $\mathfrak{u}_{X_0} \in \mathscr{T}_{X_0}$ is prescribed arbitrarily, then

$$\mathfrak{c}(X) = \Phi(X, X_0)\, \mathfrak{u}_{X_0} \tag{10.2}$$

is easily seen to define a materially constant field $\mathfrak{c}$ such that $\mathfrak{c}(X_0) = \mathfrak{u}_{X_0}$. Moreover, every materially constant field $\mathfrak{c}$ can be obtained in this fashion. Thus (10.2) describes a one-to-one correspondence between $\mathscr{T}_{X_0}$ and the set $\mathscr{T}_{\Phi}$ of all materially constant vector fields. This correspondence is actually a vector-space isomorphism, showing that $\mathscr{T}_{\Phi}$ is a three-dimensional vector-space when addition and multiplication with scalars in $\mathscr{T}_{\Phi}$ are defined pointwise.

Let $\boldsymbol{K}$ be a uniform reference (see Definition 4, Sect. 6) and $\mathfrak{c} \in \mathscr{T}_{\Phi}$. Then it follows from (10.1) and (6.4) that $\boldsymbol{K}(X)\, \mathfrak{c}(X) = \boldsymbol{K}(Y)\, \mathfrak{c}(Y)$ for all $X, Y \in \mathscr{B}$, *i.e.* that

$$\boldsymbol{K}\mathfrak{c} = \boldsymbol{c} = \text{constant}. \tag{10.3}$$

Conversely, if $c \in \mathscr{V}$, then $\mathfrak{c} = \boldsymbol{K}^{-1} c$ is easily seen to be materially constant. Thus $\mathscr{T}_\Phi$ is exactly the set of all tangent vector fields $\mathfrak{c}$ with the property (10.3).

A material uniformity Φ is said to be *of class* C^r, $r \leqq p-1$, if $\mathscr{T}_\Phi \subset \mathscr{T}_{\mathscr{B}}^r$, *i.e.*, if all tangent vector fields materially constant with respect to Φ are of class C^r. For the remainder of this paper we lay down the following:

Smoothness assumption[9]: *$\mathscr{B}$ is a materially uniform continuous body of class C^p, $p \geqq 3$*[10]*, which admits a material uniformity Φ of class C^{p-1}.*

Let Φ be a material uniformity of class C^{p-1} and let $\boldsymbol{K}$ be a uniform reference such that (6.4) holds. Then for $c \in \mathscr{V}$, $\boldsymbol{\kappa} \in \mathrm{C}$, and $\mathfrak{c} = \boldsymbol{K}^{-1} c \in \mathscr{T}_\Phi$, the vector field $((\nabla \boldsymbol{\kappa}) \boldsymbol{K}^{-1})\, c = (\nabla \boldsymbol{\kappa})\, \mathfrak{c}$ is of class C^{p-1} because $\mathfrak{c}$ is of class C^{p-1}. This is possible for all $c \in \mathscr{V}$ only if $(\nabla \boldsymbol{\kappa}) \boldsymbol{K}$ and hence $\boldsymbol{K}$ is of class C^{p-1}. Thus, if $\boldsymbol{K}$ is a uniform reference such that

$$\Phi(X, Y) = \boldsymbol{K}(X)^{-1} \boldsymbol{K}(Y), \qquad X, Y \in \mathscr{B}, \tag{10.4}$$

then $\boldsymbol{K}$ is of class C^{p-1}.

Theorem 3. *Given a material uniformity Φ of class C^{p-1}, there is a unique affine connection Γ such that $\Gamma \mathfrak{c} = 0$ holds for all materially constant tangent vector fields $\mathfrak{c} \in \mathscr{T}_\Phi$. In terms of any uniform reference $\boldsymbol{K}$ satisfying* (10.4), *Γ is given by*

$$\Gamma \mathfrak{h} = \boldsymbol{K}^{-1} \nabla_{\boldsymbol{K}} (\boldsymbol{K} \mathfrak{h}) \boldsymbol{K}, \qquad \mathfrak{h} \in \mathscr{T}_{\mathscr{B}}^1. \tag{10.5}$$

Also, Γ is of class C^{p-2}.

Proof. To prove the uniqueness, assume that Γ and $\bar{\Gamma}$ are connections such that $\Gamma \mathfrak{c} = \bar{\Gamma} \mathfrak{c}$ for all $\mathfrak{c} \in \mathscr{T}_\Phi$. Putting $\mathfrak{D}_\mathfrak{h} = \Gamma_\mathfrak{h} - \bar{\Gamma}_\mathfrak{h}$, we then have $\mathfrak{D}_\mathfrak{h} \mathfrak{c} = 0$ for all $\mathfrak{c} \in \mathscr{T}_\Phi$. We have seen at the end of the previous section that $\mathfrak{D}_\mathfrak{h}$ can be identified with an intrinsic tensor field in $\mathscr{I}_{\mathscr{B}}^0$ when $\mathfrak{h} \in \mathscr{T}_{\mathscr{B}}^0$. Hence $\mathfrak{D}_\mathfrak{h}(X_0) \mathfrak{c}(X_0) = 0$ for all $X_0 \in \mathscr{B}$ and all $\mathfrak{c} \in \mathscr{T}_\Phi$. Since for any prescribed $\mathfrak{u}_{X_0} \in \mathscr{T}_{X_0}$ the $\mathfrak{c} \in \mathscr{T}_\Phi$ given by (10.2) has the property $\mathfrak{c}(X_0) = \mathfrak{u}_{X_0}$, it follows that $\mathfrak{D}_\mathfrak{h}(X_0)\, \mathfrak{u}_{X_0} = 0$ for all $\mathfrak{u}_{X_0} \in \mathscr{T}_{X_0}$, *i.e.*, that $\mathfrak{D}_\mathfrak{h}(X_0) = 0$. Since $X_0 \in \mathscr{B}$ is arbitrary, we infer that $\mathfrak{D}_\mathfrak{h} = 0$, *i.e.* that $\Gamma = \bar{\Gamma}$.

To prove the existence of Γ we choose a uniform reference $\boldsymbol{K}$ with the property (10.4), define Γ by (10.5), and show that it has all the necessary properties. It is clear that $\Gamma \mathfrak{c} = 0$ when $\mathfrak{c} \in \mathscr{T}_\Phi$ because, by (10.3), $\nabla_{\boldsymbol{K}}(\boldsymbol{K} \mathfrak{c}) = 0$ when $\mathfrak{c} \in \mathscr{T}_\Phi$. Since $\boldsymbol{K}$ is of class C^{p-1} it follows that $\Gamma \mathfrak{h}$ is of class C^{p-2} when $\mathfrak{h} \in \mathscr{T}_{\mathscr{B}}^{p-1}$. The validity of the rules (9.2) and (9.3) follows from the validity of the analogous rules for the relative gradient $\nabla_{\boldsymbol{K}}$. Hence Γ is indeed an affine connection of class C^{p-2}. Q.E.D.

Definition 5. *The affine connection (of class C^{p-2}) with the property $\Gamma \mathfrak{c} = 0$ for all $\mathfrak{c} \in \mathscr{T}_\Phi$ is called the* **material connection** *for the material uniformity Φ (of class C^{p-1}).*

Theorem 4. *Material connections have zero Riemann-curvature.*

Proof. Let $X \in \mathscr{B}$ and $\mathfrak{u}_X \in \mathscr{T}_X$ be given. We can determine $\mathfrak{c} \in \mathscr{T}_\Phi \subset \mathscr{T}_{\mathscr{B}}^{p-1}$ such that $\mathfrak{c}(X) = \mathfrak{u}_X$. If Γ is the material connection for Φ we have $\Gamma_\mathfrak{h} \mathfrak{c} = 0$ for all $\mathfrak{h} \in \mathscr{T}_{\mathscr{B}}^0$. Hence the definition (9.15) shows that $\mathfrak{R}(\mathfrak{h}, \mathfrak{k})\, \mathfrak{c} = 0$ for all $\mathfrak{h}, \mathfrak{k} \in \mathscr{T}_{\mathscr{B}}^{p-1}$. Since $\mathfrak{R}(\mathfrak{h}, \mathfrak{k})$

[9] C. C. Wang [*8*] has recently shown that the theory given here can be extended to the case when each point has a neighborhood that admits a smooth material uniformity. This can happen even when all material uniformities for the whole body are discontinuous.

[10] For all considerations not referring to curvature, $p \geqq 2$ is actually sufficient.

can be identified with an intrinsic tensor field it follows that

$$\mathfrak{R}(\mathfrak{h},\mathfrak{k})\,\mathfrak{c}\,|_X=\mathfrak{R}(\mathfrak{h},\mathfrak{k})|_X\,\mathfrak{c}(X)=\mathfrak{R}(\mathfrak{h},\mathfrak{k})|_X\,\mathfrak{u}_X=0\,.$$

This can be valid for all $X\in\mathscr{B}$ and all $\mathfrak{u}_X\in\mathscr{T}_X$ only if $\mathfrak{R}(\mathfrak{h},\mathfrak{k})=0$. Hence, since $\mathfrak{h},\mathfrak{k}\in\mathscr{T}_{\mathscr{B}}^{p-1}$ are arbitrary, we must have $\mathfrak{R}=0$. Q.E.D.

11. Inhomogeneity

Let Φ be a material uniformity of class C^{p-1}, let Γ be the associated material connection (of class C^{p-2}) with torsion $\mathfrak{S}$, and let $\boldsymbol{K}$ be a uniform reference (of class C^{p-1}) such that (10.4) holds. We can define a field $\boldsymbol{S}$ of class C^{p-2} with values $\boldsymbol{S}(X)\colon\mathscr{V}\to\mathscr{L}$ by the condition

$$(\boldsymbol{S}\,\boldsymbol{u})\,\boldsymbol{v}=\boldsymbol{K}\,\mathfrak{S}(\boldsymbol{K}^{-1}\boldsymbol{u},\boldsymbol{K}^{-1}\boldsymbol{v}) \tag{11.1}$$

for all $\boldsymbol{u},\boldsymbol{v}\in\mathscr{V}$. In view of the linearity of the values $\boldsymbol{S}(X)$, $\mathfrak{S}(X)$, and $\boldsymbol{K}(X)$, (11.1) continues to hold if the fixed vectors $\boldsymbol{u}$ and $\boldsymbol{v}$ in (11.1) are replaced by vector fields $\boldsymbol{h}$ and $\boldsymbol{k}$. The following theorem shows how $\boldsymbol{S}$ and hence $\mathfrak{S}$ can be expressed directly in terms of $\boldsymbol{K}$:

Theorem 5. *Let γ be an arbitrary configuration of $\mathscr{B}$ and*

$$\boldsymbol{F}=(\nabla\gamma)\,\boldsymbol{K}^{-1}\in\mathscr{L}_{\mathscr{B}}^{p-1}\,. \tag{11.2}$$

Then $\boldsymbol{S}$ is given by

$$(\boldsymbol{S}\,\boldsymbol{u})\,\boldsymbol{v}=\boldsymbol{F}^{-1}[((\nabla_{\boldsymbol{K}}\boldsymbol{F})\,\boldsymbol{v})\,\boldsymbol{u}-((\nabla_{\boldsymbol{K}}\boldsymbol{F})\,\boldsymbol{u})\,\boldsymbol{v}]\,, \tag{11.3}$$

$$\boldsymbol{S}\,\boldsymbol{u}=\boldsymbol{F}^{-1}[\nabla_{\boldsymbol{K}}(\boldsymbol{F}\,\boldsymbol{u})-(\nabla_{\boldsymbol{K}}\boldsymbol{F})\,\boldsymbol{u}]\,, \tag{11.4}$$

or

$$(\boldsymbol{S}\,\boldsymbol{u})\,\boldsymbol{v}=((\nabla_{\gamma}\boldsymbol{F}^{-1})\,\boldsymbol{h})\,\boldsymbol{k}-((\nabla_{\gamma}\boldsymbol{F}^{-1})\,\boldsymbol{k})\,\boldsymbol{h}\,, \tag{11.5}$$

where $\boldsymbol{u},\boldsymbol{v}\in\mathscr{V}$ and $\boldsymbol{h}=\boldsymbol{F}\,\boldsymbol{u}$, $\boldsymbol{k}=\boldsymbol{F}\,\boldsymbol{v}$.

Proof. The tangent vector fields $\boldsymbol{K}^{-1}\boldsymbol{u}$ and $\boldsymbol{K}^{-1}\boldsymbol{v}$ are materially constant and hence are annihilated by Γ. Hence, the definitions (11.1) and (9.11) of $\boldsymbol{S}$ and $\mathfrak{S}$ yield

$$(\boldsymbol{S}\,\boldsymbol{u})\,\boldsymbol{v}=-\boldsymbol{K}[\boldsymbol{K}^{-1}\boldsymbol{u},\boldsymbol{K}^{-1}\boldsymbol{v}]=-\boldsymbol{K}[(\nabla\gamma)^{-1}\boldsymbol{h},(\nabla\gamma)^{-1}\boldsymbol{k}]\,. \tag{11.6}$$

Using Proposition 4, (8.15), we find

$$(\boldsymbol{S}\,\boldsymbol{u})\,\boldsymbol{v}=\boldsymbol{F}^{-1}[(\nabla_{\gamma}\boldsymbol{h})\,\boldsymbol{k}-(\nabla_{\gamma}\boldsymbol{k})\,\boldsymbol{h}]\,. \tag{11.7}$$

The formula (8.8), with the choices $\psi=\boldsymbol{h}$ and $\boldsymbol{G}=\nabla\gamma$, gives

$$(\nabla_{\gamma}\boldsymbol{h})\,\boldsymbol{k}=(\nabla_{\gamma}\boldsymbol{h})\,\boldsymbol{F}\,\boldsymbol{v}=(\nabla_{\boldsymbol{K}}\boldsymbol{h})\,\boldsymbol{v}=\nabla_{\boldsymbol{K}}(\boldsymbol{F}\,\boldsymbol{u})\,\boldsymbol{v}=((\nabla_{\boldsymbol{K}}\boldsymbol{F})\,\boldsymbol{v})\,\boldsymbol{u}\,. \tag{11.8}$$

Substituting (11.8) and the formula obtained from (11.8) by interchanging $\boldsymbol{h}$ and $\boldsymbol{k}$ into (11.7), we obtain (11.3) and (11.4).

To prove (11.5) we note that $\nabla_{\gamma}\boldsymbol{u}=\boldsymbol{0}$ for constant $\boldsymbol{u}\in\mathscr{V}$. Using one of the product rules for gradient operators we find

$$\boldsymbol{0}=(\nabla_{\gamma}\boldsymbol{u})\,\boldsymbol{k}=\nabla_{\gamma}(\boldsymbol{F}^{-1}\boldsymbol{h})\,\boldsymbol{k}=((\nabla_{\gamma}\boldsymbol{F}^{-1})\,\boldsymbol{k})\,\boldsymbol{h}+\boldsymbol{F}^{-1}(\nabla_{\gamma}\boldsymbol{h})\,\boldsymbol{k}\,. \tag{11.9}$$

Of course, (11.9) remains valid if we interchange $\boldsymbol{h}$ and $\boldsymbol{k}$. The formula (11.5) follows from (11.7) and (11.9). Q.E.D.

Recall that the body $\mathscr{B}$ is homogeneous if it admits a configuration gradient $\boldsymbol{K}=\nabla\boldsymbol{\kappa}$ as a uniform reference. We can then choose $\gamma=\boldsymbol{\kappa}$ in (11.2), obtaining $\boldsymbol{F}=\mathbf{1}$, which is constant and hence has gradient zero. Thus, Theorem 5 shows that $\boldsymbol{S}=\mathbf{0}$ and hence $\mathfrak{S}=0$ for suitable uniform references if the body is homogeneous. The converse of this result is not true, but it becomes true if "homogeneous" is replaced by "locally homogeneous" in a sense we shall now make precise.

Let $\mathscr{N}$ be an open subset of a simple body $\mathscr{B}$ of class C^p. We can give $\mathscr{N}$ the structure of a continuous body of class C^p by letting $\gamma\colon \mathscr{N}\to\mathscr{E}$ be a configuration of $\mathscr{N}$ if $\boldsymbol{\kappa}\circ\overset{-1}{\gamma}\colon \gamma(\mathscr{N})\to\mathscr{E}$ is of class C^p for all configurations $\boldsymbol{\kappa}\in\mathsf{C}$ of $\mathscr{B}$. We denote the set of all configurations of $\mathscr{N}$ by $\mathsf{C}_{\mathscr{N}}$. If $\boldsymbol{\kappa}\in\mathsf{C}$, then the restriction of $\boldsymbol{\kappa}$ to $\mathscr{N}$ belongs to $\mathsf{C}_{\mathscr{N}}$. However, not all configurations $\gamma\in\mathsf{C}_{\mathscr{N}}$ of $\mathscr{N}$ can be obtained in this manner. Still, given any $X\in\mathscr{N}$ and any configuration γ of $\mathscr{N}$, one can easily construct a configuration $\boldsymbol{\kappa}$ of $\mathscr{B}$ such that (3.1) holds. Therefore, an equivalence class $\boldsymbol{K}_X$ which defines a local configuration at X relative to $\mathscr{N}$ can be made to correspond to the non-empty set $\{\boldsymbol{\kappa}\in\mathsf{C}\,|\,\nabla(\boldsymbol{\kappa}\circ\overset{-1}{\gamma})|_{\gamma(X)}=\mathbf{1}$ for all $\gamma\in\boldsymbol{K}_X\}$, which is a local configuration at X relative to $\mathscr{B}$. This correspondence is one-to-one and can be used to identify local configurations at X relative to $\mathscr{N}$ with local configurations at X relative to $\mathscr{B}$. Using this identification, we can endow $\mathscr{N}$ with the structure of a simple body by using the restriction to $\mathscr{N}$ of the function $\mathfrak{G}$ which defines the simple body structure on $\mathscr{B}$ according to Definition 1 (Sect. 5). Thus, every open subset $\mathscr{N}$ of $\mathscr{B}$ has a natural structure of a simple body of class C^p, *i.e.*, every open subset $\mathscr{N}$ of $\mathscr{B}$ can be regarded as a simple body of class C^p. Such a subset is called a *neighborhood* of a material point if it contains that point.

A simple body $\mathscr{B}$ is called *locally homogeneous* if every $X\in\mathscr{B}$ has a neighborhood $\mathscr{N}$ that is homogeneous. A body can be locally homogeneous without being homogeneous, even if it is simply connected.

Definition 6. *The Cartan torsion $\mathfrak{S}$ of the material connection Γ associated with a material uniformity Φ of class C^{p-1} is called the* ***inhomogeneity*** *of Φ.*

The field $\boldsymbol{S}$ defined by (11.1) *is called the* ***inhomogeneity***[11] *relative to the reference $\boldsymbol{K}$.*

This definition finds its motivation in the result already mentioned:

Theorem 6[12]. *If $\mathscr{B}$ is homogeneous, then it admits a material uniformity of class C^{p-1} with zero inhomogeneity. If $\mathscr{B}$ admits a material uniformity of class C^{p-1} with zero inhomogeneity, then it is locally homogeneous.*

Proof. Only the second part of the theorem remains to be proved. Assume, therefore, that Φ is a material uniformity of class C^{p-1} with zero inhomogeneity. Using the same notation as before, we then have $\mathfrak{S}=0$ and hence $\boldsymbol{S}=\mathbf{0}$. Theorem 5, (11.5), shows that if $\gamma\in\mathsf{C}$ is arbitrary and $\boldsymbol{F}$ defined by (11.2), $\nabla_\gamma\boldsymbol{F}^{-1}$ has the symmetry property

$$((\nabla_\gamma\boldsymbol{F}^{-1})\,\boldsymbol{u})\,\boldsymbol{v}=((\nabla_\gamma\boldsymbol{F}^{-1})\,\boldsymbol{v})\,\boldsymbol{u} \tag{11.10}$$

[11] It corresponds to what is called "dislocation density" in the theory of continuous distributions of dislocations (*cf.* [*4*]).

[12] The theorem stated in the middle of p. 90 in reference [*2*] is incorrect and should be replaced by Theorem 6.

for all $\boldsymbol{u}, \boldsymbol{v} \in \mathscr{V}$. Let $X \in \mathscr{B}$ be given and let $\mathscr{N}'$ be a simply connected neighborhood of X. By a classical theorem of analysis, the symmetry (11.10) implies the existence of a mapping $\boldsymbol{\lambda}: \gamma(\mathscr{N}') \to \mathscr{E}$ such that $\boldsymbol{F}^{-1} = (\nabla \boldsymbol{\lambda}) \circ \gamma$ holds in $\mathscr{N}'$. Moreover, $\boldsymbol{K}, \nabla_\gamma$ and hence $\boldsymbol{F}$ and $\boldsymbol{F}^{-1}$ being of class C^{p-1}, $\boldsymbol{\lambda}$ is of class C^p. Since $\nabla \boldsymbol{\lambda} = \boldsymbol{F}^{-1} \circ \gamma^{-1}$ is invertible, it follows by the inverse function theorem that $\boldsymbol{\lambda}$ is locally (but not necessarily globally) invertible, *i.e.* that X has a neighborhood $\mathscr{N} \subset \mathscr{N}'$ on which $\boldsymbol{\lambda}$ is invertible. The mapping $\boldsymbol{\kappa} = \boldsymbol{\lambda} \circ \gamma$, when restricted to $\mathscr{N}$, is therefore a configuration of $\mathscr{N}$ with gradient $\nabla \boldsymbol{\kappa} = (\nabla \boldsymbol{\lambda} \circ \gamma)\, \nabla \gamma = \boldsymbol{F}^{-1}\, \nabla \gamma = \boldsymbol{K}$ on $\mathscr{N}$. Hence the uniform reference $\boldsymbol{K}$ on $\mathscr{N}$ is the gradient of the configuration $\boldsymbol{\kappa}$ of $\mathscr{N}$, *i.e.*, $\mathscr{N}$ is homogeneous. Q.E.D.

The first Bianchi identity gives rise to the following identity for the relative inhomogeneity $\boldsymbol{S}$ and its gradient $\nabla_{\boldsymbol{K}} \boldsymbol{S}$ relative to $\boldsymbol{K}$:

$$\sum [(((\nabla_{\boldsymbol{K}} \boldsymbol{S})\, \boldsymbol{u})\, \boldsymbol{v})\, \boldsymbol{w}) - (\boldsymbol{S}\, \boldsymbol{u})(\boldsymbol{S}\, \boldsymbol{v})\, \boldsymbol{w}] = \boldsymbol{0}. \tag{11.11}$$

To prove (11.11), substitute $\mathfrak{l} = \boldsymbol{K}^{-1} \boldsymbol{u}$, $\mathfrak{k} = \boldsymbol{K}^{-1} \boldsymbol{w}$, $\mathfrak{h} = \boldsymbol{K}^{-1} \boldsymbol{v}$ into (9.18), observe that $\mathfrak{R} = 0$ (Theorem 4, Sect. 10), and make use of $(11.6)_1$.

12. Relative Riemannian Structures, Contortion

Let $\boldsymbol{K}$ be a uniform reference of class C^{p-1}. If we choose $\boldsymbol{K}_X = \boldsymbol{K}(X)$ in (4.5), then this equation defines an inner product $*$ on each of the tangent spaces $\mathscr{T}_X$, $X \in \mathscr{B}$. The structure on $\mathscr{B}$ defined by these inner products will be called the *Riemannian structure of $\mathscr{B}$ relative to the uniform reference $\boldsymbol{K}$*. If $\mathfrak{h}$ and $\mathfrak{k}$ are tangent vector fields, we define $\mathfrak{h} * \mathfrak{k}$ pointwise. For such fields, (4.5) then yields

$$\mathfrak{h} * \mathfrak{k} = (\boldsymbol{K} \mathfrak{h}) \cdot (\boldsymbol{K} \mathfrak{k}). \tag{12.1}$$

It is clear that $\mathfrak{h} * \mathfrak{k} \in \mathscr{F}_{\mathscr{B}}^r$ if $\mathfrak{h}, \mathfrak{k} \in \mathscr{T}_{\mathscr{B}}^r$ for $0 \leqq r \leqq p-1$. This fact is expressed by saying that the Riemannian structure relative to $\boldsymbol{K}$ is *of class C^{p-1}*.

Although the following proposition is one of the basic facts of Riemannian geometry, we shall give an independent proof:

Proposition 6. *There is a unique affine connection $\overset{*}{\Gamma}$ of class C^{p-2} with the following properties:*

(a) *The torsion $\overset{*}{\mathfrak{S}}$ of $\overset{*}{\Gamma}$ vanishes.*

(b) *For any $\mathfrak{h}, \mathfrak{k}, \mathfrak{l} \in \mathscr{T}_{\mathscr{B}}^{r-1}$ the relation*

$$\mathfrak{h}(\mathfrak{k} * \mathfrak{l}) = \mathfrak{k} * \overset{*}{\Gamma}_{\mathfrak{h}} \mathfrak{l} + \mathfrak{l} * \overset{*}{\Gamma}_{\mathfrak{h}} \mathfrak{k} \tag{12.2}$$

is valid.

Proof. First we assume the existence of $\overset{*}{\Gamma}$. Let Γ be the material connection associated with $\boldsymbol{K}$ and consider the difference

$$\mathfrak{D}_{\mathfrak{h}} = \Gamma_{\mathfrak{h}} - \overset{*}{\Gamma}_{\mathfrak{h}}, \qquad \mathfrak{h} \in \mathscr{T}_{\mathscr{B}}^{p-1}. \tag{12.3}$$

According to the results given at the end of Sect. 9, $\mathfrak{D}$ can be identified with a field of class C^{p-2} on $\mathscr{B}$ whose values are linear transformations from $\mathscr{T}_X$ into $\mathscr{I}_X$.

Therefore, we can define a field $\boldsymbol{D}$ on $\mathscr{B}$ of class C^{p-2} with values $\boldsymbol{D}(X)\colon \mathscr{V}\to\mathscr{L}$ by the condition

$$\boldsymbol{D}\,\boldsymbol{u}=\boldsymbol{K}\,\mathfrak{D}_{\boldsymbol{K}^{-1}\boldsymbol{u}}\,\boldsymbol{K}^{-1} \tag{12.4}$$

for all $\boldsymbol{u}\in\mathscr{V}$. Since $\overset{*}{\mathfrak{S}}=0$, the relation (9.21) and the definitions (12.4) and (11.1) give

$$(\boldsymbol{S}\,\boldsymbol{u})\,\boldsymbol{v}=(\boldsymbol{D}\,\boldsymbol{u})\,\boldsymbol{v}-(\boldsymbol{D}\,\boldsymbol{v})\,\boldsymbol{u}\,,\qquad \boldsymbol{u},\boldsymbol{v}\in\mathscr{V}. \tag{12.5}$$

By the Definition 5 (Sect. 10), we have $\overset{*}{\Gamma}_{\mathfrak{h}}\mathfrak{c}=0$ whenever $\mathfrak{c}$ is materially constant. Hence, since $\mathfrak{c}=\boldsymbol{K}^{-1}\boldsymbol{c}$ is materially constant when $\boldsymbol{c}\in\mathscr{V}$, we infer from (12.3) and (12.4) that

$$\boldsymbol{K}\overset{*}{\Gamma}_{\mathfrak{h}}(\boldsymbol{K}^{-1}\,\boldsymbol{c})=-(\boldsymbol{D}\,\boldsymbol{K}\,\mathfrak{h})\,\boldsymbol{c} \tag{12.6}$$

when $\boldsymbol{c}\in\mathscr{V}$. Now let $\boldsymbol{u},\boldsymbol{v},\boldsymbol{w}\in\mathscr{V}$. If we substitute $\mathfrak{h}=\boldsymbol{K}^{-1}\boldsymbol{u}$, $\mathfrak{k}=\boldsymbol{K}^{-1}\boldsymbol{v}$, and $\mathfrak{l}=\boldsymbol{K}^{-1}\boldsymbol{w}$ into (12.2) and observe (12.6) and (12.1) we obtain

$$\mathfrak{h}(\boldsymbol{v}\cdot\boldsymbol{w})=-\boldsymbol{v}\cdot(\boldsymbol{D}\,\boldsymbol{u})\,\boldsymbol{w}-\boldsymbol{w}\cdot(\boldsymbol{D}\,\boldsymbol{u})\,\boldsymbol{v}\,. \tag{12.7}$$

Since $\boldsymbol{v}\cdot\boldsymbol{w}$ is constant, we have $\mathfrak{h}(\boldsymbol{v}\cdot\boldsymbol{w})=0$ (see the definition (8.10)). Therefore (12.7) states that $\boldsymbol{D}\,\boldsymbol{u}\in\mathscr{L}$ is skew for all $\boldsymbol{u}\in\mathscr{V}$:

$$\boldsymbol{D}\,\boldsymbol{u}=-(\boldsymbol{D}\,\boldsymbol{u})^{T},\qquad \boldsymbol{u}\in\mathscr{V}. \tag{12.8}$$

The equations (12.5) and (12.8) enable us to express $\boldsymbol{D}$ in terms of $\boldsymbol{S}$. Indeed, if we take the inner product of (12.5) with $\boldsymbol{w}\in\mathscr{V}$, subtract from the resulting equation the two equations obtained from it by cyclic permutations of $\boldsymbol{u}$, $\boldsymbol{v}$, $\boldsymbol{w}$, and observe (12.8), we find

$$2\boldsymbol{u}\cdot(\boldsymbol{D}\,\boldsymbol{w})\,\boldsymbol{v}=\boldsymbol{w}\cdot(\boldsymbol{S}\,\boldsymbol{u})\,\boldsymbol{v}-\boldsymbol{u}\cdot(\boldsymbol{S}\,\boldsymbol{v})\,\boldsymbol{w}-\boldsymbol{v}\cdot(\boldsymbol{S}\,\boldsymbol{w})\,\boldsymbol{u}\,. \tag{12.9}$$

Since $(\boldsymbol{S}\,\boldsymbol{u})\,\boldsymbol{v}=-(\boldsymbol{S}\,\boldsymbol{v})\,\boldsymbol{u}$, (12.9) is equivalent to

$$(\boldsymbol{D}\,\boldsymbol{u})\,\boldsymbol{v}=\tfrac{1}{2}\{[(\boldsymbol{S}\,\boldsymbol{u})-(\boldsymbol{S}\,\boldsymbol{u})^{T}]\,\boldsymbol{v}-(\boldsymbol{S}\,\boldsymbol{v})^{T}\,\boldsymbol{u}\}. \tag{12.10}$$

Now, since $\boldsymbol{S}$ is determined by the uniform reference $\boldsymbol{K}$, it follows from (12.10), (12.4), and (12.3) that $\overset{*}{\Gamma}$ is uniquely determined by $\boldsymbol{K}$.

To prove the existence of a connection $\overset{*}{\Gamma}$ with the properties (a) and (b), one can define $\overset{*}{\Gamma}$ by (12.3), (12.4), and (12.10) and verify that it has all the required properties. Q.E.D.

Definition 7. *The connection $\overset{*}{\Gamma}$ determined by the conditions* (a) *and* (b) *of Proposition 6 is called the* ***Riemannian connection*** *relative to the uniform reference* $\boldsymbol{K}$. *The field* $\boldsymbol{D}$ *determined by* (12.3) *and* (12.4) *or* (12.5) *and* (12.8) *is called the* ***contortion***[13] *of* $\boldsymbol{K}$.

The term "contortion" will be motivated in Section 13.

[13] It corresponds to what is called "Cosserat structure curvature" or "Nye curvature" in the theory of continuous distributions of dislocations (*cf.* [4]).

In general, the curvature $\overset{*}{\mathfrak{R}}$ of the Riemannian connection is not zero. We can define a field $\overset{*}{\boldsymbol{R}}$ with values $\overset{*}{\boldsymbol{R}}(X)\colon \mathscr{V}\times\mathscr{V}\to\mathscr{L}$ by the condition

$$\overset{*}{\boldsymbol{R}}(\boldsymbol{u},\boldsymbol{v})=\boldsymbol{K}\,\overset{*}{\mathfrak{R}}(\boldsymbol{K}^{-1}\boldsymbol{u},\boldsymbol{K}^{-1}\boldsymbol{v})\,\boldsymbol{K}^{-1} \tag{12.11}$$

for all $\boldsymbol{u},\boldsymbol{v}\in\mathscr{V}$. Let $\boldsymbol{u},\boldsymbol{v},\boldsymbol{w}\in\mathscr{V}$ and put $\mathfrak{h}=\boldsymbol{K}^{-1}\boldsymbol{u}$, $\mathfrak{k}=\boldsymbol{K}^{-1}\boldsymbol{v}$, $\mathfrak{l}=\boldsymbol{K}^{-1}\boldsymbol{w}$. The fields $\mathfrak{h},\mathfrak{k},\mathfrak{l}$ are then materially constant and hence are annihilated by Γ. Recalling that the curvature $\mathfrak{R}$ of Γ vanishes (Theorem 4, Sect. 10), we then infer from (9.22), (10.5), $(11.6)_1$, (12.4), and (12.11) that

$$\begin{aligned}\overset{*}{\boldsymbol{R}}(\boldsymbol{u},\boldsymbol{v})\,\boldsymbol{w}=&-\big(\nabla_{\boldsymbol{K}}(\boldsymbol{D}\,\boldsymbol{v})\,\boldsymbol{u}\big)\,\boldsymbol{w}+\big(\nabla_{\boldsymbol{K}}(\boldsymbol{D}\,\boldsymbol{u})\,\boldsymbol{v}\big)\,\boldsymbol{w}\\&+(\boldsymbol{D}\,\boldsymbol{u})(\boldsymbol{D}\,\boldsymbol{v})\,\boldsymbol{w}-(\boldsymbol{D}\,\boldsymbol{v})(\boldsymbol{D}\,\boldsymbol{u})\,\boldsymbol{w}-\boldsymbol{D}\big((\boldsymbol{S}\,\boldsymbol{u})\,\boldsymbol{v}\big)\,\boldsymbol{w}\end{aligned}$$

and hence

$$\overset{*}{\boldsymbol{R}}(\boldsymbol{u},\boldsymbol{v})=\big((\nabla_{\boldsymbol{K}}\boldsymbol{D})\,\boldsymbol{v}\big)\,\boldsymbol{u}-\big((\nabla_{\boldsymbol{K}}\boldsymbol{D})\,\boldsymbol{u}\big)\,\boldsymbol{v}+(\boldsymbol{D}\,\boldsymbol{u})(\boldsymbol{D}\,\boldsymbol{v})-(\boldsymbol{D}\,\boldsymbol{v})(\boldsymbol{D}\,\boldsymbol{u})-\boldsymbol{D}\big((\boldsymbol{S}\,\boldsymbol{u})\,\boldsymbol{v}\big). \tag{12.12}$$

In view of (12.5) and (12.10), equation (12.12) shows that $\overset{*}{\boldsymbol{R}}$ can be expressed in terms of the contortion $\boldsymbol{D}$ and its gradient relative to $\boldsymbol{K}$ or in terms of the inhomogeneity $\boldsymbol{S}$ and its gradient relative to $\boldsymbol{K}$.

13. Contorted Aeolotropy

Definition 8. *A uniform reference $\boldsymbol{K}$ of class C^{p-1} is called a* ***state of contorted aeolotropy*** *if there exists a configuration $\boldsymbol{\kappa}$ such that the tensor field*

$$\boldsymbol{Q}=(\nabla\boldsymbol{\kappa})\,\boldsymbol{K}^{-1}\in\mathscr{L}_{\mathscr{B}}^{p-1} \tag{13.1}$$

has orthogonal values ($\boldsymbol{Q}(X)\in\mathscr{o}$ for all $X\in\mathscr{B}$).

Assume that $\boldsymbol{K}$ is such a state of contorted aeolotropy. Since the inner product in $\mathscr{V}$ is preserved under orthogonal transformations, the Riemannian structure (12.1) relative to $\boldsymbol{K}$ satisfies.

$$\mathfrak{h}*\mathfrak{k}=\boldsymbol{Q}(\boldsymbol{K}\,\mathfrak{h})\cdot\boldsymbol{Q}(\boldsymbol{K}\,\mathfrak{k})=(\nabla\boldsymbol{\kappa})\,\mathfrak{h}\cdot(\nabla\boldsymbol{\kappa})\,\mathfrak{k} \tag{13.2}$$

for all tangent vector fields $\mathfrak{h}$ and $\mathfrak{k}$. It follows from (13.2) that the Riemannian connection $\overset{*}{\Gamma}$ relative to $\boldsymbol{K}$ is obtained by transporting the gradient operator ∇ from $\boldsymbol{\kappa}(\mathscr{B})$ into $\mathscr{B}$ via $(\nabla\boldsymbol{\kappa})^{-1}$, so that

$$\overset{*}{\Gamma}\,\mathfrak{k}=(\nabla\boldsymbol{\kappa})^{-1}\,\nabla_{\boldsymbol{\kappa}}\big((\nabla\boldsymbol{\kappa})\,\mathfrak{k}\big)(\nabla\boldsymbol{\kappa}),\qquad \mathfrak{k}\in\mathscr{T}_{\mathscr{B}}^{p-1}. \tag{13.3}$$

Indeed, if $\overset{*}{\Gamma}$ is defined by (13.3), condition (a) of Proposition 6 follows from the symmetry of the second gradient and condition (b) from the rule for the differentiation of inner products. By virtue of (13.1), (13.3) is equivalent to

$$\boldsymbol{K}\overset{*}{\Gamma}_{\mathfrak{h}}\,\mathfrak{k}=\boldsymbol{Q}^T\,\nabla_{\boldsymbol{\kappa}}(\boldsymbol{Q}\,\boldsymbol{K}\,\mathfrak{k})\,\boldsymbol{Q}\,\boldsymbol{K}\,\mathfrak{h},\qquad \mathfrak{h},\mathfrak{k}\in\mathscr{T}_{\mathscr{B}}^{p-1}. \tag{13.4}$$

Now let $\boldsymbol{u},\boldsymbol{v}\in\mathscr{V}$. If we substitute $\mathfrak{h}=\boldsymbol{K}^{-1}\boldsymbol{u}$, $\mathfrak{k}=\boldsymbol{K}^{-1}\boldsymbol{v}$ into (13.4) and observe (12.6), we obtain

$$\boldsymbol{Q}^T\,\nabla_{\boldsymbol{\kappa}}(\boldsymbol{Q}\,\boldsymbol{v})\,\boldsymbol{Q}\,\boldsymbol{u}=-(\boldsymbol{D}\,\boldsymbol{u})\,\boldsymbol{v},$$

which, by (13.1) and (8.8), is equivalent to

$$\boldsymbol{D}\,\boldsymbol{u} = -\boldsymbol{Q}^T(\nabla_{\boldsymbol{K}}\boldsymbol{Q})\,\boldsymbol{u}\,, \qquad \boldsymbol{u}\in\mathscr{V}. \tag{13.5}$$

This equation shows that the skew transformation $-\boldsymbol{D}\,\boldsymbol{u}|_X \in \mathscr{L}$ is the instantaneous rate of change of $\boldsymbol{Q}$ at X in the direction of $\boldsymbol{u}$, if viewed in any configuration belonging to $\boldsymbol{K}(X)$. In other words, $\boldsymbol{D}$ describes the local behavior of the rotation field $\boldsymbol{Q}$, which changes the given state of contorted aeolotropy $\boldsymbol{K}$ into the gradient of a global configuration $\boldsymbol{\kappa}$. It is this property that the term "contortion" for $\boldsymbol{D}$ is meant to express.

Theorem 7. *If $\boldsymbol{K}$ is a state of contorted aeolotropy, then the curvature of the Riemannian connection relative to $\boldsymbol{K}$ vanishes. Conversely, if the curvature of the Riemannian connection relative to $\boldsymbol{K}$ vanishes, then $\boldsymbol{K}$ is locally a state of contorted aeolotropy* (*i.e.*, every point in $\mathscr{B}$ has a neighborhood $\mathscr{N}$ such that the restriction of $\boldsymbol{K}$ to $\mathscr{N}$ is a state of contorted aeolotropy for $\mathscr{N}$).

Proof. Assume first that (13.1) holds. It follows from (13.3) that $\overset{*}{\Gamma}\mathfrak{k}=0$ if and only if $(\nabla\boldsymbol{\kappa})\mathfrak{k}$ is constant. Hence we could give a simple direct proof of $\overset{*}{\mathfrak{R}}=0$ by using the same argument as we used in the proof of Theorem 4 (Sect. 10). Another proof can be obtained on the basis of (13.5) as follows:

If γ is an arbitrary configuration and $\boldsymbol{F}=(\nabla\gamma)\boldsymbol{K}^{-1}$, then (13.5) is equivalent to

$$(\nabla_\gamma\boldsymbol{Q})\,\boldsymbol{h} = -\boldsymbol{Q}\,\boldsymbol{D}(\boldsymbol{F}^{-1}\boldsymbol{h})\,, \qquad \boldsymbol{h}\in\mathscr{V}. \tag{13.6}$$

If we take the gradient ∇_γ of (13.6) in the direction of $\boldsymbol{k}\in\mathscr{V}$, we find

$$\big((\nabla_\gamma^{(2)}\boldsymbol{Q})\,\boldsymbol{k}\big)\,\boldsymbol{h} = -\boldsymbol{Q}\,\{((\nabla_{\boldsymbol{K}}\boldsymbol{D})\,\boldsymbol{v})\,\boldsymbol{u} - (\boldsymbol{D}\,\boldsymbol{v})(\boldsymbol{D}\,\boldsymbol{u}) + \boldsymbol{D}\big(((\nabla_\gamma\boldsymbol{F}^{-1})\,\boldsymbol{k})\,\boldsymbol{h}\big), \tag{13.7}$$

where $\boldsymbol{u}=\boldsymbol{F}^{-1}\boldsymbol{h}$, $\boldsymbol{v}=\boldsymbol{F}^{-1}\boldsymbol{k}$. Of course, because of the linearity of the values of the fields $\boldsymbol{D}$, $\nabla_{\boldsymbol{K}}\boldsymbol{D}$, $\nabla_\gamma^{(2)}\boldsymbol{Q}$, etc., (13.6) and (13.7) remain valid if $\boldsymbol{h}$ and $\boldsymbol{k}$ are not fixed vectors but vector fields. In particular, they remain valid when $\boldsymbol{u}$ and $\boldsymbol{v}$ are fixed. If we interchange $\boldsymbol{u}$ and $\boldsymbol{v}$ and hence $\boldsymbol{h}$ and $\boldsymbol{k}$ in (13.7) and subtract the resulting formula from (13.7), we obtain, after observing (11.5) and (12.12),

$$\big((\nabla_\gamma^{(2)}\boldsymbol{Q})\,\boldsymbol{k}\big)\,\boldsymbol{h} - \big((\nabla_\gamma^{(2)}\boldsymbol{Q})\,\boldsymbol{h}\big)\,\boldsymbol{k} = -\boldsymbol{Q}\,\overset{*}{\boldsymbol{R}}(\boldsymbol{u},\boldsymbol{v})\,. \tag{13.8}$$

Thus, $\overset{*}{\boldsymbol{R}}=\boldsymbol{0}$ and hence $\overset{*}{\mathfrak{R}}=0$ follows also from the symmetry of the second gradient $\nabla_\gamma^{(2)}\boldsymbol{Q}$.

Assume now that $\boldsymbol{K}$ is a uniform reference such that $\overset{*}{\boldsymbol{R}}=\boldsymbol{0}$. Let γ be an arbitrary configuration and put $\boldsymbol{F}=(\nabla\gamma)\boldsymbol{K}^{-1}$, as before. We can then regard (13.6) as a differential equation for the determination of $\boldsymbol{Q}$. As we have seen, $\overset{*}{\boldsymbol{R}}=\boldsymbol{O}$ is an integrability condition necessary for the existence of a solution. According to a classical theorem, $\overset{*}{\boldsymbol{R}}=\boldsymbol{0}$ is also sufficient for the existence of a solution that is valid in a simply connected neighborhood $\mathscr{N}'$ of a given point $X_0\in\mathscr{B}$. The solution can be chosen so that for $X_0\in\mathscr{B}$, $\boldsymbol{Q}(X_0)$ has a prescribed value, which we take to be the identity $\boldsymbol{1}$. Since $\boldsymbol{D}\,\boldsymbol{u}$ is skew for all $\boldsymbol{u}\in\mathscr{V}$, it follows from (13.6) that $\boldsymbol{Q}\boldsymbol{Q}^T$ has gradient zero and hence must be equal to $\boldsymbol{1}$ everywhere in $\mathscr{N}'$. Hence $\boldsymbol{Q}$ has orthogonal values. To summarize: If $\overset{*}{\boldsymbol{R}}=\mathbf{0}$, every point in $\mathscr{B}$ has a neighborhood $\mathscr{N}'$ on which we can find an orthogonal-valued tensor field $\boldsymbol{Q}$ (of class C^{p-1}) such that (13.5) holds.

Assume, then, that (13.5) holds on $\mathscr{N}'$. Combining (12.6) with (13.5) we obtain

$$\boldsymbol{K}\overset{*}{\Gamma}_{\mathfrak{h}}\,\mathfrak{c}=\boldsymbol{Q}^T((\nabla_{\boldsymbol{K}}\boldsymbol{Q})\boldsymbol{K}\mathfrak{h})\,\boldsymbol{c}=\boldsymbol{Q}^T\,\nabla_{\boldsymbol{K}}(\boldsymbol{Q}\,\boldsymbol{c})\,\boldsymbol{K}\mathfrak{h}, \tag{13.9}$$

which is valid when $\mathfrak{c}=\boldsymbol{K}^{-1}\boldsymbol{c}$ is materially constant. Consider the affine connection $\bar{\Gamma}$ of class C^{p-1} defined by

$$\bar{\Gamma}\mathfrak{k}=\bar{\boldsymbol{K}}^{-1}\,\nabla_{\bar{\boldsymbol{K}}}(\bar{\boldsymbol{K}}\mathfrak{k})\,\bar{\boldsymbol{K}}, \qquad \mathfrak{k}\in\mathscr{T}_{\mathscr{B}}^{1}, \tag{13.10}$$

where

$$\bar{\boldsymbol{K}}=\boldsymbol{Q}\boldsymbol{K}. \tag{13.11}$$

It is easily seen that (13.9) is equivalent to the statement that

$$\overset{*}{\Gamma}\mathfrak{c}=\bar{\Gamma}\mathfrak{c} \tag{13.12}$$

holds for all materially constant tangent vector fields $\mathfrak{c}$. Using the same argument as in the uniqueness proof of Theorem 3 (Sect. 10) we conclude that $\overset{*}{\Gamma}=\bar{\Gamma}$. Since $\overset{*}{\Gamma}$ has zero torsion, an analogue of Theorem 6 (Sect. 10) shows that every point in $\mathscr{N}'$ must have a neighborhood $\mathscr{N}$ such that $\bar{\boldsymbol{K}}=\nabla\boldsymbol{\kappa}$ for some configuration $\boldsymbol{\kappa}$ of $\mathscr{N}$. Hence, by (13.11), we have $\boldsymbol{Q}=(\nabla\boldsymbol{\kappa})\boldsymbol{K}^{-1}$ on $\mathscr{N}$, *i.e.*, $\boldsymbol{K}$ is a state of contorted aeolotropy on $\mathscr{N}$. Q.E.D.

A special case of contorted aeolotropy is *curvilinear aeolotropy*. It corresponds to the case when there exists an orthogonal coordinate system on $\boldsymbol{\kappa}(\mathscr{B})\subset\mathscr{E}$ with the following property: If $(\boldsymbol{e}_1(X), \boldsymbol{e}_2(X), \boldsymbol{e}_3(X))$ is the orthonormal basis which consists of the unit vectors that point in the direction of the coordinate lines at $\boldsymbol{\kappa}(X)$, then $\boldsymbol{Q}(X)^T\,\boldsymbol{e}_i(X)$ does not depend on $X\in\mathscr{B}$.

14. Special Types of Materially Uniform Bodies

We consider first the case when the isotropy group $g_{\boldsymbol{K}}$ of $\mathscr{B}$ relative to some – and hence every – uniform reference $\boldsymbol{K}$ is *discrete*. Suppose that $\boldsymbol{K}$ and $\hat{\boldsymbol{K}}$ are two *continuous* uniform references. They must be related by (6.12), where $\boldsymbol{P}(X)\in g_{\boldsymbol{K}}$ must depend continuously on X. Since $g_{\boldsymbol{K}}$ is discrete, this is possible only when $\boldsymbol{P}$ is constant. Thus we can absorb $\boldsymbol{P}$ into $\boldsymbol{L}$, and (6.12) becomes $\hat{\boldsymbol{K}}=\boldsymbol{L}\boldsymbol{K}$, with $\boldsymbol{L}=$const. If we write (6.4) for both $\boldsymbol{K}$ and $\hat{\boldsymbol{K}}$, we see that they correspond to the same continuous material uniformity. Since every continuous material uniformity must be of the form (6.4), where $\boldsymbol{K}$ is a continuous uniform reference, we have the following result:

Theorem 8. *If the isotropy groups of a materially uniform simple body $\mathscr{B}$ are discrete, then $\mathscr{B}$ has at most one continuous material uniformity Φ. Any two continuous uniform references $\boldsymbol{K}$, $\hat{\boldsymbol{K}}$ are related by*

$$\hat{\boldsymbol{K}}=\boldsymbol{L}\boldsymbol{K}, \qquad \boldsymbol{L}=\text{const}\in\ell. \tag{14.1}$$

Since material connections are only associated with differentiable uniformities Φ and not with discontinuous ones, it follows from the uniqueness assertion of

Theorem 8 that in the case when the isotropy groups are discrete, the inhomogeneity $\mathfrak{S}$ is a characteristic of the body. If the isotropy groups are non-discrete Lie groups, however, and if there are any material uniformities of class C^{p-1} at all, there will be many, and hence also many inhomogeneities $\mathfrak{S}$ for one and the same body. This is the case, in particular, for uniform isotropic bodies.

Next, we consider a uniform *isotropic* body $\mathscr{B}$ with an undistorted uniform reference $\boldsymbol{K}$ of class C^{p-1}, so that $\mathscr{o} \subset \mathscr{g}_{\boldsymbol{K}}$. If $\boldsymbol{K}$ is a state of contorted aeolotropy, so that (13.1) holds, it follows from Theorem 2, (6.12), that $\nabla_{\boldsymbol{K}}$ is again a uniform reference and that $\mathscr{g}_{\boldsymbol{K}}$ is also the isotropy group relative to $\nabla_{\boldsymbol{K}}$:

Theorem 9. *If a uniform isotropic body has an undistorted state of contorted aeolotropy, it is homogeneous.*

The conclusion of the theorem becomes false when the qualifier "undistorted" is ommitted; *i.e.*, there are inhomogeneous isotropic bodies with distorted states of contorted aeolotropy.

Finally, suppose there is a natural way to single out, among all uniform references for $\mathscr{B}$, a particular class U with the following property (P): All members of U are of class C^{p-1} and differ from one another by a field of similarity transformations with constant ratio, so that $\boldsymbol{K}, \boldsymbol{K}' \in \mathsf{U}$ implies

$$\boldsymbol{K}' = a\,\boldsymbol{Q}\,\boldsymbol{K}, \tag{14.2}$$

where a is a real constant and $\boldsymbol{Q}$ an orthogonal valued tensor field on $\mathscr{B}$. For example, if $\mathscr{B}$ is a uniform solid body that is either isotropic or has cubic symmetry then the class U of all undistorted references has the property (P). This follows from results proved in reference [9]. Other examples are obtained by letting U be the class of all uniform references $\boldsymbol{K}$ such that the corresponding response functions $\mathfrak{H}_{\boldsymbol{K}}$ satisfy a certain special condition such as $\mathfrak{H}_{\boldsymbol{K}}(\mathbf{1}) = 0$. Such references are often called *natural references*. The nature of the response function is often such that the class U of natural references has the property (P).

If (14.2) holds, it follows from (12.1) and the fact that orthogonal transformations preserve inner products that the two Riemannian inner products corresponding to $\boldsymbol{K}'$ and $\boldsymbol{K}$ differ from one another only by the constant factor a^2. Therefore, Proposition 6 shows that the Riemannian connections relative to $\boldsymbol{K}$ and $\boldsymbol{K}'$ are the same, and we have the following result:

Theorem 10. *If $\mathscr{B}$ is a uniform simple body with a distinguished class U of uniform references with the property (P), then the Riemannian connection $\overset{*}{\Gamma}$ and its curvature $\overset{*}{\mathfrak{R}}$ are characteristics of the body.*

The assertion of Theorem 10 applies, in particular, to uniform isotropic solid bodies, for which the curvature, $\overset{*}{\mathfrak{R}}$, defined by the class of undistorted uniform references, is an intrinsic measure of deviation from homogeneity.

15. Cauchy's Equation of Balance

We now derive a new version of CAUCHY's equation of balance, which expresses the fact that the forces acting on every part of a given body $\mathscr{B}$ must add to zero. In order to do so, we first derive a lemma, Proposition 7 below.

Let $\boldsymbol{h}$ be a vector field and $\boldsymbol{T}$ a tensor field of class C^1 on $\mathscr{B}$. We define the *divergence* of these fields *relative to* some *reference* $\boldsymbol{K}$ by

$$\operatorname{div}_{\boldsymbol{K}} \boldsymbol{h} = \operatorname{tr}(\nabla_{\boldsymbol{K}} \boldsymbol{h}) \quad \text{and} \quad (\operatorname{div}_{\boldsymbol{K}} \boldsymbol{T}) \cdot \boldsymbol{u} = \operatorname{div}_{\boldsymbol{K}}(\boldsymbol{T}^T \boldsymbol{u}), \qquad \boldsymbol{u} \in \mathscr{V}, \tag{15.1}$$

respectively. The following product rules are valid for $f \in \mathscr{F}_{\mathscr{B}}^1$, $\boldsymbol{h} \in \mathscr{V}_{\mathscr{B}}^1$, $\boldsymbol{T} \in \mathscr{L}_{\mathscr{B}}^1$:

$$\begin{aligned} \operatorname{div}_{\boldsymbol{K}}(f\, \boldsymbol{h}) &= (\nabla_{\boldsymbol{K}} f) \cdot \boldsymbol{h} + f \operatorname{div}_{\boldsymbol{K}} \boldsymbol{h}, \\ \operatorname{div}_{\boldsymbol{K}}(\boldsymbol{T} \boldsymbol{h}) &= (\operatorname{div}_{\boldsymbol{K}} \boldsymbol{T}^T) \cdot \boldsymbol{h} + \operatorname{tr}(\boldsymbol{T} \nabla_{\boldsymbol{K}} \boldsymbol{h}). \end{aligned} \tag{15.2}$$

If $\boldsymbol{K} = \nabla \kappa$ is a configuration gradient, we write $\operatorname{div}_\kappa$ instead of $\operatorname{div}_{\boldsymbol{K}}$.

Suppose that a uniform reference $\boldsymbol{K}$ of class C^{p-1} for $\mathscr{B}$ and a tensor field $\boldsymbol{T}$ of class C^1 on $\mathscr{B}$ are given. For any configuration γ of $\mathscr{B}$ we then define another tensor field $\boldsymbol{T}_\gamma$ of class C^1 by

$$\boldsymbol{T}_\gamma = \frac{1}{J} \boldsymbol{T} \boldsymbol{F}^T, \tag{15.3}$$

where

$$\boldsymbol{F} = (\nabla \gamma) \boldsymbol{K}^{-1}, \qquad J = |\det \boldsymbol{F}|. \tag{15.4}$$

Proposition 7. *If*

$$\boldsymbol{K}(X) = \nabla \kappa(X) \tag{15.5}$$

for some $X \in \mathscr{B}$, then the divergences at X relative to $\boldsymbol{K}$ of the tensor field $\boldsymbol{T}$ and of the tensor field $\boldsymbol{T}_\kappa$ defined by (15.3) *and* (15.4) *are related by*

$$\operatorname{div}_{\boldsymbol{K}} \boldsymbol{T}_\kappa|_X = (\operatorname{div}_{\boldsymbol{K}} \boldsymbol{T} + \boldsymbol{T} \boldsymbol{s})|_X, \tag{15.6}$$

where the vector field $\boldsymbol{s}$ is defined, in terms of the inhomogeneity $\boldsymbol{S}$ relative to $\boldsymbol{K}$, by

$$\boldsymbol{s} \cdot \boldsymbol{u} = \operatorname{tr}(\boldsymbol{S} \boldsymbol{u}), \qquad \boldsymbol{u} \in \mathscr{V}. \tag{15.7}$$

Proof. We make use of (15.3) and (15.4) with γ replaced by κ. We then have

$$J(\boldsymbol{T}_\kappa^T \boldsymbol{u}) = \boldsymbol{F}(\boldsymbol{T}^T \boldsymbol{u}), \qquad \boldsymbol{u} \in \mathscr{V}. \tag{15.8}$$

Using the rule

$$\boldsymbol{u} \cdot \nabla_{\boldsymbol{K}}(\det \boldsymbol{F}) = (\det \boldsymbol{F}) \operatorname{tr}[\boldsymbol{F}^{-1}((\nabla_{\boldsymbol{K}} \boldsymbol{F}) \boldsymbol{u})], \qquad \boldsymbol{u} \in \mathscr{V} \tag{15.9}$$

for the differentiation of a determinant, the product rules (15.2), and the definition $(15.1)_2$, we see that taking $\operatorname{div}_{\boldsymbol{K}}$ of (15.8) yields

$$J \operatorname{tr}[\boldsymbol{F}^{-1}((\nabla_{\boldsymbol{K}} \boldsymbol{F})(\boldsymbol{T}_\kappa^T \boldsymbol{u}))] + J(\operatorname{div}_{\boldsymbol{K}} \boldsymbol{T}_\kappa) \cdot \boldsymbol{u} = (\operatorname{div}_{\boldsymbol{K}} \boldsymbol{F}^T) \cdot \boldsymbol{T}^T \boldsymbol{u} + \operatorname{tr}[\boldsymbol{F} \nabla_{\boldsymbol{K}}(\boldsymbol{T}^T \boldsymbol{u})]. \tag{15.10}$$

Since $\boldsymbol{F}(X) = \mathbf{1}$, $J(X) = 1$, and $\boldsymbol{T}_\kappa(X) = \boldsymbol{T}(X)$ by (15.5), (15.4), and (15.3), evaluation of (15.10) at $X \in \mathscr{B}$ gives

$$\{\operatorname{tr}[(\nabla_{\boldsymbol{K}} \boldsymbol{F})(\boldsymbol{T}^T \boldsymbol{u})] + (\operatorname{div}_{\boldsymbol{K}} \boldsymbol{T}_\kappa) \cdot \boldsymbol{u} - (\operatorname{div}_{\boldsymbol{K}} \boldsymbol{F}^T) \cdot \boldsymbol{T}^T \boldsymbol{u} - (\operatorname{div}_{\boldsymbol{K}} \boldsymbol{T}) \cdot \boldsymbol{u}\}_X = 0. \tag{15.11}$$

Using the rule $\operatorname{tr}[(\nabla_{\boldsymbol{K}} \boldsymbol{F}) \boldsymbol{v}] = \nabla_{\boldsymbol{K}}(\operatorname{tr} \boldsymbol{F}) \cdot \boldsymbol{v}$, we see that (15.11) can hold for all $\boldsymbol{u} \in \mathscr{V}$ only if

$$\{\boldsymbol{T}[\nabla_{\boldsymbol{K}}(\operatorname{tr} \boldsymbol{F}) - \operatorname{div}_{\boldsymbol{K}} \boldsymbol{F}^T] + \operatorname{div}_{\boldsymbol{K}} \boldsymbol{T}_\kappa - \operatorname{div}_{\boldsymbol{K}} \boldsymbol{T}\}_X = \mathbf{0}. \tag{15.12}$$

On the other hand, if we evaluate (11.4) at $X \in \mathscr{B}$ and take the trace, we obtain

$$\operatorname{tr}(\boldsymbol{S}\,\boldsymbol{u})|_X = \{(\operatorname{div}_{\boldsymbol{K}} \boldsymbol{F}^T) \cdot \boldsymbol{u} - \nabla_{\boldsymbol{K}}(\operatorname{tr} \boldsymbol{F}) \cdot \boldsymbol{u}\}_X,$$

which, in view of (15.7), is equivalent to

$$\boldsymbol{s}|_X = \{\operatorname{div}_{\boldsymbol{K}} \boldsymbol{F}^T - \nabla_{\boldsymbol{K}}(\operatorname{tr} \boldsymbol{F})\}_X. \tag{15.13}$$

The desired result (15.6) follows from (15.12) and (15.13). Q.E.D.

Let us assume now that the body $\mathscr{B}$ is subject to internal contact forces and external body forces[14]. If the forces acting on every part of $\mathscr{B}$ are balanced and if suitable regularity assumptions are satisfied one can prove the following results (*cf.* [*10*]):

(i) With every configuration κ of $\mathscr{B}$ one can associate a stress tensor field $\boldsymbol{T}_\kappa$ of class C^1 and a body force field $\boldsymbol{b}_\kappa$ of class C^0 such that the force $\boldsymbol{f}$ exerted on a part $\mathscr{P}$ of $\mathscr{B}$ by the combined action of a separate part $\mathscr{P}'$ of $\mathscr{B}$ and the external world is given by

$$\boldsymbol{f} = \int_{\kappa(\mathscr{P})} \boldsymbol{b}_\kappa \, dV + \int_{\mathscr{C}} \boldsymbol{T}_\kappa \boldsymbol{n} \, dS, \tag{15.14}$$

where $\mathscr{C}$ is the surface of contact between $\mathscr{P}$ and $\mathscr{P}'$ in the configuration κ and where $\boldsymbol{n}$ is the unit normal to $\mathscr{C}$ directed away from $\kappa(\mathscr{P})$.

(ii) CAUCHY's equation of balance

$$\operatorname{div}_\kappa \boldsymbol{T}_\kappa + \boldsymbol{b}_\kappa = \boldsymbol{0} \tag{15.15}$$

is valid on $\mathscr{B}$ for every configuration κ.

(iii) If κ and γ are two configurations of $\mathscr{B}$, then the stress fields $\boldsymbol{T}_\kappa$ and $\boldsymbol{T}_\gamma$ and the body force fields $\boldsymbol{b}_\kappa$ and $\boldsymbol{b}_\gamma$ are related by

$$\boldsymbol{T}_\gamma = \frac{1}{J} \boldsymbol{T}_\kappa \boldsymbol{F}^T, \qquad \boldsymbol{b}_\gamma = \frac{1}{J} \boldsymbol{b}_\kappa, \tag{15.16}$$

where

$$\boldsymbol{F} = (\nabla\gamma)(\nabla\kappa)^{-1}, \qquad J = |\det \boldsymbol{F}| \tag{15.17}$$

(*cf.* equation (43A.3) of reference [*2*]).

Let $\boldsymbol{K}$ be a uniform reference of class C^{p-1} for $\mathscr{B}$. Let a particular point $X \in \mathscr{B}$ be given. It is clear from (15.16) and (15.17) that $\boldsymbol{T}_\gamma(X) = \boldsymbol{T}_\kappa(X)$ and $\boldsymbol{b}_\gamma(X) = \boldsymbol{b}_\kappa(X)$ hold whenever both κ and γ belong to the equivalence class by which the local configuration $\boldsymbol{K}(X)$ is defined. Thus, we can define fields $\boldsymbol{T}_{\boldsymbol{K}}$ and $\boldsymbol{b}_{\boldsymbol{K}}$ by the condition that for each $X \in \mathscr{B}$,

$$\boldsymbol{T}_{\boldsymbol{K}}(X) = \boldsymbol{T}_\kappa(X), \qquad \boldsymbol{b}_{\boldsymbol{K}}(X) = \boldsymbol{b}_\kappa(X) \tag{15.18}$$

hold whenever $\nabla\kappa(X) = \boldsymbol{K}(X)$. We call $\boldsymbol{T}_{\boldsymbol{K}}$ and $\boldsymbol{b}_{\boldsymbol{K}}$ the *stress tensor field* and *body force field relative to the reference* $\boldsymbol{K}$. It is clear from (15.16) and (15.17) that

$$\boldsymbol{T}_\gamma = \frac{1}{J} \boldsymbol{T}_{\boldsymbol{K}} \boldsymbol{F}^T, \qquad \boldsymbol{b}_\gamma = \frac{1}{J} \boldsymbol{b}_{\boldsymbol{K}} \tag{15.19}$$

when $\boldsymbol{F}$ and J are given by (15.4), where γ is an arbitrary configuration.

[14] Inertial forces should be regarded as part of the body forces.

Since $\operatorname{div}_\kappa T|_X = \operatorname{div}_K T|_X$ whenever $\nabla \kappa(X) = K(X)$, it follows from (15.15) that

$$[\operatorname{div}_K T_\kappa + b_K]_X = 0 \tag{15.20}$$

holds whenever $\nabla \kappa(X) = K(X)$. On the other hand, Proposition 7 applies when we choose $T = T_K$. Thus, by substituting (15.6) with the choice $T = T_K$ into (15.20) we obtain the following result:

Theorem 11. *The stress tensor field* $\mathbf{T}_K$ *and the body force field* $\mathbf{b}_K$ *relative to a uniform reference* $\mathbf{K}$ *satisfy the modified equation of balance*

$$\operatorname{div}_K T_K + T_K s + b_K = 0, \tag{15.21}$$

where s *is defined in terms of the inhomogeneity* $\mathbf{S}$ *relative to* $\mathbf{K}$ *by* (15.7).

The equation (15.21) is much more useful than (15.15) for dealing with inhomogeneous materially uniform bodies. Consider, for example, an elastic body $\mathscr{B}$, for which the set of response descriptors is the set $\mathscr{S} \subset \mathscr{L}$ of symmetric linear transformations. According to Theorem 1 (Sect. 6) we can associate with a given uniform reference K a relative elastic response function $\mathfrak{H}_K : \ell \to \mathscr{S}$. In order that a configuration γ be compatible with a given force system, the constitutive equation

$$\mathfrak{H}_K(F) = T_\gamma, \qquad F = (\nabla\gamma)\, K^{-1} \tag{15.22}$$

must be satisfied on $\mathscr{B}$, where T_γ is the stress tensor field for γ. In view of $(15.19)_1$, (15.22) is equivalent to

$$\mathfrak{h}_K(F) = T_K, \qquad F = (\nabla\gamma)\, K^{-1}, \tag{15.23}$$

where $\mathfrak{h}_K : \ell \to \mathscr{L}$ is defined by

$$\mathfrak{h}_K(F) = |\det F|\, \mathfrak{H}_K(F)\, F^{T^{-1}}, \qquad F \in \ell. \tag{15.24}$$

Assume that $\mathfrak{h}_K$ is of class C^1, and denote its gradient by H_K. For each $F \in \ell$ the value $\mathsf{H}_K(F)$ is then a linear transformation from $\mathscr{L}$ into $\mathscr{L}$. If we take the gradient of (15.23) relative to K, the chain rule yields

$$(\nabla_K T_K)\, u = \mathsf{H}_K(F)\,[(\nabla_K F)\, u], \qquad u \in \mathscr{V}. \tag{15.25}$$

It follows that

$$\operatorname{div}_K T_K = \mathsf{A}_K(F)\,[(\nabla_K F)], \tag{15.26}$$

where A_K is that function on ℓ whose values $\mathsf{A}_K(F)$: $\mathscr{L}(\mathscr{V}, \mathscr{L}) \to \mathscr{V}$ are determined by the property that $\mathsf{A}_K(F)[Z] \cdot w$ is the trace of the linear transformation $u \to \{\mathsf{H}_K(F)\,[Z u]\}^T w$ for all $w \in \mathscr{V}$ and all $Z \in \mathscr{L}(\mathscr{V}, \mathscr{L})$. Of course, A_K is determined by the response function $\mathfrak{h}_K$. If we substitute (15.26) and (15.23) into (15.21), we obtain[15]

$$\mathsf{A}_K(F)\,[\nabla_K F] + \mathfrak{h}_K(F)\, s + b_K = 0, \qquad F = (\nabla\gamma)^{-1} K, \tag{15.27}$$

which is *the differential equation for the determination of configurations* γ *possible in a materially uniform elastic body.* If the body is homogeneous, we can choose

[15] This result, in terms of coordinates, was announced two years ago in reference [2] as equation (44.7).

$K = \nabla \kappa$. Then s vanishes and (15.27) reduces to the classical differential equation of finite elasticity.

Finally, we give another application of Proposition 7. Using the fact that $\mathscr{V}$ is three-dimensional (which was irrelevant up to now), we choose an orientation in $\mathscr{V}$ and consider the associated cross product $\times$. The curl of a vector field h and a tensor field T on $\mathscr{B}$ relative to some configuration γ are defined by

$$\begin{aligned} u \cdot (\nabla_\gamma h) v - v \cdot (\nabla_\gamma h) u &= (\mathrm{curl}_\gamma h) \cdot (u \times v), \\ (\mathrm{curl}_\gamma T)^T u &= \mathrm{curl}_\gamma (T u), \end{aligned} \tag{15.28}$$

where $u, v \in \mathscr{V}$. It follows from (15.28) that

$$(\mathrm{curl}_\gamma T^T)(u \times v) = ((\nabla_\gamma T) v) u - ((\nabla_\gamma T) u) v. \tag{15.29}$$

Also, we have the rule

$$\mathrm{div}_\gamma \mathrm{curl}_\gamma T = 0. \tag{15.30}$$

The inhomogeneity S reltive to a uniform reference K has the skew symmetry $(S u) v = -(S v) u$, $u, v \in \mathscr{V}$. Therefore, S determines and is determined by a tensor field A on $\mathscr{B}$ such that[16]

$$(S u) v = A(u \times v), \qquad u, v \in \mathscr{V}. \tag{15.31}$$

If we substitute (15.31) into (11.5) and observe the rule (15.29), we see that

$$A(u \times v) = [\mathrm{curl}_\gamma (F^{T^{-1}})] (F v \times F u). \tag{15.32}$$

Hence, since $F^T(F v \times F u) = (\det F)(v \times u)$ and since $u \times v$ is arbitrary, (15.32) yields

$$-\mathrm{curl}_\gamma (F^{T^{-1}}) = \frac{1}{J} A F^T, \qquad J = \det F. \tag{15.33}$$

Thus, the tensor field $-\mathrm{curl}_\gamma (F^{T^{-1}})$ is obtained from A by the rules (15.3), (15.4), except that the absolute value signs are omitted in the definition of J, which does not affect the validity of Proposition 7. Since by rule (15.30) we have

$$[\mathrm{div}_K \mathrm{curl}_\kappa (F^{T^{-1}})]_X = [\mathrm{div}_\kappa \mathrm{curl}_\kappa (F^{T^{-1}})]_X = 0,$$

(15.6) yields

$$\mathrm{div}_K A + A s = 0, \tag{15.34}$$

where s is determined by A through

$$-(A - A^T) u = s \times u, \qquad u \in \mathscr{V}. \tag{15.35}$$

Thus, (15.34) is a differential identity for A and K. One can show that it is equivalent to the Bianchi identity (11.11).

Acknowledgement. The research leading to this paper was supported by the Office of Naval Research under contract NONR-760 (30).

[16] The field A here corresponds to what was denoted by A^T in [2], Sect. 34.

References

[1] NOLL, W., A mathematical theory of the mechanical behavior of continuous media. Arch. Rational Mech. Anal. **2**, 197—226 (1958).

[2] TRUESDELL, C., & W. NOLL, The Non-linear Field Theories of Mechanics. Encyclopedia of Physics, Vol. III/3. Berlin-Heidelberg-New York: Springer 1965.

[3] LANG, S., Introduction to Differentiable Manifolds. New York-London-Sydney: Interscience 1962.

[4] KRÖNER, E.: Allgemeine Kontinuumstheorie der Versetzungen und Eigenspannungen. Arch. Rational Mech. Anal. **4**, 273—334 (1960).

[5] SEEGER, A.: Recent advances in the theory of defects in crystals. Physica status solidi **1**, 669—698 (1961).

[6] NOLL, W., Space-time structures in classical mechanics. Delaware Seminar in the Foundations of Physics, pp. 28—34. New York: Springer 1967.

[7] NOLL, W., Euclidean geometry and Minkowskian chronometry. Amer. Math. Monthly **71**, 129—144 (1964).

[8] WANG, C.-C., On the geometric structures of simple bodies, a mathematical foundation for the theory of continuous distributions of dislocations. Arch. Rational Mech. Anal., following in this issue.

[9] COLEMAN, B. D., & W. NOLL, Material symmetry and thermostatic inequalities in finite elastic deformations. Arch. Rational Mech. Anal. **15**, 87—111 (1964).

[10] NOLL, W., The foundations of classical mechanics in the light of recent advances in continuum mechanics. The Axiomatic Method, with special reference to geometry and physics, pp. 266—281. Amsterdam: North Holland 1959.

Carnegie-Mellon University
Pittsburgh, Pennsylvania

(Received July 22, 1967)

A New Mathematical Theory of Simple Materials

Walter Noll

Dedicated to Clifford Truesdell, friend and mentor of the author

Contents

1. Introduction

The original theory of *simple materials* was formulated by me in 1958 in reference [N2]. I proposed this theory in an attempt to unify and clarify the confusing variety of theories of mechanical behavior of materials that had been proposed in the literature up to that time. One can perhaps say that the attempt was moderately successful, considering that this first theory of simple materials has served as a foundation for a large part of the research in continuum physics since 1958, and considering that the concepts and even the notations of [N2] are now used routinely and without reference in textbooks on continuum mechanics.

Unfortunately, the success of the first theory of simple materials obscured the fact that this theory has at least three severe defects:

(i) In the original theory it is assumed, roughly, that the present stress is determined by the infinite past history of deformation. If one assumes that the memory of the material is limited, only a finite part of the history will have an appreciable effect on the present stress. There are materials, however, for which such an assumption is not appropriate. For such materials the original theory brings the infinite past into play. This is philosophically unacceptable, because real material bodies (perhaps even the entire universe) have not existed for an infinite time. Even if one imagines the existence of a material body for the infinite past, only a finite portion of that past is knowable.

(ii) The original theory has failed to give an adequate conceptual framework for the mathematical description of such phenomena as *plasticity*, *yield*, and *hysteresis*.

(iii) The definition of a *material of the rate type* in the framework of the first theory (*cf.* [N2], Sect. 24 or [TN], Sect. 36) was artificial and fraught with difficulties. BERNSTEIN ([B1] and [B2]) gave another and much more natural definition, but his definition did not fit into the framework of the first theory.

The new theory of simple materials presented here is free from the three defects just described. Only deformation processes of finite duration, not infinite histories, occur in the description of the response of a *material element.* I believe the new theory is suited very well as a basis for mathematical theories describing phenomena of plasticity, although such theories have not yet been developed on this basis. Finally, BERNSTEIN's definition of a material of the rate type fits naturally into the new framework.

The idea of a *state of a physical system* is used, informally or formally, in almost all branches of physics. It is used formally as a primitive notion, for example, in most axiomatic descriptions of quantum mechanics. Informally, *i.e.* without a precise mathematical meaning, the term "state" has also been used in much work on continuum mechanics. In the axiomatic definition of a material element presented here, the notion of a *state of the element* is used formally as a primitive notion with a precise mathematical meaning.

An important difference between the study presented here and all previous work in continuum physics (as far as I know) is that in the present study no *a priori* existence of *frames of reference* or *physical space* is assumed and no *a priori coordinate systems* are used for the description of the response of material bodies. In the past literature, an inordinate effort has been expended to derive the consequences of the *principle of frame-indifference* for whatever constitutive assumptions were considered. This was the case even before the principle of frame-indifference was formulated with precision and generality in [N3]; it was the case, for example, in the pioneering work of CAUCHY and STOKES. The advantage of not using frames of reference is, of course, that the principle of frame-indifference is vacuously satisfied and need not be considered explicitly at all. An increasing number of researchers recognize (mainly since the publication of my thesis [N1] in 1955) that *a priori* coordinate systems and hence considerations of coordinate invariance are redundant and an impediment to understanding.

The intrinsic, frame-free description of material elements has a second advantage: no *a priori* commitment need be made about the space-time structure used when the interaction of an element with an environment is considered. The description given here is compatible with classical, neo-classical (*cf.* [N6]), as well as relativistic spacetime. The imbedding of material bodies in relativistic space-time will be investigated in a future paper.

We now give a brief survey of the contents of this paper.

In Section 2 we assemble the mathematical notation and terminology used later. The basic vocabulary that mathematicians have adopted in the last few years is included in order to aid those readers whose primary interest is in mechanics and who may not yet be familiar with this vocabulary. It is assumed that the reader has a good grasp of the concepts of elementary abstract linear algebra and is comfortable with vector spaces that have no inner product.

In Section 3 we introduce, somewhat informally, the notions of *intrinsic configuration* of a continuous body and of an infinitesimal body element. This intrinsic notion is different from the extrinsic concept of "configuration" I have used in past work. I now prefer the term "placement in a frame of reference" for the extrinsic concept.

Section 4 contains the formal definition of a *body element* and a rigorous discussion of (intrinsic) *deformation processes* for a body element.

In Section 5 we describe the connection between (intrinsic) deformation processes and *motions in a frame space* introduced *a posteriori*.

In Section 6 we explain the concept of *intrinsic stress* and we show how work can be computed in terms of intrinsic stresses and deformations.

Section 7 contains the formal definition for the basic concept of the present paper, that of a *material element*. The structure of a material element is governed by six axioms, of which only the first three are stated in Section 7.

In Section 8 we discuss *internal constraints* and demonstrate that no formal modifications are necessary when such constraints are present.

Section 9 deals with *material isomorphisms*, which are defined in accordance with the notion of isomorphism for arbitrary mathematical structures. A *material* is defined as an equivalence class of material elements, the equivalence being material isomorphy. Included in Section 9 are also a few remarks intended to clarify the nature of the theory of materially uniform but inhomogeneous material bodies.

In Section 10, the *symmetry group* of a material element is defined as its group of automorphisms. We also give a precise definition for the *symmetry group of a state*. States whose symmetry group is an appropriate orthogonal group are called *isotropic states*, and *isotropic material elements* are defined as those that possess isotropic states.

In Section 11 we define natural uniform structures on sections of the state space and a natural topology on the state space. A *completeness axiom* postulates that the sections be complete uniform spaces. Readers who are not familiar with the elements of abstract general topology can skip this section and just assume that a notion of convergence for states is somehow defined.

In Section 12 we state and discuss the *relaxation axiom*, which postulates that if a material element is frozen in a definite configuration, its state will approach a *relaxed state*.

In Section 13 we say what it means that one state is *accessible* from another and we state the *accessibility axiom*, which postulates that all states must be accessible from a certain relaxed starting state.

In Section 14 an *elastic* element is defined by the condition that the states be in one-to-one correspondence with the configurations. A *semi-elastic* element is defined by the condition that the *relaxed* states be in one-to-one correspondence with the configurations. Solid and fluid materials are defined as semi-elastic materials with special symmetry. Notions of *undistorted state* and *undistorted configuration* are defined so as to apply to all semi-elastic materials (not just to solids and isotropic materials, as in the original theory).

In Section 15 we show that, for semi-elastic elements, a dense set of states can be described by deformation histories of infinite duration.

In Section 16 we establish a precise position for the original theory of simple materials. It turns out that simple materials in the original sense are equivalent to semi-elastic materials in the new sense. The proof of this equivalence is by no means simple. The history functional of a semi-elastic element, which relates infinite deformation histories to stresses, must satisfy two convergence conditions. These conditions express the fact that semi-elastic materials have limited memory, as is to be expected from the discussion about the first defect of the original theory.

The considerations of Section 17 are purely kinematical. A *monotonous process* is defined by the condition that all of its segments of equal duration are congruent. A non-trivial theorem, whose proof will be given elsewhere [N8], shows that the monotonous processes are related to what have been called "substantially stagnant motions" by Coleman [C] and "motions with constant stretch history" by myself [N4].

In Section 18 we explain why the dynamical analysis of monotonous processes is often not much more complicated than a statical analysis. The reason is that, if the material has a sufficiently large symmetry group and if certain other conditions are satisfied, such processes leave the reduced state unchanged. The states occurring in such processes are called *states of monotonous flow*.

Materials (as distinct from material elements) can often be characterized by numerical functionals, functions, or constants. In Section 19 such a characterization is carried out in detail for incompressible fluids.

In Section 20 we define and discuss materials of the rate type. There is no evidence, conceptual or experimental, that any real material behavior fits the theory of materials of the rate type, but this theory gives mathematically concrete examples of material elements that are not semi-elastic and hence do not fit the original theory of simple materials. (The definition of a material of the rate type given in [N2] and [TN] applies, in the language of the present paper, only to materials that are both semi-elastic and of the rate type.)

In Section 21 we discuss possible generalizations, further development, and applications of the theory presented here.

2. Mathematical Preliminaries

(a) Basic Notations. The set of real numbers is denoted by $\mathbb{R}$ and the set of non-negative reals by $\mathbb{R}^+$. If $t_1, t_2 \in \mathbb{R}$ and $t_1 \leqq t_2$, the closed interval from t_1 to t_2 is defined by $[t_1, t_2] = \{t \mid t_1 \leqq t \leqq t_2\}$. In particular, $[t, t] = \{t\}$ is a degenerate interval that reduces to a singleton. The set of natural numbers 1, 2, 3 ... is denoted by $\mathbb{N}$.

A mapping $f: \mathscr{D} \to \mathscr{C}$ assigns to each element x in the set $\mathscr{D}$ a *value* $f(x)$ in the set $\mathscr{C}$. The set $\mathscr{D}$ is called the *domain* of f and $\mathscr{C}$ is called the *codomain* of f. One expresses this situation by saying that f *maps* $\mathscr{D}$ *into* $\mathscr{C}$. The set

$$\text{Range} f = \{f(x) \mid x \in \mathscr{D}\}$$

is called the *range* (or "image") of f. The range is a subset of the codomain. If the range coincides with the codomain, one says that the mapping is *surjective* (or "onto") and calls it a *surjection.* A mapping f is said to be *injective* (or "one-to-one") and is called an *injection* if $f(x_1) = f(x_2)$ is possible only when $x_1 = x_2$.

The *composition* $g \circ f: \mathscr{D} \to \mathscr{B}$ of two mappings $f: \mathscr{D} \to \mathscr{C}$ and $g: \mathscr{C} \to \mathscr{B}$ is defined by $(g \circ f)(x) = g(f(x))$ for all $x \in \mathscr{D}$. The composition $g \circ f$ is meaningful only if the codomain of f coincides with (or at least is contained in) the domain of g. The *identity* mapping $1_{\mathscr{D}}: \mathscr{D} \to \mathscr{D}$ of a set $\mathscr{D}$ is defined by $1_{\mathscr{D}}(x) = x$ for all $x \in \mathscr{D}$. A mapping $f: \mathscr{D} \to \mathscr{C}$ is said to be *invertible* (or "bijective" or "one-to-one and onto") if it has an *inverse* $f^{-1}: \mathscr{C} \to \mathscr{D}$ with the properties $f^{-1} \circ f = 1_{\mathscr{D}}$ and $f \circ f^{-1} = 1_{\mathscr{C}}$. A mapping is invertible if and only if it is both injective and surjective. If this is the case, the mapping is also called a *bijection* (or "one-to-one correspondence").

If the values of a mapping f are given by a formula and if no special symbol for the mapping itself is available, one uses the notation $x \mapsto f(x)$ to denote the mapping. (Note the distinction between $\mapsto$ and $\to$.) For example, $t \mapsto t^2$ denotes the "squaring" mapping from $\mathbb{R}$ into $\mathbb{R}$ (or into $\mathbb{R}^+$). Another possible notation for $x \mapsto f(x)$ is $f(\cdot)$. For example, if the domain of $g: \mathscr{D}_1 \times \mathscr{D}_2 \to \mathscr{C}$ is a Cartesian product $\mathscr{D}_1 \times \mathscr{D}_2 = \{(x_1, x_2) \mid x_1 \in \mathscr{D}_1, x_2 \in \mathscr{D}_2\}$, one can obtain, for every $x_1 \in \mathscr{D}_1$, a mapping $g(x_1, \cdot): \mathscr{D}_2 \to \mathscr{C}$.

It often happens that the codomains $\mathscr{C}_1$ and $\mathscr{C}_2$ of two mappings $f_1: \mathscr{D} \to \mathscr{C}_1$ and $f_2: \mathscr{D} \to \mathscr{C}_2$ are such that there is an algebraic operation which assigns to each $y_1 \in \mathscr{C}_1$ and each $y_2 \in \mathscr{C}_2$ a $y_1 y_2$ in a set $\mathscr{C}_3$. One then defines $f_1 f_2: \mathscr{D} \to \mathscr{C}_3$ by value-wise application of the operation, *i.e.* by $(f_1 f_2)(x) = f_1(x) f_2(x)$ for all $x \in \mathscr{D}$. For example, if $f: \mathscr{D} \to \mathbb{R}$ and $g: \mathscr{D} \to \mathbb{R}$ then $f+g: \mathscr{D} \to \mathbb{R}$ and $fg: \mathscr{D} \to \mathbb{R}$ are defined by value-wise addition and multiplication, *i.e.* by $(f+g)(x) = f(x) + g(x)$ and $(fg)(x) = f(x) g(x)$ for all $x \in \mathscr{D}$.

The terms "function", "functional", and "transformation" are some of many synonyms for "mapping".

(b) Vector Spaces. We deal only with finite-dimensional real vector spaces. If $\mathscr{T}_1$ and $\mathscr{T}_2$ are such spaces, we denote by

$$\text{Lin}(\mathscr{T}_1, \mathscr{T}_2) = \{A: \mathscr{T}_1 \to \mathscr{T}_2 \mid A \text{ linear}\}$$

the set of all linear mappings from $\mathscr{T}_1$ into $\mathscr{T}_2$. The composition of two linear mappings A_1 and A_2, meaningful if the domain of A_2 coincides with the codomain of A_1, is again linear and is denoted by $A_2 A_1$. (The symbol $\circ$ is omitted in this case.) $\mathrm{Lin}(\mathscr{T}_1, \mathscr{T}_2)$ is again a vector space, its structure being defined by taking value-wise sums and scalar multiples. The notation

$$\mathrm{Invlin}(\mathscr{T}_1, \mathscr{T}_2) = \{A \in \mathrm{Lin}(\mathscr{T}_1, \mathscr{T}_2) \mid A^{-1} \text{ exists}\}$$

is used for the set of invertible linear mappings from $\mathscr{T}_1$ into $\mathscr{T}_2$. $\mathrm{Invlin}(\mathscr{T}_1, \mathscr{T}_2)$ is non-empty only if $\dim \mathscr{T}_1 = \dim \mathscr{T}_2$. We use the abbreviations

$$\mathrm{Lin}(\mathscr{T}) = \mathrm{Lin}(\mathscr{T}, \mathscr{T}), \qquad \mathrm{Invlin}(\mathscr{T}) = \mathrm{Invlin}(\mathscr{T}, \mathscr{T}).$$

$\mathrm{Lin}(\mathscr{T})$ is not only a vector space, but an algebra with multiplication defined by composition. $\mathrm{Invlin}(\mathscr{T})$ is not a vector space, but it is a group under composition, and it is often called the *general linear group* of $\mathscr{T}$. It is the group of automorphisms of the vector space $\mathscr{T}$.

The *dual* $\mathscr{T}^*$ of a vector space $\mathscr{T}$ is defined by

$$\mathscr{T}^* = \mathrm{Lin}(\mathscr{T}, \mathbb{R}).$$

It is a vector space of the same dimension as $\mathscr{T}$. The elements of $\mathscr{T}^*$ are called *linear forms* or *covectors*. The value of $u^* \in \mathscr{T}^*$ at $v \in \mathscr{T}$ is denoted by $\langle u^*, v\rangle \in \mathbb{R}$. The dual $\mathscr{T}^{**}$ of the dual $\mathscr{T}^*$ is identified with the original space $\mathscr{T}$ in such a way that the value at $u^* \in \mathscr{T}^*$ of the element of $\mathscr{T}^{**} = \mathrm{Lin}(\mathscr{T}^*, \mathbb{R})$ corresponding to $v \in \mathscr{T}$ is $\langle u^*, v\rangle$. If $v \in \mathscr{T}$ and $u^* \in \mathscr{T}^*$, we define $v \otimes u^* \in \mathrm{Lin}(\mathscr{T})$ by

$$(v \otimes u^*)(w) = \langle u^*, w\rangle v \qquad \text{for all } w \in \mathscr{T}.$$

The dual of $\mathrm{Lin}(\mathscr{T})$ contains a distinguished element $\mathrm{tr} \in (\mathrm{Lin}(\mathscr{T}))^*$, called the *trace*. It is characterized by the condition that

$$\mathrm{tr}(v \otimes u^*) = \langle u^*, v\rangle \qquad \text{for all } v \in \mathscr{T} \text{ and all } u^* \in \mathscr{T}^*.$$

Another distinguished function on $\mathrm{Lin}(\mathscr{T})$ is the *determinant* $\det\colon \mathrm{Lin}(\mathscr{T}) \to \mathbb{R}$. It is not linear but preserves products, so that $\det(A_1 A_2) = (\det A_1)(\det A_2)$. Also, one has $\det(1_{\mathscr{T}}) = 1$. We use the notations

$$\mathrm{Lin}_0(\mathscr{T}) = \{A \in \mathrm{Lin}(\mathscr{T}) \mid \mathrm{tr}\, A = 0\}$$

and

$$\mathrm{Unim}(\mathscr{T}) = \{A \in \mathrm{Invlin}(\mathscr{T}) \mid (\det A)^2 = 1\}.$$

$\mathrm{Lin}_0(\mathscr{T})$ is a subspace of $\mathrm{Lin}(\mathscr{T})$ and $\mathrm{Unim}(\mathscr{T})$ is a subgroup of $\mathrm{Invlin}(\mathscr{T})$. The group $\mathrm{Unim}(\mathscr{T})$ is called the *unimodular group* of $\mathscr{T}$.

To every $A \in \mathrm{Lin}(\mathscr{T}_1, \mathscr{T}_2)$ one can associate a unique *adjoint* $A^* \in \mathrm{Lin}(\mathscr{T}_2^*, \mathscr{T}_1^*)$ characterized by the condition that $\langle A^* w^*, v\rangle = \langle w^*, Av\rangle$ for all $v \in \mathscr{T}_1$ and all $w^* \in \mathscr{T}_2^*$. If $A \in \mathrm{Lin}(\mathscr{T})$, then $A^* \in \mathrm{Lin}(\mathscr{T}^*)$ and we have $\mathrm{tr}\, A = \mathrm{tr}\, A^*$, $\det A^* = \det A$.

(c) Bilinear Forms, Orthogonal Groups, Inner Product Spaces. Let $\mathscr{T}$ be a finite-dimensional real vector space. The space $\mathrm{Lin}(\mathscr{T}, \mathscr{T}^*)$ can be identified with the space of all *bilinear forms* on $\mathscr{T}$, because every such bilinear form is a mapping from $\mathscr{T} \times \mathscr{T}$ into $\mathbb{R}$ of the type

$$(v, w) \mapsto \langle Gv, w\rangle, \quad \text{with} \quad G \in \mathrm{Lin}(\mathscr{T}, \mathscr{T}^*).$$

The subspace

$$\mathrm{Sym}(\mathscr{T},\mathscr{T}^*)=\{G\in\mathrm{Lin}(\mathscr{T},\mathscr{T}^*)\,|\,\langle Gv,w\rangle=\langle Gw,v\rangle \text{ for all } v,w\in\mathscr{T}\}$$

of $\mathrm{Lin}(\mathscr{T},\mathscr{T}^*)$ is then identified with the space of *symmetric bilinear forms* on $\mathscr{T}$. The subset

$$\mathrm{Sym}^+(\mathscr{T},\mathscr{T}^*)=\{G\in\mathrm{Sym}(\mathscr{T},\mathscr{T}^*)\,|\,\langle Gv,v\rangle>0 \text{ if } v\in\mathscr{T}, v\neq 0\}$$

of $\mathrm{Sym}(\mathscr{T},\mathscr{T}^*)$ is identified with the set of all *positive definite symmetric bilinear forms* on $\mathscr{T}$. We note that

$$\mathrm{Sym}^+(\mathscr{T},\mathscr{T}^*)\subset\mathrm{Invlin}(\mathscr{T},\mathscr{T}^*),$$

i.e., G^{-1} exists for every $G\in\mathrm{Sym}^+(\mathscr{T},\mathscr{T}^*)$. The following fact will be needed: For every $G_0\in\mathrm{Sym}^+(\mathscr{T},\mathscr{T}^*)$ we have

$$\mathrm{Sym}^+(\mathscr{T},\mathscr{T}^*)=\{A^*G_0A\,|\,A\in\mathrm{Invlin}\,\mathscr{T}\}.$$

For any $G\in\mathrm{Sym}^+(\mathscr{T},\mathscr{T}^*)$ we put

$$\mathrm{Orth}(G)=\{A\in\mathrm{Lin}(\mathscr{T})\,|\,A^*GA=G\}.$$

For every $G\in\mathrm{Sym}^+(\mathscr{T},\mathscr{T}^*)$, the set $\mathrm{Orth}(G)$ is a proper subgroup of $\mathrm{Unim}(\mathscr{T})$. We call $\mathrm{Orth}(G)$ the *orthogonal group of* G. The following facts concerning orthogonal groups will be needed:

(i) *For all* $G_1, G_2\in\mathrm{Sym}^+(\mathscr{T},\mathscr{T}^*)$ *we have*

$$\mathrm{Orth}(G_1)\subset\mathrm{Orth}(G_2)\Leftrightarrow\mathrm{Orth}(G_1)=\mathrm{Orth}(G_2)$$
$$\Leftrightarrow G_2=c\,G_1 \quad \textit{for some } c>0.$$

(ii) *The groups* $\mathrm{Orth}(G)$ *are maximal subgroups of* $\mathrm{Unim}(\mathscr{T})$, *i.e., if* $\mathscr{g}$ *is a proper subgroup of* $\mathrm{Unim}(\mathscr{T})$ *such that* $\mathscr{g}\supset\mathrm{Orth}(G)$, *then* $\mathscr{g}=\mathrm{Orth}(G)$.[1]

An *inner product space* $\mathscr{V}$ is a vector space equipped with an additional structure by singling out a particular member $I\in\mathrm{Sym}^+(\mathscr{V},\mathscr{V}^*)$, which is called the *inner product of* $\mathscr{V}$. One uses the notation

$$u\cdot v=\langle Iu,v\rangle \quad \text{for } y, v\in\mathscr{V}$$

and calls $u\cdot v$ the *inner product of* u *and* v. Also, since $I\colon\mathscr{V}\to\mathscr{V}^*$ is a bijection, it is used to identify $\mathscr{V}$ with $\mathscr{V}^*$. Thus, when $u\in\mathscr{V}$ is regarded as a linear form on $\mathscr{V}$, its value at $v\in\mathscr{V}$ is $u\cdot v$. $\mathrm{Lin}(\mathscr{V}^*)$ is identified with $\mathrm{Lin}(\mathscr{V})$ and hence the adjoint of $A\in\mathrm{Lin}(\mathscr{V})$ is identified with another element of $\mathrm{Lin}(\mathscr{V})$. This element is denoted by A^T instead of A^* and called the *transpose* of A. The space $\mathrm{Sym}(\mathscr{V},\mathscr{V}^*)$ is identified with the subspace

$$\mathrm{Sym}(\mathscr{V})=\{A\in\mathrm{Lin}(\mathscr{V})\,|\,A=A^T\}$$

of $\mathrm{Lin}(\mathscr{V})$. The orthogonal group of the inner product I is also called the *orthogonal group of the inner product space* $\mathscr{V}$ and denoted by

$$\mathrm{Orth}(\mathscr{V})=\{Q\in\mathrm{Lin}(\mathscr{V})\,|\,Q^TQ=1_{\mathscr{V}}\}.$$

$\mathrm{Orth}(\mathscr{V})$ is the group of automorphisms of the inner product space $\mathscr{V}$.

[1] This is a rather non-trivial theorem. The simplest proof I know is the one I gave in [N5].

Let $\mathscr{V}$ be an inner product space and $\mathscr{T}$ a vector space. If $K \in \mathrm{Lin}(\mathscr{T}, \mathscr{V})$, then $K^* \subset \mathrm{Lin}(\mathscr{V}^*, \mathscr{T}^*) = \mathrm{Lin}(\mathscr{V}, \mathscr{T}^*)$ and hence $K^* K \in \mathrm{Lin}(\mathscr{T}, \mathscr{T}^*)$. Actually, we always have $K^* K \in \mathrm{Sym}(\mathscr{T}, \mathscr{T}^*)$ and

$$\mathrm{Sym}^+(\mathscr{T}, \mathscr{T}^*) = \{K^* K \mid K \in \mathrm{Invlin}(\mathscr{T}, \mathscr{V})\}$$

if $\dim \mathscr{T} = \dim \mathscr{V}$.

A special inner product space is the numerical space $\mathbb{R}^n$ consisting of n-tuples of real numbers. Actually, $\mathbb{R}^n$ is equipped with even more structure than an inner product space, because $\mathbb{R}^n$ possesses a distinguished orthonormal basis, while for inner product spaces without additional structure any orthonormal basis is as good as any other.

3. Continuous Bodies, Intrinsic Configurations

A *continuous body* $\mathscr{B}$ is usually defined[1] as a set endowed with a structure described by a class Φ of mappings $\kappa: \mathscr{B} \to \mathscr{E}_\kappa$ whose codomains $\mathscr{E}_\kappa$ are Euclidean spaces. The elements $X, Y, \ldots$ of $\mathscr{B}$ are called *material points*. The spaces $\mathscr{E}_\kappa$ have an intuitive interpretation as *frames of reference*, and a mapping $\kappa \in \Phi$ assigns to each material point $X \in \mathscr{B}$ a *place* $\kappa(X)$ in the frame of reference $\mathscr{E}_\kappa$. We will refer to the mappings in Φ as the possible **placements**[2] *of the body* $\mathscr{B}$.

The class Φ of placements of $\mathscr{B}$ is subject to a number of axioms which we will not repeat here. For our purposes it is enough to know that the axioms are such as to determine, on $\mathscr{B}$, a unique structure of a differentiable manifold. Therefore one can associate with each material point $X \in \mathscr{B}$ a tangent space $\mathscr{T}_X$, which is a finite-dimensional real vector space (see [N7], Sect. 4). Intuitively, the members of $\mathscr{T}_X$ describe the material points in an infinitesimal neighborhood of X. Therefore, we call $\mathscr{T}_X$ the (infinitesimal) *body element* at X of the (global) body $\mathscr{B}$. If $\kappa \in \Phi$ is a placement of $\mathscr{B}$, its gradient $\nabla\kappa(X): \mathscr{T}_X \to \mathscr{V}_\kappa$ is an invertible linear mapping from $\mathscr{T}_X$ onto the translation space $\mathscr{V}_\kappa$ of $\mathscr{E}_\kappa$ (see, *e.g.*, [N7], Sects. 3 and 4). Note that $\mathscr{V}_\kappa$ is an inner product space, while $\mathscr{T}_X$ is a vector space without inner product.

We thus see that localizations of placements of a body to one of its body elements $\mathscr{T}_X$ are described by invertible linear mappings from $\mathscr{T}_X$ onto inner product spaces. We therefore call such mappings *placements of the body element* $\mathscr{T}_X$.

We observe that one can associate with each placement $\kappa: \mathscr{B} \to \mathscr{E}_\kappa$ a distance function $\delta_\kappa: \mathscr{B} \times \mathscr{B} \to \mathbb{R}^+$ defined by

$$\delta_\kappa(X, Y) = d_\kappa(\kappa(X), \kappa(Y)), \tag{3.1}$$

where d_κ denotes the Euclidean distance in the frame space $\mathscr{E}_\kappa$. Two placements $\kappa_1, \kappa_2 \in \Phi$ determine the same distance function if and only if $\kappa_2 \circ \kappa_1^{-1}$ is an isometry between a subset of $\mathscr{E}_{\kappa_1}$ and a subset of $\mathscr{E}_{\kappa_2}$.

Let

$$\Delta = \{\delta_\kappa \mid \kappa \in \Phi\} \tag{3.2}$$

---

[1] Although the concept has been used in the literature on continuum mechanics since time immemorial, the first precise mathematical definition was given in [N3] (see also [N7], Sect. 3).

[2] In earlier work I used the term "configuration". I now believe "placement" is more apt. Also, I would like to reserve "configuration" for the intrinsic concept to be introduced below.

be the class of all distance functions of the form (3.1). Instead of using Φ to describe the structure on $\mathscr{B}$ and then defining Δ by (3.2) in terms of Φ, one can also use a suitable class Δ of mappings $\delta: \mathscr{B} \times \mathscr{B} \to \mathbb{R}^+$ to describe the structure of a continuous body on $\mathscr{B}$ and then define Φ to be the class of mappings $\kappa: \mathscr{B} \to \mathscr{E}_\kappa$ for which $(X, Y) \mapsto d_\kappa(\kappa(X), \kappa(Y))$ belongs to Δ. The advantage of describing the body structure in terms of the class Δ is that such a description is completely intrinsic. The elements of Δ do not involve extrinsic frames of reference, as do the elements of Φ.

We shall refer to the distance functions in Δ as the possible *(intrinsic)* **configurations** *of the body* $\mathscr{B}$. A configuration $\delta \in \Delta$ assigns to any two material points $X, Y \in \mathscr{B}$ a number $\delta(X, Y) \in \mathbb{R}^+$, which we call the *distance between X and Y in the configuration* δ.

A configuration $\delta: \mathscr{B} \times \mathscr{B} \to \mathbb{R}^+$ induces, by localization, on each of the tangent spaces $\mathscr{T}_X$ a positive definite symmetric bilinear form $\delta_X: \mathscr{T}_X \times \mathscr{T}_X \to \mathbb{R}^+$. In precise terms, δ_X is the mixed second gradient of δ with respect to its two point variables, evaluated at (X, X). It is natural, therefore, to call such bilinear forms *(intrinsic) configurations of the body element* $\mathscr{T}_X$. Recalling that $\mathrm{Sym}^+(\mathscr{T}_X, \mathscr{T}_X^*)$ can be identified with the set of all positive definite symmetric bilinear forms on $\mathscr{T}_X$ (see Section 2), we see that the configurations of $\mathscr{T}_X$ can be regarded as elements of $\mathrm{Sym}^+(\mathscr{T}_X, \mathscr{T}_X^*)$.

A *motion* of a body $\mathscr{B}$ is a one-parameter family of placements of $\mathscr{B}$ (see [N2]). By a *deformation process* for a body $\mathscr{B}$ we mean a one-parameter family of configurations of $\mathscr{B}$. Motions and deformation processes of body elements are defined in the same way.

From now on we shall deal almost exclusively with body elements rather than with (global) bodies. We think of a body element as a physical system in itself, regarding the rest of the continuous body to which the element belongs merely as part of the environment.

Precise formal definitions, motivated by the considerations of this section, for the concepts of a body element and of configurations, deformation processes, placements, and motions of a body element will be given in the next two sections.

4. Deformation Processes, Body Elements

Let $\mathscr{G}$ be any set. A function P of the type

$$P: [0, d_P] \to{}' \mathscr{G}, \qquad d_P \in \mathbb{R}^+ \tag{4.1}$$

will be called a **process** *with values in* $\mathscr{G}$. The number d_P will be called the *duration* of the process and the values

$$P^i = P(0), \qquad P^f = P(d_P) \tag{4.2}$$

will be referred to as the *initial and final values* of the process.

If $G \in \mathscr{G}$ and $t \in \mathbb{R}^+$, we define a process $G_{(t)}: [0, t] \to \mathscr{G}$ by

$$G_{(t)}(r) = G \quad \text{for } r \in [0, t] \tag{4.3}$$

and call it the **freeze** *of duration t at G*.

Let P be a process and let $t_1, t_2 \in [0, d_P]$, with $t_1 \leqq t_2$. We then define a new process $P_{[t_1, t_2]}$, of duration $t_2 - t_1$, by

$$P_{[t_1, t_2]}(t) = P(t + t_1) \quad \text{for } t \in [0, t_2 - t_1]. \tag{4.4}$$

Such a process will be called a **segment** of the given process P.

Let P_1 and P_2 be processes such that $P_1^f = P_2^i$. We then define a new process $P_1 * P_2$, of duration $d_{P_1} + d_{P_2}$, by

$$(P_1 * P_2)(t) = \begin{cases} P_1(t) & \text{if } t \in [0, d_{P_1}] \\ P_2(t - d_{P_1}) & \text{if } t \in [d_{P_1}, d_{P_1} + d_{P_2}]. \end{cases} \tag{4.5}$$

We call $P_1 * P_2$ the **continuation** of P_1 with P_2. Continuation is an associative operation, *i.e.*, we have

$$(P_1 * P_2) * P_3 = P_1 * (P_2 * P_3) \tag{4.6}$$

if $P_1^f = P_2^i = (P_2 * P_3)^i$ and $P_2^f = (P_1 * P_2)^f = P_3^i$.

We note that processes of duration zero are not excluded. Of course, all such processes are of the form $G_{(0)}$, *i.e.* they are freezes of duration zero. It is clear that for any process P

$$P = P^i_{(0)} * P = P * P^f_{(0)}. \tag{4.7}$$

Definition 4.1. *A* **body element** *is a triple* $(\mathscr{T}, \mathscr{G}, \Pi)$, *where* $\mathscr{T}$ *is a finite-dimensional real vector space,* $\mathscr{G}$ *is a closed and connected subset of* $\operatorname{Sym}^+(\mathscr{T}, \mathscr{T}^*)$, *and* Π *is a class of processes with values in* $\mathscr{G}$ *which satisfies the following conditions:*

(P1) *Any freeze at any* $G \in \mathscr{G}$ *belongs to* Π.

(P2) *If* P *belongs to* Π, *so does every segment of* P.

(P3) Π *is closed under continuation, i.e., if* $P_1, P_2 \in \Pi$ *and* $P_1^f = P_2^i$, *then* $P_1 * P_2 \in \Pi$.

(P4) *Any two elements* $G_1, G_2 \in \mathscr{G}$ *can be connected by a process in* Π, *i.e., there is at least one* $P \in \Pi$ *such that* $G_1 = P^i$ *and* $G_2 = P^f$.

When we speak about a body element $\mathscr{T}$, we consider it understood that $\mathscr{T}$ is endowed with a structure defined by the prescription of $\mathscr{G}$ and Π. The elements of $\mathscr{G}$ will be called the **configurations** of $\mathscr{T}$ and the processes in Π the **deformation processes** for $\mathscr{T}$.

Examples of classes Π that satisfy (P1)–(P4) are: (a) the class of all processes with values in $\mathscr{G}$, (b) the class of all continuous processes with values in $\mathscr{G}$, (c) if $\mathscr{G}$ is arcwise connected, the class of all processes P that are continuous and have a piecewise continuous derivative $\dot{P}$ in the sense that $\dot{P}(t)$ exist for all but a finite number of $t \in [0, d_P]$ and $\dot{P}$ has left and right limits for all $t \in [0, d_P]$.

Remark. It cannot happen that the class Π contains only processes that are everywhere differentiable, except in the trivial case when $\mathscr{G}$ is a singleton. To see this, we note that (P4) implies the existence of a non-constant $P \in \Pi$. If P is everywhere differentiable, there must be a $t > 0$ such that $\dot{P}(t) \neq 0$. By (P1)–(P3), the process $P_{[0,t]} * P(t)_{(t')}$, where $t' > 0$, must also belong to Π. This process does not have a derivative at t.

5. Motions in a Frame-Space

Let $(\mathcal{T}, \mathcal{G}, \Pi)$ be a body element and let $\mathcal{V}$ be a real inner product space having the same dimension as $\mathcal{T}$. An element $K \in \mathrm{Invlin}(\mathcal{T}, \mathcal{V})$ will be called a **placement** of the element $\mathcal{T}$ in the *frame-space* $\mathcal{V}$ if $K^* K \in \mathcal{G}$. The set of all placements of $\mathcal{T}$ is denoted by

$$\mathcal{P}_{\mathcal{V}} = \{K \in \mathrm{Invlin}(\mathcal{T}, \mathcal{V}) \mid K^* K \in \mathcal{G}\}. \tag{5.1}$$

We call $K^* K$ the *configuration determined by the placement K*. It is easily seen that two placements $K_1, K_2 \in \mathcal{P}_{\mathcal{V}}$ determine the same configuration if and only if $K_2 K_1^{-1} \in \mathrm{Orth}(\mathcal{V})$ and that

$$\mathcal{G} = \{K^* K \mid K \in \mathcal{P}_{\mathcal{V}}\}. \tag{5.2}$$

A process M: $[0, d_M] \to \mathcal{P}_{\mathcal{V}}$ with values in $\mathcal{P}_{\mathcal{V}}$ will be called a **motion** *of $\mathcal{T}$ in the frame-space $\mathcal{V}$* if $M^* M \in \Pi$. We call $M^* M$ the *deformation process determined by the motion M.*

The notions of *duration, initial and final placements, freezes, segments, and continuation* apply also to motions since they apply to processes of any type. Two motions M_1 and M_2 determine the same deformation process if and only if they have the same duration and if $M_2 M_1^{-1}$ has values in $\mathrm{Orth}(\mathcal{V})$.

It is often useful to fix a particular *reference placement* $K_R \in \mathcal{P}_{\mathcal{V}}$ and to describe the possible motions of the body-element with reference to K_R. In this way one obtains all the standard tensor functions connected with a given motion M (see [TN], Chapter CI). For example,

$$F = M K_R^{-1} \colon [0, d_M] \to \mathrm{Lin}(\mathcal{V}) \tag{5.3}$$

defines the *displacement tensor*[1] of M. If $P = M^* M$ is the deformation process determined by M, then

$$C = K_R^{*-1} P K_R^{-1} = F^T F \colon [0, d_M] \to \mathrm{Lin}(\mathcal{V}) \tag{5.4}$$

defines the *right Cauchy-Green tensor.* If $t \in [0, d_M]$ and if we take $M(t)$, rather than K_R, as reference placement, we obtain the *relative displacement tensor*

$$F_{(t)} = M M(t)^{-1} \colon [0, d_M] \to \mathrm{Lin}(\mathcal{V}) \tag{5.5}$$

and the *relative right Cauchy-Green tensor*

$$C_{(t)} = M(t)^{*-1} G M(t)^{-1} = F_{(t)}^T F_{(t)} \colon [0, d_M] \to \mathrm{Lin}(\mathcal{V}). \tag{5.6}$$

If the deformation process $P = M^* M$ is of class C^k, we can define the kth *Rivlin-Ericksen tensor* by

$$A_k = M^{*-1} \overset{(k)}{P} M^{-1} \colon [0, d_M] \to \mathrm{Lin}(\mathcal{V}). \tag{5.7}$$

6. Intrinsic Stress, Work

It follows from the basic principles of continuum mechanics that, roughly speaking, the contact forces exerted on a body element $\mathcal{T}$ in a placement $K \in P_{\mathcal{V}}$

[1] In [TN] the term "deformation gradient" is used. For the present context, I prefer the term I used in [N1], p. 14.

can be described by a (Cauchy-) *stress tensor* $T \in \mathrm{Sym}(\mathscr{V})$ (see Section 16 of [TN]). We define a corresponding **intrinsic stress** $S \in \mathrm{Sym}(\mathscr{T}^*, \mathscr{T})$ by

$$S = K^{-1} T K^{*-1}. \tag{6.1}$$

It describes the contact forces acting on the element $\mathscr{T}$ intrinsically, without reference to any frame space.

A *mechanical process* for a body element $\mathscr{T}$ is defined to be a pair (P, S), where $P \in \Pi$ is a deformation process for $\mathscr{T}$ and S is a process with values in $\mathrm{Sym}(\mathscr{T}^*, \mathscr{T})$ and of the same duration as P, *i.e.* S: $[0, d_P] \to \mathrm{Sym}(\mathscr{T}^*, \mathscr{T})$. The values of S are to be interpreted as intrinsic stresses. If P is determined by a motion M of $\mathscr{T}$ in the frame space $\mathscr{V}$, then

$$T = MSM^*: [0, d_P] \to \mathrm{Sym}(\mathscr{V}) \tag{6.2}$$

gives the Cauchy-stress corresponding to the intrinsic stress S as a function on $[0, d_P]$.

If S is continuous and P continuously differentiable, then $S\dot{P}$: $[0, d_P] \to \mathrm{Lin}(\mathscr{T})$ is meaningful and one can easily show (see (5.7)) that

$$\tfrac{1}{2}\mathrm{tr}(TA_1) = \tfrac{1}{2}\mathrm{tr}(S\dot{P}): [0, d_P] \to \mathbb{R} \tag{6.3}$$

is the *stress power*. Its value at t measures the rate at which the contact forces exerted on the element do work, per unit volume in the configuration $P(t)$. The total work done on the element during the mechanical process (P, S), per unit volume in the initial configuration $P^i = P(0)$, is given by

$$w = \tfrac{1}{2}\int_0^{d_P} \mathrm{tr}(S\dot{P})\, j, \tag{6.4}$$

where j: $[0, t_P] \to \mathbb{R}^+$ is determined by

$$j^2 = \det((P^i)^{-1} P). \tag{6.5}$$

The meaning of $j(t)$ is that of the ratio of the volume of the configuration $P(t)$ to the initial volume.

7. Material Elements

We are now ready to introduce the main concept of the present paper, the concept of a material element. A material element will be a body element endowed with additional structure that is designed to describe physical material properties. For definiteness we confine ourselves to mechanical material properties, although it is not difficult to include non-mechanical physical phenomena (see Section 21). The principal feature of our definition is that it associates with the element a space of possible physical *states*. The underlying intuitive idea is that when a material element is given in a concrete physical situation, it is given in a definite state. The state determines everything about the element: its configuration, its stress, and, most importantly, the response of the element in every possible test. We imagine that a test consists of subjecting the element to a deformation process and measuring the stress at the end.

Definition 7.1. *A* **material element** *is a septuple* $(\mathscr{T}, \mathscr{G}, \Pi, \Sigma, \hat{G}, \hat{S}, \hat{\rho})$ *which satisfies Axioms I–VI stated below. The nature of the objects of the septuple is as follows:*

(a) $(\mathscr{T}, \mathscr{G}, \Pi)$ *is a body element as in Definition 4.1. It is called the* **underlying body element** *of the material element.*

(b) Σ *is a set, called the* **state space** *of the material element.*

(c) $\hat{G}$ *is a mapping* $\hat{G}: \Sigma \to \mathscr{G}$ *from the state space* Σ *into the configuration space* $\mathscr{G}$.

(d) $\hat{S}$ *is a mapping* $\hat{S}: \Sigma \to \mathscr{S}$ *from the state space* Σ *into the space*

$$\mathscr{S} = \mathrm{Sym}(\mathscr{T}^*, \mathscr{T}), \tag{7.1}$$

which we call the **stress space** *of the element.*

(c) $\hat{\rho}$ *is a mapping* $\hat{\rho}: (\Sigma \times \Pi)_{\mathrm{fit}} \to \Sigma$, *whose domain*

$$(\Sigma \times \Pi)_{\mathrm{fit}} = \{(\sigma, P) \mid \sigma \in \Sigma,\ P \in \Pi,\ P^i = \hat{G}(\sigma)\} \tag{7.2}$$

is the set of all state-process pairs such that the state "fits" the initial configuration of the process.

When we speak about a material element $\mathscr{T}$, we consider it understood that $\mathscr{T}$ is endowed with a structure defined by the prescription of $\mathscr{G}$, Π, Σ, $\hat{G}$, $\hat{S}$, and $\hat{\rho}$. The elements σ of Σ are the possible *states* of $\mathscr{T}$. The value $\hat{G}(\sigma)$ is the *configuration* and $\hat{S}(\sigma)$ is the (intrinsic) *stress* determined by the state σ. The function $\hat{\rho}$ will be called the **evolution function** of $\mathscr{T}$. The value $\hat{\rho}(\sigma, P)$ is the *state reached* by the element if, starting from the state σ, it is subjected to the deformation process P.

In addition to (7.1) and (7.2), we shall use the following notations throughout the rest of the paper:

(i) The set of all deformation processes beginning at the configuration $G \in \mathscr{G}$ is denoted by

$$\Pi_G = \{P \in \Pi \mid P^i = G\}. \tag{7.3}$$

(ii) The set of all states that fit the configuration $G \in \mathscr{G}$ is denoted by

$$\Sigma_G = \{\sigma \in \Sigma \mid \hat{G}(\sigma) = G\} \tag{7.4}$$

and called the G-**section** of Σ.

(iii) The **response functional** $\tilde{S}$ of the element is defined by

$$\tilde{S} = \hat{S} \circ \hat{\rho}: (\Sigma \times \Pi)_{\mathrm{fit}} \to \mathscr{S}. \tag{7.5}$$

If $\sigma \in \Sigma$ and $P \in \Pi_{\hat{G}(\sigma)}$, then $\tilde{S}(\sigma, P)$ is the *stress produced* by the process P when the initial state was σ.

It is evident that Π is the disjoint union of the Π_G, $G \in \mathscr{G}$, that Σ is the disjoint union of the sections Σ_G, $G \in \mathscr{G}$, and that $(\Sigma \times \Pi)_{\mathrm{fit}}$ is the disjoint union of the Cartesian products $\Sigma_G \times \Pi_G$, $G \in \mathscr{G}$.

The first two axioms flow directly from the intuitive meaning of the evolution function $\hat{\rho}$. One states that the state reached after a process must fit the final configuration. The other expresses the fact that when the element is first subjected to a process P_1 and then to a process P_2, it must reach the same state as when it is subjected to the continuation $P_1 * P_2$ of P_1 by P_2.

Axiom I. *For all* $(\sigma, P)\in(\Sigma\times\Pi)_{\text{fit}}$,

$$\hat{\rho}(\sigma, P)\in\Sigma_{P^f}, \quad \text{i.e.,} \quad \hat{G}(\hat{\rho}(\sigma, P))=P^f. \tag{7.6}$$

It follows from Axiom I that $\hat{G}: \Sigma\to\mathcal{G}$ must be surjective, because, by condition (P4) of Definition 4.1, for any $\sigma\in\Sigma$ and any $G\in\mathcal{G}$, there must be a $P\in\Pi$ such that $(\sigma, P)\in(\Sigma\times\Pi)_{\text{fit}}$ and $P^f=G$.

Axiom II. *If* $P_1, P_2\in\Pi$, $\sigma\in\Sigma_{P_1^i}$ *and* $P_1^f=P_2^i$, *then*

$$\hat{\rho}(\sigma, P_1 * P_2)=\hat{\rho}(\hat{\rho}(\sigma, P_1), P_2). \tag{7.7}$$

An immediate consequence of (7.7) is

$$\tilde{S}(\sigma, P_1 * P_2)=\tilde{S}(\hat{\rho}(\sigma, P_1), P_2). \tag{7.8}$$

The next axiom is basic. It expresses the assumption that there must be some operational way to distinguish between states. Specifically, if two states are different but fit the same configuration, there must be some process which produces different stresses with the two states as initial states.

Axiom III. *For any* $G\in\mathcal{G}$, *if* $\sigma_1, \sigma_2\in\Sigma_G$ *and*

$$\tilde{S}(\sigma_1, P)=\tilde{S}(\sigma_2, P)$$

for all $P\in\Pi_G$, *then* $\sigma_1=\sigma_2$.

A simple application of Axiom III gives the following result, which states that processes of zero duration leave states unchanged.

Proposition 7.1. *For all* $\sigma\in\Sigma$ *we have*

$$\hat{\rho}(\sigma, \hat{G}(\sigma)_{(0)})=\sigma, \qquad \tilde{S}(\sigma, \hat{G}(\sigma)_{(0)})=\hat{S}(\sigma). \tag{7.9}$$

Proof. Given any $P\in\Pi_{\hat{G}(\sigma)}$, we apply (7.8) with the choice $P_1=\hat{G}(\sigma)_{(0)}=P^i_{(0)}$ and $P_2=P$. In view of (4.7), we get

$$\tilde{S}(\sigma, P)=\tilde{S}(\hat{\rho}(\sigma, \hat{G}(\sigma)_{(0)}), P).$$

Since P is arbitrary, Axiom III yields the desired result $(7.9)_1$, and $(7.9)_2$ is an immediate consequence. Q.E.D.

There will be three more axioms. They all involve a certain topology on the state space Σ, a topology which we will define in Section 11. Axiom IV of Section 11 requires that the state space Σ should be in a certain sense complete. Axiom V, to be stated in Section 12, postulates the existence of "relaxed states". Axiom VI, to be stated in Section 13, requires that all states must be "accessible" in an appropriate sense.

In the next section we shall describe how our conceptual framework applies even when the material is subject to "internal constraints". Sections 9 and 10 deal with material isomorphisms and material symmetry, for which Axioms IV–VI are not needed.

8. Internal Constraints

We recall (Section 4) that in order to define the structure of a body element $\mathcal{T}$, one must prescribe the set $\mathcal{G}$ of its possible configurations, and $\mathcal{G}$ must be a closed connected subset of $\mathrm{Sym}^+(\mathcal{T}, \mathcal{T}^*)$. If every neighborhood in $\mathcal{G}$ of every $G \in \mathcal{G}$ has a non-empty interior in $\mathrm{Sym}^+(\mathcal{T}, \mathcal{T}^*)$, we say that the element is *free of internal constraints.*

On the other hand, for a given configuration G, it may happen that some neighborhood of G in $\mathcal{G}$ is contained in a proper submanifold $\mathcal{C}_G$ of $\mathrm{Sym}^+(\mathcal{T}, \mathcal{T}^*)$. If this is the case, we say that the element is subject to an *internal constraint* at G with *constraint manifold* $\mathcal{C}_G$. For example, if $\mathcal{G}$ is contained in a proper submanifold $\mathcal{C}$ of $\mathrm{Sym}^+(\mathcal{T}, \mathcal{T}^*)$, then $\mathcal{C}$ is a constraint manifold for every $G \in \mathcal{G}$.

The most important internal constraint is *incompressibility*. It is defined by the condition that any two possible configurations $G_1, G_2 \in \mathcal{G}$ must have the same volume. Although it is easy to give a precise mathematical meaning to the notion of "volume of a configuration",[1] we shall not do so here but merely remark that the square of the ratio of the volume of G_2 to the volume of G_1 is given by $\det(G_1^{-1} G_2)$ (*cf.* [TN], (23.11)). We say that $\mathcal{T}$ is an **incompressible element** if for any $G_1, G_2 \in \mathcal{G}$ we have $\det G_1^{-1} G_2 = 1$. This means that $\mathcal{G}$ is contained in

$$\mathcal{C} = \{G \in \mathrm{Sym}^+(\mathcal{T}, \mathcal{T}^*) \mid \det(G_0^{-1} G) = 1\}, \tag{8.1}$$

where $G_0 \in \mathcal{G}$ is arbitrary. Thus, for an incompressible element, (8.1) is a constraint manifold for all $G \in \mathcal{G}$.

The severest possible constraint is *rigidity*. A rigid body element has only one possible configuration, *i.e.* $\mathcal{G}$ reduces to a singleton $\mathcal{G} = \{G_0\}$. The only possible deformation processes in Π are the freezes at G_0.

Another internal constraint is *inextensibility* in some direction. We say that $\mathcal{T}$ is *inextensible* in the direction of $v \in \mathcal{T}$ if for any $G_1, G_2 \in \mathcal{G}$ we have $\langle G_1 v, v \rangle = \langle G_2 v, v \rangle$. This means that $\mathcal{G}$ is contained in

$$\mathcal{C} = \{G \in \mathrm{Sym}^+(\mathcal{T}, \mathcal{T}^*) \mid \langle G v, v \rangle = c\}, \tag{8.2}$$

where $c > 0$. For an inextensible element, (8.2) is a constraint manifold for all $G \in \mathcal{G}$.

The definition of a material element given in the previous sections applies whether or not internal constraints are present. However, if $\mathcal{T}$ is subject to internal constraints, the values of $\hat{S}: \Sigma \to \mathcal{S}$ must be reinterpreted as being *intrinsic extra stresses* (*cf.* [TN], p. 71) rather than intrinsic stresses. The formulas (6.3) and (6.4) remain valid if $\hat{S}$ is so reinterpreted. The indeterminacy of the extra stress must be removed by normalization. Such normalization is somewhat arbitrary, but the following condition seems to be the simplest. We note that if $\mathcal{C}$ is a submanifold of $\mathrm{Sym}^+(\mathcal{T}, \mathcal{T}^*)$, then the tangent spaces of $\mathcal{C}$ are subspaces of $\mathrm{Sym}(\mathcal{T}, \mathcal{T}^*)$. Also, if $S \in \mathrm{Sym}(\mathcal{T}^*, \mathcal{T})$ and $G \in \mathcal{G}$, then $GSG \in \mathrm{Sym}(\mathcal{T}, \mathcal{T}^*)$.

Normalization Condition. *If $\mathcal{T}$ is subject to an internal constraint at $G \in \mathcal{G}$ with constraint manifold $\mathcal{C}_G$, then, for every $\sigma \in \Sigma_G$, $G\hat{S}(\sigma)G$ must belong to the tangent space of $\mathcal{C}_G$ at G.*

[1] As an element of the space $\bigwedge^n \mathcal{T}^*$ of n-covectors, with $n = \dim \mathcal{T}$.

One can show that the tangent space of the manifold (8.1) at $G\in\mathcal{G}$ consists of all $D\in\operatorname{Sym}(\mathcal{T},\mathcal{T}^*)$ such that $\operatorname{tr}(G^{-1}D)=0$. Hence, for incompressible body elements, the Normalization Condition reduces to

$$\operatorname{tr}(\hat{S}(\sigma)\,G)=0 \quad \text{for all } G\in\mathcal{G} \text{ and all } \sigma\in\Sigma_G. \tag{8.3}$$

For a rigid material element, the constraint manifold is the zero-dimensional manifold $\mathcal{C}=\mathcal{G}=\{G_0\}$. Its tangent-space at G_0 is the zero-dimensional space whose only member is the zero element of $\operatorname{Sym}(\mathcal{T},\mathcal{T}^*)$. Hence the Normalization Condition gives $\hat{S}(\sigma)=0$ for all $\sigma\in\Sigma$. Using Axiom III, one can see immediately that the state space Σ must also reduce to a singleton $\Sigma=\{\sigma_0\}$, and the entire material element structure becomes trivial.

9. Material Isomorphisms, Materially Uniform Bodies

Isomorphisms between body elements (Definition 4.1) and between material elements (Definition 7.1) will be defined in accordance with the general notion of isomorphisms for arbitrary mathematical structures.

We note that the set $\operatorname{Invlin}(\mathcal{T}_1,\mathcal{T}_2)$ is the set of vector-space isomorphisms from $\mathcal{T}_1$ onto $\mathcal{T}_2$. Each $A\in\operatorname{Invlin}(\mathcal{T}_1,\mathcal{T}_2)$ induces an isomorphism $G\mapsto A^*GA$ from $\operatorname{Sym}(\mathcal{T}_2,\mathcal{T}_2^*)$ onto $\operatorname{Sym}(\mathcal{T}_1,\mathcal{T}_1^*)$ and an isomorphism $S\mapsto ASA^*$ from $\operatorname{Sym}(\mathcal{T}_1^*,\mathcal{T}_1)$ onto $\operatorname{Sym}(\mathcal{T}_2^*,\mathcal{T}_2)$.

Definition 9.1. *Let $(\mathcal{T}_1,\mathcal{G}_1,\Pi_1)$ and $(\mathcal{T}_2,\mathcal{G}_2,\Pi_2)$ be body elements. A mapping $A:\mathcal{T}_1\to\mathcal{T}_2$ is called an isomorphism between the body elements $\mathcal{T}_1$ and $\mathcal{T}_2$ if* (i) $A\in\operatorname{Invlin}(\mathcal{T}_1,\mathcal{T}_2)$, (ii) $A^*\mathcal{G}_2A=\mathcal{G}_1$, *and* (iii) $A^*\Pi_2A=\Pi_1$.

Definition 9.2. *Let $(\mathcal{T}_1,\mathcal{G}_1,\Pi_1,\Sigma_1,\hat{G}_1,\hat{S}_1,\hat{\rho}_1)$ and $(\mathcal{T}_2,\mathcal{G}_2,\Pi_2,\Sigma_2,\hat{G}_2,\hat{S}_2,\hat{\rho}_2)$ be material elements. A mapping $A:\mathcal{T}_1\to\mathcal{T}_2$ is called a* **material isomorphism** *between the material elements $\mathcal{T}_1$ and $\mathcal{T}_2$ if A is an isomorphism between the underlying body elements of $\mathcal{T}_1$ and $\mathcal{T}_2$ and if there exists a bijection*

$$\iota_A:\Sigma_1\to\Sigma_2 \tag{9.1}$$

with the following properties:

(a) *for all $\sigma\in\Sigma_1$,*

$$\hat{G}_1(\sigma)=A^*\,\hat{G}_2(\iota_A(\sigma))\,A, \tag{9.2}$$

(b) *for all $\sigma\in\Sigma_1$,*

$$\hat{S}_2(\iota_A(\sigma))=A\hat{S}_1(\sigma)\,A^*, \tag{9.3}$$

(c) *for all $\sigma\in\Sigma_1$ and all $P\in\Pi_2$ such that $\hat{G}_1(\sigma)=A^*P^{\iota}A$,*

$$\iota_A(\hat{\rho}_1(\sigma,A^*PA))=\hat{\rho}_2(\iota_A(\sigma),P). \tag{9.4}$$

Proposition 9.1. *The bijection ι_A of* (9.1)–(9.4) *is uniquely determined by the material isomorphism A.*

Proof. Application of (9.3) and (9.4) shows that

$$A\hat{S}_1(\hat{\rho}_1(\sigma,A^*PA))\,A^*=\hat{S}_2(\hat{\rho}_2(\iota_A(\sigma),P)) \tag{9.5}$$

for all $\sigma\in\Sigma_1$ and all $P\in\Pi_2$ with $\hat{G}_1(\sigma)=A^*P^iA$. Hence, if ι_A and ι'_A are both bijections satisfying (9.1)–(9.4), we have

$$\hat{S}_2(\hat{\rho}_2(\iota_A(\sigma),P))=\hat{S}_2(\hat{\rho}_2(\iota'_A(\sigma),P))$$

for all $P\in\Pi_2$ with $P^i=A^{*-1}\hat{G}_1(\sigma)A^{-1}=\hat{G}_2(\iota_A(\sigma))=\hat{G}_2(\iota'_A(\sigma))$. It follows from Axiom III that $\iota_A(\sigma)=\iota'_A(\sigma)$. Since $\sigma\in\Sigma_1$ is arbitrary, we obtain $\iota_A=\iota'_A$. Q.E.D.

If A is a material isomorphism, so is its inverse A^{-1}, and the corresponding bijections ι_A and $\iota_{A^{-1}}$ are related by

$$\iota_{A^{-1}}=(\iota_A)^{-1}. \tag{9.6}$$

If A_1 and A_2 are material isomorphisms and if their composition A_2A_1 makes sense, it is again a material isomorphism. The bijections ι_{A_1}, ι_{A_2}, and $\iota_{A_2A_1}$ are related by

$$\iota_{A_2A_1}=\iota_{A_2}\circ\iota_{A_1}. \tag{9.7}$$

If two material elements $\mathscr{T}_1$ and $\mathscr{T}_2$ are materially isomorphic we also say that they *consist of the same material*. We give a precise meaning to the term "material" by saying that a **material** is an equivalence class of material elements, the equivalence being material isomorphy.

Suppose that we have defined, on one and the same vector space $\mathscr{T}$, two material element structures in such a way that the identity $1_{\mathscr{T}}$ is a material isomorphism between them. By Definition 9.1, the sets $\mathscr{G}$ and Π must be the same for both structures. Although the state spaces need not be the same *a priori*, we shall always employ the bijection $\iota_{1_{\mathscr{T}}}$ to identify one with the other. The functions $\hat{G}$, $\hat{S}$, and $\hat{\rho}$ then become the same for both structures, *i.e.* the two structures become the same. Thus, we do not distinguish between material element structures on $\mathscr{T}$ that are isomorphic via the identity mapping of $\mathscr{T}$.

Remark. In Section 3 we have indicated how a continuous body $\mathscr{B}$ can be defined and how a body element $\mathscr{T}_X$ can be associated with each of the material points $X\in\mathscr{B}$. We say that $\mathscr{B}$ has the structure of a **material body** if each of its body elements $\mathscr{T}_X$ has been endowed with the structure of a material element in accordance with Definition 7.1. We say that a material body is **materially uniform** if any two of its material elements consist of the same material. (This definition is consistent with the one I gave in [TN], p. 59 and [N7], p. 7.)

Let $\mathscr{B}$ be a materially uniform material body, and let $\mathscr{A}$ be the set of all material isomorphisms between material elements of $\mathscr{B}$. It is easily seen that this set $\mathscr{A}$ has the following properties:

(I1) Every member of $\mathscr{A}$ is a vector-space isomorphism between two (not necessarily distinct) tangent spaces of $\mathscr{B}$.

(I2) If $A\in\mathscr{A}$, then $A^{-1}\in\mathscr{A}$.

(I3) If $A_1, A_2\in\mathscr{A}$ and Domain $A_2=$ Range A_1, then $A_2\circ A_1\in\mathscr{A}$.

(I4) For any $X, Y\in\mathscr{B}$, there is at least one $A\in\mathscr{A}$ such that $A:\mathscr{T}_X\to\mathscr{T}_Y$.

The set $\mathscr{A}$ endows the body-manifold $\mathscr{B}$ with a certain "geometrical" structure. This structure is closely related to what, in the literature on abstract differential geometry, is called a G-structure (see [S], Chap. VII). The theory of inhomogeneous but materially uniform bodies as developed by me in [N7] and C.-C. WANG in [W2] is essentially nothing more than a theory of the geometrical structure on a manifold $\mathscr{B}$ defined by a set $\mathscr{A}$ with the properties (I1)–(I4). For example, a *material connection* Γ on $\mathscr{B}$ is defined to be an affine connection on $\mathscr{B}$ such that every parallel transport with respect to Γ belongs to $\mathscr{A}$. (This concept of material connection is the one introduced by C.-C. WANG in [W2], p. 66. It is a more inclusive concept than the original concept introduced by me in [TN], p. 90 and [N7], p. 19.)

10. Material Symmetry, Isotropy

Let $\mathscr{T}$ be a material element. A material isomorphism (see Definition 9.2) between $\mathscr{T}$ and $\mathscr{T}$ itself is called a *material automorphism* or a **symmetry** of $\mathscr{T}$. The set

$$g = \mathrm{Aut}(\mathscr{T}) \subset \mathrm{Invlin}(\mathscr{T}) \tag{10.1}$$

of all material automorphisms is obviously a subgroup of the general linear group Invlin($\mathscr{T}$). We call this group g the **symmetry group** of material element $\mathscr{T}$.[1]

The following result is an immediate corollary of Prop. 9.1 and (9.6), (9.7):

Proposition 10.1. *There is a unique homomorphism*

$$\imath: g \to \mathrm{Perm}(\Sigma) \tag{10.2}$$

from the symmetry group g into the group Perm(Σ) *of all permutations of the state space Σ such that for every $A \in g$ the value $\imath_A$ of $\imath$ at A satisfies the following conditions*:

(i) *for all $\sigma \in \Sigma$*

$$\hat{G}(\sigma) = A^* \hat{G}(\imath_A(\sigma))A, \tag{10.3}$$

and

$$\hat{S}(\imath_A(\sigma)) = A\hat{S}(\sigma)A^*, \tag{10.4}$$

(ii) *for all $\sigma \in \Sigma$ and all $P \in \Pi$ such that $\hat{G}(\sigma) = A^* P^i A$,*

$$\imath_A(\hat{\rho}(\sigma, A^* P A)) = \hat{\rho}(\imath_A(\sigma), P). \tag{10.5}$$

This proposition asserts that the symmetry group g acts as a transformation group on the state space Σ. Therefore, we can define the *orbit* Ω_σ of $\sigma \in \Sigma$ under the action of g by

$$\Omega_\sigma = \{\imath_A(\sigma) \mid A \in g\}. \tag{10.6}$$

The orbits under the action of g form a partition of the state space Σ. We call these orbits the **reduced states** of the given material element. Every state determines a unique reduced state. Two states that determine the same reduced state are physically equivalent in an intuitively obvious sense.

Let $\sigma \in \Sigma$ be any state. The set

$$g_\sigma = \{A \in g \mid \imath_A(\sigma) = \sigma\} \tag{10.7}$$

[1] In [N2], Sect. 19, and in later work (notably [TN]) I used the term "isotropy group". Some of my colleagues convinced me that "symmetry group" is preferable.

of all symmetries that leave the state σ invariant will be called the **symmetry group of the state** σ. Of course, $\mathcal{g}_\sigma$ is a subgroup of $\mathcal{g}$.

Proposition 10.2. *The symmetry group $\mathcal{g}_\sigma$ of a state σ is a subgroup of the orthogonal group of $\hat{G}(\sigma)$ and hence*

$$\mathcal{g}_\sigma \subset \mathcal{g} \cap \operatorname{Orth}(\hat{G}(\sigma)). \tag{10.8}$$

Proof. It follows from (10.3) and (10.7) that $\hat{G}(\sigma) = A^* \hat{G}(\sigma) A$ for all $A \in \mathcal{g}_\sigma$. This means (see Section 2) that $A \in \operatorname{Orth}(\hat{G}(\sigma))$. Q.E.D.

We say that a state σ is **isotropic** if

$$\mathcal{g}_\sigma = \operatorname{Orth}(\hat{G}(\sigma)). \tag{10.9}$$

We say that $\mathscr{T}$ is an *isotropic material element* if it has isotropic states.

Proposition 10.3. *The symmetry group $\mathcal{g}$ of an isotropic material element $\mathscr{T}$ is such that either $\mathcal{g} = \mathcal{g}_\sigma$ if and only if σ is isotropic or else $\mathcal{g} \supset \operatorname{Unim}(\mathscr{T})$.*

Proof. By the definition (10.9), σ is isotropic if and only if $\mathcal{g}_\sigma = \operatorname{Orth}(\hat{G}(\sigma))$. Since $\mathcal{g} \supset \mathcal{g}_\sigma$ we can apply the theorem on the maximality of an orthogonal group in the unimodular group (see Section 2) and conclude that we must have either $\mathcal{g} = \operatorname{Orth} \hat{G}(\sigma) = \mathcal{g}_\sigma$ or $\mathcal{g} \supset \operatorname{Unim}(\mathscr{T})$. Since these are mutually exclusive possibilities, the assertion follows. Q.E.D.

Proposition 10.4. *The stress $\hat{S}(\sigma)$ determined by a state σ is such that*

$$A\hat{S}(\sigma)\hat{G}(\sigma) = \hat{S}(\sigma)\hat{G}(\sigma)A \qquad \text{for all } A \in \mathcal{g}_\sigma. \tag{10.10}$$

In particular, if σ is an isotropic state, then

$$\hat{S}(\sigma) = -p\,\hat{G}(\sigma)^{-1}, \tag{10.11}$$

where $p \in \mathbb{R}$.

Proof. (10.10) is an immediate consequence of (10.4), (10.7), and (10.3). If σ is isotropic, *i.e.* $\mathcal{g}_\sigma = \operatorname{Orth}(\hat{G}(\sigma))$, then (10.10) states that $\hat{S}(\sigma)\,\hat{G}(\sigma)$ commutes with all $A \in \operatorname{Orth}(\hat{G}(\sigma))$. It is well known that this can happen only if $\hat{S}(\sigma)\,\hat{G}(\sigma)$ is a multiple of the identity $1_{\mathscr{T}}$, *i.e.* if (10.11) holds. Q.E.D.

The number p of (10.11) is called the *pressure* corresponding to the isotropic state σ. For incompressible material elements, it follows from Prop. 10.4 and the normalization (8.3) that the extra-stress is zero for all isotropic states.

We recall that two states σ, σ' determine the same reduced state if one is in the orbit of the other under the action of $\mathcal{g}$, *i.e.* if $\sigma' = \iota_A(\sigma)$ for some $A \in \mathcal{g}$. The following proposition shows that if two states determine the same reduced state, then the symmetry groups of the two states are conjugate in $\mathcal{g}$ and hence isomorphic.

Proposition 10.5. *If $\sigma \in \Sigma$ and $A \in \mathcal{g}$, then*

$$\mathcal{g}_{\iota_A(\sigma)} = A\,\mathcal{g}_\sigma A^{-1}. \tag{10.12}$$

Proof. In view of (9.6), (9.7), and the definition (10.7), we have $B \in \mathcal{g}_\sigma$ if and only if

$$\iota_{ABA^{-1}}(\iota_A(\sigma)) = (\iota_A \circ \iota_B \circ \iota_{A^{-1}} \circ \iota_A)(\sigma) = \iota_A(\iota_B(\sigma)) = \iota_A(\sigma),$$

which is valid if and only if $ABA^{-1} \in \mathcal{g}_{\iota_A(\sigma)}$. Q.E.D.

We say that a state σ is a **state of maximal symmetry** if $\mathcal{g}_\sigma$ is maximal in the class of all symmetry groups of states, *i.e.* if $\mathcal{g}_\sigma \subset \mathcal{g}_\tau$ is possible for $\tau \in \Sigma$ only if $\mathcal{g}_\sigma = \mathcal{g}_\tau$.

Proposition 10.6. *For an isotropic material element, a state is of maximal symmetry if and only if it is isotropic.*

Proof. In view of Prop. 10.3 it is trivial that isotropic states are always of maximal symmetry.

On the other hand, let σ be a state of maximal symmetry. By Prop. 10.3, we have two possibilities for $\mathcal{g}$:

(a) $\mathcal{g} = \mathcal{g}_\tau$, with τ isotropic. In this case, (10.8) becomes

$$\mathcal{g}_\sigma \subset \mathcal{g}_\tau \cap \operatorname{Orth}(\hat{G}(\sigma)) \subset \mathcal{g}_\tau.$$

Since $\mathcal{g}_\sigma$ is maximal, the inclusions must reduce to equalities, which is possible only when $\mathcal{g}_\sigma = \mathcal{g}_\tau = \mathcal{g}$. By Prop. 10.3, σ must be isotropic.

(b) $\mathcal{g} \supset \operatorname{Unim}(\mathcal{T})$. Let τ be an isotropic state. We can always find a $A \in \operatorname{Unim}(\mathcal{T})$ such that $\hat{G}(\sigma) = cA^* \hat{G}(\tau) A$, with $c > 0$ (see Section 2). Since $A \in \mathcal{g}$, we have, by (10.3), $\hat{G}(\sigma) = c\hat{G}(\tau')$, where $\tau' = \iota_A(\tau)$ is another isotropic state (see Prop. 10.5). It follows that $\mathcal{g}_{\tau'} = \operatorname{Orth}(\hat{G}(\tau')) = \operatorname{Orth}(\hat{G}(\sigma))$. Using (10.8), we obtain

$$\mathcal{g}_\sigma \subset \operatorname{Orth}(\hat{G}(\sigma)) = \operatorname{Orth}(\hat{G}(\tau')) = \mathcal{g}_{\tau'}.$$

Since $\mathcal{g}_\sigma$ is maximal, the inclusion must reduce to equality, so that $\mathcal{g}_\sigma = \operatorname{Orth}(\hat{G}(\sigma))$. This means that σ is isotropic. Q.E.D.

It is clear from Prop. 10.5 that if a state σ is of maximal symmetry or is isotropic, so are all states that determine the same reduced state as σ. Thus, it is meaningful to speak about *reduced states of maximal symmetry* and *isotropic reduced states.*

The conditions (10.3)–(10.5) are obviously satisfied for $A = 1_{\mathcal{T}}$ and $A = -1_{\mathcal{T}}$ if we put

$$\iota_{1_{\mathcal{T}}} = \iota_{(-1_{\mathcal{T}})} = 1_\Sigma. \tag{10.13}$$

Hence the symmetry groups $\mathcal{g}_\sigma$, $\sigma \in \Sigma$, and $\mathcal{g}$ always contain the two-element group $\{1_{\mathcal{T}}, -1_{\mathcal{T}}\}$.

11. Uniformity and Topology in the State Space

We now introduce natural uniformities on the sections of the state space of a material element and a natural topology on the whole state space. For the explanation of the concepts occurring in the following definition we refer to [BK], Chapt. II, §2, No. 3, and Chapt. I, §2, No. 4, or to [FT], Chapt. 24. Recall that the stress space $\mathcal{S} = \operatorname{Sym}(\mathcal{T}^*, \mathcal{T})$ is a finite-dimensional real vector space and hence has a natural uniformity.

Definition 11.1. *Let $\mathcal{T}$ be a material element with response functional $\tilde{S}$ (see (7.5)). For every $G \in \mathcal{G}$, we call* **natural uniformity** *of the G-section Σ_G (see (7.4)) the coarsest uniformity on Σ_G which renders the mappings*

$$\tilde{S}(\cdot, P) : \Sigma_G \to \mathcal{S}$$

uniformly continuous for all $P \in \Pi_G$ ((see 7.3)).[1] *We call* **natural topology** *of the state space $\Sigma = \bigcup\{\Sigma_G \mid G \in \mathscr{G}\}$ the sum of the topologies on the sections Σ_G induced by the natural uniformities of these sections.*

The following result, a special case of [BK], Prop. 4 of §2, Chap. II, characterizes the natural uniformity of Σ_G.

Proposition 11.1. *A mapping $\varphi\colon \mathscr{U} \to \Sigma_G$ from a uniform space into the G-section Σ_G is uniformly continuous if and only if the mappings*

$$\tilde{S}(\varphi(\cdot), P)\colon \mathscr{U} \to \mathscr{S}$$

are uniformly continuous for all $P \in \Pi_G$.

This proposition is used to prove the following result.

Proposition 11.2. *For all $P \in \Pi$, the mapping*

$$\hat{\rho}(\cdot, P)\colon \Sigma_{P_i} \to \Sigma_{P_f}$$

is uniformly continuous.

Proof. It follows from (7.8) that

$$\tilde{S}(\hat{\rho}(\sigma, P), P') = \tilde{S}(\sigma, P * P')$$

for all $P' \in \Pi_{P_f}$. Since $\tilde{S}(\cdot, P * P')\colon \Sigma_{P_i} \to \mathscr{S}$ is uniformly continuous by definition of the natural uniformity on Σ_{P_i}, it follows that $\tilde{S}(\hat{\rho}(\cdot, P), P')\colon \Sigma_{P_i} \to \mathscr{S}$ is uniformly continuous for all $P' \in \Pi_{P_f}$. Application of Prop. 11.1 with the choice $\mathscr{U} = \Sigma_{P_i}$, $\varphi = \hat{\rho}(\cdot, P)$ yields the assertion. Q.E.D.

The following result, a special case of [BK] Prop. 10 of §8, Chap. I, characterizes the natural topology of Σ. (For the explanation of convergence of nets we refer to [K], Chapt. 2.)

Proposition 11.3. *A net $(i \mapsto \sigma_i)$ in Σ converges to $\sigma \in \Sigma$ if and only if* (i) *σ_i belongs eventually to some fixed section Σ_G,* (ii) *$\sigma \in \Sigma_G$, and* (iii)

$$\lim_i \tilde{S}(\sigma_i, P) = \tilde{S}(\sigma, P) \tag{11.1}$$

for all $P \in \Pi_G$.

Proposition 11.4. *The state space Σ, with its natural topology, is a Hausdorff space, i.e., nets in Σ cannot have more than one limit.*

Proof. If σ_1 and σ_2 are both limits of the net $(i \mapsto \sigma_i)$, we must have, by Prop. 11.3, $\hat{G}(\sigma_1) = \hat{G}(\sigma_2) = G$ and $\tilde{S}(\sigma_1, P) = \tilde{S}(\sigma_2, P)$ for all $P \in \Pi_G$. By Axiom III, Sect. 7, it follows that $\sigma_1 = \sigma_2$. Q.E.D.

The natural topology on Σ is such that all sections Σ_G, $G \in \mathscr{G}$, are both open and closed in Σ. Hence $\hat{G}\colon \Sigma \to \mathscr{G}$ is trivially continuous, because it is constant on each section. Also, $\hat{S}\colon \Sigma \to \mathscr{S}$ is continuous, because by $(7.9)_2$ and Definition 11.1, the restriction of $\hat{S}$ to Σ_G is the same as $\tilde{S}(\cdot, G_{(0)})$ and hence (uniformly) continuous.

[1] This uniformity is also called the "initial uniformity" on Σ_G associated with the mappings $\tilde{S}(\cdot, P)$, $P \in \Pi_G$.

If $\mathscr{T}_1$ and $\mathscr{T}_2$ are material elements with state spaces Σ_1 and Σ_2 and $A: \mathscr{T}_1 \to \mathscr{T}_2$ a material isomorphism between $\mathscr{T}_1$ and $\mathscr{T}_2$, then the bijection $\iota_A: \Sigma_1 \to \Sigma_2$ induced by A (see Prop. 9.1) is a homeomorphism when Σ_1 and Σ_2 are endowed with their natural topologies. This is but a special instance of the general fact that isomorphisms preserve all structures induced from the given structure. In particular, if A is a symmetry of an element $\mathscr{T}$, then ι_A is an auto-homeomorphism of the state space Σ of $\mathscr{T}$.

The sections Σ_G, with their natural uniformity described by Definition 11.1, may or may not be complete spaces. If they are not, we may form the completions $\bar{\Sigma}_G$ of the Σ_G (*cf.* [K], p. 196 or [BK] Chapt. II, §3) and define the *completed states space* $\bar{\Sigma}$ by $\bar{\Sigma} = \bigcup\{\bar{\Sigma}_G \mid G \in \mathscr{G}\}$. Then Σ is dense in $\bar{\Sigma}$ and, in view of the continuity of $\hat{G}$ and $\hat{S}$, we can extend the domains of these functions from Σ to $\bar{\Sigma}$, preserving continuity. Also, in view of Prop. 11.2, one can extend the domain of $\hat{\rho}$ from $(\Sigma \times \Pi)_{\text{fit}}$ to $(\bar{\Sigma} \times \Pi)_{\text{fit}}$, preserving uniform continuity of $\hat{\rho}(\cdot, P)$ for all $P \in \Pi$. After this has been done, one obtains what is technically a new structure having $(\mathscr{T}, \mathscr{G}, \Pi)$ as the underlying body element but $\bar{\Sigma}$ instead of Σ as the state space. However, the distinction between the original structure and the new one with completed state space is physically insignificant, and artificial distinctions are avoided by means of the following postulate:

Axiom IV. *The sections Σ_G, $G \in \mathscr{G}$, with the natural uniformity described by Definition 11.1, are complete spaces.*

12. Relaxed States

In Axiom V, below, the topology of Definition 11.1 on the state space Σ of the element $\mathscr{T}$ plays an essential role. The axiom has the following interpretation: If the element is initially in the state σ and then frozen in the corresponding configuration, its state will approach, in the limit of infinite time, a "relaxed" state.

Axiom V. *For all $\sigma \in \Sigma$, the limit*

$$\lim_{t \to \infty} \hat{\rho}(\sigma, \hat{G}(\sigma)_{(t)}) = \hat{\lambda}(\sigma) \tag{12.1}$$

exists (it is unique by Prop. 11.4).

The mapping $\hat{\lambda}: \Sigma \to \Sigma$ determined by (12.1) will be called the *state relaxation mapping*, and the members of its range

$$\Sigma_{\text{rel}} = \hat{\lambda}(\Sigma) \tag{12.2}$$

will be called **relaxed states.**[1]

It is clear from (7.6) that

$$\hat{G}(\hat{\lambda}(\sigma)) = \hat{G}(\sigma). \tag{12.3}$$

Hence, since $\hat{G}$ is surjective, the restriction $\hat{G}|_{\Sigma_{\text{rel}}}$ of $\hat{G}$ to Σ_{rel} is also surjective. In other terms, $\Sigma_G \cap \Sigma_{\text{rel}}$ is non-empty for all $G \in \mathscr{G}$.

[1] This concept of a relaxed state is similar to what I have called, in a very special and now obsolete context, a "body at ease" in [N1], p. 26.

If we apply Prop. 11.3 to the net $t \mapsto \hat{\rho}(\sigma, \hat{G}(\sigma)_{(t)})$ and observe (7.8), we see that (12.1) is equivalent to

$$\lim_{t\to\infty} \tilde{S}(\sigma, G_{(t)} * P) = \tilde{S}(\hat{\lambda}(\sigma), P) \quad \text{for all } P \in \Pi_G, \tag{12.4}$$

where $G = \hat{G}(\sigma)$.

We now give several characterizations of relaxed states.

Proposition 12.1. *Let $\lambda \in \Sigma$ and $G = \hat{G}(\lambda)$. The following four conditions are equivalent:*

(i) $\lambda \in \Sigma_{\text{rel}}$,

(ii) *for all* $t \in \mathbb{R}^+$, $\hat{\rho}(\lambda, G_{(t)}) = \lambda$,

(iii) *for some* $t_0 > 0$, $\hat{\rho}(\lambda, G_{(t_0)}) = \lambda$,

(iv) $\hat{\lambda}(\lambda) = \lambda$.

Proof. First we note that, in view of (4.3) and (4.5),

$$G_{(t+r)} = G_{(r)} * G_{(t)} \tag{12.5}$$

holds for any $r, t \in \mathbb{R}^+$. For every $\sigma \in \Sigma_G$, it follows from (12.5) and (7.7) that

$$\hat{\rho}(\sigma, G_{(t+r)}) = \hat{\rho}(\hat{\rho}(\sigma, G_{(r)}), G_{(t)}). \tag{12.6}$$

We now prove (i)$\Rightarrow$(ii): $\lambda \in \Sigma_{\text{rel}}$ means that $\lambda = \hat{\lambda}(\sigma)$ for some $\sigma \in \Sigma$. Since (12.3) shows that $\sigma \in \Sigma_G$ we see that (12.6) is valid. Taking the limit $r \to \infty$ in (12.6) and observing Prop. 11.2 as well as Axiom V, we obtain

$$\lambda = \hat{\lambda}(\sigma) = \hat{\rho}(\hat{\lambda}(\sigma), G_{(t)}) = \hat{\rho}(\lambda, G_{(t)}),$$

i.e., condition (ii) is satisfied.

The implication (ii)$\Rightarrow$(iii) is trivial.

To prove (iii)$\Rightarrow$(iv), we assume that (iii) holds and apply (12.6) with the choice $\sigma = \lambda$, $r = t_0$, $t = (n-1)t_0$, where $n \in \mathbb{N}$:

$$\hat{\rho}(\lambda, G_{(n t_0)}) = \hat{\rho}(\hat{\rho}(\lambda, G_{(t_0)}), G_{(n-1)t_0}) = \hat{\rho}(\lambda, G_{(n-1)t_0}).$$

Since $\hat{\rho}(\lambda, G_{(0)}) = \lambda$ by (7.9), we see that induction over n yields $\lambda = \hat{\rho}(\lambda, G_{(n t_0)})$ for all $n \in \mathbb{N}$. Taking the limit $n \to \infty$ and using Axiom V, we get $\lambda = \hat{\lambda}(\lambda)$.

The implication (iv)$\Rightarrow$(i) is trivial. Q.E.D.

Prop. 12.1 has the following interpretation: Suppose a material element is initially in a state λ and then frozen in the corresponding configuration. The state will remain unchanged if and only if λ is relaxed. The material cannot return to its initial state λ in a finite time or in the limit of infinite time unless λ is relaxed.

Proposition 12.2. *Relaxation cannot decrease the symmetry group of a state, i.e., for all $\sigma \in \Sigma$, we have*

$$\mathcal{g}_\sigma \subset \mathcal{g}_{\hat{\lambda}(\sigma)}. \tag{12.7}$$

Proof. Let $G = \hat{G}(\sigma) = \hat{G}(\hat{\lambda}(\sigma))$ (see (12.3)). If $A \in \mathcal{g}_\sigma$, i.e. $\iota_A(\sigma) = \sigma$, we have, by (10.3), $G = A^* G A$ and hence, for all $t \in \mathbb{R}^+$, $G_{(t)} = A^* G_{(t)} A$. Writing (10.5) with

$P=G_{(t)}$, we get

$$\iota_A(\hat{\rho}(\sigma, G_{(t)}))=\hat{\rho}(\sigma, G_{(t)}).$$

Taking the limit $t\to\infty$ and observing Axiom V, we obtain $\iota_A(\hat{\lambda}(\sigma))=\hat{\lambda}(\sigma)$ and hence $A\in\mathcal{g}_{\hat{\lambda}(\sigma)}$. Q.E.D.

The following result is a corollary to Prop. 12.2.

Proposition 12.3. *If σ is an isotropic state or a state of maximal symmetry, so is $\hat{\lambda}(\sigma)$.*

If λ is a relaxed state and A a material isomorphism, it is clear that $\iota_A(\lambda)$ is again a relaxed state. This applies, in particular, when $A\in\mathcal{g}$ is a material symmetry. It follows that if λ is relaxed, so are all states that determine the same reduced state as λ. Thus it is meaningful to speak about *relaxed reduced states.*

13. Accessibility

Let σ be a state of a material element with state space Σ. The set of all states that can be reached from σ is $\{\hat{\rho}(\sigma, P) \mid P\in\Pi_{\hat{G}(\sigma)}\}$. The closure of this set, in the natural topology of Σ described in Definition 11.1, will be denoted by

$$\Sigma_\sigma=\mathrm{cl}\{\hat{\rho}(\sigma, P)\mid P\in\Pi_{\hat{G}(\sigma)}\}. \tag{13.1}$$

If $\tau\in\Sigma_\sigma$, we say that the state τ is **accessible** from the state σ. Thus, τ is accessible from σ if every neighborhood of τ contains states that can actually be reached from σ. We note that, by (7.9), we have $\sigma=\hat{\rho}(\sigma, \hat{G}(\sigma)_{(0)})\in\Sigma_\sigma$, showing that every state can be reached from itself.

The following proposition states that accessibility is a transitive relation, *i.e.*, if ν is accessible from τ and τ is accessible from σ, then ν is also accessible from σ.

Proposition 13.1. *If $\tau\in\Sigma_\sigma$, then $\Sigma_\tau\subset\Sigma_\sigma$.*

Proof. First we recall that for any $\nu\in\Sigma$, $\Sigma_{\hat{G}(\nu)}$ is open in Σ and hence a neighborhood of ν. Thus, the neighborhoods of ν that are contained in $\Sigma_{\hat{G}(\nu)}$ form a neighborhood basis for ν.

Now let $\nu\in\Sigma_\tau$ and let Υ_ν be an arbitrary neighborhood of ν with $\Upsilon_\nu\subset\Sigma_{\hat{G}(\nu)}$. By the definition of Σ_τ, there is a process $P_1\in\Pi_{\hat{G}(\tau)}$ such that $\hat{\rho}(\tau, P_1)\in\Upsilon_\nu$. By (7.6) we have $P_1^f=\hat{G}(\nu)$. The continuity of $\hat{\rho}(\cdot, P_1)$: $\Sigma_{\hat{G}(\tau)}\to\Sigma_{\hat{G}(\nu)}$ (see Prop. 11.2) insures that there is a neighborhood $\Upsilon_\tau\subset\Sigma_{\hat{G}(\tau)}$ of τ such that $\hat{\rho}(\tau', P_1)\in\Upsilon_\nu$ for all $\tau'\in\Upsilon_\tau$. The condition $\tau\in\Sigma_\sigma$ implies the existence of a process $P_2\in\Pi_{\hat{G}(\sigma)}$ such that $\tau'=\hat{\rho}(\sigma, P_2)\in\Upsilon_\tau$. Hence we have $\hat{\rho}(\hat{\rho}(\sigma, P_2), P_1)\in\Upsilon_\nu$, *i.e.*, by (7.7),

$$\hat{\rho}(\sigma, P_2 * P_1)\in\Upsilon_\nu.$$

Thus we have found a state in Υ_ν which can be reached from σ. Since the neighborhood Υ_ν was arbitrary, it follows that $\nu\in\Sigma_\sigma$. Q.E.D.

The next result is a direct consequence of (12.1) and Prop. 13.1:

Proposition 13.2. *For every $\sigma\in\Sigma$, if $\tau\in\Sigma_\sigma$, then $\hat{\lambda}(\tau)\in\Sigma_\sigma$, i.e., if τ is accessible from σ, so is $\hat{\lambda}(\tau)$. In particular, $\hat{\lambda}(\sigma)\in\Sigma_\sigma$.*

We are now ready to state our final axiom. It postulates that every state must be accessible from some relaxed starting state λ_0. The axiom thus limits the possible size of the state space Σ. If an axiom of this type were not imposed, there could exist two states which had no connection whatever. Physically, these states would have to be regarded as states of distinct materials.

Axiom VI. *There is at least one* $\lambda_0 \in \Sigma_{\text{rel}}$ *such that* $\Sigma_{\lambda_0} = \Sigma$.

If the septuple $(\mathcal{T}, \mathcal{G}, \Pi, \Sigma, \hat{G}, \hat{S}, \hat{\rho})$ satisfies Axioms I–V but not Axiom VI, one can use the following procedure to define material element structures on $\mathcal{T}$: We select $\lambda_0 \in \Sigma_{\text{rel}}$ arbitrarily and put $\Sigma' = \Sigma_{\lambda_0}$. Let G' and $\hat{S}'$ be the restrictions of $\hat{G}$ and $\hat{S}$ to Σ', and let $\hat{\rho}'$ be the restriction of $\hat{\rho}$ to $(\Sigma' \times \Pi_G)_{\text{fit}}$. It is not hard to verify that the septuple $(\mathcal{T}, \mathcal{G}, \Pi, \Sigma', \hat{G}', \hat{S}', \hat{\rho}')$ thus obtained again satisfies Axioms I–V. It obviously also satisfies Axiom VI and hence defines a structure of a material element on $\mathcal{T}$. This structure may depend, of course, on the particular choice of λ_0, and hence the original septuple may give rise to not one, but many material element structures on $\mathcal{T}$.

The procedure just described will be illustrated by an example in Section 20.

14. Semi-Elastic and Elastic Materials

Definition 14.1. *A material element* $\mathcal{T}$ *is said to be* **semi-elastic** *if the restriction* $\hat{G}|_{\Sigma_{\text{rel}}}$ *of its configuration function* $\hat{G}$ *to the set* Σ_{rel} *of relaxed states is injective. The material element is said to be* **elastic** *if it is semi-elastic and if, in addition, all of its states are relaxed, so that* $\Sigma = \Sigma_{\text{rel}}$.

We have seen in Section 12 that $\hat{G}|_{\Sigma_{\text{rel}}}$ is always surjective. Hence, semi-elastic materials are characterized by the condition that $\hat{G}|_{\Sigma_{\text{rel}}}$ has an inverse

$$\bar{\lambda}: \mathcal{G} \to \Sigma_{\text{rel}}. \tag{14.1}$$

The value $\bar{\lambda}(G)$ is the relaxed state uniquely determined by the configuration G. Observing (12.3), we see that $\bar{\lambda}$ satisfies

$$\hat{\lambda}(\sigma) = \bar{\lambda}(\hat{G}(\sigma)) \tag{14.2}$$

for all $\sigma \in \Sigma$. Hence (12.1) reduces to

$$\lim_{t \to \infty} \hat{\rho}(\sigma, G_{(t)}) = \bar{\lambda}(G) \quad \text{if } \hat{G}(\sigma) = G. \tag{14.3}$$

For semi-elastic elements with a known symmetry group, one can determine the symmetry groups of its relaxed states:

Proposition 14.1. *Let* $\mathcal{T}$ *be a semi-elastic element with symmetry group* g. *For all* $\lambda \in \Sigma_{\text{rel}}$ *we then have*

$$g_\lambda = g \cap \operatorname{Orth}(\hat{G}(\lambda)), \tag{14.4}$$

where g_λ, *the symmetry group of* λ, *is defined by* (10.7).

Proof. In view of (10.8) we need only prove that $A \in g \cap \operatorname{Orth}(\hat{G}(\lambda))$ implies $\iota_A(\lambda) = \lambda$. Now, $A \in \operatorname{Orth}(\hat{G}(\lambda))$ means that $A^* \hat{G}(\lambda) A = \hat{G}(\lambda)$. On the other hand, by (10.3), $A \in g$ implies that $\hat{G}(\lambda) = A^* \hat{G}(\iota_A(\lambda)) A$. Thus we have $\hat{G}(\lambda) = \hat{G}(\iota_A(\lambda))$. Since λ is relaxed, so is $\iota_A(\lambda)$, and the injectivity of $\hat{G}|_{\Sigma_{\text{rel}}}$ yields $\lambda = \iota_A(\lambda)$. Q.E.D.

We say that a state of a semi-elastic element is **undistorted** if it is relaxed and of maximal symmetry. We say that the configuration $G \in \mathscr{G}$ is *undistorted* if the corresponding relaxed state $\bar{\lambda}(G)$ is undistorted.[1] Undistorted states always exist, for if σ is any state of maximal symmetry, so is the relaxed state $\hat{\lambda}(\sigma)$, as is evident from Prop. 12.2. It follows from Prop. 10.6 that for an isotropic semi-elastic element, a state is undistorted if and only if it is both relaxed and isotropic.

We call a semi-elastic element a **solid element**[2] if its symmetry group is contained in the orthogonal group of some configuration.

Proposition 14.2. *In a solid element a state λ is undistorted if and only if it is relaxed and satisfies $g_\lambda = g$.*

Proof. We need only show that for any $\lambda \in \Sigma_{\text{rel}}$, if λ is of maximal symmetry then $g_\lambda = g$. Since the element is solid, we have $g \subset \text{Orth}(G)$ for some $G \in \mathscr{G}$. Let $\lambda_0 = \bar{\lambda}(G)$, so that $G = \hat{G}(\lambda_0)$. By (14.4) we have $g_{\lambda_0} = g \cap \text{Orth}(G) = g$ and hence $g_\lambda = g \cap \text{Orth}(\hat{G}(\lambda)) \subset g = g_{\lambda_0}$. Since λ is of maximal symmetry, it follows that $g_\lambda = g_{\lambda_0} = g$. Q.E.D.

Proposition 14.3. *If the symmetry group of a semi-elastic element is $g = \text{Orth}(G)$ for some $G \in \mathscr{G}$, then the element is isotropic (and solid). Moreover, a state σ is isotropic if and only if $g_\sigma = g$. If σ is isotropic, then $\hat{G}(\sigma) = cG$ for some $c > 0$.*

Proof. We must prove that isotropic states exist. Since undistorted states exist and since the element is solid if $g = \text{Orth}(G)$, it follows from Prop. 14.2 that there are states σ such that $g_\sigma = g = \text{Orth}(G)$. Now, in view of (10.8), $g_\sigma = g = \text{Orth}(G)$ implies $\text{Orth}(G) \subset \text{Orth}(G) \cap \text{Orth}(\hat{G}(\sigma))$, and hence $\text{Orth}(G) \subset \text{Orth}(\hat{G}(\sigma))$. This is possible (see Section 2) only when $\text{Orth}(\hat{G}(\sigma)) = \text{Orth}(G) = g = g_\sigma$, which shows that σ is isotropic.

If, conversely σ is isotropic, we have $\text{Orth}(\hat{G}(\sigma)) = g_\sigma \subset g = \text{Orth}(G)$. Again, this is possible only if $g_\sigma = g$ and $\hat{G}(\sigma) = cG$ with some $c > 0$ (see Section 2). Q.E.D.

We call a semi-elastic element a **fluid element**[3] if its symmetry group g contains the unimodular group $\text{Unim}(\mathscr{T})$. It follows from Prop. 14.1 that all relaxed states of a fluid element are isotropic. The converse is also valid, except when the set $\mathscr{G}$ of configurations is *degenerate* in the sense that all $G \in \mathscr{G}$ are scalar multiples of one of them. Degeneracy is a severe constraint and of little if any physical interest.

Proposition 14.4. *If all relaxed states of a semi-elastic element are isotropic and if $\mathscr{G}$ is non-degenerate, then the element is fluid.*

Proof. To say that all relaxed states are isotropic means that $g_\lambda = \text{Orth}(\hat{G}(\lambda))$ for all $\lambda \in \Sigma_{\text{rel}}$. Hence, since $\hat{G}|_{\Sigma_{\text{rel}}}$ is surjective, it follows that $g \supset \text{Orth}(G)$ for all

[1] In [TN], Sects. 31 and 33 and in earlier work ([N2], [CN1], [CN2]) I have used a related notion of "undistorted configuration", applicable only to isotropic materials and to solids. The concept introduced here applies also to materials that are neither one nor the other.

[2] See [N2], Sect. 20 or [TN], Sect. 33.

[3] See [N2], Sect. 21 or [TN], Sect. 32.

$G \in \mathcal{G}$. Since $\mathcal{G}$ is non-degenerate, the groups Orth(G) are not all the same (see Section 2). Therefore, we must have $g \neq$ Orth(G) for some $G \in \mathcal{G}$. By the theorem on the maximality of an orthogonal group in the unimodular group it follows that $g \supset$ Unim$(\mathcal{T})$. Q.E.D.

For a fluid element, all configurations are undistorted and, by Prop. 10.4, the stress in all relaxed states reduces to a pressure. For an isotropic solid element, by Prop. 14.3, all undistorted configurations are scalar multiples of any one of them. An incompressible isotropic solid element has exactly one undistorted configuration.

Remark 1. For material elements that are not semi-elastic, the conclusions of Props. 14.1–4 need no longer be valid. It is conceivable, for example, that the symmetry group is an orthogonal group or even the unimodular group and yet the element is not isotropic, because there are no isotropic states.

For semi-elastic materials, Axiom VI of Section 13 yields the following strong conclusion.

Proposition 14.5. *In a semi-elastic element, every state is accessible from every other state, i.e.* $\Sigma_\sigma = \Sigma$ *for all* $\sigma \in \Sigma$.

Proof. Let $\tau \in \Sigma_\sigma$. By Prop. 13.2 we then also have $\hat{\lambda}(\tau) \in \Sigma_\sigma$ and hence, by Prop. 13.1, $\Sigma_{\hat{\lambda}(\tau)} \subset \Sigma_\sigma$.

Now choose a process $P \in \Pi$ such that $P^i = \hat{G}(\hat{\lambda}(\tau))$ and $P^f = \hat{G}(\lambda_0)$, where $\lambda_0 \in \Sigma_{\text{rel}}$ satisfies $\Sigma_{\lambda_0} = \Sigma$ (see Axiom VI). The existence of such a process is assured by the condition (P4) of Definition 4.1. We have

$$\hat{\rho}(\hat{\lambda}(\tau), P) \in \Sigma_{\hat{\lambda}(\tau)}, \qquad P^f = \hat{G}(\hat{\rho}(\hat{\lambda}(\tau), P)) = \hat{G}(\lambda_0). \tag{14.5}$$

Using (12.3), we obtain from $(14.5)_2$ that $\hat{G}(\hat{\lambda}(\hat{\rho}(\hat{\lambda}(\tau), P))) = \hat{G}(\lambda_0)$. The injectivity of $\hat{G}|_{\Sigma_{\text{rel}}}$ yields $\lambda_0 = \hat{\lambda}(\hat{\rho}(\hat{\lambda}(\tau), P))$. Hence, by $(14.5)_1$ and Prop. 13.2, we conclude that $\lambda_0 \in \Sigma_{\hat{\lambda}(\tau)} \subset \Sigma_\sigma$. Using Prop. 13.1 again, we find $\Sigma = \Sigma_{\lambda_0} \subset \Sigma_\sigma$, *i.e.* $\Sigma_\sigma = \Sigma$. Q.E.D.

Remark 2. Prop. 14.5 would remain valid if Axiom VI were replaced by the weaker requirement that every state σ be accessible from some relaxed state λ_σ which may depend on σ.

The results of this section are valid, *a fortiori*, for elastic material elements. For such elements, $\hat{G}: \Sigma \to \mathcal{G}$ is a bijection, and the sections $\Sigma_G = \{\hat{\lambda}(G)\}$ are singletons. The natural topology on Σ is then the discrete topology. One can use $\hat{G}: \Sigma \to \mathcal{G}$ and its inverse $\hat{\lambda}: \mathcal{G} \to \Sigma$ to identify the state space Σ with the set $\mathcal{G}$ of configurations. If this is done one must remember, however, that $\mathcal{G}$ is equipped with the discrete topology when considered as the state space.

In order to define the structure of an elastic element on a given body element $(\mathcal{T}, \mathcal{G}, \Pi)$, it is sufficient to prescribe a *stress function* $\hat{S}: \mathcal{G} \to \mathcal{S} = \text{Sym}(\mathcal{T}^*, \mathcal{T})$. For the state space one can then take $\Sigma = \mathcal{G}$, for $\hat{G}$ the identity mapping of $\mathcal{G}$, and for $\hat{\rho}$ the trivial mapping given by $\hat{\rho}(P^i, P) = P^f$.

15. Deformation Histories for Semi-Elastic Elements

As in Section 4, let $\mathscr{G}$ be any set. We observe the notation and terminology of Section 4. A function H of the type

$$H: \mathbb{R}^+ \to \mathscr{G} \tag{15.1}$$

will be called a **history** *with values in* $\mathscr{G}$. Its value

$$H(0) = H^f \tag{15.2}$$

will be called the *final value* of H.

If $G \in \mathscr{G}$ is given, we can define a history $G_{(\infty)}: \mathbb{R}^+ \to \mathscr{G}$ by

$$G_{(\infty)}(s) = G \quad \text{for } s \in \mathbb{R}^+; \tag{15.3}$$

we call this history the *constant history* at G.

For any $s \in \mathbb{R}^+$, we define the **s-section** $H^{(s)}: \mathbb{R}^+ \to \mathscr{G}$ of a history H by

$$H^{(s)}(r) = H(s+r) \quad \text{for } r \in \mathbb{R}^+. \tag{15.4}$$

It is again a history. The s'-section of the s-section of H is the same as the $(s+s')$-section of H, *i.e.*,

$$H^{(s)(s')} = H^{(s+s')} \quad \text{for } s, s' \in \mathbb{R}^+. \tag{15.5}$$

Of course, we have $H^{(0)} = H$.

If H is a history and $s \in \mathbb{R}^+$, we can define a process $H_{[s]}$, of duration s, by

$$H_{[s]}(t) = H(s-t) \quad \text{for } t \in [0, s]. \tag{15.6}$$

We call $H_{[s]}$ the **s-segment** of H. We have

$$(H_{[s]})^i = H(s), \quad (H_{[s]})^f = H^f = H(0). \tag{15.7}$$

The s-segment $(H^{(r)})_{[s]}$ of the r-section $H^{(r)}$ of H is a segment of the segment $H_{[s+r]}$, namely

$$(H^{(r)})_{[s]} = (H_{[s+r]})_{[0,s]}. \tag{15.8}$$

If H is a history and P a process, of duration d_P, such that $H^f = P^i$, we define the **continuation** $H * P: \mathbb{R}^+ \to \mathscr{G}$ *of H with P* by

$$(H * P)(s) = \begin{cases} H(s-d_P) & \text{if } s \geqq d_P \\ P(d_P - s) & \text{if } s \in [0, d_P]. \end{cases} \tag{15.9}$$

We have, for all $s \in \mathbb{R}^+$,

$$H = H^{(s)} * H_{[s]}. \tag{15.10}$$

The continuation is associative, *i.e.*

$$(H * P_1) * P_2 = H * (P_1 * P_2) \quad \text{if } H^f = P_1^i, \quad P_1^f = P_2^i.$$

The s-section of a continuation $H * P$ is given by

$$(H * P)^{(s)} = \begin{cases} H^{(s-d_P)} & \text{if } s \geqq d_P \\ H * P_{[0, d_P - s]} & \text{if } s \in [0, d_P]. \end{cases} \tag{15.11}$$

For the s-segment of $H * P$ we have

$$(H * P)_{[s]} = \begin{cases} P_{[d_P - s,\, d_P]} & \text{if } s \in [0, d_P] \\ H_{[s-d_P]} * P & \text{if } s \geqq d_P. \end{cases} \tag{15.12}$$

Definition 15.1. *Let $(\mathscr{T}, \mathscr{G}, \Pi, \Sigma, \hat{G}, \hat{S}, \hat{\rho})$ be a semi-elastic material element, so that $\hat{G}|_{\Sigma_{rel}}$ is invertible with inverse $\bar{\lambda}: \mathscr{G} \to \Sigma_{rel}$. A history $H: \mathbb{R}^+ \to \mathscr{G}$ is called a* **deformation history** *for the element $\mathscr{T}$ if*

(i) *for all $s \in R^+$, the segment $H_{[s]}$ belongs to Π* (in view of (15.8) and (P2) of Definition 4.1, we then also have $(H^{(r)})_{[s]} \in \Pi$ for all $s, r \in R^+$), *and*

(ii) *the limit*

$$\lim_{s \to \infty} \hat{\rho}(\bar{\lambda}(H^{(r)}(s)), (H^{(r)})_{[s]}) \in \Sigma_{H(r)}$$

exists for all $r \in \mathbb{R}^+$.

The set of all deformation histories for $\mathscr{T}$ will be denoted by Φ and, in view of condition (ii), *we can define $\bar{\rho}: \Phi \to \Sigma$ by*

$$\bar{\rho}(H) = \lim_{s \to \infty} \hat{\rho}(\bar{\lambda}(H(s)), H_{[s]}). \tag{15.13}$$

We use the notation

$$\Phi_G = \{H \in \Phi \mid H^f = G\} \tag{15.14}$$

for the set of all deformation histories for $\mathscr{T}$ that end at the configuration $G \in \mathscr{G}$.

We note that if H belongs to Φ, so do all of its sections $H^{(s)}$, $s \in \mathbb{R}^+$. Also, in view of (12.1), all constant histories (15.3) belong to Φ, and by Prop. 12.1 (iv), we have

$$\bar{\rho}(G_{(\infty)}) = \bar{\lambda}(G) \tag{15.15}$$

for all $G \in \mathscr{G}$. The next observation is somewhat less trivial:

Proposition 15.1. *If $H \in \Phi$ and $P \in \Pi_{H^f}$, then $H * P \in \Phi$ and we have*

$$\bar{\rho}(H * P) = \hat{\rho}(\bar{\rho}(H), P). \tag{15.16}$$

Proof. By (15.9) and (15.12) we have, for $s \geqq d_P$,

$$\hat{\rho}(\bar{\lambda}((H * P)(s)), (H * P)_{[s]}) = \hat{\rho}(\bar{\lambda}(H(s - d_P)), H_{[s-d_P]} * P).$$

Application of (7.7) yields

$$\hat{\rho}(\bar{\lambda}((H * P)(s)), (H * P)_{[s]}) = \hat{\rho}(\hat{\rho}(\bar{\lambda}(H(s - d_P)), H_{[s-d_P]}), P).$$

Taking the limit $s \to \infty$ and using the continuity of $\hat{\rho}(\cdot, P)$ (see Prop. 11.2), we see that (15.13) gives the desired result (15.16). Q.E.D.

Observing (15.10) and (15.15), we obtain the following special cases of (15.16):

$$\bar{\rho}(H) = \hat{\rho}(\bar{\rho}(H^{(s)}), H_{[s]}), \qquad s \in \mathbb{R}^+, \ H \in \Phi, \tag{15.17}$$

$$\bar{\rho}(P^i_{(\infty)} * P) = \hat{\rho}(\bar{\lambda}(P^i), P), \qquad P \in \Pi. \tag{15.18}$$

Proposition 15.2. *The function* $\bar{\rho}\colon \Phi \to \Sigma$ *satisfies*

$$\lim_{s\to\infty} \bar{\rho}(H(s)_{(\infty)} * H_{[s]}) = \bar{\rho}(H) \qquad \textit{for all } H \in \Phi \tag{15.19}$$

and

$$\lim_{t\to\infty} \bar{\rho}(H * H^f_{(t)}) = \bar{\rho}(H^f_{(\infty)}) \qquad \textit{for all } H \in \Phi. \tag{15.20}$$

Proof. If we write (15.18) with the choice $P = H_{[s]}$ and substitute the result into (15.13), we obtain (15.19).

To prove (15.20), we apply (15.16) with the choice $P = H^f_{(t)}$ and take the limit $t \to \infty$. Observing (14.3), we obtain

$$\lim_{t\to\infty} \bar{\rho}(H * H^f_{(t)}) = \lim_{t\to\infty} \hat{\rho}(\bar{\rho}(H), H^f_{(t)}) = \bar{\lambda}(H^f).$$

In view of (15.15) this is the desired result. Q.E.D.

The next result states that arbitrarily close to any state there is a state that can be described by a deformation history.

Proposition 15.3. *The range of* $\bar{\rho}$ *is dense in the state space* Σ, *i.e.*,

$$\Sigma = \mathrm{cl}\,\Sigma_{\mathrm{hist}}, \quad \textit{where} \quad \Sigma_{\mathrm{hist}} = \mathrm{Range}\,\bar{\rho}. \tag{15.21}$$

Proof. Let λ be any relaxed state. It follows from the definition of Σ_λ (*cf.* (13.1)) and from (15.18) that

$$\Sigma_\lambda = \mathrm{cl}\{\bar{\rho}(P^i_{(\infty)} * P) \mid P \in \Pi_{\hat{G}(\lambda)}\}.$$

Since this set is the closure of the range of the restriction of $\bar{\rho}$ to histories of the form $P^i_{(\infty)} * P$ with $P^i = \hat{G}(\lambda)$, it follows *a fortiori* that $\Sigma_\lambda = \mathrm{cl}\,\Sigma_{\mathrm{hist}}$. By Prop. 14.5, this is the desired result. Q.E.D.

For semi-elastic elements, it is often sufficient to confine one's attention to the space Σ_{hist} of states that correspond to deformation histories. The remaining states, which are limits of states determined by histories, are of little physical interest.

Finally, we describe how $\bar{\rho}(H)$ changes under a symmetry.

Proposition 15.4. *If* $A \in \mathscr{g}$ *is a symmetry of a semi-elastic element, then* $A^* \Phi A = \Phi$ *and*

$$\bar{\rho}(A^* H A) = \iota_{A^{-1}}(\bar{\rho}(H)) \qquad \textit{for all } H \in \Phi. \tag{15.22}$$

Proof. Let $A \in \mathscr{g}$. It follows from (10.3) and the definition of $\bar{\lambda}$ as the inverse $\hat{G}|_{\Sigma_{\mathrm{rel}}}$ that

$$\bar{\lambda}(A^* G A) = \iota_A^{-1}(\bar{\lambda}(G)) \tag{15.23}$$

for all $G \in \mathscr{g}$. Now let $H \in \Phi$. Using (15.23) with $G = H(s)$ and then (10.5) with the choice $P = H_{[s]}$ and $\sigma = \iota_A^{-1}(\bar{\lambda}(H(s)))$, we obtain

$$\begin{aligned}\hat{\rho}(\bar{\lambda}((A^* H A)(s)), (A^* H A)_{[s]}) &= \hat{\rho}(\iota_A^{-1}(\bar{\lambda}(H(s))), A^* H_{[s]} A)\\ &= \iota_A^{-1}(\hat{\rho}(\bar{\lambda}(H(s)), H_{[s]})).\end{aligned}$$

Recalling that $\iota_A^{-1} = \iota_{A^{-1}}$ (*cf.* (9.6)) is a homeomorphism and remembering the definition (15.13) of $\bar{\rho}$, we see, by taking the limit $s \to \infty$, that $A^* H A \in \Phi$ and that (15.22) holds. Q.E.D.

16. The History Functional of a Semi-Elastic Element

Let $\mathscr{T}$ be a semi-elastic element, Φ the set of its deformation histories, and $\bar{\rho}: \Phi \to \Sigma$ the function defined by (15.13). We define the **history functional** $\bar{S}$ of the element by

$$\bar{S} = \hat{S} \circ \bar{\rho}: \Phi \to \mathscr{S}, \tag{16.1}$$

where $\mathscr{S} = \mathrm{Sym}(\mathscr{T}^*, \mathscr{T})$ is the stress space. If we apply $\hat{S}$ to (15.16), we see that $\bar{S}$ is related to the response functional $\tilde{S} = \hat{S} \circ \hat{\rho}$ by

$$\tilde{S}(\bar{\rho}(H), P) = \bar{S}(H * P) \qquad \text{for } H \in \Phi,\ P \in \Pi_{H^f}. \tag{16.2}$$

A knowledge of the history functional is sufficient for deciding whether two deformation histories determine the same state. Indeed, it follows from Axiom III, Section 7, and (16.2) that

$$\bar{\rho}(H_1) = \bar{\rho}(H_2) \Leftrightarrow \left\{ \begin{matrix} H_1^f = H_2^f \text{ and } \bar{S}(H_1 * P) = \bar{S}(H_2 * P) \\ \text{for all } P \in \Pi \text{ with } P^i = H_1^f = H_2^f \end{matrix} \right\}. \tag{16.3}$$

We shall show in Theorem 16.1, below, that a knowledge of the set Φ of deformation histories and of the history functional $\bar{S}: \Phi \to \mathscr{S}$ characterizes the semi-elastic element structure on $\mathscr{T}$. We first give a complete list of conditions that are satisfied by Φ. They are easy consequences of the results of the previous section.

(H1) *Every* $H \in \Phi$ *is a history* $H: \mathbb{R}^+ \to \mathscr{G}$ *with values in a closed and connected subset* $\mathscr{G}$ *of* $\mathrm{Sym}^+(\mathscr{T}, \mathscr{T}^*)$.

(H2) *For all* $G \in \mathscr{G}$ *we have* $G_{(\infty)} \in \Phi$.

(H3) *If* $H \subset \Phi$ *and* $s \in \mathbb{R}^+$, *then* $H^{(s)} \in \Phi$.

(H4) *If* $H, K \in \Phi$ *and* $s \in \mathbb{R}^+$ *with* $K^f = H(s)$, *then* $K * H_{[s]} \in \Phi$.

(H5) *For any* $G_1, G_2 \in \mathscr{G}$ *there is a* $H \in \Phi$ *such that* $G_1, G_2 \in \mathrm{Range}\, H$.

The set Π of all deformation processes for $\mathscr{T}$ is determined by Φ as follows:

$$\Pi = \{H_{[s]} \mid H \in \Phi,\ s \in \mathbb{R}^+\}. \tag{16.4}$$

Next, we show that the history functional $\bar{S}$ has the following two convergence properties:

(F1) *For every* $H \in \Phi$ *we have*

$$\lim_{s \to \infty} \bar{S}(H(s)_{(\infty)} * H_{[s]}) = \bar{S}(H). \tag{16.5}$$

(F2) *If* $H \in \Phi$ *and* $P \in \Pi_{H^f}$, *so that* $H^f = P^i$, *then*

$$\lim_{t \to \infty} \bar{S}(H * H^f_{(t)} * P) = \bar{S}(H^f_{(\infty)} * P). \tag{16.6}$$

Property (F1) is an immediate consequence of (15.19) and the continuity of $\hat{S}$. If we apply the convergence criterion for states given by Prop. 11.3 to (15.20) and observe (16.2), we see that (F2) must be valid. Applying the same convergence criterion to the condition (ii) of Definition 15.1 and observing (15.15) and (15.16), we see that the class Φ must satisfy the following *saturation condition* with respect to $\bar{S}$:

(S) *If H is a history with values in $\mathscr{G}$ such that $H_{[s]}\in\Pi$ for every $s\in\mathbb{R}^+$ and such that*

$$\lim_{s\to\infty}\bar{S}((H^{(r)}(s))_{(\infty)}*(H^{(r)})_{[s]}*P)$$

exists whenever $r\in\mathbb{R}^+$ and $P^i=H(r)$, then H belongs to Φ.

Theorem 16.1. *Let $\mathscr{T}$ be a finite-dimensional real vector space, and let Φ be a class of histories with the properties* (H1)–(H5). *Let Π be defined by* (16.4). *Finally, let $\bar{S}:\Phi\to\mathscr{S}$ be a functional with the properties* (F1), (F2) *and assume that Φ satisfies the saturation condition* (S) *with respect to $\bar{S}$. Then there is a unique semi-elastic element structure on $\mathscr{T}$ for which Φ is the class of deformation histories and $\bar{S}$ the history functional.*

Proof. First we assume the existence of a semi-elastic element structure ($\mathscr{T}$, $\mathscr{G}$, Π, Σ, $\hat{G}$, $\hat{S}$, $\hat{\rho}$) whose set of deformation histories is Φ and whose history functional is $\bar{S}$, and we prove the uniqueness of this structure. It is clear that Π must be the set defined by (16.4) in terms of Φ and that $\mathscr{G}$ must be the set mentioned in (H1). Hence the underlying body element ($\mathscr{T}$, $\mathscr{G}$, Π) is determined by Φ.

Guided by (16.3), we now introduce an equivalence relation $\sim$ on Φ by

$$H_1\sim H_2\Leftrightarrow\left\{\begin{matrix}H_1^f=H_2^f \text{ and } \bar{S}(H_1*P)=\bar{S}(H_2*P)\\ \text{for all } P\in\Pi \text{ with } P^i=H_1^f=H_2^f\end{matrix}\right\}. \tag{16.7}$$

Let Σ' be the quotient of Φ with respect to this equivalence relation. In other words, the elements of Σ' are the equivalence classes determined by $\sim$. We use the notation $\bar{\rho}':\Phi\to\Sigma'$ for the quotient mapping, which assigns to each $H\in\Phi$ the equivalence class $\bar{\rho}'(H)\in\Sigma'$ to which H belongs. It follows from (16.3) that there is a bijection $\iota:\Sigma'\to\Sigma_{\text{hist}}=\text{Range}\,\bar{\rho}$ such that $\bar{\rho}=\iota\circ\bar{\rho}'$. The bijection ι associates with every equivalence class $\sigma'\in\Sigma$ the common value of $\bar{\rho}$ at all members $H\in\sigma'$. We now define $\hat{G}':\Sigma'\to\mathscr{G}$ by $\hat{G}'=\hat{G}\circ\iota$, $\hat{S}':\Sigma'\to\mathscr{S}$ by $\hat{S}'=\hat{S}\circ\iota$, and $\hat{\rho}':(\Sigma'\times\Pi)_{\text{fit}}\to\Sigma'$ by $\iota(\hat{\rho}'(\cdot,\cdot))=\hat{\rho}(\iota(\cdot),\cdot)$. We then have

$$\hat{G}'(\sigma')=H^f,\qquad \hat{S}'(\sigma')=\bar{S}(H)\qquad \text{if } H\in\sigma' \tag{16.8}$$

and

$$\hat{\rho}'(\sigma',P)=\bar{\rho}'(H*P)\qquad \text{if } H\in\sigma',\ P^i=H^f. \tag{16.9}$$

It is clear from (16.8) and (16.9) that $\hat{G}'$, $\hat{S}'$, and $\hat{\rho}'$ are completely determined by Φ and $\bar{S}$.

We have obtained a structure ($\mathscr{T}$, $\mathscr{G}$, Π, Σ', $\hat{G}'$, $\hat{S}'$, $\hat{\rho}'$) which is determined by Φ and $\bar{S}$ alone. It is easily seen that this structure satisfies Axioms I–III. Hence we can apply Definition 11.1 to it and endow each section $\Sigma'_G=\{\sigma'\in\Sigma'\mid\hat{G}'(\sigma')=G\}$ with a natural uniformity and $\Sigma'=\bigcup\{\Sigma'_G\mid G\in\mathscr{G}\}$ with a natural topology. The functions $\hat{G}'$, $\hat{S}'$, and $\hat{\rho}'(\cdot,P)$ are continuous with respect to this topology. Moreover, it is easily seen that the bijection $\iota:\Sigma'\to\Sigma_{\text{hist}}$ is a homeomorphism when Σ_{hist} is endowed with the topology induced by the natural topology of Σ.

Next, we consider the completions $\bar{\Sigma}'_G$ and $\Sigma^*=U\{\bar{\Sigma}'_G\mid G\in\mathscr{G}\}$ as described at the end of Section 11, and we extend the functions $\hat{G}'$, $\hat{S}'$, and $\hat{\rho}'$ so as to obtain a structure $(\mathscr{T}, \mathscr{G}, \Pi, \Sigma^*, \hat{G}^*, \hat{S}^*, \hat{\rho}^*)$ which satisfies not only Axioms I–III but also Axiom IV. Since Axiom IV is satisfied for $(\mathscr{T}, \mathscr{G}, \Pi, \Sigma, \hat{G}, \hat{S}, \hat{\rho})$, we can also extend the homeomorphism $\iota: \Sigma' \to \Sigma_{\text{hist}}$ to a homeomorphism $\iota^*: \Sigma^* \to \operatorname{cl} \Sigma_{\text{hist}} = \Sigma$ (see (15.21)). The relations $\hat{G}' = \hat{G} \circ \iota$, $\hat{S}' = \hat{S} \circ \iota$, and $\iota(\hat{\rho}'(\cdot, \cdot)) = \hat{\rho}(\iota(\cdot), \cdot)$ remain satisfied after $\hat{G}'$, $\hat{S}'$, $\hat{\rho}'$, and ι have been extended to $\hat{G}^*$, $\hat{S}^*$, $\hat{\rho}^*$, and ι^*. It follows that the identity $1_{\mathscr{T}}$ is a material isomorphism between the material element $(\mathscr{T}, \mathscr{G}, \Pi, \Sigma, \hat{G}, \hat{S}, \hat{\rho})$ and the structure $(\mathscr{T}, \mathscr{G}, \Pi, \Sigma^*, \hat{G}^*, \hat{S}^*, \hat{\rho}^*)$ determined by Φ and $\bar{S}$, the bijection $\iota_{1_{\mathscr{T}}}$ being the same as ι^*. Since we have agreed, under these circumstances (see Section 9), to employ ι^* to identify Σ with Σ^* and not to distinguish between the two structures, we have the following conclusion: *If there is a semi-elastic element structure on $\mathscr{T}$ whose set of deformation histories is Φ and whose history functional is $\bar{S}$, it must be given by $(\mathscr{T}, \mathscr{G}, \Pi, \Sigma^*, \hat{G}^*, \hat{S}^*, \hat{\rho}^*)$.*

We now prove the existence of a semi-elastic element structure on $\mathscr{T}$ whose set of deformation histories is Φ and whose history functional is $\bar{S}$. If $\mathscr{G}$ is the set mentioned in (H 1) and if Π is defined by (16.4), it is easily inferred from (H 1)–(H 5) that Π satisfies the conditions (P 1)–(P 4) of Definition 4.1 and hence that $(\mathscr{T}, \mathscr{G}, \Pi)$ endows $\mathscr{T}$ with the structure of a body element.

Consider the quotient set Σ' of Φ with respect to the equivalence relation defined by (16.7) and the quotient mapping $\bar{\rho}': \Phi \to \Sigma'$. It follows from (16.7) that $H_1 \sim H_2$ implies $H_1^f = H_2^f$, $\bar{S}(H_1) = \bar{S}(H_2)$, and $H_1 * P \sim H_2 * P$ when $P^i = H_1^f = H_2^f$. Thus, it is meaningful to *define* $\hat{G}'$, $\hat{S}'$ by (16.8) and $\hat{\rho}'$ by (16.9). We introduce a corresponding response functional $\tilde{S}' = \hat{S}' \circ \hat{\rho}'$, so that

$$\tilde{S}'(\sigma', P) = \bar{S}(H * P) \quad \text{if } H \in \sigma', \ P^i = H^f. \tag{16.10}$$

It is easily verified that the septuple $(\mathscr{T}, \mathscr{G}, \Pi, \Sigma', \hat{G}', \hat{S}', \hat{\rho}')$ thus constructed satisfies Axioms I–III. Observing (16.10), it follows from Condition (F2) for $\bar{S}$ that $\tilde{S}'$ satisfies (12.4) when $\hat{\lambda}': \Sigma' \to \Sigma'_{\text{rel}}$ is defined by

$$\hat{\lambda}'(\sigma') = \bar{\rho}'(H^f_{(\infty)}) \quad \text{if } H \in \sigma'. \tag{16.11}$$

Therefore Axiom V is also satisfied, because it is equivalent to (12.4). The set Σ'_{rel} consists of the equivalence classes that contain constant histories. The restriction of $\hat{G}'$ to Σ'_{rel} has an inverse $\bar{\lambda}': \mathscr{G} \to \Sigma'_{\text{rel}}$ given by

$$\bar{\lambda}'(G) = \bar{\rho}'(G_{(\infty)}), \tag{16.12}$$

so that $\bar{\lambda}'(G)$ is the equivalence class determined by the constant history $G_{(\infty)}$. Thus, the defining condition of Definition 14.1 for a semi-elastic element is satisfied.

Next we construct the completion $\Sigma^* = \bigcup \{\bar{\Sigma}'_G \mid G \in \mathscr{G}\}$ and the extensions $\hat{G}^*$, $\hat{S}^*$, and $\hat{\rho}^*$ as described in the uniqueness proof above. The septuple $(\mathscr{T}, \mathscr{G}, \Pi, \Sigma^*, \hat{G}^*, \hat{S}^*, \hat{\rho}^*)$ thus constructed satisfies not only Axioms I, II, III, and V, but also Axiom IV. Note that the completion introduces no new relaxed states, so that $\Sigma^*_{\text{rel}} = \Sigma'_{\text{rel}}$. To show that the accessibility axiom, Axiom VI, is satisfied, let $G \in \mathscr{G}$ and $H \in \Phi$ be given. For each $s \in \mathbb{R}^+$ we select a process $P_s \in \Pi$ such that $P_s^i = G$, $P_s^f = H(s)$. This is possible because Π satisfies (P4) of Definition 4.1. Using the assumption that $\bar{S}$ satisfies both (F 1) and (F 2), we easily infer that

$$\lim_{s \to \infty} \lim_{t \to \infty} \bar{S}(G_{(\infty)} * P_s * H(s)_{(t)} * H_{[s]} * P') = \bar{S}(H * P') \tag{16.13}$$

holds for all $P'\in\Pi_{H^t}$. Observing (16.9), (16.10), (16.11), and using the convergence criterion of Prop. 11.3, we see that (16.13) is equivalent to

$$\lim_{s\to\infty}\lim_{t\to\infty}\hat{\rho}'(\bar{\lambda}'(G), P_s * H(s)_{(t)} * H_{[s]}) = \bar{\rho}'(H)$$

in the natural topology of $\Sigma' = \text{Range}\,\bar{\rho}'$. It follows that $\bar{\rho}'(H)$ is accessible from $\bar{\lambda}'(G)$. Since H was arbitrary, we conclude that $\Sigma'\subset\Sigma^*_{\bar{\lambda}'(G)}$, and since Σ' is dense in Σ^*, we find $\Sigma^*=\Sigma^*_{\bar{\lambda}'(G)}$, *i.e.* that Axiom VI is satisfied. Therefore $(\mathcal{T}, \mathcal{G}, \Pi, \Sigma^*, \hat{G}^*, \hat{S}^*, \hat{\rho}^*)$ is a semi-elastic material element structure.

It remains to be shown that Φ is the set of deformation histories and $\bar{S}$ the history functional of the semi-elastic element just constructed. It follows from (H 2), (H 3), (H 4), and (F 1), using also (16.10) and (16.12), that

$$\lim_{s\to\infty}\bar{S}'(\bar{\lambda}'(H^{(r)}(s)), (H^{(r)})_{[s]} * P) = \bar{S}(H^{(r)} * P) \tag{16.14}$$

for all $H\in\Phi$, all $r\in\mathbb{R}^+$, and $P\in\Pi_{H(r)}$. From the convergence criterion of Prop. 11.3 we see that condition (ii) as well as condition (i) of Definition 15.1 is satisfied for all $H\in\Phi$ and hence that Φ is contained in the set of deformation histories for the constructed element. The fact that Φ is actually equal to this set of deformation histories is a consequence of the saturation condition (S). It is not hard to verify, finally, that $\bar{S}$ is the history functional of the constructed element. Q.E.D.

Remark. If Φ and $\bar{S}: \Phi\to\mathcal{S}$ are such that (H 1)–(H 5) and (F 1), (F 2), but not necessarily the saturation condition (S), are satisfied, the construction of the semi-elastic element structure $(\mathcal{T}, \mathcal{G}, \Pi, \Sigma^*, \hat{G}^*, \hat{S}^*, \hat{\rho}^*)$ described in the proof above can still be carried out. However, the set Φ^* of deformation histories for the element so constructed will then be larger than the original set Φ. To obtain Φ^* from Φ, one must saturate Φ by joining to it all histories for which the limits of condition (S) exist.

Theorem 16.1 states, in essence, that a semi-elastic material element structure on $\mathcal{T}$ is the same as a structure defined by a triple $(\mathcal{T}, \Phi, \bar{S})$ subject to the conditions (H 1)–(H 5), (F 1), (F 2), and (S) as axioms. It follows that material isomorphisms between semi-elastic elements can be described as isomorphisms between structures of the type $(\mathcal{T}, \Phi, \bar{S})$. More precisely, we have the following corollary to Theorem 16.1.

Proposition 16.2. *Let $\mathcal{T}_1$ and $\mathcal{T}_2$ be semi-elastic elements whose sets of deformation histories are Φ_1 and Φ_2 and whose history functionals are $\bar{S}_1$ and $\bar{S}_2$, respectively. A mapping $A\in\text{Invlin}(\mathcal{T}_1, \mathcal{T}_2)$ is a material isomorphism between $\mathcal{T}_1$ and $\mathcal{T}_2$ if and only if $\Phi_1 = A^*\Phi_2 A$ and*

$$\bar{S}_1(A^* H A) = A^{-1}\bar{S}_2(H) A^{*-1} \quad \textit{for all } H\in\Phi_2. \tag{16.15}$$

Applying this proposition to material automorphism, we obtain the following corollary.

Proposition 16.3. *Let $\mathcal{T}$ be a semi-elastic element whose set of deformation histories is Φ and whose history functional is $\bar{S}$. A mapping $A\in\text{Invlin}(\mathcal{T})$ belongs to the symmetry group g of $\mathcal{T}$ if and only if $\Phi = A^*\Phi A$ and*

$$\bar{S}(A^* H A) = A^{-1}\bar{S}(H) A^{*-1} \quad \textit{for all } H\in\Phi. \tag{16.16}$$

17. Monotonous Processes

We deal with processes and histories as described in Sections 4 and 15. Let $\mathcal{T}_1$ and $\mathcal{T}_2$ be two real vector spaces of equal finite dimension. We say that a process P_1 with values in $\mathrm{Sym}^+(\mathcal{T}_1, \mathcal{T}_1^*)$ is *congruent* to a process P_2 with values in $\mathrm{Sym}^+(\mathcal{T}_2, \mathcal{T}_2^*)$ if P_1 and P_2 have the same duration and if there is an $A \in \mathrm{Invlin}(\mathcal{T}_1, \mathcal{T}_2)$ such that

$$P_1 = A^* P_2 A. \tag{17.1}$$

For example, if $(\mathcal{T}_1, \mathcal{G}_1, \Pi_1)$ and $(\mathcal{T}_2, \mathcal{G}_2, \Pi_2)$ are body elements (*cf.* Definition 4.1) and if A is an isomorphism between these elements (*cf.* Definition 9.1), then every process in Π_1 is of the form $A^* P_2 A$ with $P_2 \in \Pi_2$ and hence congruent to a process in Π_2. It is clear that congruence is an equivalence relation.

Definition 17.1. *We say that a process P with values in* $\mathrm{Sym}^+(\mathcal{T}, \mathcal{T}^*)$ *is* **monotonous** *if P is continuous and if any two segments of P of equal duration are congruent. We say that a history H with values in* $\mathrm{Sym}^+(\mathcal{T}, \mathcal{T}^*)$ *is* **monotonous** *if all of its segments $H_{[s]}$, $s \in R^+$, are monotonous.*

Proposition 17.1. *A process P with values in* $\mathrm{Sym}^+(\mathcal{T}, \mathcal{T}^*)$ *is monotonous if and only if there is a function A:* $[0, d_P] \to \mathrm{Invlin}(\mathcal{T})$ *such that*

$$P(r+t) = A(r)^* P(t) A(r) \quad \textit{for } r \in [0, d_P],\ t \in [0, d_P - r]. \tag{17.2}$$

Proof. If P is monotonous, then $P_{[0, d_P - r]}$ and $P_{[r, d_P]}$ are congruent, and (17.2) expresses this congruence. If (17.2) holds, then $P_{[r, r+d]}$ and $P_{[0, d]}$ are congruent if $r \leqq d_P - d$. Hence all segments of P of duration d are congruent to $P_{[0, d]}$ and hence to one another, *i.e.*, P is monotonous. Q.E.D.

Every freeze is obviously monotonous. A complete characterization of all monotonous processes is given in the following theorem.

Theorem 17.1. *A process P with values in* $\mathrm{Sym}^+(\mathcal{T}, \mathcal{T}^*)$ *is monotonous if and only if there is an element $E \in \mathrm{Lin}(\mathcal{T})$ such that*[1]

$$P(t) = \exp(tE^*)\, P^i \exp(tE) \quad \textit{for } t \in [0, d_P]. \tag{17.3}$$

Moreover, if P is monotonous, then (17.3) *holds for exactly one E in the subalgebra of* $\mathrm{Lin}(\mathcal{T})$ *that is generated by the subset* $\{P(t_1)^{-1} P(t_2) \mid t_1, t_2 \in [0, d_P]\}$ *of* $\mathrm{Lin}(\mathcal{T})$.

The proof of this rather non-trivial theorem will be given in a separate paper [N8]. If it were known in advance that both P and A of (17.2) are differentiable, one could prove the assertion by the method used in the proof of Theorem 1 of [N4]. The hard part of the proof of Theorem 17.1 is to show that *a priori* continuity and differentiability assumptions on A are unnecessary.

If (17.3) holds, we call P the **monotonous process of exponent** E, *initial configuration P^i, and duration d_P.* It is clear that every segment of a monotonous process of exponent E is again a monotonous process of exponent E.

[1] If $E \in \mathrm{Lin}(\mathcal{T})$, then $\exp E \in \mathrm{Invlin}(\mathcal{T})$ is defined by $\exp E = \sum_{n=1}^{\infty} \frac{1}{n!} E^n$ or as the value $A(1)$ at $t=1$ of the solution of the initial value problem $\dot{A}(t) = EA(t)$, $A(0) = 1$.

In general, the exponent is not uniquely determined by the monotonous process P. For example, if $P=G_{(d)}$ is a freeze, then (17.3) holds not only when $E=0$, but also when E is skew-symmetric relative to G, *i.e.* when $E^*G=-GE$. In this case, the subalgebra mentioned in Theorem 17.1 consists of the scalar multiples of the identity of $\mathcal{T}$, and $E=0$ is the only exponent in that subalgebra for which (17.3) holds.

Proposition 17.2. *A history H with values in* $\mathrm{Sym}^+(\mathcal{T},\mathcal{T}^*)$ *is monotonous if and only if it is of the form*

$$H(s)=\exp(-sE^*)\,H^f\exp(-sE),\qquad s\in\mathbb{R}^+, \tag{17.4}$$

with $E\in\mathrm{Lin}(\mathcal{T})$. *If* (17.4) *holds, then every segment of H is a monotonous process of exponent E.*

Proof. If (17.4) holds, it is easily seen (*cf.* (15.6) and (15.7)) that

$$H_{[s]}(t)=\exp(tE^*)\,H(s)\exp(tE),\qquad t\in[0,s]$$

and hence that $H_{[s]}$ is monotonous with exponent E for each $s\in\mathbb{R}^+$, which means that H is monotonous.

Now assume that H is monotonous, and choose $r\in\mathbb{R}^+$, $r>0$. Since $H_{[r]}$ is then monotonous, we can apply Theorem 17.1 to $H_{[r]}$ obtaining

$$H(r-t)=H_{[r]}(t)=\exp(tE^*)\,H(r)\exp(tE)\qquad\text{for } t\in[0,r]. \tag{17.5}$$

The choice $t=r$ gives $H^f_{[r]}=H^f=\exp(rE^*)\,H(r)\exp(rE)$. Solving this equation for $H(r)$, substituting the result into (17.5) and then putting $s=r-t$, we derive

$$H(s)=\exp(-sE^*)\,H^f\exp(-sE)\qquad\text{for } s\in[0,r]. \tag{17.6}$$

We must show that (17.6) holds not only for $s\in[0,r]$ but also when $s>r$. Now, if $s>r$, we select $r'>s$ and observe that an equation of the form (17.6), but possibly with an exponent E' different from E, holds for all $s'\in[0,r']$. It follows that H is analytic on $[0,r']$. By the principle of analytic continuation, (17.6) must also hold for s. Q.E.D.

If (17.4) holds, we say that H is the *monotonous history of exponent E* and final configuration H^f. If H is a monotonous history of exponent E, so are all its sections $H^{(s)}$ (see (15.4)) and, since $\exp(r+s)E=(\exp sE)\,(\exp rE)$, we have

$$H^{(s)}=\exp(-sE^*)\,H\exp(-sE). \tag{17.7}$$

We recall the definitions and terminology of Section 5 and consider a body element $(\mathcal{T},\mathcal{G},\Pi)$ as well as a frame-space $\mathcal{V}$. We say that a motion $M\colon[0,d_M]\to\mathcal{P}_{\mathcal{V}}$ is *monotonous* if the deformation process $M^*M\colon[0,d_M]\to\mathcal{G}$ determined by it is monotonous.

Proposition 17.3. *Let M be a motion of $\mathcal{T}$ in $\mathcal{V}$ and $F_{(0)}=MM_i^{-1}$ its displacement tensor relative to the initial placement $M_i=M(0)$. The motion M is monotonous if and only if $F_{(0)}$ is of the form*[1]

[1] Comparison of this result with Theorem I of [N4] shows that the class of monotonous motions is identical to the class of motions introduced by COLEMAN [C] under the name of "substantially stagnant motions", which I later called "motions with constant stretch history" in [N4] and [TN], Sect. 109.

$$F_{(0)}(t) = Q(t)\exp(tB), \tag{17.8}$$

where $B \in \mathrm{Lin}(\mathscr{V})$ *and* Q: $[0, d_M] \to \mathrm{Orth}(\mathscr{V})$, $Q(0) = 1_{\mathscr{V}}$.

Proof. By Theorem 17.1, $M^* M$ and hence M is monotonous if and only if

$$M^*(t)\,M(t) = \exp(tE^*)\,M^{i*}\,M^i \exp(tE). \tag{17.9}$$

If (17.9) holds, then M and $t \mapsto M^i \exp(tE)$ determine the same process. Hence, as was remarked in Section 5,

$$M(t)\,(M^i \exp(tE))^{-1} = Q(t) \tag{17.10}$$

belongs to $\mathrm{Orth}(\mathscr{V})$. If we put $B = M_i E M_i^{-1}$ so that $M_i \exp(tE)\,M_i^{-1} = \exp(tB)$, we see that (17.10) implies (17.8). Conversely, if (17.8) holds, it is easily seen that (17.9) is valid with $E = M_i^{-1} B M_i$, and hence that M is monotonous. Q.E.D.

An important special type of monotonous deformation process is one whose exponent E satisfies $E^2 = 0$ but $E \neq 0$. We call such a process a **simple shearing.** Since $\exp(tE) = 1 + tE$ when $E^2 = 0$, (17.3) shows that a simple shearing P has the form

$$P(t) = P^i + t(E^* P^i + P^i E) + t^2 E^* P^i E, \qquad t \in [0, d_P]. \tag{17.11}$$

A motion which determines a simple shearing is called a *viscometric motion* (*cf.* [CMN]).

18. Processes of Constant Reduced State, States of Monotonous Flow

Definition 18.1. *Let* $\mathscr{T}$ *be a material element with symmetry group* $\mathscr{g}$. *We say that a deformation process* $P \in \Pi$ *is of* **constant reduced state** *relative to a state* $\sigma \in \Sigma_{P^i}$ *if the states* $\hat{\rho}(\sigma, P_{[0,t]})$, $t \in [0, d_P]$, *all belong to the reduced state* Ω_σ *determined by* σ.

To say that P is of constant reduced state relative to $\sigma \in \Sigma_{P^i}$ means (*cf.* (10.6)) that there is a function A: $[0, d_P] \to \mathscr{g}$ such that

$$\hat{\rho}(\sigma, P_{[0,t]}) = \iota_{A(t)}(\sigma), \qquad t \in [0, d_P]. \tag{18.1}$$

Since $\hat{\rho}(\sigma, P_{[0,0]}) = \sigma$ by Prop. 7.1, we may assume that $A(0) = 1_{\mathscr{T}}$.

It is clear from Prop. 12.1 (ii) that all freezes $G_{(t)}$ are of constant reduced state (actually of constant state) relative to any relaxed state λ with $\hat{G}(\lambda) = G$. If the symmetry group $\mathscr{g}$ is large enough, one can expect to find non-constant processes of constant reduced state among the monotonous processes in Π, as is shown in the proposition below. We say that $\sigma \in \Sigma$ is a **state of monotonous flow** *with exponent* E if $\exp(sE) \in \mathscr{g}$ for all $s \in \mathbb{R}^+$ and if every monotonous process $P \in \Pi$ of exponent E and initial configuration $P^i = \hat{G}(\sigma)$ satisfies

$$\hat{\rho}(\sigma, P_{[0,t]}) = \iota_{\exp(-tE)}(\sigma) \tag{18.2}$$

and hence is of constant reduced state relative to σ. The states of monotonous flow with exponent 0 are identical to the relaxed states (*cf.* Prop. 12.1 (ii)). If $\mathscr{g}$

is a Lie group, then the condition that $\exp(sE) \in g$ for all $s \in \mathbb{R}^+$ means that E belongs to the Lie algebra of g.

Proposition 18.1. *Let H be a monotonous history, with values in $\mathcal{G}$ and of exponent E, such that*

(i) *the segments $H_{[s]}$, $s \in \mathbb{R}^+$, belong to Π,*

(ii) *for all $s \in \mathbb{R}^+$, $\exp(sE)$ belongs to the symmetry group g,*

(iii) *for some $\sigma_0 \in \Sigma_{H^f}$, the limit*

$$\lim_{s\to\infty} \hat{\rho}\left(\iota_{\exp(sE)}(\sigma_0), H_{[s]}\right) = \sigma \tag{18.3}$$

exists.

Then σ is a state of monotonous flow with exponent E.

Proof. We use the abbreviation

$$A(t) = \exp(-tE), \qquad t \in \mathbb{R}. \tag{18.4}$$

It follows from (17.4) and (17.5) that

$$A(r)^* H_{[t]}(r) A(r) = A(t)^* H^f A(t), \qquad r \in [0, t]. \tag{18.5}$$

On the other hand, if P is a monotonous process of exponent E and initial configuration H^f, it follows from (17.3) that $A(r)^* P_{[0,t]}(r) A(r) = H^f$, $r \in [0, t]$, $t \in [0, d_P]$. Substituting this expression for H^f into (18.5) and using the fact that $A(r)$ and $A(t)$ commute, we obtain

$$H_{[t]} = A(t)^* P_{[0,t]} A(t), \qquad t \in [0, d_P]. \tag{18.6}$$

Since $H_{[t]} \in \Pi$ and $A(t)^{-1} \in g$, we infer that $P_{[0,t]}$ belongs to Π. Moreover, by (15.10) and (17.7), (18.6) gives

$$A(t)^* (H * P_{[0,t]}) A(t) = H \tag{18.7}$$

and hence

$$A(t)^* (H_{[s]} * P_{[0,t]}) A(t) = H_{[t+s]}. \tag{18.8}$$

By (10.5), it follows from (18.8) that

$$\iota_{A(t)} \hat{\rho}\left(\iota_{A(t)}^{-1}(\iota_{A(-s)}(\sigma_0)), H_{[t+s]}\right) = \hat{\rho}\left(\iota_{A(-s)}(\sigma_0), H_{[s]} * P_{[0,t]}\right).$$

If we apply (7.7) and observe $\iota_{A(-(t+s))} = \iota_{A(t)}^{-1} \circ \iota_{A(-s)}$, we obtain

$$\iota_{A(t)}\left(\hat{\rho}\left(\iota_{A(-(t+s))}(\sigma_0), H_{[t+s]}\right)\right) = \hat{\rho}\left(\hat{\rho}\left(\iota_{A(-s)}(\sigma_0), H_{[s]}\right), P_{[0,t]}\right).$$

Taking the limit $s \to \infty$ and observing (18.3), we get

$$\iota_{A(t)}(\sigma) = \hat{\rho}(\sigma, P_{[0,t]}),$$

which shows that (18.2) holds and hence that σ is a state of monotonous flow. Q.E.D.

The following result is a corollary to Prop. 18.1.

Proposition 18.2. *Let $\mathcal{T}$ be a semi-elastic element and Φ the set of deformation histories for $\mathcal{T}$ (cf. Definition 15.1). If H belongs to Φ and is monotonous of ex-*

ponent E such that $\exp(sE)\in g$ *for all* $s\in \mathbb{R}^+$, *then* $\bar{p}(H)$ (*see* (15.13)) *is a state o monotonous flow with exponent E.*

Proof. H satisfies the condition (i) of Prop. 18.1 because H is in Φ, the condition (ii) by hypothesis, and the condition (iii) with $\sigma=\bar{p}(H)$ and $\sigma_0=\bar{\lambda}(H^f)$ because $\iota_{\exp(sE)}(\bar{\lambda}(H^f))=\bar{\lambda}(H(s))$. Q.E.D.

For a process of constant reduced state it is possible to determine the stress at time t from a knowledge of the initial stress and the symmetry-valued function A of (18.1) alone. Indeed, it follows from (18.1) and (10.4) that the stress $\tilde{S}(\sigma, P_{[0,t]})$ at time t is given by

$$\tilde{S}(\sigma, P_{[0,t]})=A(t)\,\hat{S}(\sigma)\,A(t)^*. \tag{18.9}$$

Remark. It is (18.9) that often makes the dynamical analysis of processes of constant reduced state as simple as a statical analysis. For example, (18.9) is at the root of the dynamical analysis of viscometric flows of incompressible fluids (*cf.* [CMN]) as well as of the incompressible "subfluids" of WANG [W1].

For semi-elastic elements, it is possible to give a lower bound on the size of the symmetry group of a state of monotonous flow of the form $\bar{p}(H)$.

Proposition 18.3. *Let H be a monotonous history of exponent E that belongs to the set* Φ *of deformation histories of a semi-elastic element* $\mathscr{T}$. *Let*

$$\mathrm{Comm}(E)=\{B\in \mathrm{Invlin}(\mathscr{T})\,|\,BE=EB\} \tag{18.10}$$

be the group of all invertible transformations that commute with E. The symmetry group $g_{\bar{p}(H)}$ *then satisfies*

$$g_{\bar{p}(H)}\supset g\cap \mathrm{Orth}(H^f)\cap \mathrm{Comm}(E). \tag{18.11}$$

Proof. If $B\in\mathrm{Orth}(H^f)\cap\mathrm{Comm}(E)$, it is clear from (17.4) that $B^{*-1}HB^{-1}=H$. If also $B\in g$, it follows from Prop. 15.4 that $\bar{p}(H)=\bar{p}(B^{*-1}HB^{-1})=\iota_B(\bar{p}(H))$, *i.e.* that $B\in g_{\bar{p}(H)}$. Q.E.D.

The Propositions 18.2 and 18.3 apply, in particular, to fluid elements, for which $g\supset \mathrm{Unim}(\mathscr{T})$ (see Section 14). Since $\exp(sE)\in\mathrm{Unim}(\mathscr{T})$ if and only if $\mathrm{tr}\,E=0$, we see that for fluids all monotonous deformation histories H in Φ with traceless exponent define a state $\bar{p}(H)$ of monotonous flow. The symmetry group of this state satisfies

$$g_{\bar{p}(H)}\supset \mathrm{Orth}(H^f)\cap \mathrm{Comm}(E). \tag{18.12}$$

By Prop. 10.4 the stress $\bar{S}(H)=\hat{S}(\bar{p}(H))$ is such that $\bar{S}(H)H^f$ commutes with all transformations in $\mathrm{Orth}(H^f)\cap\mathrm{Comm}(E)$.

19. Material Functionals and Functions for Incompressible Fluids

We recall (Section 9) that a material is defined to be an equivalence class of material elements, the equivalence being material isomorphy. It is often possible to characterize such an equivalence class by means of numerical functionals. We call such functionals *material functionals* for the material in question. Numerical functions and numbers derived from a material functional are called *material functions* and *material constants*. For simplicity we consider only the case when the material is an incompressible fluid.

Proposition 19.1. *The set $\mathscr{G}$ of configurations of an incompressible fluid element $\mathscr{T}$ has the form*

$$\begin{aligned}\mathscr{G} &= \{G \in \mathrm{Sym}^+(\mathscr{T}, \mathscr{T}^*) \mid \det G_0^{-1} G = 1\} \\ &= \{A^* G_0 A \mid A \in \mathrm{Unim}\,\mathscr{T}\},\end{aligned} \tag{19.1}$$

where $G_0 \in \mathrm{Sym}^+(\mathscr{T}, \mathscr{T}^*)$. The symmetry group g of $\mathscr{T}$ is $g = \mathrm{Unim}(\mathscr{T})$.

Proof. Let $G_0 \in \mathscr{G}$. By definition of a fluid element (Section 14) we have $\mathrm{Unim}(\mathscr{T}) \subset g$. Hence, since $A^* \mathscr{G} A = \mathscr{G}$ for all $A \in g$ (see Definition 9.1 (ii)), $\mathscr{G}$ must contain the set $\mathscr{G}_1 = \{A^* G_0 A \mid A \in \mathrm{Unim}(\mathscr{T})\}$. On the other hand, since the element is incompressible, $\mathscr{G}$ must be contained in the set $\mathscr{C}$ given by (8.1). Now, since $\det(G_0^{-1}(A^* G_0 A)) = (\det A)^2 = 1$ holds if and only if $A \in \mathrm{Unim}(\mathscr{T})$ and since $\mathrm{Sym}^+(\mathscr{T}, \mathscr{T}^*) = \{A^* G_0 A \mid A \in \mathrm{Invlin}(\mathscr{T})\}$ (see Section 2), it follows that $\mathscr{G}_1 = \mathscr{C}$ and hence $\mathscr{G} = \mathscr{G}_1 = \mathscr{C}$, which is (19.1). Also, $A^* \mathscr{G} A = \mathscr{G}$ can hold only if $A \in \mathrm{Unim}(\mathscr{T})$, which implies $g = \mathrm{Unim}(\mathscr{T})$. Q.E.D.

We assume now that $\mathscr{G}$ is given by (19.1) and that a frame space $\mathscr{V}$ has been selected (see Section 5). The set $\mathscr{P}_{\mathscr{V}}$ of all placements corresponding to $\mathscr{G}$, as defined by (5.1), is easily seen to have the form

$$\mathscr{P}_{\mathscr{V}} = \{K_0 A \mid A \in \mathrm{Unim}(\mathscr{T})\}, \tag{19.2}$$

where $K_0 \in \mathrm{Lin}(\mathscr{T}, \mathscr{V})$ is such that $K_0^* K_0 \in \mathscr{G}$.

Given any set Φ of histories, with values in $\mathscr{G}$ and such that $A^* \Phi A = \Phi$ for all $A \in \mathrm{Unim}(\mathscr{T})$, we define

$$\Phi_{\mathscr{V}} = \{K^{*-1} H K^{-1} \mid H \in \Phi,\ K \in \mathscr{P}_{\mathscr{V}},\ K^* K = H(0)\}. \tag{19.3}$$

This set $\Phi_{\mathscr{V}}$ consists of histories with values in $\mathrm{Sym}^+(\mathscr{V}) \cap \mathrm{Unim}(\mathscr{V})$ and all $C \in \Phi_{\mathscr{V}}$ satisfy $C(0) = 1_{\mathscr{V}}$. Moreover, $\Phi_{\mathscr{V}}$ satisfies $Q^T \Phi_{\mathscr{V}} Q = \Phi_{\mathscr{V}}$ for all $Q \in \mathrm{Orth}(\mathscr{V})$, because if $K \in \mathscr{P}_{\mathscr{V}}$ and $K^* K = H(0)$, then also $QK \in \mathscr{P}_{\mathscr{V}}$ and $(QK)^*(QK) = H(0)$ for all $Q \in \mathrm{Orth}(\mathscr{V})$.

Conversely, suppose that $\Phi_{\mathscr{V}}$ is a set of histories with values in $\mathrm{Sym}^+(\mathscr{V}) \cap \mathrm{Unim}(\mathscr{V})$ and such that $C(0) = 1_{\mathscr{V}}$ for all $C \in \Phi_{\mathscr{V}}$ and $Q^T \Phi_{\mathscr{V}} Q = \Phi_{\mathscr{V}}$ for all $Q \in \mathrm{Orth}(\mathscr{V})$. Then

$$\Phi = \{K^* C K \mid C \in \Phi_{\mathscr{V}},\ K \in \mathscr{P}_{\mathscr{V}}\} \tag{19.4}$$

defines a set of histories with values in $\mathscr{G}$ and such that $A^* \Phi A = \Phi$ for all $A \in \mathrm{Unim}(\mathscr{T})$.

We say that a mapping $\overline{T}\colon \Phi_{\mathscr{V}} \to \mathrm{Sym}(\mathscr{V})$ is an *isotropic functional* if

$$\overline{T}(Q^T C Q) = Q^T\, \overline{T}(C)\, Q \qquad \text{for all } C \in \Phi_{\mathscr{V}} \text{ and all } Q \in \mathrm{Orth}(\mathscr{V}). \tag{19.5}$$

The following lemma is the key for the construction of material functionals.

Lemma 19.2. *Let Φ be a set of histories with values in $\mathscr{G}$ such that $A^* \Phi A = \Phi$ for all $A \in \mathrm{Unim}(\mathscr{T})$, and let $\Phi_{\mathscr{V}}$ be defined by* (19.3). *If $\bar{S}\colon \Phi \to \mathrm{Sym}(\mathscr{T}^*, \mathscr{T})$ satisfies*

$$\bar{S}(A^* H A) = A^{-1}\, \bar{S}(H)\, A^{*-1} \qquad \textit{for all } H \in \Phi \textit{ and all } A \in \mathrm{Unim}(\mathscr{T}), \tag{19.6}$$

then there is a unique isotropic functional $\bar{T}: \Phi_{\mathscr{V}} \to \mathrm{Sym}(\mathscr{V})$ *such that*

$$\bar{T}(C) = K\bar{S}(K^* CK)K^* \quad \textit{for all } C \in \Phi_{\mathscr{V}} \textit{ and all } K \in \mathscr{P}_{\mathscr{V}}. \tag{19.7}$$

Conversely, let $\Phi_{\mathscr{V}}$ *be a set of histories with values in* $\mathrm{Sym}^+(\mathscr{V}) \cap \mathrm{Unim}(\mathscr{V})$ *such that* $C(0) = 1_{\mathscr{V}}$ *for all* $C \in \Phi_{\mathscr{V}}$ *and* $Q^T \Phi_{\mathscr{V}} Q = \Phi_{\mathscr{V}}$ *for all* $Q \in \mathrm{Orth}(\mathscr{V})$, *and let* Φ *be defined by* (19.4). *If* $\bar{T}: \Phi_{\mathscr{V}} \to \mathrm{Sym}(\mathscr{V})$ *is an isotropic functional, then there is a unique functional* $\bar{S}: \Phi \to \mathrm{Sym}(\mathscr{T}^*, \mathscr{T})$ *which satisfies* (19.6) *and* (19.7).

Proof. Assume Φ and $\bar{S}$ are given and that (19.6) holds. Select $K_0 \in \mathscr{P}_{\mathscr{V}}$ and define $\bar{T}$ by

$$\bar{T}(C) = K_0 \bar{S}(K_0^* CK_0) K_0^*. \tag{19.8}$$

Now let $K \in \mathscr{P}_{\mathscr{V}}$ be arbitrary. By (19.2), there is a $A \in \mathrm{Unim}(\mathscr{T})$ such that $K = K_0 A$. Using (19.6) and (19.8), we get

$$K\bar{S}(K^* CK)K^* = K_0 A\bar{S}(A^* K_0^* CK_0 A) A^* K_0^* = \bar{T}(C),$$

i.e., (19.7) is valid. Since $QK_0 \in \mathscr{P}_{\mathscr{V}}$ whenever $Q \in \mathrm{Orth}(\mathscr{V})$, we can substitute QK_0 for K in (19.7). Using (19.8) again, we obtain

$$\bar{T}(C) = QK_0 \bar{S}(K_0^* Q^T CQK_0) K_0^* Q^T = Q\bar{T}(Q^T CQ) Q^T.$$

Hence $\bar{T}$ satisfies (19.5) and hence is an isotropic functional.

Assume now, conversely, that $\Phi_{\mathscr{V}}$ and an isotropic functional $\bar{T}: \Phi_{\mathscr{V}} \to \mathrm{Sym}(\mathscr{V})$ are given. Let $H \in \Phi$, with Φ defined by (19.4). If $K_1, K_2 \in \mathscr{P}_{\mathscr{V}}$ satisfy $K_1^* K_1 = K_2^* K_2 = H(0)$, we have $K_1^{*-1} HK_1^{-1}$, $K_2^{*-1} HK_2^{-1} \in \Phi_{\mathscr{V}}$ and there is a $Q \in \mathrm{Orth}(\mathscr{V})$ such that $K_2 = QK_1$. Since $\bar{T}$ is isotropic, we obtain

$$\begin{aligned} K_2^{-1} \bar{T}(K_2^{*-1} HK_2^{-1}) K_2^{*-1} &= K_1^{-1} Q^T \bar{T}(QK_1^{*-1} HK_1^{-1} Q^T) QK_1^{*-1} \\ &= K_1^{-1} \bar{T}(K_1^{*-1} HK_1^{-1}) K_1^{*-1}. \end{aligned}$$

Therefore it is meaningful to define $\bar{S}$ by

$$\bar{S}(H) = K^{-1} \bar{T}(K^{*-1} HK^{-1}) K^{*-1} \quad \text{if } K^* K = H(0), \tag{19.9}$$

because the right side depends only on H, not on the particular choice of K. It is clear that (19.7) holds when $\bar{S}$ is defined by (19.9). For any $A \in \mathrm{Unim}(\mathscr{T})$ we have $(KA)^*(KA) = A^* K^* KA = A^* H(0) A$. Hence (19.9) remains valid when H is replaced by $A^* HA$ and K by KA. It follows that $\bar{S}$ also satisfies (19.6). Q.E.D.

We assume now that $\mathscr{T}$ is an incompressible fluid element, that Φ is the set of all deformation histories for $\mathscr{T}$ (see Definition 15.1), and that $\bar{S}$ is the history functional of $\mathscr{T}$ (see Section 16). It follows from Prop. 16.3 and Prop. 19.1 that $\bar{S}$ satisfies (19.6) and hence that Lemma 19.2 can be applied to it. Thus, given any frame space $\mathscr{V}$, we can associate with the element $\mathscr{T}$ a set $\Phi_{\mathscr{V}}$ of histories with values in $\mathrm{Sym}^+(\mathscr{V}) \cap \mathrm{Unim}(\mathscr{V})$ and an isotropic functional $\bar{T}$ with domain $\Phi_{\mathscr{V}}$. Since $\hat{S}$ satisfies the normalization condition (8.3) for incompressible elements, it easily follows from (19.7) that the values of $\bar{T}$ must belong to the space

$$\mathrm{Sym}_0(\mathscr{V}) = \{T \in \mathrm{Sym}(\mathscr{V}) \mid \mathrm{tr}\, T = 0\} \tag{19.10}$$

of traceless symmetric tensors.

If we take for $\mathscr{V}$ the numerical space $\mathbb{R}^n$, we denote the resulting isotropic functional by $\mathfrak{T}$ and call it the **material functional** of the element. The domain $\Phi_{\mathbb{R}^n}$ of $\mathfrak{T}$ consists of histories whose values are unimodular and positive-definite symmetric matrices. The values of $\mathfrak{T}$ are symmetric matrices with zero trace. By (19.7), the material functional $\mathfrak{T}$ is related to the history functional $\bar{S}$ in such a way that

$$\mathfrak{T}(C) = K\bar{S}(K^* C K) K^* \quad \text{for all } C \in \Phi_{\mathbb{R}^n} \tag{19.11}$$

and all placements K of $\mathscr{T}$ into the numerical space $\mathbb{R}^n$.

Proposition 19.3. *Two incompressible fluid elements are materially isomorphic if and only if they have the same material functional.*

Proof. Recall that for semi-elastic elements $\mathscr{T}$, the structure is determined by the triple $(\mathscr{T}, \Phi, \bar{S})$ (see Section 16). Now, if $\mathscr{T}_1$ and $\mathscr{T}_2$ are two incompressible fluid elements with the same material functional $\mathfrak{T}$ and if K_1 and K_2 are arbitrary placements of $\mathscr{T}_1$ and $\mathscr{T}_2$ into $\mathbb{R}^n$, then it follows from (19.11) and Prop. 16.2 that $K_2^{-1} K_1$ is a material isomorphism. Conversely, if $A \in \operatorname{Invlin}(\mathscr{T}_1, \mathscr{T}_2)$ is a material isomorphism and K_1 a placement of $\mathscr{T}_1$ into $\mathbb{R}^n$, then $K_2 = A K_1$ is a placement of $\mathscr{T}_2$ and Prop. 16.2 shows that (19.11) defines the same material functional for both $\mathscr{T}_1$ and $\mathscr{T}_2$. Q.E.D.

Using the properties (H 1)–(H 5) of Φ listed in Section 16, one can easily show that the domain $\Phi_{\mathbb{R}^n}$ of a material functional of an incompressible fluid must have the following properties:

(D 1) *Every $C \in \Phi_{\mathbb{R}^n}$ is a history with values in* $\operatorname{Sym}^+(\mathbb{R}^n) \cap \operatorname{Unim}(\mathbb{R}^n)$.

(D 2) *For every $C \in \Phi_{\mathbb{R}^n}$, $C(0) = 1_{\mathbb{R}^n}$ and $(1_{\mathbb{R}^n})_{(\infty)} \in \Phi_{\mathbb{R}^n}$.*

(D 3) *If $C \in \Phi_{\mathbb{R}^n}$, $s \in \mathbb{R}^+$ and $U \in \operatorname{Unim}(\mathbb{R}^n)$ such that $U^T U = C(s)$, then*

$$U^{T-1} C^{(s)} U^{-1} \in \Phi_{\mathbb{R}^n}.$$

(D 4) *If $C, D \in \Phi_{\mathbb{R}^n}$, $s \in \mathbb{R}^+$ and $U \in \operatorname{Unim}(\mathbb{R}^n)$ such that $U^T U = C(s)$, then*

$$U^T D U * C_{[s]} \in \Phi_{\mathbb{R}^n}.$$

(D 5) *For all $C_0 \in \operatorname{Sym}^+(\mathbb{R}^n) \cap \operatorname{Unim}(\mathbb{R}^n)$ there is a $C \in \Phi_{\mathbb{R}^n}$ such that $C_0 \in \operatorname{Range}(C)$.*

The properties (F 1), (F 2), of Section 16, for the history functionals imply that the material functional $\mathfrak{T}\colon \Phi_{\mathbb{R}^n} \to \operatorname{Sym}_0(\mathbb{R}^n)$, in addition to being isotropic, must have the following convergence properties.

(M 1) *For all $C \in \Phi_{\mathbb{R}^n}$ we have*

$$\lim_{s \to \infty} \mathfrak{T}(C(s)_{(\infty)} * C_{[s]}) = \mathfrak{T}(C). \tag{19.12}$$

(M 2) *For all $C, D \in \Phi_{\mathbb{R}^n}$, $s \in \mathbb{R}^n$ and $U \in \operatorname{Unim}(\mathbb{R}^n)$ such that $U^T U = C(s)$ we have*

$$\lim_{t \to \infty} \mathfrak{T}(U^t D U * C(s)_{(t)} * C_{[s]}) = \mathfrak{T}(C(s)_{(\infty)} * C_{[s]}). \tag{19.13}$$

Prop. 19.3 states, in essence, that every incompressible fluid is characterized by an isotropic material functional $\mathfrak{T}$ whose domain $\Phi_{\mathbb{R}^n}$ satisfies (D 1)–(D 5) and which has the convergence properties (M 1) and (M 2). Conversely, every functional $\mathfrak{T}: \Phi_{\mathbb{R}^n} \to \mathrm{Sym}_0(\mathbb{R}^n)$ of this type determines an incompressible fluid material. Indeed, if we select a vector space $\mathcal{T}$ of dimension n and any $G_0 \in \mathrm{Sym}^+(\mathcal{T}, \mathcal{T}^*)$, we can define $\mathcal{G}$ by (19.1), then $\mathcal{P}_{\mathbb{R}^n}$ by (19.2), then Φ by (19.4), and finally $\bar{S}$ by (19.9). The triple $(\mathcal{T}, \Phi, \bar{S})$ then satisfies (H 1)–(H 5) and (F 1), (F 2) of Section 16. As remarked there, one can then construct a structure of a semi-elastic material element on $\mathcal{T}$ whose set Φ^* of deformation histories is obtained by saturation of Φ with respect to $\bar{S}$. The element is incompressible and fluid, and its material functional is obtained from the given $\mathfrak{T}$ by extension from $\Phi_{\mathbb{R}^n}$ to an appropriate saturation of $\Phi_{\mathbb{R}^n}$.

Let $\mathfrak{T}$ be a material functional of an incompressible fluid and assume that the domain of $\mathfrak{T}$ contains all histories of the form $s \mapsto \exp(-sB^T)\exp(-sB)$ with B in the space $\mathrm{Lin}_0(\mathbb{R}^n)$ of all traceless matrices. We then define $\mathrm{t}: \mathrm{Lin}_0(\mathbb{R}^n) \to \mathrm{Sym}_0(\mathbb{R}^n)$ by

$$\mathrm{t}(B) = \mathfrak{T}\big(s \mapsto \exp(-sB^T)(\exp - sB)\big). \tag{19.14}$$

Since $\mathfrak{T}$ is an isotropic functional, t is an isotropic function in the sense that

$$\mathrm{t}(Q^T B Q) = Q\,\mathrm{t}(B)\,Q^T \quad \text{for all } B \in \mathrm{Lin}_0(\mathbb{R}^n) \text{ and all } Q \in \mathrm{Orth}(\mathbb{R}^n). \tag{19.15}$$

The following result shows that the material function t derived from $\mathfrak{T}$ by (19.14) describes the behavior of the incompressible fluid determined by $\mathfrak{T}$ in all monotonous flow states (see Section 18).

Proposition 19.4. *Let $\mathcal{T}$ be any incompressible fluid element with material functional $\mathfrak{T}$. Then for every $G \in \mathcal{G}$ and every $E \in \mathrm{Lin}_0(\mathcal{T})$ there is a unique state σ of monotonous flow with exponent E and such that $\hat{G}(\sigma) = G$. The intrinsic stress $\hat{S}(\sigma)$ satisfies*

$$\hat{S}(\sigma) - K^{-1}\,\mathrm{t}(KEK^{-1})\,K^{*-1} \tag{19.16}$$

for all $K \in \mathcal{P}_{\mathbb{R}^n}$ such that $K^* K = G$.

Proof. Let $G \in \mathcal{G}$ and $E \in \mathrm{Lin}_0(\mathcal{T})$ be given. Let H be the monotonous history with exponent E and final value $H^f = G$. This history is given by (17.4). We now choose $K \in \mathcal{P}_{\mathbb{R}^n}$ such that $K^* K = G$. It is easily seen that we then have

$$H = K^* C K, \quad \text{where} \quad C = \big(s \mapsto \exp(-sB^T)\exp(-sB)\big) \quad \text{with} \quad B = KEK^{-1}. \tag{19.17}$$

Since $C \in \Phi_{\mathbb{R}^n}$ by assumption, we infer from (19.4) that $H \in \Phi$. As we have seen at the end of Section 18, it follows that $\sigma = \bar{p}(H)$ is the state of monotonous flow with exponent E and configuration $\hat{G}(\sigma) = H^f = G$. The relation (19.16) is an immediate consequence of (19.11), (19.14), and (19.17). Q.E.D.

The material function t determines, in particular, the *viscometric functions* τ, σ_1, σ_2. These are given by

$$\tau(\kappa) = \mathrm{t}_{12}(\kappa B_0), \qquad \sigma_1(\kappa) = (\mathrm{t}_{11} - \mathrm{t}_{33})(\kappa B_0), \qquad \sigma_2(\kappa) = (\mathrm{t}_{22} - \mathrm{t}_{33})(\kappa B_0), \tag{19.20}$$

for $\kappa \in \mathbb{R}$. Here t_{ij}, $i, j = 1, 2, 3$ denote the components of t and B_0 is the matrix $B_0 = \begin{bmatrix} 0 & 0 & 0 \\ 1 & 0 & 0 \\ 0 & 0 & 0 \end{bmatrix}$. As is well known (see, *e.g.*, [N4], Sect. 4), the viscometric functions describe the behavior of the fluid for all simple shearings (see end of Section 17).

20. Materials of the Rate Type

We consider a body element $(\mathscr{T}, \mathscr{G}, \Pi)$ such that Π is the class of piecewise continuously differentiable processes in the sense described in example (c) after Definition 4.1. We use the notation

$$\dot{\mathscr{G}}_G = \{\dot{P}(t) \,|\, P \in \Pi, t \in [0, d_P], \dot{P}(t) \text{ exists}, P(t) = G\} \tag{20.1}$$

for the class of all possible derivatives of processes at a time at which they have the value G.

We assume that with each $G \in \mathscr{G}$, there is associated a closed subset $\mathscr{S}_G$ of the stress space $\mathscr{S} = \mathrm{Sym}^+(\mathscr{T}^*, \mathscr{T})$ and that we are given a function

$$\mathfrak{H}: \mathscr{D} \to \mathscr{S}$$

whose domain $\mathscr{D}$ consists of triples $(G, \dot{G}, S)$ as follows:

$$\mathscr{D} = \{(G, \dot{G}, S) \,|\, G \in \mathscr{G}, \dot{G} \in \dot{\mathscr{G}}_G, S \in \mathscr{S}_G\}. \tag{20.2}$$

We assume that $\mathfrak{H}$ satisfies the following conditions:

(i) If P is continuously differentiable and $S^i \in \mathscr{S}_{P^i}$, the initial value problem

$$\dot{S}(t) = \mathfrak{H}(P(t), \dot{P}(t), S(t)), \qquad S(0) = S^i \tag{20.3}$$

has a unique solution S: $[0, d_P] \to \mathscr{S}$. We denote the dependence of the final value $S(d_P) = S^f$ on P and on the initial value S^i by

$$S^f = \mathfrak{S}(P, S^i), \tag{20.4}$$

so that $t \mapsto \mathfrak{S}(P_{[0,t]}, S^i)$ is the solution of (20.3).

(ii) The final value $S^f = \mathfrak{S}(P, S^i)$ belongs to $\mathscr{S}_{P^f}$ and the dependence of the final value on the initial value is uniformly continuous, *i.e.* the function

$$\mathfrak{S}(P, \cdot): \mathscr{S}_{P^i} \to \mathscr{S}_{P^f} \tag{20.5}$$

is uniformly continuous for each continuously differentiable $P \in \Pi$.

(iii) If $G \in \mathscr{G}$ and $S \in \mathscr{S}_G$, then

$$\lim_{t \to \infty} \mathfrak{S}(G_{(t)}, S) = \hat{\mathfrak{S}}(G, S) \tag{20.6}$$

exists.

If $\mathfrak{H}$ is a reasonably regular function, then one can apply general theorems on ordinary differential equations to verify Conditions (i) and (ii). Condition (iii) says that the solutions of $\dot{S}(t) = \mathfrak{H}(G, 0, S(t))$ have limits as $t \to \infty$ for every (constant) $G \in \mathscr{G}$. This condition is closely related to one I have described in [N1], pp. 52–53.

We now attempt to use the function $\mathfrak{H}$ for endowing $\mathscr{T}$ with the structure of a material element. For the state space Σ we take

$$\Sigma=\{(G, S) \mid G\in\mathscr{G}, S\in\mathscr{S}_G\}. \tag{20.7}$$

Thus, the states are configuration-stress pairs. The functions $\hat{G}$ and $\hat{S}$ merely assign to each pair its first and second member, so that

$$\hat{G}(G, S)=G, \quad \hat{S}(G. S)=S. \tag{20.8}$$

In order to define the evolution function $\hat{\rho}$, we first note that there is a natural way to extend the solution functional $\mathfrak{S}$ of (20.3) in such a way that $\mathfrak{S}(P, S^i)$ becomes meaningful for *all* $P\in\Pi$ with $S^i\in\mathscr{S}_{P^i}$, not just those that have a continuous derivative for all $t\in[0, d_P]$. The general member $P\in\Pi$ is of the form

$$P=P_1 * P_2 *\cdots * P_m,$$

where the P_k are continuously differentiable. We then define $\mathfrak{S}(P, S^i)$ inductively by

$$\mathfrak{S}(P_1 *\cdots * P_{m-1} * P_m, S^i)=\mathfrak{S}(P_m, \mathfrak{S}(P_1 *\cdots * P_{m-1}, S^i). \tag{20.9}$$

With $\mathfrak{S}$ thus extended, it makes sense to define $\hat{\rho}$ by

$$\hat{\rho}\big((G, S), P\big)=\big(P^f, \mathfrak{S}(P, S)\big), \qquad P\in\Pi_G, \; S\in\mathscr{S}_G. \tag{20.10}$$

The response functional $\tilde{S}=\hat{S}\circ\hat{\rho}$ satisfies

$$\tilde{S}\big((G, S), P\big)=\mathfrak{S}(P, S) \quad \text{if } P^i=G \tag{20.11}$$

and hence is essentially the same as the solution functional of (20.3).

It is very easy to verify that the septuple $(\mathscr{T}, \mathscr{G}, \Pi, \Sigma, \hat{G}, \hat{S}, \hat{\rho})$ thus constructed satisfies Axioms I–III. We note that for all $G\in\mathscr{G}$, there is a natural one-to-one correspondence $S\leftrightarrow(G, S)$ between $\mathscr{S}_G$ and the G-section Σ_G of the state space. It is not hard to prove, using Condition (ii) and (20.11), that the natural uniformity on Σ_G (see Definition 11.1) is identical to the uniformity on Σ_G induced by the natural uniformity of $\mathscr{S}_G\subset\mathscr{S}$. Since $\mathscr{S}_G$ was assumed to be closed in $\mathscr{S}$ and since $\mathscr{S}$ is complete, it follows that $\mathscr{S}_G$ and hence Σ_G is complete. Thus, Axiom IV, Section 11, is satisfied. The validity of the relaxation axiom, Axiom V, Section 12, is now an immediate consequence of Condition (iii), and it follows from (20.6) that the relaxation mapping $\hat{\lambda}$ is given by

$$\hat{\lambda}(G, S)=\big(G, \hat{\mathfrak{S}}(G, S)\big). \tag{20.12}$$

We have proved that $(\mathscr{T}, \mathscr{G}, \Pi, \Sigma, \hat{G}, \hat{S}, \hat{\rho})$ satisfies Axioms I–V. The accessibility axiom (Axiom VI, Section 13), however, will *not* be satisfied unless the sets $\mathscr{S}_G$ of stresses compatible with a given configuration are properly selected. Indeed Axiom VI is equivalent to the requirement that the $\mathscr{S}_G$ be of the form

$$\mathscr{S}_G=\mathrm{cl}\{\mathfrak{S}(P, S_0) \mid P^i=G_0, P^f=G\}, \tag{20.13}$$

where $G_0\in\mathscr{G}$ and $S_0\in\mathscr{S}_{G_0}$ such that $\hat{\mathfrak{S}}(G_0, S_0)=S_0$.

If Axiom VI is not satisfied, one can apply the procedure described at the end of Section 13 to construct material element structures on $\mathscr{T}$. In the present case, this procedure goes as follows: Select $G_0 \in \mathscr{G}$ and $S_1 \in \mathscr{S}_{G_1}$ arbitrarily and compute $S_0 = \hat{\mathfrak{S}}(G_0, S_1)$ using (20.7). Define $\mathscr{S}'_G$ to be the right side of (20.13) and then determine $(\Sigma', \hat{G}', \hat{S}', \hat{\rho}')$ in terms of the $\mathscr{S}'_G$ in the same way as $(\Sigma, \hat{G}, \hat{S}, \hat{\rho})$ were determined in terms of the $\mathscr{S}_G$. The septuple $(\mathscr{T}, \mathscr{G}, \Pi, \Sigma', \hat{G}', \hat{S}', \hat{\rho}')$ then satisfies all axioms and hence defines the structure of a material element on $\mathscr{T}$.

A material element obtained from a function $\mathfrak{H}$ satisfying Conditions (i)–(iii) in the manner just described is called a **material element of the rate type.** As we have seen, one and the same function $\mathfrak{H}$ may define more than one material element structure on $\mathscr{T}$.

We now consider the case when $\dot{\mathscr{G}}_G$ is a subspace (rather than merely a subset) of $\mathrm{Sym}(\mathscr{T}, \mathscr{T}^*)$ and when the given function $\mathfrak{H}$ is linear in its second variable, *i.e.* if

$$\mathfrak{H}(G, \cdot, S) \in \mathrm{Lin}(\dot{\mathscr{G}}_G, \mathscr{S}) \tag{20.14}$$

for every $G \in \mathscr{G}$ and $S \in \mathscr{S}_G$. A material element whose structure is defined by such a function will be called a **hypo-elastic element**[1]. Since (20.14) implies that $\mathfrak{H}(G, 0, S) = 0$, it follows that the solutions of $\dot{S}(t) = \mathfrak{H}(G, 0, S(t))$ are constants and hence that Condition (iii) is automatically satisfied with

$$\hat{\mathfrak{S}}(G, S) = S \tag{20.15}$$

for all $G \in \mathscr{G}$ and $S \in \mathscr{S}_G$. In view of (20.12) it follows that *for a hypo-elastic element all states are relaxed.* Thus, a hypo-elastic element cannot be semi-elastic unless it is elastic. If the element is elastic, then the sets $\mathscr{S}_G$ must be singletons $\mathscr{S}_G = \{\hat{S}(G)\}$. It can be proved that the elastic elements which are also hypo-elastic are those whose stress function $\hat{S}$ satisfies certain differentiability and invertibility conditions[2].

When (20.14) holds, then the solution functional $\mathfrak{S}$ of (20.4) is easily seen to have the property

$$S^f = \mathfrak{S}(P, S^i) \Leftrightarrow S^i = \mathfrak{S}(P_{\mathrm{rev}}, S^f), \tag{20.16}$$

where the *reversed process* P_{rev} is defined by $P_{\mathrm{rev}}(t) = P(d_P - t)$, $t \in [0, d_P]$. It follows from (20.16), as is easy to show, that accessibility (see Section 13) becomes an equivalence relation on Σ when Σ is defined by (20.7). Even when $\mathfrak{H}$ is a very simple function there can be more than one equivalence class, as is shown by an example due to BERNSTEIN ([B1]; see also [TN], Sect. 100). Axiom V is satisfied only after the $\mathscr{S}_G$ are replaced by subsets $\mathscr{S}'_G$ which consist of all those $S \in \mathscr{S}_G$ for which (G, S) belongs to a particular equivalence class in Σ. The state space Σ' defined by (20.7) in terms of the $\mathscr{S}'_G$ then is the same as this particular equivalence class. Thus, each of the equivalence classes in Σ defines on $\mathscr{T}$ the structure of a hypo-elastic element[3].

Remark 1. In the past literature (*cf.* [N2], Sect. 24 or [TN], Sect. 36) materials of the rate type have been defined by differential equations involving Cauchy-stresses and motions in a frame space rather than intrinsic stresses and deformation

[1] This notion of hypo-elasticity is more inclusive than the original one of TRUESDELL [T].

[2] The method of proof is the same as the one used, in a special situation, in Sect. 15b of [N1].

[3] This procedure was first used by BERNSTEIN [B2] to define a concept of "material" in hypo-elasticity.

processes. It is instructive to consider the following special non-intrinsic constitutive equation of the rate type:

$$\overset{*}{T}=\mathfrak{G}(T, A_1). \tag{20.17}$$

Here, T is the Cauchy-stress, $\overset{*}{T}$ any invariant time flux of T, and A_1 the first Rivlin-Ericksen tensor. Apart from a possible dependence on density, (20.17) is the constitutive equation for what I called "hygrosteric" materials in [N1]. If $\mathfrak{G}$ is linear in its second variable, then (20.17) is the constitutive equation of hypo-elasticity in the original sense of TRUESDELL ([T]; see also [TN], Chapter DIV).

It will be convenient to choose for $\overset{*}{T}$ the time flux given by

$$\overset{*}{T}=\dot{T}-LT-TL^T, \tag{20.18}$$

where $L(t)=\dot{F}_{(t)}(t)$, with $F_{(t)}$ defined by (5.5). Using (5.7), (6.2), (5.5), and (20.18), we can express A_1, T, and $\overset{*}{T}$ in terms of the motion M, the intrinsic stress S, and the process $P=M^*M$:

$$A_1=M^{*-1}\dot{P}M^{-1}, \quad T=MSM^*, \quad \overset{*}{T}=M\dot{S}M^*. \tag{20.19}$$

Now, (20.17) can serve to define material elements of the rate type in the intrinsic sense only if, after substitution of (20.19), equation (20.17) becomes a special case of $(20.3)_1$. This is the case if and only if there is a function $\mathfrak{H}$ of the type described at the start of this section such that

$$\mathfrak{H}(G, \dot{G}, S)=K^{-1}\mathfrak{G}(KSK^*, K^{*-1}\dot{G}K^{-1})K^{*-1} \tag{20.20}$$

holds for all $(G, \dot{G}, S)$ in the domain of $\mathfrak{H}$ and all placements K such that $K^*K=G$. Since $K^*K=K_1^*K_1=G$ is valid if and only if $K_1=QK$ for some $Q\in\mathrm{Orth}(\mathscr{V})$, it follows that (20.20) can be valid for *all* K with $K^*K=G$ only if $\mathfrak{G}$ is an isotropic function, *i.e.* if

$$\mathfrak{G}(QTQ^T, QA_1Q^T)=Q\,\mathfrak{G}(T, A_1)Q^T \tag{20.21}$$

for all $Q\in\mathrm{Orth}(\mathscr{V})$ and all (T, A_1) in the domain of $\mathfrak{G}$. We have thus recovered the well known result ([N1], Sect. 7) that (20.17) cannot define a material unless $\mathfrak{G}$ is isotropic.

If (20.20) holds, then $\mathfrak{H}$ satisfies

$$\mathfrak{H}(A^*GA, A^*\dot{G}A, A^{-1}SA^{*-1})=A^{-1}\mathfrak{H}(G, \dot{G}, S)A^{*-1} \tag{20.22}$$

for all $(G, \dot{G}, S)$ in the domain $\mathscr{D}$ of $\mathfrak{H}$ and all A in the set

$$\mathscr{g}=\{A\in\mathrm{Invlin}(\mathscr{T})\,|\,(G, \dot{G}, S)\in\mathscr{D}\Rightarrow(A^*GA, A^*\dot{G}A, A^{-1}SA^{*-1})\in\mathscr{D}\}. \tag{20.23}$$

Indeed, if we replace K by KA in (20.20) so that $G=K^*K$ becomes $(KA)^*(KA)=A^*K^*KA=A^*GA$, if we replace also $\dot{G}$ by $A^*\dot{G}A$ and S by $A^{-1}SA^{*-1}$, we see that (20.22) must hold. Now, we recall that in order to define a material element structure by means of $\mathfrak{H}$, one may have to replace the sets $\mathscr{S}_G$ by suitable subsets. In view of (20.2), this means that $\mathscr{D}$ has to be replaced by a suitable subset. After this has been done, (20.23) can easily be shown to be the symmetry group of the element.

One and the same constitutive equation (20.17) may determine several material element structures on $\mathscr{T}$, and these structures need not all have the same symmetry group.

Remark II. More general materials of the rate type are obtained if the initial value problem (20.3) is replaced by an initial value problem for a differential equation of higher order. No new insight is gained by considering such materials.

21. Outlook

We give here a list of possible generalizations, further developments, and applications of the ideas presented in this paper.

(a) Abstract theory of systems with memory. The basic features of the theory presented here can be used to define a very general concept of *physical systems with memory*. In the definition of such a system, one uses as primitive notions a set $\mathscr{G}$ of *abstract configurations* and an *action space* $\mathscr{S}$, whose elements are assumed to describe the possible actions of the system on the environment. A system with memory is then defined to be a septuple $(\mathscr{G}, \Pi, \Sigma, \hat{G}, \mathscr{S}, \hat{S}, \hat{\rho})$ whose entries Π, Σ, $\hat{G}$, $\hat{S}$, and $\hat{\rho}$ are as described in Definitions 4.1 and 7.1. The only difference is that $\mathscr{G}$ and $\mathscr{S}$ are now abstract sets which need not be defined concretely in terms of a vector space $\mathscr{T}$.

Axioms I–VI remain in force. It must be assumed that the action space $\mathscr{S}$ is endowed *a priori* with a uniform topology. This is necessary in order that Definition 11.1 and hence Axioms IV–VI remain meaningful.

A material element in the sense of Definition 7.1 is a very special kind of physical system with memory.

Note Added in Proof. The structure of an abstract system with memory described here has many features in common with the structure used in "mathematical systems theory". (See, *e.g.*, WILLEMS, J. C.: Dissipative dynamical systems I. Arch. Rational Mech. Anal. **45**, 322–351 (1972), and the literature cited there.)

(b) Thermodynamic theory of material elements. To include thermal and energetic phenomena, one should modify Definition 4.1 by assuming that $\mathscr{G}$ is a closed and connected subset of $\mathrm{Sym}^+(\mathscr{T}, \mathscr{T}^*) \times \mathbb{R}^{++} \times \mathscr{T}^*$, where $\mathbb{R}^{++}$ denotes the set of all positive reals. The members of $\mathscr{G}$ are triples $G = (C, \theta, g)$, where C is interpreted as a configuration, θ as a temperature of the material element, and g as a temperature gradient across the element.

In Definition 7.1, Equation (7.1) should be replaced by

$$\mathscr{S} = \mathrm{Sym}(\mathscr{T}^*, \mathscr{T}) \times \mathbb{R} \times \mathscr{T} \times \overset{n}{\wedge} \mathscr{T}^*, \tag{21.1}$$

where $\overset{n}{\wedge} \mathscr{T}^*$, the nth exterior power of $\mathscr{T}^*$ $(n = \dim \mathscr{T})$, is the one-dimensional space of possible volumes of the element. The members of $\mathscr{S}$ are quadruples (S, η, h, ψ), where S is interpreted as an intrinsic stress, η as an entropy per unit volume, h as an intrinsic heat flux, and ψ as a free energy.

Let $P=(\bar{C}, \bar{\theta}, \bar{g})$: $[0, d_P] \to \mathcal{G}$ be a process (*i.e.* $P \in \Pi$), let $\sigma \in \Sigma$ be a state such that $P^i = \hat{G}(\sigma)$ and let

$$(\bar{S}, \bar{\eta}, \bar{h}, \bar{\psi})\colon [0, d_P] \to \mathscr{S} \tag{21.2}$$

be defined with the help of the response functional $\tilde{S}$ (see (7.5)) by

$$(\bar{S}(t), \bar{\eta}(t), \bar{h}(t), \bar{\psi}(t)) = \tilde{S}(\sigma, P_{[0,t]}), \qquad t \in [0, d_P]. \tag{21.3}$$

One then defines the *rate of entropy production* $\bar{\gamma}$: $[0, d_P] \to \overset{n}{\wedge} \mathscr{T}^*$ corresponding to σ and P by

$$\bar{\theta}\bar{\gamma} = v_{\bar{C}} \left(\frac{1}{2} \operatorname{tr}(\bar{S}\dot{\bar{C}}) - \bar{\eta}\dot{\bar{\theta}} + \frac{1}{\bar{\theta}} \langle \bar{g}, \bar{h} \rangle \right) - \dot{\bar{\psi}}, \tag{21.4}$$

where $v_{\bar{C}}(t) = v_{\bar{C}(t)}$ is the volume in $\overset{n}{\wedge} \mathscr{T}^*$ corresponding to the configuration $\bar{C}(t)$.

In a detailed development of a thermodynamic theory of material elements, the Definition (21.4) should play a fundamental role.

I believe, also, that the framework just described can serve to simplify and clarify the axiomatic approach to thermodynamics developed by W. A. DAY [D].

(c) Theories of plastic behavior. D. R. OWEN ([O1], [O2]) has developed, within the framework of the old theory of simple materials, concepts and results intended to describe in mathematical terms what physicists usually call plastic behavior. I believe that the new theory presented here provides a more natural setting for such concepts and that it can be used to simplify and clarify OWEN's ideas.

We note that the theory of states of monotonous flow given in Section 18 applies not only to semi-elastic materials, but also to materials that exhibit plastic behavior, provided the materials possess sufficient symmetry. I expect, therefore, that one can develop concrete descriptions of plastic monotonous flows, in particular viscometric flows.

(d) Theories of fading memory. In the framework of the old concept of simple material, several theories of fading memory are known (see [TN], Chapter CV; also [CM] and the literature cited there). Analogous theories for the new concept given here remain to be created. Such theories should involve continuity assumptions imposed upon the response functional

$$\tilde{S}(\sigma, \cdot)\colon \Pi_G \to \mathscr{S} \qquad (G = \hat{G}(\sigma)). \tag{21.5}$$

Suitable topologies on Π_G, which are needed to make continuity meaningful, can be defined in terms of a natural metric on the configuration space $\mathcal{G}$. I shall demonstrate the existence of such a natural metric in a future paper.

For semi-elastic materials, it should not be too difficult to translate the known theories of fading memory into the intrinsic setting of this paper. It would be interesting if one could also create theories of fading memory for non-semi-elastic materials.

Acknowledgement. The research leading to this paper was supported by the National Science Foundation under Grant GP-28430. I am grateful to Dr. F. BLOOM for help in proofreading.

References

[B1] BERNSTEIN, B., Hypo-elasticity and elasticity. Arch. Rational Mech. Anal. **6**, 89–104 (1960).

[B2] BERNSTEIN, B., Remarks on materials of the rate type and the principle of determinism. Trans. Soc. Rheol. **5**, 35–40 (1961).

[BK] BOURBAKI, N., Topologie Générale, Chapitres 1 et 2, quatrième édition. Paris: Hermann 1965.

[C] COLEMAN, B. D., Kinematical concepts with applications in the mechanics and thermodynamics of incompressible viscoelastic fluids. Arch. Rational Mech. Anal. **9**, 273–300 (1962).

[CM] COLEMAN, B. D., & V. J. MIZEL, On the general theory of fading memory. Arch. Rational Mech. Anal. **29**, 18–31 (1968).

[CMN] COLEMAN, B. D., H. MARKOVITZ, & W. NOLL, Viscometric Flows of Non-Newtonian Fluids. New York: Springer 1966.

[CN1] COLEMAN, B. D., & W. NOLL, On the thermostatics of continuous media. Arch. Rational Mech. Anal. **4**, 97–128 (1959).

[CN2] COLEMAN, B. D., & W. NOLL, Material symmetry and thermostatic inequalities in finite elastic deformations. Arch. Rational Mech. Anal. **15**, 87–111 (1964).

[D] DAY, W. A., Thermodynamics based on a work axiom. Arch. Rational Mech. Anal. **31**, 1–34 (1968).

[FT] FAIRCHILD, W. W., & C. IONESCU-TULCEA, Topology. Philadelphia-London-Toronto: W. B. Saunders 1971.

[K] KELLEY, J. L., General Topology. Princeton: Van Nostrand 1955.

[N1] NOLL, W., On the continuity of the solid and fluid states. J. Rational Mech. Anal. **4**, 3–81 (1955).

[N2] NOLL, W., A mathematical theory of the mechanical behavior of continuous media. Arch. Rational Mech. Anal. **2**, 197–226 (1958).

[N3] NOLL, W., The foundations of classical mechanics in the light of recent advances in continuum mechanics. Proc. of the Berkeley Symposium on the Axiomatic Method, p. 266–281, Amsterdam 1959.

[N4] NOLL, W., Motions with constant stretch history. Arch. Rational Mech. Anal. **11**, 97–105 (1962).

[N5] NOLL, W., Proof of the maximality of the orthogonal group in the unimodular group. Arch. Rational Mech. Anal. **18**, 100–102 (1965).

[N6] NOLL, W., Space-time structures in classical mechanics. Delaware Seminar in the Foundations of Physics, p. 28–34. Berlin-Heidelberg-New York: Springer 1967.

[N7] NOLL, W., Materially uniform simple bodies with inhomogeneities. Arch. Rational Mech. Anal. **27**, 1–32 (1967).

[N8] NOLL, W., The representation of monotonous processes by exponentials. To appear.

[O1] OWEN, D. R., Thermodynamics of materials with elastic range. Arch. Rational Mech. Anal. **32**, 91–112 (1968).

[O2] OWEN, D. R., A mechanical theory of materials with elastic range. Arch. Rational Mech. Anal. **37**, 85–110 (1970).

[T] TRUESDELL, C., Hypo-elasticity. J. Rational Mech. Anal. **4**, 83–133 (1955).

[TN] TRUESDELL, C., & W. NOLL, The Non-linear Field Theories of Mechanics. Encyclopedia of Physics, Vol. III/3. Berlin-Heidelberg-New York: Sringer 1965.

[S] STERNBERG, S., Lectures on Differential Geometry. Englewood Cliffs: Prentice Hall 1964.

[W1] WANG, C. C., A general theory of subfluids. Arch. Rational Mech. Anal. **20**, 1–40 (1965).

[W2] WANG, C. C., On the geometric structures of simple bodies, a mathematical foundation for the theory of continuous distributions of dislocations. Arch. Rational Mech. Anal. **27**, 33–94 (1967).

Department of Mathematics
Carnegie-Mellon University
Pittsburgh

(Received May 12, 1972)

Lectures on the Foundations of Continuum Mechanics and Thermodynamics

Walter Noll

Expository Memoir, invited by the Editors

Lectures on the Foundations of Continuum Mechanics and Thermodynamics

WALTER NOLL

Expository Memoir, invited by the Editors

Contents

1. Introduction

This is an outline of a series of lectures I delivered at the Technion in Haifa, Israel, in the summer of 1972. This outline gives my view of what the basic concepts of modern continuum mechanics and thermodynamics are and how they should be presented to graduate students of mathematics and theoretical mechanics. This view evolved gradually over the past eight years. A first version was given in a series of lectures I delivered at the Summer Session in Bressanone, Italy, in 1965 (reference [1]). The material was reworked several times for an introductory graduate course repeatedly given at Carnegie-Mellon University and for lecture series given at the University of Karlsruhe, Germany, in 1968 and in Jabłonna, Poland, in 1970. This paper is the latest version, and it renders reference [1] obsolete.

The basic mathematical notations and tools used in the present paper, mainly from linear algebra, are described in Section 2 of reference [2].

I. Kinematics of Continuous Bodies

2. Euclidean Spaces

A *metric* on a set $\mathscr{E}$ is a function

$$d: \mathscr{E} \times \mathscr{E} \to \mathbb{R}^+$$

which associates with each pair (x, y) of points in $\mathscr{E}$ a number $d(x, y)$, called the *distance* from x to y. Euclidean geometry can be based entirely on an appropriate metric.

The *group of isometries* $\mathscr{I}$ for the metric d is the set of all invertible mappings of $\mathscr{E}$ onto itself that leave distances unchanged:

$$\mathscr{I} = \{\alpha: \mathscr{E} \to \mathscr{E} \mid \alpha \text{ invertible, } d(x, y) = d(\alpha(x), \alpha(y)) \text{ for all } x, y \in \mathscr{E}\}.$$

It is clear that $\mathscr{I}$ is a group under composition.

If the metric d is to describe Euclidean geometry it must have certain properties. It is intuitively clear that translations are isometries, and we characterize Euclidean metrics in terms of the translations as follows:

It may happen that $\mathscr{I}$ contains a subgroup $\mathscr{V}$ with the following properties:

(E1) $\mathscr{V}$ is commutative, *i.e.*, $\boldsymbol{u} \circ \boldsymbol{v} = \boldsymbol{v} \circ \boldsymbol{u}$ for all $\boldsymbol{u}, \boldsymbol{v} \in \mathscr{V}$.

(E2) $\mathscr{V}$ is transitive, *i.e.*, for any $x, y \in \mathscr{E}$ there is a $\boldsymbol{u} \in \mathscr{V}$ such that $\boldsymbol{u}(x) = y$.

(E3) $\mathscr{V}$ acts freely, *i.e.*, if $\boldsymbol{v}(x) = x$ for some $x \in \mathscr{E}$, then $\boldsymbol{v} = 1_{\mathscr{E}}$.

(E4) There is a scalar multiplication $\mathbb{R} \times \mathscr{V} \to \mathscr{V}$ which makes $\mathscr{V}$ a vector space when composition in $\mathscr{V}$ is taken as addition, and there is an inner product I on the vector space $\mathscr{V}$ such that

$$d(x, y)^2 = \boldsymbol{u} \cdot \boldsymbol{u} = \langle I\boldsymbol{u}, \boldsymbol{u}\rangle \quad \text{when } \boldsymbol{u}(x) = y.$$

The properties (E1)–(E4) are intuitively plausible when $\mathscr{V}$ is the group of translations.

Uniqueness Theorem. *There is at most one subgroup $\mathscr{V}$ of $\mathscr{I}$ with the properties* (E1)–(E4). *Moreover, if $\mathscr{V}$ is such a subgroup, the scalar multiplication and the inner product on $\mathscr{V}$ required for* (E4) *are uniquely determined.*

The proof of this theorem will not be given; it is not trivial but not too difficult (see, *e.g.*, reference [3]).

Definition. The function $d: \mathscr{E} \times \mathscr{E} \to \mathbb{R}^+$ is called a *Euclidean metric* on $\mathscr{E}$ if $\mathscr{I}$ has a subgroup $\mathscr{V}$ with the properties (E1)–(E4). If a Euclidean metric d is prescribed, we say that d endows $\mathscr{E}$ with the structure of a *Euclidean space.*

The inner product space $\mathscr{V}$ determined by d is then called the *translation space* of $\mathscr{E}$.

Let $\mathscr{E}$ be a Euclidean space with translation space $\mathscr{V}$. The following notations are useful:

(i) if $\boldsymbol{u}, \boldsymbol{v} \in \mathscr{V}$, write $\boldsymbol{u} + \boldsymbol{v}$ for $\boldsymbol{v} \circ \boldsymbol{u}$,

(ii) write $\boldsymbol{0} \in \mathscr{V}$ for $1_{\mathscr{E}}$,

(iii) if $\boldsymbol{u} \in \mathscr{V}$, write $-\boldsymbol{u}$ for $\boldsymbol{u}^{-1}$.

(iv) if $x \in \mathscr{E}$ and $\boldsymbol{u} \in \mathscr{V}$, write $x+\boldsymbol{u}$ for $\boldsymbol{u}(x)$.

(v) if $x, y \in \mathscr{E}$, write $\boldsymbol{u}=y-x$ for the unique $\boldsymbol{u} \in \mathscr{V}$ for which $\boldsymbol{u}(x)=y$.

If these notations are used, the rules of the ordinary algebra of addition and subtraction are valid.

Let $\mathscr{E}_1$ and $\mathscr{E}_2$ be Euclidean spaces, with metrics d_1 and d_2 and with translation spaces $\mathscr{V}_1$ and $\mathscr{V}_2$. An invertible mapping $\alpha: \mathscr{E}_1 \to \mathscr{E}_2$ is called an *isomorphism* from $\mathscr{E}_1$ onto $\mathscr{E}_2$ if it preserves distance, *i.e.* if

$$d_1(x, y)=d_2(\alpha(x), \alpha(y))$$

for all $x, y \in \mathscr{E}_1$. An isomorphism from $\mathscr{V}_1$ onto $\mathscr{V}_2$ is a linear mapping which preserves the inner product. The following result describes the isomorphisms of Euclidean spaces:

Representation Theorem. *Every isomorphism $\alpha: \mathscr{E}_1 \to \mathscr{E}_2$ from one Euclidean space onto another induces an isomorphism $\bar{\alpha}: \mathscr{V}_1 \to \mathscr{V}_2$ between the corresponding translation spaces in such a way that*

$$\alpha(x)-\alpha(q)=\bar{\alpha}(x-q) \tag{1}$$

holds for all $x, q \in \mathscr{E}_1$.

Proof. We define $\mathscr{V}_2'=\{\alpha \circ \boldsymbol{u} \circ \alpha^{-1} \mid \boldsymbol{u} \in \mathscr{V}_1\}$. It is easily seen that $\mathscr{V}_2'$ is a subgroup of the group $\mathscr{I}_2$ of isometries of $\mathscr{E}_2$. It is also easily verified that $\mathscr{V}_2'$ satisfies the conditions (E1)–(E3). We define a scalar multiplication on $\mathscr{V}_2'$ by putting $a(\alpha \circ \boldsymbol{u} \circ \alpha^{-1})=\alpha \circ (a\boldsymbol{u}) \circ \alpha^{-1}$ for $a \in \mathbb{R}$, $\boldsymbol{u} \in \mathscr{V}_1$ and an inner product on $\mathscr{V}_2'$ by putting

$$(\alpha \circ \boldsymbol{u} \circ \alpha^{-1}) \cdot (\alpha \circ \boldsymbol{v} \circ \alpha^{-1})=\boldsymbol{u} \cdot \boldsymbol{v} \quad \text{when } \boldsymbol{u}, \boldsymbol{v} \in \mathscr{V}_1.$$

It then becomes clear that (E4) is also satisfied by $\mathscr{V}_2'$. By the uniqueness theorem, it follows that $\mathscr{V}_2=\mathscr{V}_2'$. Moreover, the mapping $\bar{\alpha}: \mathscr{V}_1 \to \mathscr{V}_2$ given by

$$\bar{\alpha}(\boldsymbol{u})=\alpha \circ \boldsymbol{u} \circ \alpha^{-1}, \quad \boldsymbol{u} \in \mathscr{V}_1 \tag{2}$$

is an isomorphism from $\mathscr{V}_1$ onto $\mathscr{V}_2$. Equation (2) is equivalent to

$$\bar{\alpha}(\boldsymbol{u})(\alpha(q))=\alpha(\boldsymbol{u}(q)) \quad \text{for all } q \in \mathscr{E}_1,\ \boldsymbol{u} \in \mathscr{V}_1.$$

If we put $\boldsymbol{u}=x-q$ and use the notations (i)–(v) above, we get the desired result. Q.E.D.

The case when $\mathscr{E}_1=\mathscr{E}_2=\mathscr{E}$ gives rise to the following

Corollary. *To every isometry $\alpha \in \mathscr{I}$ of a Euclidean space $\mathscr{E}$ with translation space $\mathscr{V}$ corresponds a unique $Q_\alpha \in \operatorname{Orth}(\mathscr{V})$ such that*

$$\alpha(x)-\alpha(q)=Q_\alpha(x-q) \tag{3}$$

for all $x, q \in \mathscr{E}$.

The dimension of a Euclidean space is defined to be the dimension of its translation space. Finite-dimensional Euclidean spaces are isomorphic if and only if they have the same dimension.

3. Neo-Classical Spacetime

A basic primitive notion of any physical theory is an *event-world* $\mathscr{W}$. It is a set whose elements $e, f, \ldots$ are called *events*. Intuitively, events are actual or possible "atoms of experience" and the event world is a mathematical idealization of experience.

The classical time concept can be described by a *time-lapse function*

$$\hat{t}: \mathscr{W} \times \mathscr{W} \to \mathbb{R},$$

whose value $\hat{t}(e,f)$ gives the time-lapse between the events e and f, as measured by a stop watch. We agree to count $\hat{t}(e,f)$ as positive when e is earlier than f, negative when e is later than f. Our intuitive idea of time requires that $\hat{t}$ have the following properties:

(T1) $\hat{t}(e,f) = -\hat{t}(f,e)$ for all $e, f \in \mathscr{W}$.

(T2) $\hat{t}(e,f) + \hat{t}(f,g) = \hat{t}(e,g)$ for all $e, f, g \in \mathscr{W}$.

(T3) For every $e \in \mathscr{W}$ and every $t \in \mathbb{R}$ there is a $f \in \mathscr{W}$ such that $\hat{t}(e,f) = t$.

We say that e is *earlier* than, *later* than, or *simultaneous* with f according to whether $\hat{t}(e,f) > 0$, < 0, or $= 0$. Simultaneity is easily seen to be an equivalence relation on $\mathscr{W}$. Its graph is the set

$$\mathscr{S} = \{(e,f) \in \mathscr{W} \times \mathscr{W} \mid \hat{t}(e,f) = 0\} \tag{1}$$

of all pairs of simultaneous events. The simultaneity relation determines a partition Γ of $\mathscr{W}$ into classes τ of simultaneous events so that

$$\mathscr{W} = \bigcup \{\tau \mid \tau \in \Gamma\}, \qquad \mathscr{S} = \bigcup \{\tau \times \tau \mid \tau \in \Gamma\}, \tag{2}$$

the unions being disjoint.

The classes $\tau \in \Gamma$ are called *instants*. If $e \in \tau$ we say that *the event e happens at the instant τ*.

The time-lapse $\hat{t}(e,f)$ depends only on the instants τ and σ at which e and f happen. Therefore one can define a *time-lapse function*

$$\bar{t}: \Gamma \times \Gamma \to \mathbb{R}$$

between instants by

$$\bar{t}(\tau, \sigma) = \hat{t}(e,f) \quad \text{if } e \in \tau,\ f \in \sigma.$$

The absolute value $|\bar{t}|$ can easily be shown to be a Euclidean metric on Γ. The corresponding translation space of Γ can be identified with the reals $\mathbb{R}$. If the notations of Section 2 are used, we then have

$$\tau - \sigma = \bar{t}(\sigma, \tau), \qquad \sigma + t = \tau \quad \text{if } t = \bar{t}(\sigma, \tau)$$

for $\tau, \sigma \in \Gamma$, $t \in \mathbb{R}$.

Distances between events, as measured by measuring rods, are meaningful only when the events are simultaneous, because a measurement has to take place at a particular instant. Thus, distance measurements are described by a function

$$\hat{d}: \mathscr{S} \to \mathbb{R}^+,$$

where $\mathscr{S}$ is the set (1) of all simultaneous pairs of events. Experience with distance measurements shows that $\hat{d}$ should have the following basic properties:

(D 1) For each instant $\tau \in \Gamma$, the restriction $d_\tau\colon \tau \times \tau \to \mathbb{R}^+$ of $\hat{d}$ to $\tau \times \tau \subset \Gamma$ is a Euclidean metric and hence endows τ with the structure of a Euclidean space.

(D 2) For each instant $\tau \in \Gamma$ the dimension of the translation space $\mathscr{V}_\tau$ of τ is 3.

Definition. We say that a set $\mathscr{W}$ is a *neo-classical event-world* if it is endowed with the structure defined by a prescribed time-lapse function $\hat{t}$ with the properties (T 1)–(T 3) and a prescribed distance function $\hat{d}$ with the properties (D 1), (D 2).

The internal symmetry of a neo-classical event-world $\mathscr{W}$ is described by its group of automorphisms. An automorphism α of $\mathscr{W}$ is an invertible mapping of $\mathscr{W}$ onto itself which preserves time lapses and distances, i.e. satisfies

$$t(\alpha(e), \alpha(f)) = \hat{t}(e,f) \quad \text{for all } e, f \in \mathscr{W},$$

$$\hat{d}(\alpha(e), \alpha(f)) = \hat{d}(e,f) \quad \text{for all } (e,f) \in \mathscr{S}.$$

The following result is an easy consequence of the Representation Theorem of Section 2.

Theorem. *To every automorphism α of a neo-classical event world $\mathscr{W}$ corresponds a unique number $t_\alpha \in \mathbb{R}$ and to each instant $\tau \in \Gamma$ a unique isomorphism $\bar{\alpha}_\tau\colon \mathscr{V}_\tau \to \mathscr{V}_{\tau + t_\alpha}$ such that $\alpha(e) \in \tau + t_\alpha$ if $e \in \tau$ and*

$$\alpha(e) - \alpha(f) = \bar{\alpha}_\tau(e - f) \quad \text{if } e, f \in \tau.$$

4. Kinematical Processes, Frames, Motions

A *material system* $\mathscr{B}$ is a set of material points $X, Y, \ldots$. These are idealizations of material objects or markings in or on material objects.

From now on we assume a neo-classical event-world $\mathscr{W}$ given once and for all. By a *time-interval* $\mathscr{T}$ we mean a connected subset of the set Γ of all instants, connected in the sense that if $\tau \in \mathscr{T}$ and $\tau + t \in \mathscr{T}$, then also $\tau + s \in \mathscr{T}$ for all $s \in [0, t]$.

A mapping

$$\chi\colon \mathscr{B} \times \mathscr{T}_\chi \to \mathscr{W},$$

where $\mathscr{T}_\chi$ is a time-interval, will be called a **kinematical process** if

$$\chi(X, \tau) \in \tau \quad \text{for all } X \in \mathscr{B} \text{ and all } \tau \in \mathscr{T}_\chi.$$

The event $\chi(X, \tau)$ is interpreted to be the event experienced by the material point X at the instant τ.

The set

$$\{\chi(X, \tau) \mid \tau \in \mathscr{T}_\chi\}$$

of all events experienced by X is called the *worldline of* X in the kinematical process χ.

Let $f: \mathcal{T} \to \mathscr{E}$ be a function on a time-interval $\mathcal{T}$ with values in some Euclidean space $\mathscr{E}$. The limit

$$\dot{f}(\tau) = \lim_{s \to 0} \frac{f(\tau+s) - f(\tau)}{s} = \frac{\mathrm{d}}{\mathrm{d}\tau} f(\tau),$$

if it exists, is called the *derivative* of f at $\tau \in \mathcal{T}$. The value $\dot{f}(\tau)$ belongs to the translation space $\mathscr{V}$ of $\mathscr{E}$. If $\dot{f}(\tau)$ exists for all $\tau \in \mathcal{T}$ we say that f is differentiable and call $\dot{f}: \mathcal{T} \to \mathscr{V}$ its derivative. We say that f is of class C^1 if $\dot{f}$ exists and is continuous. The nth derivative $\overset{(n)}{f}$ of f is defined inductively by

$$\overset{(0)}{f} = f, \quad \overset{(1)}{f} = \dot{f}, \quad \overset{(2)}{f} = \ddot{f} = (\dot{f})^{\cdot}, \quad \overset{(n)}{f} = (\overset{(n-1)}{f})^{\cdot}.$$

We say that f is of class C^n if $\overset{(n)}{f}$ exists and is continuous.

A material system $\mathscr{B}$ is assumed to be endowed with a structure defined by the prescription of a non-empty class $\mathfrak{K}$ of admissible kinematical processes. The members of $\mathfrak{K}$ are assumed to be such that for any two material points $X, Y \in \mathscr{B}$, the function

$$\tau \mapsto \hat{d}(\chi(X, \tau), \chi(Y, \tau)$$

is of class C^2.

The derivative $\frac{\mathrm{d}}{\mathrm{d}\tau} \hat{d}(\chi(X, \tau), \chi(Y, \tau))$ is called the *relative speed* of X and Y at τ and the second derivative $\frac{\mathrm{d}^2}{\mathrm{d}\tau^2} \hat{d}(\chi(X, \tau), \chi(Y, \tau))$ the *relative scalar acceleration* of X and Y at τ. A kinematical process χ is called a *rigid process* if $\hat{d}(\chi(X, \tau), \chi(Y, \tau))$ is independent of τ for all $X, Y \in \mathscr{B}$, *i.e.*, if the relative speed of any two material points is always zero.

A material system $\mathscr{E}$ is called a *rigid system* if its class $\mathfrak{R}$ of admissible kinematical processes is defined in terms of a distance function $d: \mathscr{E} \times \mathscr{E} \to \mathbb{R}$ by $\mathfrak{R} = \{\beta \mid \beta: \mathscr{E} \times \mathcal{T}_\beta \to \mathscr{W}$ is a kinematical process and $d(x, y) = \hat{d}(\beta(x, \tau), \beta(y, \tau))$ for all $x, y \in \mathscr{E}$ and all $\tau \in \mathcal{T}_\beta\}$. If $\beta \in \mathfrak{R}$ and $\tau \in \mathcal{T}_\beta$, we define

$$\beta_\tau: \mathscr{E} \to \tau \quad \text{by} \quad \beta_\tau(x) = \beta(x, \tau). \tag{1}$$

If β_τ is invertible for some $\beta \in \mathfrak{R}$ and some $\tau \in \Gamma$ then β_τ is an isomorphism and d must be a Euclidean metric on $\mathscr{E}$. In this case, $\mathscr{E}$ is called a **frame of reference** (or simply *frame*). The processes in $\mathfrak{R}$ will then be called *reference processes*.

We now assume that a frame $\mathscr{E}$ with Euclidean metric d and class $\mathfrak{R}$ of reference processes is given. Let $\mathscr{B}$ be a material system, unrelated to $\mathscr{E}$, and let $\mathfrak{K}$ be the class of admissible processes of $\mathscr{B}$.

We assume that if $\chi \in \mathfrak{K}$ and if α is an automorphism of $\mathscr{W}$, then also $\chi^\alpha \in \mathfrak{K}$, where χ^α is defined by $\chi^\alpha(X, \tau) = \alpha(\chi(X, \tau - t_\alpha))$. For every $\chi \in \mathfrak{K}$ and every $\tau \in \mathcal{T}_\chi$, we define

$$\chi_\tau: \mathscr{B} \to \tau \quad \text{by} \quad \chi_\tau(X) = \chi(X, \tau). \tag{2}$$

If a reference process $\beta \in \mathfrak{R}$ with $\mathcal{T}_\chi = \mathcal{T}_\beta$ is given, we define the **motion** $\mu: \mathscr{B} \times \mathcal{T}_\chi \to \mathscr{E}$ *of* $\mathscr{B}$ *relative to the frame* $\mathscr{E}$, determined by $\chi \in \mathfrak{K}$, by

$$\mu(X, \tau) = \mu_\tau(X) \quad \text{with} \quad \mu_\tau = \beta_\tau^{-1} \circ \chi_\tau, \qquad \tau \in \mathcal{T}_\chi.$$

Let $\mathfrak{M}$ be the class of all motions obtained in this way, i.e.

$$\mathfrak{M}=\{\mu \mid \mu_\tau=\beta_\tau\circ\chi_\tau,\ \chi\in\mathfrak{K},\ \tau\in\mathcal{T}_\chi\}.$$

It follows easily from the invariance of $\mathfrak{K}$ under the automorphisms of $\mathscr{W}$ that $\mathfrak{M}$ is independent of the choice of $\beta\in\mathfrak{R}$. For every choice $\beta\in\mathfrak{R}$ with $\mathcal{T}_\beta=\mathcal{T}_\chi$ the relations

$$\mu_\tau=\beta_\tau^{-1}\circ\chi_\tau,\qquad \chi_\tau=\beta_\tau\circ\mu_\tau \tag{3}$$

define a one-to-one correspondence between the class $\mathfrak{K}$ of kinematical processes of $\mathscr{B}$ and the class $\mathfrak{M}$ of motions of $\mathscr{B}$ relative to the frame $\mathscr{E}$. This correspondence depends, of course, on the choice of the reference process $\beta\in\mathfrak{R}$.

Let a kinematical process $\chi\in\mathfrak{K}$ be given. If β, $\beta^*\in\mathfrak{R}$ are two reference processes (with $\mathcal{T}_\chi=\mathcal{T}_\beta=\mathcal{T}_{\beta^*}$) we obtain two motions μ, $\mu^*\in\mathfrak{M}$ corresponding to χ by

$$\mu_\tau=\beta_\tau^{-1}\circ\chi_\tau,\qquad \mu_\tau^*=\beta_\tau^{*\,-1}\circ\chi_\tau,\qquad \tau\in\mathcal{T}_\chi. \tag{4}$$

It follows from (4) that these motions μ and μ^* are related by

$$\mu_\tau^*=\alpha_\tau\circ\mu_\tau,\qquad \alpha_\tau\colon\mathscr{E}\to\mathscr{E}, \tag{5}$$

where $\alpha_\tau=\beta_\tau^{*\,-1}\circ\beta_\tau$, for each $\tau\in\mathcal{T}_\chi$, is an isometry of the frame $\mathscr{E}$. Using the representation (3) of Section 2, we infer from (5) that

$$\mu^*=c+Q(\mu-q), \tag{6}$$

where $q\in\mathscr{E}$ can be chosen arbitrarily, $c\colon\mathcal{T}_\chi\to\mathscr{E}$ is defined by $c(\tau)=\alpha_\tau(q)$, and $Q\colon\mathcal{T}_\alpha\to\mathrm{Orth}(\mathscr{V})$ is defined by $Q(\tau)=Q_{\alpha_\tau}$.

If a motion μ of $\mathscr{B}$ relative to $\mathscr{E}$ is of class C^2, we can define a corresponding *velocity field* $\dot\mu\colon\mathscr{B}\times\mathcal{T}_\mu\to\mathscr{V}$ and a corresponding *acceleration field* $\ddot\mu\colon\mathscr{B}\times\mathcal{T}_\mu\to\mathscr{V}$ by

$$\dot\mu(X,\tau)=\frac{\mathrm{d}}{\mathrm{d}\tau}\mu(X,\tau),\qquad \ddot\mu(X,\tau)=\frac{\mathrm{d}^2}{\mathrm{d}\tau^2}\mu(X,\tau).$$

Assume that the motion μ, corresponding to a given kinematical process χ relative to a reference process β, is of class C^2. Let β^* be another reference process such that the functions c and Q of (6) determined by β^* are also of class C^2. Then the motion μ^* corresponding to χ and β^*, given by (6), is also of class C^2. The velocity and acceleration fields of μ and μ^* are easily seen to be related by

$$\dot\mu^*-Q\dot\mu=\dot c+A(\mu^*-c), \tag{7}$$

$$\ddot\mu^*-Q\ddot\mu=\ddot c+2A(\dot\mu^*-\dot c)+(\dot A-A^2)(\mu^*-c), \tag{8}$$

where

$$A\colon\mathcal{T}_\chi\to\mathrm{Skew}(\mathscr{V})\quad\text{is defined by } A=\dot Q Q^T$$

$$(\mathrm{Skew}(\mathscr{V})=\{A\in\mathrm{Lin}(\mathscr{V})\mid A=-A^T\}).$$

It is possible to associate with a kinematical process χ and a reference process β an *instantaneous velocity field* v by $v_\tau=\overline{\beta}_\tau\circ\dot\mu_\tau\colon\mathscr{B}\to\mathscr{V}_\tau$, where $\mu_\tau=\beta_\tau^{-1}\circ\chi_\tau$. If the right side of (7) were zero, one could prove that v_τ is independent of the choice of the reference process. Since the right side of (7) is not zero, it follows that there is no way of associating a velocity field with the kinematical process itself, independent of any choice of reference process. We express this fact by saying that the velocity is *frame-dependent*. Of course, the acceleration is also frame-dependent. Later, we will see that there are certain tensor fields derived from the velocity that are *frame-indifferent*, *i.e.* independent of the choice of reference process.

5. Differential Calculus in Euclidean Spaces

Let $\phi: \mathscr{D} \to \mathscr{D}'$ be a mapping from an open subset $\mathscr{D}$ of a Euclidean space $\mathscr{E}$ into an open subset $\mathscr{D}'$ of a Euclidean space $\mathscr{E}'$. Let $\mathscr{V}$ and $\mathscr{V}'$ be the translation spaces of $\mathscr{E}$ and $\mathscr{E}'$, respectively. We say that ϕ is *differentiable at* $x \in \mathscr{D}$ if there is a $L \in \mathrm{Lin}(\mathscr{V}, \mathscr{V}')$ such that

$$\phi(x+\boldsymbol{u}) - \phi(x) = L\boldsymbol{u} + o(u)$$

for $\boldsymbol{u}$ in a neighborhood of $\mathbf{0} \in \mathscr{V}$. The notation $o(\boldsymbol{u})$ indicates that $\boldsymbol{u} \mapsto o(\boldsymbol{u})$ is some function of order small o. This means that for every $\varepsilon > 0$, there is a $\delta > 0$ such that

$$|o(\boldsymbol{u})| < \varepsilon |\boldsymbol{u}| \quad \text{when} \quad |\boldsymbol{u}| < \delta.$$

Roughly, $o(\boldsymbol{u})$ can be neglected compared to $L\boldsymbol{u}$ for small $\boldsymbol{u}$. The linear mapping L is called the *gradient of* ϕ *at* x and is denoted by $\nabla\phi(x)$. The gradient $\nabla\phi(x)$ is a linear approximation to the mapping $\boldsymbol{u} \mapsto \phi(x+\boldsymbol{u}) - \phi(x)$. A mapping $\phi: \mathscr{D} \to \mathscr{D}'$ is said to be of class C^1 if ϕ is differentiable at every $x \in \mathscr{D}$ and if the gradient function

$$\nabla\phi: \mathscr{D} \to \mathrm{Lin}(\mathscr{V}, \mathscr{V}')$$

is continuous. It is possible to define on $\mathrm{Lin}(\mathscr{V}, \mathscr{V}')$ a natural structure of an inner product space in the following way: If $L, M \in \mathrm{Lin}(\mathscr{V}, \mathscr{V}')$ then

$$M^* \in \mathrm{Lin}(\mathscr{V}'^*, \mathscr{V}^*) = \mathrm{Lin}(\mathscr{V}', \mathscr{V})$$

and hence $LM^* \in \mathrm{Lin}(\mathscr{V})$. If we put $L \cdot M = \mathrm{tr}(LM^*)$, then $(L, M) \mapsto L \cdot M$ is easily proved to be an inner product on $\mathrm{Lin}(\mathscr{V}, \mathscr{V}')$.

Since every inner product space can be regarded as a Euclidean space, it is meaningful to discuss the possibility that $\nabla\phi$ is of class C^1. If this is the case, we say that ϕ is of class C^2. In a similar way, one can define what is meant by saying that ϕ is of class C^n, $n = 1, 2, \ldots$.

The most important theorem of the differential calculus is the *chain rule*: Let $\phi: \mathscr{D} \to \mathscr{D}'$ and $\psi: \mathscr{D}' \to \mathscr{D}''$ be mappings of class C^1, where $\mathscr{D}, \mathscr{D}'$, and $\mathscr{D}''$ are open subsets of Euclidean spaces $\mathscr{E}, \mathscr{E}'$, and $\mathscr{E}''$. The composition $\psi \circ \phi: \mathscr{D} \to \mathscr{D}''$ is then again of class C^1 and its gradient is given by

$$\nabla(\psi \circ \phi)(x) = \nabla\psi(\phi(x))\, \nabla\phi(x), \qquad x \in \mathscr{D}.$$

Moreover, if ϕ and ψ are both of class C^n, $n > 1$, so is $\psi \circ \phi$.

Another important theorem is the *product rule*:

Let $\phi: \mathscr{D} \to \mathscr{V}_1$, $\psi: \mathscr{D} \to \mathscr{V}_2$ be mappings of class C^1, where $\mathscr{D}$ is an open subset of a Euclidean space and where $\mathscr{V}_1$ and $\mathscr{V}_2$ are inner product spaces.

Let $B: \mathscr{V}_1 \times \mathscr{V}_2 \to \mathscr{V}_3$ be a bilinear function. Then

$$\eta = B(\phi(\cdot), \psi(\cdot)): \mathscr{D} \to \mathscr{V}_3$$

is again of class C^1 and its gradient is given by

$$(\nabla\eta(x))\,\boldsymbol{u} = B((\nabla\phi(x))\,\boldsymbol{u}, \psi(x)) + B(\phi(x), (\nabla\psi(x))\,\boldsymbol{u}).$$

Finally, we have the following *symmetry theorem*:
If $\phi: \mathscr{D} \to \mathscr{D}'$ is of class C^2, then $\nabla\nabla\phi: \mathscr{D} \to \mathrm{Lin}(\mathscr{V}, \mathrm{Lin}(\mathscr{V}, \mathscr{V}'))$ has symmetric values in the sense that

$$(((\nabla\nabla\phi)(x))\,\boldsymbol{u})\,\boldsymbol{v} = (((\nabla\nabla\phi)(x))\,\boldsymbol{v})\,\boldsymbol{u}$$

for all $x \in \mathscr{D}$ and all $\boldsymbol{u}, \boldsymbol{v} \in \mathscr{V}$.

6. Continuous Bodies

We describe here precise mathematical structures intended to idealize the vague physical notion of continuous body. Such structures depend on the choice of a class $\boldsymbol{D}$ of mappings called **displacements.** This class is assumed to satisfy the following requirements:

(i) the members of $\boldsymbol{D}$ are invertible mappings whose domains and codomains are open subsets of Euclidean spaces.

(ii) Every mapping in $\boldsymbol{D}$ is of class C^1.

(iii) If $\lambda \in \boldsymbol{D}$ then $\lambda^{-1} \in \boldsymbol{D}$.

(iv) If $\lambda, \mu \in \boldsymbol{D}$ and $\mathrm{Range}(\lambda) = \mathrm{Domain}(\mu)$ then $\mu \circ \lambda \in \boldsymbol{D}$.

(v) Isometric (*i.e.* distance-preserving) bijections between open subsets of Euclidean spaces belong to $\boldsymbol{D}$.

The following classes $\boldsymbol{D}$ are important examples of classes of displacements:

(a) $\boldsymbol{D} = \boldsymbol{D}_n$ = class of all restrictions to open sets of C^n-diffeomorphisms between Euclidean spaces ($n \geq 1$).

(b) $\boldsymbol{D} = \boldsymbol{D}_n^{\mathrm{iso}}$ = class of all isochoric (i.e. volume-preserving) mappings belonging to $\boldsymbol{D}_n$.

(c) $\boldsymbol{D} = \boldsymbol{D}^{\mathrm{rig}}$ = class of all isometric bijections between open sets of Euclidean spaces.

Definition. We say that a material system $\mathscr{B}$ is a **continuous body** of type $\boldsymbol{D}$ if it is endowed with a structure defined by a non-empty class $\boldsymbol{P}$ of mappings subject to the following conditions

(B1) The members of $\boldsymbol{P}$ are invertible mappings from $\mathscr{B}$ onto open subsets of Euclidean spaces.

(B2) If $\kappa, \gamma \in \boldsymbol{P}$, then $\kappa \circ \gamma^{-1} \in \boldsymbol{D}$.

(B3) If $\kappa \in \boldsymbol{P}$, $\lambda \in \boldsymbol{D}$ and $\mathrm{Range}(\kappa) = \mathrm{Domain}(\lambda)$, then $\lambda \circ \kappa \in \boldsymbol{P}$.

The members of $\boldsymbol{P}$ are called the **placements** of $\mathscr{B}$. The range $\mathscr{R}_\kappa$ of a $\kappa \in \boldsymbol{P}$ is called the *region occupied* by $\mathscr{B}$ in the placement κ.

The Euclidean space $\mathscr{E}_\kappa$ of which the range $\mathscr{R}_\kappa$ of $\kappa\in\boldsymbol{P}$ is an open subset is called the *range space* of κ. Its translation space is denoted by $\mathscr{V}_\kappa$.

We say that $\mathscr{B}$ is a *body of class* C^n if $\boldsymbol{D}\subset\boldsymbol{D}_n$, an *incompressible body* of class C^n if $\boldsymbol{D}\subset\boldsymbol{D}_n^{\text{iso}}$, and a *rigid body* if $\boldsymbol{D}=\boldsymbol{D}^{\text{rig}}$.

Let $\mathscr{F}$ be some Euclidean space with translation space $\mathscr{U}$ and let $\mathscr{B}$ be a body of class C^n. We say that the mapping

$$f\colon \mathscr{B}\to\mathscr{F}$$

is of class C^r $(0\leqq r\leqq n)$ if $f\circ\kappa^{-1}\colon \mathscr{R}_\kappa\to\mathscr{F}$ is of class C^r for *some* $\kappa\in\boldsymbol{P}$. It follows from (B2) and the chain rule that $f\circ\kappa^{-1}$ is then of class C^r for *every* $\kappa\in\boldsymbol{P}$. If f is of class C^1 and if $\kappa\in\boldsymbol{P}$, we define

$$\nabla_\kappa f\colon \mathscr{B}\to\operatorname{Lin}(\mathscr{V}_\kappa,\mathscr{U})$$

by

$$\nabla_\kappa f(X)=\nabla(f\circ\kappa^{-1})(\kappa(X)),\qquad X\in\mathscr{B} \tag{1}$$

and call it the *gradient of f in the placement κ*.

If κ and γ are two placements, it follows from the chain rule that

$$\nabla_\gamma f=(\nabla_\kappa f)(\nabla_\gamma\kappa). \tag{2}$$

We say that $\nabla_\gamma\kappa$ is the *gradient of the displacement $\lambda=\gamma\circ\kappa^{-1}$ from κ to γ.*

The intuitive notion of an "infinitesimal element" of a continuous body can be made precise in the following way.

Let $X\in\mathscr{B}$ and let γ be a placement of $\mathscr{B}$. If $\boldsymbol{u}$ varies in a neighborhood of zero in the translation space $\mathscr{V}_\gamma$ of the range space of γ, then $\gamma(X)+\boldsymbol{u}$ varies in a neighborhood of $\gamma(X)$ in the region occupied by the body in the configuration γ, and hence $\gamma^{-1}(\gamma(X)+\boldsymbol{u})$ varies in a neighborhood of $X\in\mathscr{B}$. If κ is another placement then

$$\gamma^{-1}(\gamma(X)+\boldsymbol{u})=\kappa^{-1}(\kappa(X)+\boldsymbol{v})$$

describes the same point in the neighborhood of X if

$$\lambda(\gamma(X)+\boldsymbol{u})=\kappa(X)+\boldsymbol{v},\qquad \lambda=\kappa\circ\gamma^{-1}. \tag{3}$$

Approximating the displacement λ by its gradient in the neighborhood of $\gamma(X)$, we see that (3) can be written

$$(\nabla_\gamma\kappa(X))\,\boldsymbol{u}+o(\boldsymbol{u})=\boldsymbol{v}.$$

Thus, the two vectors $\boldsymbol{u}\in\mathscr{V}_\gamma$ and $\boldsymbol{v}\in\mathscr{V}_\kappa$ describe approximately the same material point if $(\nabla_\gamma\kappa(X))\,\boldsymbol{u}=\boldsymbol{v}$.

Definition. We say that the pairs $(\gamma,\boldsymbol{u})$ and $(\kappa,\boldsymbol{v})$, where $\kappa,\gamma\in\boldsymbol{P}$, $\boldsymbol{u}\in\mathscr{V}_\gamma$, $\boldsymbol{v}\in\mathscr{V}_\kappa$, are *equivalent at* $X\in\mathscr{B}$ and write $(\gamma,\boldsymbol{u})\sim_X(\kappa,\boldsymbol{v})$ if

$$(\nabla_\gamma\kappa(X))\,\boldsymbol{u}=\boldsymbol{v}. \tag{4}$$

The equivalence classes determined by the equivalence relation thus defined are called the *tangent vectors* at X. The set of all tangent vectors at X is called the **tangent space** at X and is denoted by $\mathscr{T}_X$.

The tangent vectors at X describe, approximately, the points in a neighborhood of X independently of any placement. Thus, the notion of tangent space is a mathematization of the intuitive notion of an "infinitesimal element about X".

Let $f\colon \mathscr{B} \to \mathscr{F}$ be a mapping of class C^1 from $\mathscr{B}$ into some Euclidean space $\mathscr{F}$ with translation space $\mathscr{U}$. It It follows from (2) and (4) that $(\nabla_\kappa f(X))\boldsymbol{v} = (\nabla_\gamma f(X))\boldsymbol{u}$ if $(\gamma, \boldsymbol{u}) \sim_X (\kappa, \boldsymbol{v})$. Hence we can define a mapping

$$\nabla f(X)\colon \mathscr{T}_X \to \mathscr{U} \tag{5}$$

by

$$(\nabla f(X))\boldsymbol{t} = (\nabla_\kappa f(X))\boldsymbol{v} \quad \text{if } (\kappa, \boldsymbol{v}) \in \boldsymbol{t} \in \mathscr{T}_X. \tag{6}$$

The following theorem justifies the term "tangent vectors" for the elements of $\mathscr{T}_X$.

Theorem. *The tangent space $\mathscr{T}_X$ has a unique natural vector space structure such that for all $f\colon \mathscr{B} \to \mathscr{F}$ of class C^1, the mapping* (5) *defined by* (6) *is linear.*

Proof. If $\kappa \in \boldsymbol{P}$, then $K = \nabla\kappa(X)\colon \mathscr{T}_X \to \mathscr{V}_\kappa$ is easily seen to be an invertible mapping. The vector-space structure on $\mathscr{T}_X$ is obtained by transporting the vector-space structure of $\mathscr{V}_\kappa$ to $\mathscr{T}_X$ by means of K^{-1}. It is easily seen that the structure on $\mathscr{T}_X$ thus obtained is independent of the choice of $\kappa \in \boldsymbol{P}$. Q.E.D.

Remark. Given $\kappa \in \boldsymbol{P}$, one can also transport the inner product of $\mathscr{V}_\kappa$ to $\mathscr{T}_X$ by means of K^{-1}. But one gets different inner products on $\mathscr{T}_X$ for different choices of $\kappa \in \boldsymbol{P}$. Hence $\mathscr{T}_X$ has no intrinsic natural inner product.

7. Configurations, Deformations

Let $\mathscr{B}$ be a continuous body. With every placement $\kappa \in \boldsymbol{P}$ we can associate a distance function

$$d_\kappa\colon \mathscr{B} \times \mathscr{B} \to \mathbb{R}^+$$

by putting

$$d_\kappa(X, Y) = d(\kappa(X), \kappa(Y)) = |\kappa(Y) - \kappa(X)|. \tag{1}$$

(If $\boldsymbol{u}$ belongs to some inner product space, we define the magnitude $|\boldsymbol{u}|$ of $\boldsymbol{u}$ by $|\boldsymbol{u}| = \sqrt{\boldsymbol{u} \cdot \boldsymbol{u}}$.)

The distance functions obtained in this manner will be called **configurations** of $\mathscr{B}$, so that

$$\boldsymbol{C} = \{d_\kappa \mid \kappa \in \boldsymbol{P}\}$$

is the set of all configurations of $\mathscr{B}$. Two placements $\kappa, \gamma \in \boldsymbol{P}$ determine the same configuration $d_\kappa = d_\gamma$ if and only if the displacement $\lambda = \gamma \circ \kappa^{-1}$ from κ to γ is an isometric bijection.

It is possible to recover the class $\boldsymbol{P}$ of placements of $\mathscr{B}$ from a knowledge of the set $\boldsymbol{C}$ of configurations of $\mathscr{B}$, because $\boldsymbol{P}$ is the class of all invertible mappings $\kappa\colon \mathscr{B} \to \mathscr{R}_\kappa$ onto open subsets $\mathscr{R}_\kappa$ of Euclidean spaces $\mathscr{E}_\kappa$ such that $d(X, Y) = |\kappa(Y) - \kappa(X)|$ holds for some $d \in \boldsymbol{C}$.

Let $d \in \boldsymbol{C}$ be a configuration of $\mathscr{B}$ and let $X \in \mathscr{B}$ be a material point. The definition (6) of Section 6 can be applied to $d^2(\cdot, Y)\colon \mathscr{B} \to \boldsymbol{R}$, where $Y \in \mathscr{B}$. We denote the gradient at $X \in \mathscr{B}$ of this function by

$$(\nabla d^2(\cdot, Y))(X) = \nabla_1 d^2(X, Y) \in \operatorname{Lin}(\mathscr{T}_X, \mathbb{R}) = \mathscr{T}_X^*.$$

Since $Y \in \mathscr{B}$ is arbitrary we thus obtain a function $\nabla_1 d^2(X, \cdot): \mathscr{B} \to \mathscr{T}_X^*$. The definition (6) of Section 6 can be applied again, with Y instead of X, and we obtain

$$\nabla_2 \nabla_1 d^2(X, Y) = \nabla(\nabla_1 d^2(X, \cdot))(Y) \in \operatorname{Lin}(\mathscr{T}_Y, \mathscr{T}_X^*).$$

Using the fact that $d^2(X, Y) = d^2(Y, X)$ and the symmetry theorem of Section 5, one can easily prove that

$$G_d = \nabla_2 \nabla_1 d^2(X, X) \in \operatorname{Sym}^+(\mathscr{T}_X, \mathscr{T}_X^*). \tag{2}$$

We say that G_d is the *configuration of the infinitesimal element* $\mathscr{T}_X$ *induced* by the configuration d of the body $\mathscr{B}$. The set

$$\boldsymbol{G} = \{G_d \mid d \in \boldsymbol{C}\} \subset \operatorname{Sym}^+(\mathscr{T}_X, \mathscr{T}_X^*)$$

is then the set of all configurations of $\mathscr{T}_X$.

There are several other ways to characterize the induced configurations G_d of $\mathscr{T}_X$:

(i) If $\boldsymbol{t} \in \mathscr{T}_X$ and $(\kappa, \boldsymbol{u}) \in \boldsymbol{t}$, then

$$\langle G_d \boldsymbol{t}, \boldsymbol{t} \rangle = \frac{\mathrm{d}^2}{\mathrm{d}s^2} d^2(X, \kappa^{-1}(\kappa(X) + s\boldsymbol{u}))|_{s=0}.$$

(ii) If $\kappa \in \boldsymbol{P}$ is a placement and $K = \nabla\kappa(X) \in \operatorname{Invlin}(\mathscr{T}_X, \mathscr{V}_\kappa)$, then

$$G_{d_\kappa} = K^* K; \tag{3}$$

hence G_{d_κ} is just the "inner product" on $\mathscr{T}_X$ obtained by transporting the inner product of $\mathscr{V}_\kappa$ to $\mathscr{T}_X$ via K^{-1}.

A **deformation process** of a body $\mathscr{B}$ is a function of the type

$$p: [0, r] \to \boldsymbol{C}, \quad r \in \mathbb{R}^+.$$

The value of p at $t \in [0, r]$ is denoted by p_t. The number r is called the *duration* of the process. With every kinematical process

$$\chi: \mathscr{B} \times \mathscr{T}_\chi \to \mathscr{W} \quad \text{of } \mathscr{B}, \quad \text{where} \quad \mathscr{T}_\chi = [\tau_0, \tau_0 + r]$$

is a closed time-interval, we can associate a deformation process $p: [0, r] \to \boldsymbol{C}$ by

$$p_t(X, Y) = \hat{d}(\chi(X, \tau_0 + t), \chi(Y, \tau_0 + t)), \quad t \in [0, r]. \tag{4}$$

Similarly, with every motion $\mu: \mathscr{B} \times \mathscr{T}_\mu \to \mathscr{E}$ of $\mathscr{B}$ relative to some frame $\mathscr{E}$, where $\mathscr{T}_\mu = [\tau_0, \tau_0 + r]$, we can associate a deformation process $p: [0, r] \to \boldsymbol{C}$ by

$$p_t(X, Y) = d(\mu(X, \tau_0 + t), \mu(Y, \tau_0 + t)), \quad t \in [0, r]. \tag{5}$$

Theorem A. *Two kinematical processes χ and χ^* of $\mathscr{B}$ determine the same deformation process if and only if there is an automorphism α of the event-world $\mathscr{W}$ such that $\chi^* = \chi^\alpha$.* (Recall that $\chi^\alpha(X, \tau) = \alpha(\chi(X, \tau - t_\alpha))$.)

Theorem B. *Two motions μ and μ^* of $\mathscr{B}$ relative to $\mathscr{E}$ determine the same deformation process if and only if there is a kinematical process χ and two reference processes β*

and β^* such that μ and μ^* are the motions corresponding to χ, β and χ, β^*, respectively. This is the case if and only if μ and μ^* are connected by a relation of the form (6) *of Section* 4.

A continuous body $\mathscr{B}$ can be endowed with the structure of a material system in the sense of Section 4 in the following manner: A class $\mathfrak{D}$ of deformation processes of $\mathscr{B}$ is prescribed. The class $\mathfrak{K}$ of admissible kinematical processes then consist of all those that determine a deformation process in $\mathfrak{D}$.

In view of Theorem A, the condition that $\chi\in\mathfrak{K}\Rightarrow\chi^\alpha\in\mathfrak{K}$ for every automorphism α of $\mathscr{W}$ is automatically satisfied. In view of Theorem B, the class $\mathfrak{M}$ of admissible motions of $\mathscr{B}$ relative to the frame $\mathscr{E}$ consists of all those that determine a deformation process in $\mathfrak{D}$.

If μ is a motion of $\mathscr{B}$, then μ_τ: $\mathscr{B}\to\mathscr{E}$ is called the *placement of $\mathscr{B}$ in $\mathscr{E}$ at time $\tau\in\mathfrak{T}_\mu$*. The gradient

$$M(\tau)=\nabla\mu_\tau(X)\in\mathrm{Lin}(\mathscr{T}_X,\mathscr{V})$$

is called the *placement of the body element $\mathscr{T}_X$ in $\mathscr{V}$ at time τ*. The function $M:\mathfrak{T}_\mu\to\mathrm{Lin}(\mathscr{T}_X,\mathscr{V})$ is called the *motion of the body element $\mathscr{T}_X$* induced by the motion μ of $\mathscr{B}$.

If p: $[0,r]\to\boldsymbol{C}$ is a deformation process of $\mathscr{B}$, then

$$P:[0,r]\to\boldsymbol{G},\qquad P(t)=G_{p_t}$$

is called the *deformation process of $\mathscr{T}_X$ induced* by the deformation process p of $\mathscr{B}$.

It is often useful to consider body elements in isolation, as if they were disconnected from the continuous body from which they are obtained.

Definition. A **body element** is a 3-dimensional vector space $\mathscr{T}$ endowed with structure defined by the prescription of a set $\boldsymbol{G}\subset\mathrm{Sym}^+(\mathscr{T},\mathscr{T}^*)$ of *configurations* and a set Π of *deformation processes* P: $[0,r]\to\boldsymbol{G}$, $r\in\mathbb{R}^+$.

A *placement* of the element $\mathscr{T}$ in a frame space $\mathscr{V}$ (inner product space) is defined to be an element $K\in\mathrm{InvLin}(\mathscr{T},\mathscr{V})$ such that $K^*K\in\boldsymbol{G}$. A *motion* of $\mathscr{T}$ in the frame space $\mathscr{V}$ is a function M: $[0,r]\to\mathrm{InvLin}(\mathscr{T},\mathscr{V})$ such that $M^*M\in\Pi$. It is often useful to fix a particular *reference placement* K_R and describe the possible motions in relation to K_R. The following functions from $[0,r]$ into $\mathrm{Lin}(\mathscr{V})$ are obtained in this fashion:

(i) *Displacement tensor:*

$$F=MK_R^{-1}.$$

(ii) *Right Cauchy-Green tensor:*

$$C=F^TF=K_R^{*-1}PK_R^{-1},\qquad P=M^*M.$$

In the case when $K_R=M(t)$ for some $t\in[0,r]$, we get:

(iii) *Relative displacement tensor:*

$$F_{(t)}=MM(t)^{-1}.$$

(iv) *Relative right Cauchy-Green tensor:*

$$C_{(t)}=F_{(t)}^T F_{(t)}=M(t)^{*-1} PM(t)^{-1}.$$

(v) k^{th} *Rivlin-Ericksen tensor:*

$$A_k=M^{*-1}\overset{(k)}{P}M^{-1}, \qquad A_k(t)=\overset{(k)}{C}_{(t)}(t).$$

If L is defined by $L(t)=\dot{F}_{(t)}(t)$, we have $\frac{1}{2}A=D=\frac{1}{2}(L+L^T)$, which is called the *stretching rate* (or *rate of deformation*). The *spin* (or *vorticity*) W is defined by $W=\frac{1}{2}(L-L^T)$.

The functions C, $C_{(t)}$, A_k, and D are frame-indifferent, F, $F_{(t)}$, L, and W are frame-dependent.

II. Thermomechanics of Continuous Bodies

8. Interactions

In this section we deal with what we call a **material universe** Ω, whose members $\mathscr{A}, \mathscr{B}, \mathscr{C}, \ldots$ we call *material objects*. We assume that Ω is endowed with a structure defined by a relation $\prec$ and we read $\mathscr{A}\prec\mathscr{B}$ as "*$\mathscr{A}$ is a part of $\mathscr{B}$*".

We assume that Ω and $\prec$ satisfies the 6 axioms stated below, which are all intuitively plausible.

(M 1) $\mathscr{A}=\mathscr{B}$ if and only if $\mathscr{A}\prec\mathscr{B}$ and $\mathscr{B}\prec\mathscr{A}$.

(M 2) $\mathscr{A}\prec\mathscr{B}$ and $\mathscr{B}\prec\mathscr{C}$ implies $\mathscr{A}\prec\mathscr{C}$.

These two axioms state that $\prec$ is a partial order on Ω. Given $\mathscr{A}, \mathscr{B}\in\Omega$, there can be at most one $\mathscr{C}\in\Omega$ such that $\mathscr{C}\prec\mathscr{A}, \mathscr{B}$ and $\mathscr{C}'\prec\mathscr{A}, \mathscr{B}\Rightarrow\mathscr{C}'\prec\mathscr{C}$. If such a $\mathscr{C}$ exists, we call it the *greatest common part* of $\mathscr{A}$ and $\mathscr{B}$ and write

$$\mathscr{C}=\mathscr{A}\wedge\mathscr{B}.$$

Given $\mathscr{A}, \mathscr{B}\in\Omega$, there can be at most one $\mathscr{D}\in\Omega$ such that $\mathscr{A}, \mathscr{B}\prec\mathscr{D}$ and $\mathscr{A}, \mathscr{B}\prec\mathscr{D}'\Rightarrow\mathscr{D}\prec\mathscr{D}'$. If such a $\mathscr{D}$ exists, we call it the *least envelope* of $\mathscr{A}$ and $\mathscr{B}$ and we write

$$\mathscr{D}=\mathscr{A}\vee\mathscr{B}.$$

(M 3) There are two elements $\emptyset, \infty\in\Omega$, called the *material nothing* and the *material all*, such that $\emptyset\prec\mathscr{A}\prec\infty$ for all $\mathscr{A}\in\Omega$.

If $\mathscr{A}\wedge\mathscr{B}=\emptyset$ we say that $\mathscr{A}$ and $\mathscr{B}$ are *separate*.

(M 4) For each $\mathscr{A}\in\Omega$ there is exactly one $\mathscr{A}^e\in\Omega$, called the *exterior* of $\mathscr{A}$, such that $\mathscr{A}\wedge\mathscr{A}^e=\emptyset$ and $\mathscr{A}\vee\mathscr{A}^e=\infty$.

(M 5) If $\mathscr{A}\wedge\mathscr{B}^e=\emptyset$, then $\mathscr{A}\prec\mathscr{B}$.

(M 6) For all $\mathscr{A}, \mathscr{B}\in\Omega$, $\mathscr{A}\wedge\mathscr{B}$ exists.

One can prove, from (M 1)–(M 6), that the operations $\wedge$ and $\vee$ define on Ω the structure of a Boolean algebra. The proof (taken from reference [1]) is given in the Appendix.

We use the notation

$$(\Omega\times\Omega)_{\text{sep}}=\{(\mathscr{A},\mathscr{B})\in\Omega\times\Omega\mid \mathscr{A}\wedge\mathscr{B}=\emptyset\}$$

for the set of all separate pairs of bodies, and the notation

$$\Omega_{\mathscr{B}}=\{\mathscr{P}\in\Omega\mid \mathscr{P}\prec\mathscr{B}\}$$

for the set of all parts of $\mathscr{B}$.

A mapping $F\colon\Omega'\to\mathscr{U}$ from some subset Ω' of Ω into some vector space $\mathscr{U}$ is said to be *additive* if

$$F(\mathscr{A}\vee\mathscr{B})=F(\mathscr{A})+F(\mathscr{B}) \tag{1}$$

whenever $\mathscr{A},\mathscr{B},\mathscr{A}\vee\mathscr{B}\in\Omega'$ and $\mathscr{A}\wedge\mathscr{B}=\emptyset$.

Definition. A mapping

$$I\colon(\Omega\times\Omega)_{\text{sep}}\to\mathscr{U},$$

where $\mathscr{U}$ is some vector space, is called an **interaction** if $I(\cdot,\mathscr{A}^e)\colon\Omega_{\mathscr{A}}\to\mathscr{U}$ and $I(\mathscr{A}^e,\cdot)\colon\Omega_{\mathscr{A}}\to\mathscr{U}$ are additive for every $\mathscr{A}\in\Omega$.

Lemma. *If I is an interaction and $\mathscr{A}\wedge\mathscr{B}=\emptyset$ then*

$$I(\mathscr{A},\mathscr{B})+I(\mathscr{B},\mathscr{A})=I(\mathscr{A},\mathscr{A}^e)+I(\mathscr{B},\mathscr{B}^e)-I\big(\mathscr{A}\vee\mathscr{B},(\mathscr{A}\vee\mathscr{B})^e\big). \tag{2}$$

Proof. Let $\mathscr{C}=\mathscr{A}\vee\mathscr{B}$. Then $\mathscr{C}^e\wedge\mathscr{A}=\emptyset$, $\mathscr{C}^e\wedge\mathscr{B}=\emptyset$, and

$$\mathscr{A}^e=\mathscr{C}^e\vee\mathscr{B},\qquad \mathscr{B}^e=\mathscr{C}^e\vee\mathscr{A}.$$

The additivity of $I(\mathscr{A},\cdot)$ and $I(\mathscr{B},\cdot)$ gives

$$I(\mathscr{A},\mathscr{A}^e)+I(\mathscr{B},\mathscr{B}^e)=I(\mathscr{A},\mathscr{C}^e)+I(\mathscr{A},\mathscr{B})+I(\mathscr{B},\mathscr{C}^e)+I(\mathscr{B},\mathscr{A}).$$

Using the additivity of $I(\cdot,\mathscr{C}^e)$, we get the desired result (2). Q.E.D.

The following theorem is an immediate consequence of the Lemma.

Theorem A. *An interaction I satisfies the "law of action and reaction"*

$$I(\mathscr{A},\mathscr{B})=-I(\mathscr{B},\mathscr{A}) \tag{3}$$

for all $\mathscr{A},\mathscr{B}\in\Omega$ with $\mathscr{A}\wedge\mathscr{B}=\emptyset$ if and only if the mapping $\mathscr{A}\mapsto I(\mathscr{A},\mathscr{A}^e)$ is additive.

An interaction I is said to be **balanced** if

$$I(\mathscr{A},\mathscr{A}^e)=0\qquad\text{for all }\mathscr{A}\in\Omega. \tag{4}$$

Theorem A implies that the law (3) of action and reaction holds for balanced interactions.

We now assume that a continuous body $\mathscr{B}$ of class C^1 is given and that for some (and hence every) placement κ of $\mathscr{B}$ the region $\mathscr{R}_\kappa$ occupied by $\mathscr{B}$ is bounded and has a piecewise C^1 boundary $\partial\mathscr{R}_\kappa$. One can then complete $\mathscr{B}$ by joining to $\mathscr{B}$ a boundary $\partial\mathscr{B}$. The placements of $\mathscr{B}$ can be extended to the completion

$$\bar{\mathscr{B}}=\mathscr{B}\cup\partial\mathscr{B}$$

of $\mathscr{B}$ in such a way that for every $\kappa \in \boldsymbol{P}$, κ is continuous in $\overline{\mathscr{B}}$ and

$$\kappa(\partial\mathscr{B}) = \partial\mathscr{R}_\kappa.$$

For every subset $\mathscr{P}$ of $\mathscr{B}$, we say that the boundary $\partial\mathscr{P} \subset \overline{\mathscr{B}}$ is piecewise C^1 if $\kappa(\partial\mathscr{P})$ is piecewise C^1 for some (and hence every) placement $\kappa \in \boldsymbol{P}$.

We assume, further, that we have singled out a certain class $\Omega_{\mathscr{B}}$ of open subsets of $\mathscr{B}$ such that

(i) Every $\mathscr{P} \in \Omega_{\mathscr{B}}$ has a piecewise C^1 boundary.

(ii) If $\mathscr{P} \in \Omega_{\mathscr{B}}$, then

$$\mathscr{P}^b = \operatorname{int}(\mathscr{B} \backslash \mathscr{P}) \in \Omega_{\mathscr{B}}. \tag{5}$$

(iii) If $\mathscr{P}, \mathscr{Q} \in \Omega_{\mathscr{B}}$, then

$$\mathscr{P} \wedge \mathscr{Q} = \mathscr{P} \cap \mathscr{Q} \in \Omega_{\mathscr{B}}. \tag{6}$$

(iv) If $\mathscr{P}, \mathscr{Q} \in \Omega_{\mathscr{B}}$, then

$$\mathscr{P} \vee \mathscr{Q} = \operatorname{int}\operatorname{cl}(\mathscr{P} \cup \mathscr{Q}) \in \Omega_{\mathscr{B}}. \tag{7}$$

(v) The class $\Omega_{\mathscr{B}}$ is as large as possible (in a sense not made precise here).

We now construct a material universe Ω from a given continuous body $\mathscr{B}$ and a prescribed class $\Omega_{\mathscr{B}}$ of *parts of* $\mathscr{B}$ as explained above, and from a given Euclidean space $\mathscr{E}$. The class Ω consists of $\Omega_{\mathscr{B}}$ and objects of the form $\mathscr{P} \vee \mathscr{E}$ with $\mathscr{P} \in \Omega_{\mathscr{B}}$. The "is a part of" relation $\prec$ is defined in such a way that $\mathscr{P} \prec \mathscr{Q}$ means $\mathscr{P} \subset \mathscr{Q}$ if $\mathscr{P}, \mathscr{Q} \in \Omega_{\mathscr{B}}$ and

$$\mathscr{E} = \mathscr{B}^e. \tag{8}$$

In words, the Euclidean space $\mathscr{E}$ is regarded as the exterior of the body, or as the *external world*. For the parts of $\mathscr{B}$, the greatest common part and the least envelope are given by (6) and (7), respectively. If $\mathscr{P} \in \Omega_{\mathscr{B}}$, we have

$$\mathscr{P}^e = \mathscr{P}^b \vee \mathscr{E}, \tag{9}$$

where $\mathscr{P}^b$ is given by (5).

Let I be an interaction in the material universe Ω just constructed with values in a normed space $\mathscr{U}$. The restriction

$$I^{\text{int}}: (\Omega_{\mathscr{B}} \times \Omega_{\mathscr{B}})_{\text{sep}} \to \mathscr{U}$$

of I to seperate pairs of parts of $\mathscr{B}$ is called the **internal interaction** determined by I.

The additive function $I^{\text{ext}}: \Omega_{\mathscr{B}} \to \mathscr{U}$ defined by $I^{\text{ext}}(\mathscr{P}) = I(\mathscr{P}, \mathscr{E})$ is called the **external action** determined by I. It is clear that for any $\mathscr{P}, \mathscr{Q} \in \Omega_{\mathscr{B}}$ with $\mathscr{P} \subset \mathscr{Q}$, we have

$$I(\mathscr{P}, \mathscr{Q}^e) = I^{\text{int}}(\mathscr{P}, \mathscr{Q}^b) + I^{\text{ext}}(\mathscr{P}). \tag{10}$$

The following two assumptions are made about most of the interactions that occur in continuum physics:

(I1) The internal interaction I^{int} is a *contact interaction* and it is *area-continuous*. More presisely, for some (and hence every) placement κ of $\mathscr{B}$, there is a $k > 0$ such that

$$|I^{\text{int}}(\mathscr{P}, \mathscr{Q})| < k \operatorname{Area}_\kappa(\partial\mathscr{P} \cap \partial\mathscr{Q})$$

holds for all $\mathscr{P}, \mathscr{Q} \in \Omega_{\mathscr{B}}$ with $\mathscr{P} \wedge \mathscr{Q} = \emptyset$.

(I2) The external action I^{ext} can be *by contact and at a distance* and it is *area-volume-continuous.* More presisely, for some (and hence every) placement κ of $\mathscr{B}$, there are $k_1, k_2>0$ such that

$$|I^{\text{ext}}(\mathscr{P})|<k_1 \operatorname{Vol}_\kappa(\mathscr{P})+k_2 \operatorname{Area}_\kappa(\partial\mathscr{P}\cap\partial\mathscr{B}).$$

for all $\mathscr{P}\in\Omega$.

These assumptions have the following consequence:

Theorem B. *Let I be interaction that satisfies the conditions* (I1) *and* (I2) *and let* κ *be a placement of* $\mathscr{B}$. *One can then find an integrable function*

$$f_\kappa: \mathscr{B}\to\mathscr{U}$$

and, for each oriented surface $\mathscr{S}\subset\overline{\mathscr{B}}$ *belonging to a suitable class, an integrable function*

$$\bar{\bar{j}}_{\mathscr{S},\kappa}: \mathscr{S}\to\mathscr{U}$$

such that for any $\mathscr{P}$, $\mathscr{Q}\in\Omega_\beta$ *with* $\mathscr{P}\subset\mathscr{Q}$, *we have*

$$I(\mathscr{P},\mathscr{Q}^e)=\int_{\mathscr{P}} f_\kappa\, d(\operatorname{Vol}_\kappa)+\int_{\mathscr{S}} \bar{\bar{j}}_{\mathscr{S},\kappa}\, d(\operatorname{Area}_\kappa), \tag{11}$$

where $\mathscr{S}$ *is the surface of contact between* $\mathscr{P}$ *and* $\mathscr{Q}$, *oriented in such a way that* $\mathscr{P}$ *and* $\mathscr{Q}$ *are on the interior side of* $\mathscr{S}$. *The functions* f_κ *and* $\bar{\bar{j}}_{\mathscr{S},\kappa}$ *are uniformly bounded.*

The proof of this theorem is not difficult but too technical to be presented here. The essential parts of this proof are contained in Section III and the Appendix of reference [4]. The following theorem is non-trivial and surprising. The original proof is contained in reference [5].

Theorem C. *Let I be an interaction that satisfies the conditions* (I1) *and* (I2) *so that the conclusions of Theorem B are valid. Assume that* $\mathscr{P}\mapsto I(\mathscr{P},\mathscr{P}^e)$ *is volume-continuous on* $\Omega_{\mathscr{B}}$ (i.e., *for some, and hence every, placement* κ *of* $\mathscr{B}$, *there is a* $k>0$ *such that* $|I(\mathscr{P},\mathscr{P}^e)|\leqq k\operatorname{Vol}_\kappa(\mathscr{P})$ *holds for all* $\mathscr{P}\in\Omega_{\mathscr{B}}$).

Then there is a function

$$\bar{j}_\kappa: \overline{\mathscr{B}}\times\mathscr{V}_\kappa^1\to\mathscr{U},$$

where $\mathscr{V}_\kappa^1=\{\boldsymbol{n}\in\mathscr{V}_\kappa \mid |\boldsymbol{n}|=1\}$, *such that*

$$\bar{\bar{j}}_{\mathscr{S},\kappa}(X)=\bar{j}_\kappa(X,\boldsymbol{n}) \tag{12}$$

whenever $\boldsymbol{n}$ *is the exterior unit normal to* $\mathscr{S}$ *at* $X\in\mathscr{S}$ *in the placement* κ.

The following theorem is a classical non-trivial result that was proved, in essense, by CAUCHY in 1823. A modern proof, which I have used in courses and lectures since 1958 but never published, is reproduced in Section VIII of reference [4].

Theorem D. *If the hypotheses of Theorem* C *are valid and if the function*

$$\bar{j}_\kappa(\cdot,\boldsymbol{n}): \overline{\mathscr{B}}\to\mathscr{U}$$

of Theorem C *is continuous for every* $\boldsymbol{n} \in \mathscr{V}_\kappa^1$, *then there is a continuous function*

$$j_\kappa : \bar{\mathscr{B}} \to \operatorname{Lin}(\mathscr{V}_\kappa, \mathscr{U})$$

such that

$$\bar{j}_\kappa(X, \boldsymbol{n}) = j_\kappa(X)\,\boldsymbol{n}. \tag{13}$$

It is unfortunate that nobody has been able, so far, to prove the conclusion of Theorem D on the basis of the hypotheses of Theorem C alone, without the *ad hoc* continuity assumption of Theorem D. A step towards filling this hole in the foundations of continuum thermomechanics has been made by GURTIN, MIZEL & WILLIAMS in 1967 (reference [6]), who proved the existence of a measurable function j_κ under the assumption that $\bar{j}_\kappa(\cdot, \boldsymbol{n})$ is measurable for each $\boldsymbol{n}$.

9. Equilibrium of Forces, Mechanical Processes

We assume in this section that $\mathscr{B}$ is a continuous body of class C^1, that $\Omega_{\mathscr{B}}$ is a class of parts of $\mathscr{B}$, that $\mathscr{E} = \mathscr{B}^e$ is a Euclidean space representing the external world, and that Ω is the material universe constructed from $\Omega_{\mathscr{B}}$ and $\mathscr{E}$ as explained in the previous section.

An interaction $\boldsymbol{f}$ in Ω with values in the translation space $\mathscr{V}$ of $\mathscr{E}$ will be called a **force system** for $\mathscr{B}$. We consider only force systems that satisfy the conditions (I 1) and (I 2) of the previous section. It is then possible to associate, with each part $\mathscr{P} \in \Omega_{\mathscr{B}}$ a bounded vector-valued Borel measure $\boldsymbol{f}_{\mathscr{P}}$ on the Borel subsets of $\bar{\mathscr{P}}$ such that

$$\boldsymbol{f}_{\mathscr{P}}(\mathscr{C}) = \int_{\mathscr{C}} \boldsymbol{b}_\kappa \,\mathrm{d}(\mathrm{Vol}_\kappa) + \int_{\mathscr{C} \cap \partial\mathscr{P}} \boldsymbol{t}_{\partial\mathscr{P},\kappa} \,\mathrm{d}(\mathrm{Area}_\kappa), \tag{1}$$

holds for all placements κ of $\mathscr{B}$ in $\mathscr{E}$, where $\boldsymbol{b}_\kappa$ and $\boldsymbol{t}_{\partial\mathscr{P},\kappa}$ correspond to f_κ and $\bar{j}_{\mathscr{S},\kappa}$, with $\mathscr{S} = \partial\mathscr{P}$, of Theorem B of the previous section. This measure describes, intuitively, the forces exerted on $\mathscr{P}$ and its parts by the exterior of $\mathscr{P}$. It is meaningful to integrate continuous functions on $\bar{\mathscr{P}}$ relative to the measure $\boldsymbol{f}_{\mathscr{P}}$. For example, if $\boldsymbol{v}: \bar{\mathscr{P}} \to \mathscr{V}$ is a continuous function, we have

$$\int_{\bar{\mathscr{P}}} \boldsymbol{v} \cdot \mathrm{d}\boldsymbol{f}_{\mathscr{B}} = \int_{\mathscr{P}} (\boldsymbol{b}_\kappa \cdot \boldsymbol{v}) \,\mathrm{d}(\mathrm{Vol}_\kappa) + \int_{\partial\mathscr{P}} (\boldsymbol{t}_{\partial\mathscr{P},\kappa} \cdot \boldsymbol{v}) \,\mathrm{d}(\mathrm{Area}_\kappa). \tag{2}$$

A vector field $\boldsymbol{k}: \mathscr{E} \to \mathscr{V}$ is called an **infinitesimal isometry** of $\mathscr{E}$ if the solution $\xi(s) = \alpha(s, x)$ of the initial value problem $\dot{\xi} = k(\xi)$, $\xi(0) = x$, determines an isometry $x \mapsto \alpha(s, x)$ for every $s \in \mathbb{R}$. The Representation Theorem for isometries (Corollary at the end of Section 2) is easily seen to imply the following result.

Representation Theorem for Infinitesimal Isometries. *To every infinitesimal isometry* $\boldsymbol{k}$ *of* $\mathscr{E}$ *corresponds a unique* $A_{\boldsymbol{k}} \in \operatorname{Skew}(\mathscr{V})$ *such that*

$$\boldsymbol{k}(x) = \boldsymbol{k}(q) + A_{\boldsymbol{k}}(x - q) \tag{3}$$

holds for all $x, q \in \mathscr{E}$.

Definition. We say that a force system $\boldsymbol{f}$ for $\mathscr{B}$ is **in equilibrium** *in the placement* κ of $\mathscr{B}$ in $\mathscr{E}$ if for every part $\mathscr{P}$ of $\mathscr{B}$ and every infinitesimal isometry $\boldsymbol{k}$ of $\mathscr{B}$

$$\int_{\bar{\mathscr{P}}} (\boldsymbol{k} \circ \kappa) \cdot \mathrm{d}\boldsymbol{f}_{\mathscr{P}} = 0 \tag{4}$$

holds.

The right hand side of (4) can be interpreted as the rate at which the forces do work in a "virtual" rigid motion (*i.e.* a motion obtained from a rigid kinematical process; see Section 4).

The following result is obtained by applying the Representation Theorem above. The details of the proof are given in reference [7], Section 7.

Theorem A. *A force system $\boldsymbol{f}$ for $\mathscr{B}$ is in equilibrium in the placement κ of $\mathscr{B}$ in $\mathscr{E}$ if and only if the following two balance laws are valid:*

I. Balance of Forces. For every part $\mathscr{P}$ of $\mathscr{B}$

$$\boldsymbol{f}(\mathscr{P}, \mathscr{P}^e)=\mathbf{0} \tag{5}$$

holds.

II. Balance of Moments. For every part $\mathscr{P}$ of $\mathscr{B}$ and every $q\in\mathscr{E}$ the tensor

$$K_\kappa(\mathscr{P}, q)=\Big(\int_{\bar{\mathscr{P}}} (\boldsymbol{r}\circ\kappa)\otimes \mathrm{d}\boldsymbol{f}_{\mathscr{P}}\Big)^T\in \mathrm{Lin}(\mathscr{V}) \tag{6}$$

is symmetric, where $\boldsymbol{r}\colon \mathscr{E}\to\mathscr{V}$ is defined by $\boldsymbol{r}(x)=x-q$.

We note that the condition I does not involve the placement κ at all. Hence, if the forces are not balanced, the force system cannot be in equilibrium in *any* placement.

Since we have assumed that the force system $\boldsymbol{f}$ satisfies the conditions (I1) and (I2) of the previous section, we can apply Theorems C and D of the previous Section to a balanced force system. If it were not for the *ad hoc* continuity condition of Theorem D, we could conclude that there exists, for every placement κ of $\mathscr{B}$ in $\mathscr{E}$ a field

$$T_\kappa\colon \bar{\mathscr{B}}\to \mathrm{Lin}(\mathscr{V})$$

such that for any $\mathscr{P}\in\Omega_{\mathscr{B}}$ and any Borel subset $\mathscr{C}$ of $\bar{\mathscr{P}}$ we have

$$\boldsymbol{f}_{\mathscr{P}}(\mathscr{C})=\int_{\mathscr{C}} \boldsymbol{b}_\kappa\, \mathrm{d}(\mathrm{Vol}_\kappa)+\int_{\mathscr{C}\cap\partial\mathscr{P}} T_\kappa \boldsymbol{n}_\kappa\, \mathrm{d}(\mathrm{Area}_\kappa), \tag{7}$$

where $\boldsymbol{n}_\kappa\colon \partial\mathscr{P}\to\mathscr{V}$ is the unit normal to $\kappa(\partial\mathscr{P})$, directed towards the exterior of $\mathscr{P}$.

The field T_κ, if it exists, is called the **stress** determined by $\boldsymbol{f}$ in the placement κ.

For an arbitrary force system $\boldsymbol{f}$, the tensor $K_\kappa(\mathscr{P}, q)$ defined by (6) is called the **astatic load** on $\mathscr{P}$ relative to q in κ. Its skew-symmetric part is called the **moment** relative to q of the forces acting on $\mathscr{P}$ in the placement κ. A proof of the following theorem is given in Section 5 of reference [8].

Theorem B. *Assume that $\boldsymbol{f}$ satisfies the balance of force condition* I *of Theorem* A *and that the stress T_κ, exists for some and hence all placements κ of $\mathscr{B}$. Then the astatic load $K_\kappa(\mathscr{P}, q)=K_\kappa(\mathscr{P})$ is independent of q and given by*

$$K_\kappa(\mathscr{P})=\int_{\bar{\mathscr{P}}} T_\kappa\, \mathrm{d}(\mathrm{Vol}_\kappa). \tag{8}$$

Since the left side of (8) cannot by symmetric for *every* part $\mathscr{P}$ of $\mathscr{B}$ unless T_κ has symmetric values, we obtain the following consequence of Theorems A and B:

Theorem C. *A force system $\boldsymbol{f}$ that satisfies the hypotheses of Theorem* B *is in equilibrium in the placement κ of $\mathscr{B}$ in $\mathscr{E}$ if and only if the stress T_κ has symmetric values.*

Assume now that f is a force system that has a representation of the form (7) for some, and hence all, placements κ. Condition I, expressing balance of forces, is then equivalent to the condition that

$$\int_{\mathscr{P}} \boldsymbol{b}_\kappa \,\mathrm{d}(\mathrm{Vol}_\kappa) + \int_{\partial\mathscr{P}} T_\kappa \boldsymbol{n}_\kappa \,\mathrm{d}(\mathrm{Area}_\kappa) = \boldsymbol{0} \tag{10}$$

hold for all parts $\mathscr{P}$ of $\mathscr{B}$. If T_κ is of class C^1, this condition is equivalent to *Cauchy's Law*:

$$\mathrm{div}_\kappa\, T_\kappa + \boldsymbol{b}_\kappa = \boldsymbol{0}. \tag{11}$$

If Cauchy's Law (11) holds for one placement κ, it holds automatically for all placements.

Definition. A **mechanical process** is a triple $(\chi, \beta, \boldsymbol{f})$ with the following properties:

(a) $\chi: \mathscr{B} \times \Upsilon \to \mathscr{W}$ is a kinematical process,

(b) $\beta: \mathscr{E} \times \Upsilon \to \mathscr{W}$ is a reference process,

(c) For each $\tau \in \Upsilon$, $\boldsymbol{f}_\tau$ is a force system in Ω that satisfies the conditions (I1) and (I2) of the previous section.

(d) For each $\tau \in \Upsilon$ and each $\mathscr{P} \in \Omega$, the rate of work

$$P(\mathscr{P}, \tau) = \int_{\mathscr{P}} \dot{\mu}_\tau^* \cdot \mathrm{d}\boldsymbol{f}_\tau, \tag{12}$$

of the forces $\boldsymbol{f}_\tau$ in a motion μ^* determined by χ and a reference process β^* is independent of the reference process β^*.

It follows from Equation (7) of Section 4 that the Condition (d) is equivalent to the following:

(d′) For each $\tau \in \Upsilon$ the force system $\boldsymbol{f}_\tau$ is in equilibrium in the placement μ_τ, where μ is the motion determined by χ and β.

If the continuity condition of Theorem D of the previous section could be eliminated, one would have the following result: *A mechanical process can be characterized by a triple* (χ, β, T), where $\chi: \mathscr{B} \times \Upsilon \to \mathscr{W}$ *is a kinematical process of* $\mathscr{B}$, $\beta: \mathscr{E} \times \Upsilon \to \mathscr{W}$ *a reference process of the external world* $\mathscr{E}$, *and* $T: \mathscr{B} \times \Upsilon \to \mathrm{Sym}(\mathscr{V})$ *a symmetric stress tensor field.*

The force-systems $\boldsymbol{f}_\tau$, $\tau \in \Upsilon$, of the process (χ, β, T) are determined as follows: If $\mathscr{P}, \mathscr{Q} \in \Omega_{\mathscr{B}}$, $\mathscr{P} \wedge \mathscr{Q} = \emptyset$, then

$$\boldsymbol{f}_\tau(\mathscr{P}, \mathscr{Q}) = \int_{\partial\mathscr{P} \cap \partial\mathscr{Q}} T_\tau \boldsymbol{n}_\tau \,\mathrm{d}(\mathrm{Area}_\tau), \tag{13}$$

where $\boldsymbol{n}_\tau$ is the exterior unit normal in the placement $\mu_\tau = \beta_\tau^{-1} \circ \chi_\tau$ and Area_τ denotes the surface area in that placement. For each $\tau \in \Upsilon$ there is a unique body force field $\boldsymbol{b}_\tau$ such that (10) holds with κ replaced by μ_τ (or simply τ). If T_τ is of class C^1, then $\boldsymbol{b}_\tau$ is actually given by $\boldsymbol{b}_\tau = -\mathrm{div}_{\mu_\tau} T_\tau$. We have

$$\boldsymbol{f}_\tau(\mathscr{P}, \mathscr{E}) = \int_{\partial\mathscr{B} \cap \partial\mathscr{P}} T_\tau \boldsymbol{n}_\tau \,\mathrm{d}(\mathrm{Area}_\tau) + \int_{\mathscr{P}} \boldsymbol{b}_\tau \,\mathrm{d}(\mathrm{Vol}_\tau). \tag{14}$$

If we put $\boldsymbol{f}_\tau(\mathscr{E}, \mathscr{P}) = -\boldsymbol{f}_\tau(\mathscr{P}, \mathscr{E})$ when $\mathscr{P} \in \Omega_{\mathscr{B}}$, then $\boldsymbol{f}_\tau$ is completely determined by (13) and (14).

10. Thermodynamic Processes, Entropy Production

As in the last section we assume that $\mathscr{B}$ is a continuous body of class C^1, and that Ω is a material universe constructed from $\mathscr{B}$ and an external world $\mathscr{E}$.

Definition. A **thermodynamic process** is a septuple $(\chi, \beta, \boldsymbol{f}, E, H, S, \theta)$ whose entries have the following nature:

(α) $(\chi, \beta, \boldsymbol{f})$ is a mechanical process on a time interval $\mathscr{T}$ and the motion μ determined by χ and β if of class C^1 with velocity $\boldsymbol{v}=\dot{\mu}$.

(β) E is a function of the type $E\colon \Omega_{\mathscr{B}} \times \mathscr{T} \to \mathbb{R}$, whose value $E(\mathscr{P}, \tau)=E_\tau(\mathscr{P})$ is called the *energy content* of $\mathscr{P}$ at time $\tau \in \mathscr{T}$. It is assumed that E_τ is additive and that E_τ and $\dot{E}_\tau$ (defined by $\dot{E}_\tau(\mathscr{P})=\dot{E}(\mathscr{P}, \tau)$) are volume continuous.

(γ) H is a function of the type $H\colon (\Omega \times \Omega)_{\text{sep}} \times \mathscr{T} \to \mathbb{R}$, whose value $H(\mathscr{P}, \mathscr{A}, \tau)=H_\tau(\mathscr{P}, \mathscr{A})$ is called the *rate of heat transfer* from $\mathscr{A}$ to $\mathscr{P}$ at time $\tau \in \mathscr{T}$. It is assumed that H_τ is an interaction that satisfies the conditions (I1) and (I2) of Sect. 8.

(δ) S is a function of the type $S\colon \Omega_{\mathscr{B}} \times \mathscr{T} \to \mathbb{R}$, whose value $S(\mathscr{P}, \tau)=S_\tau(\mathscr{P})$ is called the *entropy content* of $\mathscr{P}$ at time $\tau \in \mathscr{T}$. It is assumed that S_τ and $\dot{S}_\tau$ are volume continuous.

(ε) θ is a function of the type $\theta\colon \overline{\mathscr{B}} \times \mathscr{T} \to \mathbb{R}^{++}$ whose value $\theta(\chi, \tau)$ is called the *temperature* at $\chi \in \overline{\mathscr{B}}$ at time $\tau \in \mathscr{T}$.

The septuple is subject to the following axiom, called the **law of energy balance**: *For all parts $\mathscr{P}$ of $\mathscr{B}$ and all $\tau \in \mathscr{T}$, we have*

$$\dot{E}_\tau(\mathscr{P})=H_\tau(\mathscr{P}, \mathscr{P}^e)+P_\tau(\mathscr{P}), \tag{1}$$

where

$$P_\tau(\mathscr{P})=\int_{\mathscr{P}} \boldsymbol{v}_\tau \cdot \mathrm{d}\boldsymbol{f}_{\tau, \mathscr{P}}, \qquad \boldsymbol{v}_\tau(X)=\dot{\mu}(X, \tau), \tag{2}$$

is the rate of work of the forces acting on $\mathscr{P}$ in the motion μ determined by χ and β.

Recall that by Condition (d) of the previous section, P is independent of the reference process β and depends on μ only through χ. Using the results of the previous section, one can express P in terms of the stress as follows:

Theorem A. *Let κ be a placement of $\mathscr{B}$ in $\mathscr{E}$. If the force systems $\boldsymbol{f}_\tau$ are described by a stress $T_\kappa\colon \overline{\mathscr{B}} \times \mathscr{T} \to \operatorname{Lin}(\mathscr{V})$, then the rate of working is given by*

$$P(\mathscr{P})=\int_{\mathscr{P}} \operatorname{tr}(T_\kappa \nabla_\kappa \boldsymbol{v})\, \mathrm{d}(\mathrm{Vol}_\kappa). \tag{3}$$

From now on we omit the subscript τ and consider dependence on time as understood.

It follows from the assumed additivity and volume continuity of E_τ, $\dot{E}_\tau$, S_τ and $\dot{S}_\tau$ that for every placement κ there exists an *energy density* $\varepsilon_\kappa\colon \mathscr{B} \times \mathscr{T} \to \mathbb{R}$ and an *entropy density* $\eta_\kappa\colon \mathscr{B} \times \mathscr{T} \to \mathbb{R}$ such that

$$E(\mathscr{P})=\int_{\mathscr{P}} \varepsilon_\kappa\, \mathrm{d}(\mathrm{Vol}_\kappa), \qquad \dot{E}(\mathscr{P})=\int_{\mathscr{P}} \dot{\varepsilon}_\kappa\, \mathrm{d}(\mathrm{Vol}_\kappa) \tag{4}$$

and

$$S(\mathscr{P})=\int_{\mathscr{P}} \eta_\kappa\, \mathrm{d}(\mathrm{Vol}_\kappa), \qquad \dot{S}(\mathscr{P})=\int_{\mathscr{P}} \dot{\eta}_\kappa\, \mathrm{d}(\mathrm{Vol}_\kappa). \tag{5}$$

If the force systems can be described by a stress field, then Theorem A shows that the rate of working P is additive and volume continuous. Since $\dot{E}$ is additive and volume continuous by assumption, it follows from the law (1) of energy balance that $\mathscr{P} \mapsto H(\mathscr{P}, \mathscr{P}^e)$ is additive and volume continuous. The hypotheses of Theorem C of Section 8 are therefore satisfied for the interaction H. If it were not for the continuity condition of Theorem D of Section 9 we could conclude that with each $\mathscr{P} \in \Omega_{\mathscr{B}}$ one can associate a bounded Borel measure $H_{\mathscr{P}}$ on $\overline{\mathscr{P}}$ of the form

$$H_{\mathscr{P}}(\mathscr{C}) = \int_{\mathscr{C}} r_\kappa \, \mathrm{d}(\mathrm{Vol}_\kappa) + \int_{\mathscr{C} \cap \partial \mathscr{P}} \boldsymbol{q}_\kappa \cdot \boldsymbol{n}_\kappa \, \mathrm{d}(\mathrm{Area}_\kappa) \tag{6}$$

where κ is an arbitrary placement of $\mathscr{B}$.

We have $H_{\mathscr{Q}}(\mathscr{P}) = H(\mathscr{P}, \mathscr{Q}^e)$ whenever $\mathscr{P}, \mathscr{Q} \in \Omega_{\mathscr{B}}$, $\mathscr{P} \subset \mathscr{Q}$. The function r_κ: $\mathscr{B} \times \mathscr{T} \to \mathbb{R}$ is called the *heat supply* and $\boldsymbol{q}_\kappa$: $\mathscr{B} \times \mathscr{T} \to \mathscr{V}$ the *heat flux* determined by H in the placement κ.

If we substitute (3), (4) and (6), with $\mathscr{C} = \mathscr{P}$, into the energy balance equation (1), we obtain

$$\int_{\mathscr{P}} (\dot{\varepsilon}_\kappa - \mathrm{tr}(T_\kappa \nabla_\kappa \boldsymbol{v}) - r_\kappa) \, \mathrm{d}(\mathrm{Vol}_\kappa) - \int_{\partial \mathscr{P}} \boldsymbol{q}_\kappa \cdot \boldsymbol{n}_\kappa \, \mathrm{d}(\mathrm{Area}_\kappa) = 0. \tag{7}$$

If $\boldsymbol{q}_\kappa$ is of class C^1, then (7) is valid for all parts $\mathscr{P}$ of $\mathscr{B}$ if and only if the *local energy balance equation*

$$\dot{\varepsilon}_\kappa = \mathrm{tr}(T_\kappa \nabla_\kappa \boldsymbol{v}) + \mathrm{div}_\kappa \, \boldsymbol{q}_\kappa + r_\kappa \tag{8}$$

holds.

One can derive a result for thermodynamical process analogous to the one for mechanical processes stated at the end of the previous section: If continuity conditions could be ignored, every thermodynamical process could be characterized by a septuple $(\chi, \beta, T, \varepsilon, \boldsymbol{q}, \eta, \theta)$, where $T, \varepsilon, \boldsymbol{q}, \eta, \theta$ are functions on $\mathscr{B} \times \mathscr{T}$ representing the stress, energy density, heat flux, entropy density, and temperature corresponding to the actual placements μ_τ of the motion determined by χ and β. The local energy balance equation (8) then determines the heat supply $r(X, \tau) = r_{\mu_\tau}(X, \tau)$.

We assume now that a thermodynamical process is given. Using the measures $H_{\mathscr{Q}}$ defined by (6), we define the rate of *entropy transfer* from $\mathscr{Q}^e$ to $\mathscr{P} \in \Omega_{\mathscr{B}}$, $\mathscr{P} \subset \mathscr{Q}$, by

$$M(\mathscr{P}, \mathscr{Q}^e) = \int_{\overline{\mathscr{P}}} \frac{1}{\theta} \, \mathrm{d}H_{\mathscr{Q}} = \int_{\mathscr{P}} \frac{1}{\theta} r_\kappa \, \mathrm{d}(\mathrm{Vol}_\kappa) + \int_{\partial \mathscr{P} \cap \partial \mathscr{Q}} \frac{1}{\theta} \boldsymbol{q}_\kappa \cdot \boldsymbol{n}_\kappa \, \mathrm{d}(\mathrm{Area}_\kappa). \tag{9}$$

Definition. The *rate of* **entropy production** Γ: $\Omega_{\mathscr{B}} \times \mathscr{T} \to \mathbb{R}$ of a thermodynamic process is defined by

$$\Gamma(\mathscr{P}, \tau) = \dot{S}(\mathscr{P}, \tau) - M(\mathscr{P}, \mathscr{P}^e, \tau). \tag{10}$$

If the heat flux $\boldsymbol{q}_\kappa$ and the temperature θ are of class C^1, one can use the divergence theorem in (9). After substitution of $(5)_2$ and (9) into (10) one then obtains

$$\Gamma(\mathscr{P}) = \int_{\mathscr{P}} \gamma_\kappa \, \mathrm{d}(\mathrm{Vol}_\kappa), \tag{11}$$

where γ_κ: $\mathscr{B} \times \mathscr{T} \to \mathbb{R}$, the *local rate of entropy production*, is given by

$$\gamma_\kappa = \dot{\eta}_\kappa - \frac{1}{\theta} r_\kappa - \mathrm{div}_\kappa \left(\frac{1}{\theta} \boldsymbol{q}_\kappa \right). \tag{12}$$

Using (8), we can eliminate r_κ from (12) and obtain

$$\theta\gamma_\kappa = \theta\dot{\eta}_\kappa - \dot{\varepsilon}_\kappa + \operatorname{tr}(T_\kappa \nabla_\kappa \boldsymbol{v}) + \frac{1}{\theta}\,\boldsymbol{q}_\kappa \cdot \nabla_\kappa \theta. \tag{13}$$

In view of (11) we see that $\Gamma(\mathscr{P}) \geqq 0$ can hold for all parts $\mathscr{P}$ of $\mathscr{B}$ if and only if $\gamma_\kappa \geqq 0$. Hence (13) gives the following result:

Theorem B. *If the thermodynamic process is such that for some, and hence every, placement κ of $\mathscr{B}$ the stress T_κ and heat flux $\boldsymbol{q}_\kappa$ exist and if $\boldsymbol{q}_\kappa$ and the temperature θ are of class C^1, then the rate of entropy production Γ has non-negative values if and only if*

$$\theta\dot{\eta}_\kappa - \dot{\varepsilon}_\kappa + \operatorname{tr}(T_\kappa \nabla_\kappa \boldsymbol{v}) + \frac{1}{\theta}\,\boldsymbol{q}_\kappa \cdot \nabla_\kappa \theta \geqq 0 \tag{14}$$

holds for some, and hence every, placement κ.

Remark. Gurtin & Williams in 1967 (reference [4]) have developed a foundation of thermodynamics in which the rate of entropy transfer M is regarded as one of the primitive data in the definition of a thermodynamical process. Using certain physically well motivated axioms, they then derive from M and H not one but two temperatures, a conductive temperature and a radiative temperature. They then proceed to show that under many, but not all, circumstances, the two temperatures are the same.

11. Constitutive Assumptions, the Dissipation Principle

Constitutive assumptions are mathematical statements that describe particular physical circumstances which govern the behavior of particular bodies and their environments. A constitutive assumption is a relation involving the seven functions χ, β, $\boldsymbol{f}$, E, H, S, θ that define a thermodynamic process. Processes that satisfy this relation are said to be *admissible* for the constitutive assumption in question.

External constitutive assumptions describe the influence of the environment $\mathscr{E}$ on the body $\mathscr{B}$. We describe those assumptions that are very often made in continuum physics:

(a) There are two additive and volume-continuous functions $m_i, m_g \colon \Omega_{\mathscr{B}} \to \mathbb{R}^+$ and a scalar field $\phi \colon \mathscr{E} \times \mathscr{T} \to \mathbb{R}$ of class C^1 such that

$$\boldsymbol{f}^{\text{ext}}(\mathscr{P}) = -\int_{\mathscr{P}} \boldsymbol{a}\,\mathrm{d}m_i + \int_{\mathscr{P}} \nabla\phi \circ \mu\,\mathrm{d}m_g \quad \text{when} \quad \bar{\mathscr{P}} \subset \mathscr{B} \tag{1}$$

holds for all admissible processes. Here $\boldsymbol{f}^{\text{ext}}$ is the external force defined by $\boldsymbol{f}$, *i.e.* $\boldsymbol{f}^{\text{ext}}(\mathscr{P}) = \boldsymbol{f}(\mathscr{P}, \mathscr{E})$, and $\boldsymbol{a} = \ddot{\mu}$ is the acceleration for the motion μ defined by χ and β. m_i is called the *inertial mass* and m_g the *gravitational mass.* The first term on the right side of (1) is called the *inertial force*, the second is called the *gravitational force* acting on $\mathscr{P}$. The field ϕ is called the *gravitational potential* of the enviroment $\mathscr{E}$.

It is an experimental fact that the ratio $m_i(\mathscr{P})/m_g(\mathscr{P})$ is not only independent of $\mathscr{P} \in \Omega_{\mathscr{B}}$, but actually the same for all bodies ever encountered in nature. By suitable choice of units, one therefore can take $m_i = m_g = m$, replace (1) by

$$\boldsymbol{f}^{\text{ext}}(\mathscr{P}) = \int_{\mathscr{P}} (-\boldsymbol{a} + \nabla\phi \circ \mu)\,\mathrm{d}m \qquad \text{when} \quad \bar{\mathscr{P}} \subset \mathscr{B}, \tag{2}$$

and call the function $m\colon \Omega_{\mathscr{B}} \to \mathbb{R}^+$ the *mass*. Since m is assumed to be volume continuous, we can associate with every placement κ of $\mathscr{B}$ a *mass density* $\rho_\kappa\colon \mathscr{B} \to \mathbb{R}^+$ such that

$$m(\mathscr{P}) = \int_{\mathscr{P}} \rho_\kappa \, \mathrm{d}(\mathrm{Vol}_\kappa). \tag{3}$$

The condition (2) can be expressed in terms of ρ_κ and the body force density $\boldsymbol{b}_\kappa$ by

$$\boldsymbol{b}_\kappa = \rho_\kappa(-\boldsymbol{a} + \nabla\phi \circ \mu). \tag{4}$$

(b) If $\overline{\mathscr{P}} \not\subset \mathscr{B}$, *i.e.* if $\partial\mathscr{P} \cap \partial\mathscr{B} \neq \emptyset$, a surface integral over $\partial\mathscr{P} \cap \partial\mathscr{B}$ is added to the right side of (2). This integral represents the *surface traction* acting on $\mathscr{B}$. A variety of assumptions on this surface traction represent a variety of physical circumstances.

(c) It is usually assumed that there is no heat transfer at a distance from the environment, *i.e.* that

$$H^{\mathrm{ext}}(\mathscr{P}) = H(\mathscr{P}, \mathscr{E}) = 0 \qquad \text{when } \overline{\mathscr{P}} \subset \mathscr{B}. \tag{5}$$

In terms of the heat supply, (5) means that

$$r_\kappa = 0 \tag{6}$$

for some, and hence all, placements κ of $\mathscr{B}$.

(d) If $\partial\mathscr{P} \cap \partial\mathscr{B} \neq \emptyset$, then $H^{\mathrm{ext}}(\mathscr{P})$ is set equal to a surface integral over $\partial\mathscr{P} \cap \partial\mathscr{B}$. This integral represents the heat supplied to $\mathscr{B}$ by contact with the environment. A variety of assumptions on this integral represent a variety of physical conditions.

Internal constitutive assumptions describe the internal physical constitution of the body $\mathscr{B}$. They are relations that involve only the internal portion (χ, $\beta\boldsymbol{f}^{\mathrm{int}}$, E, H^{int}, S, θ) of the thermodynamic process (χ, β, $\boldsymbol{f}$, E, H, S, θ) considered. The kinematical process β of the enviroment $\mathscr{E}$ should not occur and only the internal parts $\beta\boldsymbol{f}^{\mathrm{int}}$ and H^{int} of the interactions $\boldsymbol{f}$ and H should be involved. It is $\beta_\tau \boldsymbol{f}_\tau^{\mathrm{int}}$ rather than $\boldsymbol{f}_\tau^{\mathrm{int}}$ that describes the internal forces at time τ intrinsically, without reference to the environment $\mathscr{E}$.

A fundamental restriction on the possible internal consititutive assumptions must be imposed:

Dissipation Principle. *For all members of the class of thermodynamic processes that are admissible for an internal constitutive assumption, the rate of entropy production must have non-negative values.*

A body $\mathscr{B}$ is said to consist of a **simple material** if its physical behaviour is completely determined by the physical properties of its infinitesimal elements. In order to make this idea more precise, we introduce the concept of a thermodynamic process of a body element as follows:

Let (χ, β, $\boldsymbol{f}$, E, H, S, Θ) be a thermodynamic process, let μ be the motion determined by χ and β, and let $X \in \mathscr{B}$ be a material point. Assume that the process is defined on the closed time interval $\Upsilon = [\tau_0, \tau_0 + r]$. The function M: $[0, r] \to$

Invlin $(\mathscr{T}_X, \mathscr{V})$ defined by

$$M(t)=\nabla\mu_{\tau+t}(X), \qquad t\in[0,r] \tag{7}$$

describes the motion of the body element $\mathscr{T}_X$ (see Sect. 7). The deformation process P of $\mathscr{T}_X$ induced by the deformation process p induced by the kinematical process χ is

$$P=M^* M:[0,r]\to \boldsymbol{G}\subset \mathrm{Sym}^+(\mathscr{T}_X,\mathscr{T}_X^*). \tag{8}$$

We define the *intrinsic stress*

$$S:[0,r]\to \mathrm{Sym}(\mathscr{T}_X^*,\mathscr{T}_X)$$

in terms of the stress T_{μ_τ} (assumed to exist) by

$$S(t)=M(t)^{-1}\, T_{\mu_{\tau_0+t}}(X,\tau_0+t)\, M(t)^{*\,-1}. \tag{9}$$

Also, we define the *intrinsic heat flux*

$$\boldsymbol{q}:[0,r]\to\mathscr{T}_X$$

in terms of the heat fluxes $\boldsymbol{q}_{\mu_\tau}$ (assumed to exist) by

$$\boldsymbol{q}(t)=M(t)^{-1}\,\boldsymbol{q}_{\mu_{\tau_0+t}}(X,\tau_0+t). \tag{10}$$

The *intrinsic temperature gradient* $\boldsymbol{g}$: $[0,r]\to\mathscr{T}_X^*$ is defined by

$$\boldsymbol{g}(t)=\nabla\theta(X,\tau_0+t)=\left(\nabla_{\mu_{\tau_0+t}}\theta(X,\tau_0+t)\right)M(t). \tag{11}$$

It is a useful to choose a *reference configuration* $G_R\in\boldsymbol{G}$ of $\mathscr{T}_X$ and to introduce an energy density ε_R: $[0,r]\to\mathbb{R}$ and an entropy density η_R: $[0,r]\to\mathbb{R}$ by

$$\varepsilon_R(t)=\varepsilon_\kappa(X,\tau_0+t), \qquad \eta_R(t)=\eta_\kappa(X,\tau_0+t), \tag{12}$$

where $\kappa\in\boldsymbol{P}$ is such that $G_R=\nabla\kappa(X)^*\,\nabla\kappa(X)$. It is easily seen that ε_R and η_R depend on κ only through G_R.

The septuple $(P, S, \varepsilon_R, \boldsymbol{q}, \eta_R, \theta, g)$ of functions on $[0,r]$ with values in $\boldsymbol{G}$, $\mathrm{Sym}(\mathscr{T}_X^*, \mathscr{T}_X)$, $\mathbb{R}$, $\mathscr{T}_X$, $\mathbb{R}$, $\mathbb{R}^{++}$, and $\mathscr{T}_X^*$, respectively, is called the *thermodynamic process of the element* $\mathscr{T}_X$ induced by the thermodynamic process $(\chi, \beta, \boldsymbol{f}, E, H, S, \theta)$ of $\mathscr{B}$. It is easily seen that the induced process depends only on the internal portion $(\chi, \beta\boldsymbol{f}^{\mathrm{int}}, E, H^{\mathrm{int}}, S, \theta)$ of the given process.

We say that $\mathscr{B}$ is a *simple body* if the internal constitutive assumptions for $\mathscr{B}$ are relations involving only the thermodynamical processes of the body elements.

We introduce a *local volume* j_R: $[0,r]\to\mathbb{R}^{++}$ by

$$j_R(t)^2=\det\left(G_R^{-1}P(t)\right), \qquad t\in[0,r], \tag{13}$$

and a *local entropy production* γ_R: $[0,r]\to\mathbb{R}$ by

$$\gamma_R(t)=\gamma_\kappa(X,\tau_0+t), \qquad t\in[0,r], \tag{14}$$

when $G_R=\nabla\kappa(X)^*\,\nabla\kappa(X)$. It follows from Equation (13) of the previous section that

$$\theta\gamma_R=\theta\dot\eta_R-\dot\varepsilon_R+j_R\left\{\frac{1}{2}\operatorname{tr}(S\dot P)+\frac{1}{\theta}\langle \boldsymbol{g},\boldsymbol{q}\rangle\right\}, \tag{15}$$

which shows that γ_R depends only on the thermodynamic process of the element $\mathcal{T}_X$ under consideration. Theorem *B* of the previous Section implies the following result:

Theorem A. *The dissipation principle is satisfied for a material element* $\mathcal{T}$ *if and only if for every admissible thermodynamical process* $(P, S, \varepsilon_R, \boldsymbol{q}, \eta_R, \theta, g)$ *of* $\mathcal{T}$, *the local entropy production* γ_R *defind by* (15) *has non-negative values.*

We now show by an example how Theorem A can be applied. A *Navier-Stokes fluid with Fourier heat conduction* is described by 6 material functions $\bar{p}, \bar{\lambda}, \bar{\mu}, \bar{\varepsilon}_R, \bar{\theta}$, and $\bar{\kappa}$ from $\mathbb{R}^{++} \times \mathbb{R}$ into $\mathbb{R}$. The constitutive relations are the following:

(i) Navier-Stokes relation:

$$PSP = (-p + \lambda \operatorname{tr}(P^{-1}\dot{P}))P + \mu\dot{P}, \tag{16}$$

where

$$p = \bar{p}(j_R, \eta_R), \qquad \lambda = \bar{\lambda}(j_R, \eta_R), \qquad \mu = \bar{\mu}(j_R, \eta_R). \tag{17}$$

(ii) Equations of state:

$$\varepsilon_R = \bar{\varepsilon}_R(j_R, \eta_R), \qquad \theta = \bar{\theta}(j_R, \eta_R). \tag{18}$$

(iii) Fourier's law of heat conduction:

$$P\boldsymbol{q} = \kappa g, \quad \text{where} \quad \kappa = \bar{\kappa}(j_R, \eta_R). \tag{19}$$

The following result is a consequence of Theorem A, but the proof, given in reference [9], is not trivial.

Theorem B. *The constitutive relations* (16)–(19) *for a Navier-Stokes fluid with Fourier heat conduction satisfy the dissipation principle if and only if the following conditions are satisfied*:

(a) $\bar{\theta} = \bar{\varepsilon}_{R,2}$ (temperature relation),
(b) $\bar{p} = -\bar{\varepsilon}_{R,1}$ (pressure relation),
(c) $\bar{\mu} \geqq 0,\ \bar{\lambda} + \frac{2}{3}\bar{\mu} \geqq 0$ (viscosity inequalities),
(d) $\bar{\kappa} \geqq 0$ (heat conductivity inequality).

Remark. The formulation of the dissipation principle given here is due to Coleman & Noll (1963) (reference [9]), who also gave the first non-trivial application, of which Theorem B is a special case. Since then, the principle has been applied by many authors to a variety of constitutive relations, often with suprising results.

Appendix: The Concept of a Material Universe

A *material universe* Ω is a set endowed with a structure defined by a relation $\prec$, subject to the axioms (B1)–(B6) below. The elements $A, B, C, \ldots$ are called *material objects* or *bodies.*

(B1) *Certain pairs* (A, B) *of bodies are related by* $\prec$; *we write* $A \prec B$ *if this is the case and say that* A *is a part of* B.

(B2) $A = B$ *if and only if both* $A \prec B$ *and* $B \prec A$.

(B3) *If* $A \prec B$ *and* $B \prec C$, *then* $A \prec C$.

The axioms (B1)–(B3) state that $\prec$ is a *partial ordering* of Ω.
If $B \prec A$ and $C \prec A$ we call A an *envelope* of $\{B, C\}$ and write $\{B, C\} \prec A$. If $\{B, C\} \prec D$ and if $\{B, C\} \prec A$ implies $D \prec A$ we say that D is the *least envelope* of $\{B, C\}$ and we write $D = B \vee C$, so that

$$\{B, C\} \prec A \quad \text{implies} \quad \{B, C\} \prec B \vee C \prec A. \tag{1}$$

A least envelope may or may not exist, but if it does, it is uniquely determined by B and C. If $A \prec B$ and $A \prec C$ we say that A is a *common part* of $\{B, C\}$ and write $A \prec \{B, C\}$. If $D \prec \{B, C\}$ and if $A \prec \{B, C\}$ implies $A \prec D$ we say that D is the *greatest common part* of $\{B, C\}$ and write $D = B \vee C$, so that

$$A \prec \{B, C\} \quad \text{implies} \quad A \prec B \wedge C \prec \{B, C\}. \tag{2}$$

If there is a greatest common part it is unique. Envelopes, the greatest envelope, common parts, and the greatest common part of an arbitrary collection of bodies are defined in a similar manner. Let $\{A_i \mid i \in I\}$ be such a collection, its members A_i marked with indices i taken from an index set I. If the least envelope of the collection exists, we denote it by $\bigvee_{i \in I} A_i$; if the greatest common part exists, we denote it by $\bigwedge_{i \in I} A_i$.

Two bodies A and B are said to be *separate* if they have no common part. It will be convenient to adjoin to the material universe Ω two *improper bodies*, the *null-body* $\emptyset$ and the *universal body* ∞. We extend the relation $\prec$ to the extended universe

$$\Omega' = \Omega \cup \{\emptyset, \infty\} \tag{3}$$

by putting

$$\emptyset \prec A \prec \infty \qquad \text{for all } A \in \Omega'. \tag{4}$$

Of course, the relation $\prec$, when extended by (4), remains a partial ordering. We have

$$A \wedge B = \emptyset, \qquad A, B \in \Omega \tag{5}$$

if and only if A and B are separate.

The following rules are easily established:

I. $A \prec B$ if and only if $A \wedge B = A$ or $A \vee B = B$.

II. If $A \prec B$ and if $A \wedge C$, $B \wedge C$ exist, then

$$A \wedge C \prec B \wedge C. \tag{6}$$

III. If $A \prec B$ and if $A \vee C$, $B \vee C$ exist, then

$$A \vee C \prec B \vee C. \tag{7}$$

IV. If $A \prec B$ and if $B \wedge C = \emptyset$, then $A \wedge C = \emptyset$.

V. If $A \wedge B$ and $B \wedge C$ exist, then

$$(A \wedge B) \wedge C = A \wedge (B \wedge C) = A \wedge B \wedge C, \tag{8}$$

provided either $(A \wedge B) \wedge C$ or $A \wedge (B \wedge C)$ exist.

It is possible that $A \wedge B \wedge C$ exists but $A \wedge B$ or $B \wedge C$ do not.

We are now ready to state the next axiom:

(B4) *For each body $A \in \Omega$ there is exactly one body $A^e \in \Omega$ with the property that $\{A, A^e\}$ have neither a common part nor an envelope; i.e.,*

$$A \wedge A^e = \emptyset, \qquad A \vee A^e = \infty. \tag{9}$$

The body A^e is called *the exterior* of the body A. If we put $\emptyset^e = \infty$, $\infty^e = \emptyset$, then (9) is valid for all $A \in \Omega'$. It follows directly from the definition of A^e that

$$(A^e)^e = A \tag{10}$$

holds for every $A \in \Omega'$. Also,

$$A \prec B \quad \text{implies} \quad A \wedge B^e = \emptyset, \tag{11}$$

which is a consequence of $(9)_1$ and rule IV. The next axiom postulates that the converse of (11) is valid:

(B5) *If A and the exterior of B have no common part, then A is a part of B.*

It follows from (B5) and (11) that

$$A \prec B \quad \text{if and only if} \quad A \wedge B^e = \emptyset. \tag{12}$$

By (12) and (10) we have $A \prec B$ if and only if $B^e \wedge (A^e)^e = \emptyset$. Hence, using (12) again, we find that

$$A \prec B \quad \text{if and only if} \quad B^e \prec A^e. \tag{13}$$

The following propositions are corollaries of (13): Let $\{A_i \mid i \in I\}$ be a collection of bodies. If $\bigvee_{i \in I} A_i$ exists, so does $\bigwedge_{i \in I} (A_i^e)$ and

$$\bigwedge_{i \in I} (A_i^e) = \left(\bigvee_{i \in I} A_i\right)^e. \tag{14}$$

If $\bigwedge_{i \in I} A_i$ exists, so does $\bigvee_{i \in I} (A_i^e)$ and

$$\bigvee_{i \in I} (A_i^e) = \left(\bigwedge_{i \in I} A_i\right)^e. \tag{15}$$

The following result is basic to the theory of bodies:

Lemma. *Assume that A_1, A_2, and B are bodies and that $A_1 \wedge B$, $A_2 \wedge B$, and $A_1 \vee A_2$ exist. Then we have*

$$\{A_1 \wedge B, A_2 \wedge B\} \prec \{B, A_1 \vee A_2\}; \tag{16}$$

furthermore, if

$$\{A_1 \wedge B, A_2 \wedge B\} \prec C \tag{17}$$

and

$$D \prec \{B, A_1 \vee A_2\}, \tag{18}$$

then

$$D \prec C. \tag{19}$$

Proof. It follows from the definitions (1) and (2) that $A_i \wedge B \prec B$, and $A_i \wedge B \prec A_i \prec A_1 \vee A_2$, $i = 1, 2$, which immediately gives (16).

Assume now that (17) and (18) hold. It follows from (17) and (11) that

$$\{A_i \wedge B) \wedge C^e = \emptyset, \quad i=1,2. \tag{20}$$

Suppose that E is a common part of $\{D, C^e\}$, *i.e.*

$$E \prec \{D, C^e\}. \tag{21}$$

By (18) we have $E \prec D \prec B$, and hence

$$E \prec \{B, C^e\}. \tag{22}$$

Now, if G_i is a common part of $\{E, A_i\}$,

$$G_i \prec \{E, A_i\}, \tag{23}$$

then $G_i \prec E \prec B$ by (22) and hence $G_i \prec \{B, A_i\}$, which is equivalent to

$$G_i \prec A_i \wedge B, \quad i=1,2. \tag{24}$$

Using rule IV, we infer from (24) and (20) that $G_i \wedge C^e = \emptyset$. But by (23) and (22) we also have $G_i \prec E \prec C^e$ and hence, by rule IV, $G_i = G_i \wedge G_i \prec G_i \wedge C^e = \emptyset$, *i.e.* $G_i = \emptyset$. We have shown that (23) implies $G_i = \emptyset$, which means that

$$E \wedge A_i = \emptyset, \quad i=1,2. \tag{25}$$

Applying (10) and (12), we infer from (25) that $A_i \prec E^e$, $i=1, 2$, which is equivalent to $A_1 \vee A_2 \prec E^e$, or, by (13), to

$$E \prec (A_1 \vee A_2)^e. \tag{26}$$

On the other hand, it follows from (21) and (18) that

$$E \prec D \prec A_1 \vee A_2. \tag{27}$$

The relations (26) and (27) can both hold only if $E = \emptyset$. We have shown that (21) implies $E = \emptyset$, which means that $D \wedge C^e = \emptyset$. But by (12) this is equivalent to (19). Q.E.D.

The following four propositions are easy consequences of the Lemma:

(A) *If $A_1 \wedge B$, $A_2 \wedge B$, and $A_1 \vee A_2$ exist, then*

$$(A_1 \vee A_2) \wedge B = (A_1 \wedge B) \vee (A_2 \wedge B), \tag{28}$$

provided either the left hand side or the right hand side exists.

(B) *If $A_1 \vee B$, $A_2 \vee B$ and $A_1 \wedge A_2$ exist, then*

$$(A_1 \wedge A_2) \vee B = (A_1 \vee B) \wedge (A_2 \vee B), \tag{29}$$

provided either the left hand side or the right hand side exists.

(C) *If $A \wedge B$ and $A^e \wedge B$ exist, then*

$$B = (A \wedge B) \vee (A^e \wedge B). \tag{30}$$

(D) *We have*

$$A = B \vee C, \quad B \wedge C = \emptyset \tag{31}$$

if and only if

$$B \prec A, \quad C = A \wedge B^e. \tag{32}$$

Proposition (A) is an immediate corollary of the Lemma; (B) results from applying (A) to A_1^e, A_2^e, and B^e and using (14) and (15); (C) corresponds to the special case $A_1 = A$, $A_2 = A^e$ of (A). To prove (D), assume that (31) holds. Then $B \prec A$ follows from $(31)_1$. Using (12) and (10) we infer from $(31)_2$ that $C \prec B^e$, *i.e.* that $C \wedge B^e = C$. Using $(31)_1$ and (A), we obtain

$$A \wedge B^e = (B \vee C) \wedge B^e = (B \wedge B^e) \vee (C \wedge B^e) = \emptyset \vee C = C,$$

which proves (32). Assume, conversely, that (32) holds. We then have $A \wedge B = B$ and hence, by (C),

$$A = (B \wedge A) \vee (B^e \wedge A) = B \vee C.$$

Also, $\emptyset = \emptyset \wedge A = (B \wedge B^e) \wedge A = B \wedge (B^e \wedge A) = B \wedge C$, which proves (31).

We now state the final axiom:

(B6) *If $A, B \in \Omega$ are not separate, then the greatest common part $A \wedge B \in \Omega$ exists.*

It follows from (B6) that in the extended universe defined by (3), $A \wedge B$ exists for all $A, B \in \Omega'$. By (15), the least envlope $A \vee B = (A^e \wedge B^e)^e \in \Omega'$ exists always also. Equations (28) and (29) state that the operations $\wedge$ and $\vee$ are distributive with respect to one another. This fact, together with axiom (B4), can be expressed by saying that Ω' is a *complemented distributive lattice.*

Suppose that a body B is given. A finite collection $\mathfrak{P} = \{P_i \mid i = 1, \ldots, n\}$ is called a *partition of B* if

$$B = \bigvee_{i=1}^{n} P_i \quad \text{and} \quad P_i \wedge P_j = 0 \quad \text{if } i \neq j. \tag{33}$$

We say that the partition $\mathfrak{P}' = \{P_i' \mid i = 1, \ldots, n'\}$ is a *refinement* of $\mathfrak{P}$ if for each $P_i' \in \mathfrak{P}'$ there is a $P_j \in \mathfrak{P}$ such that $P_i' \prec P_j$.

Theorem. *Any two partitions $\mathfrak{P} = \{P_i \mid i = 1, \ldots, n\}$, $\mathfrak{Q} = \{Q_j \mid j = 1, \ldots, m\}$ have a common refinement; i.e., there is a partition which is a refinement of both $\mathfrak{P}$ and $\mathfrak{Q}$.*

Proof. The required refinement consists of the nm parts $P_i \wedge Q_j$, $i = 1, \ldots, n$, $j = 1, \ldots, m$. The fact that these parts do indeed form a refinement of both $\mathfrak{P}$ and $\mathfrak{Q}$ follows easily from the distributive laws (28), (29), and from $P_i \wedge Q_j \prec \{P_i, Q_j\}$. Q.E.D.

The axioms (B1)–(B6) and the theorems derived from them reflect our common sense experiences with physical bodies. It is necessary here to make the mathematical description of bodies independent of any imbedding in "space", simply because there is no such thing as "space" in our development.

The following purely mathematical special examples illustrate the concept of a material universe:

(α) Let Ω' consist in all subsets of an arbitrary set X and let $\prec$ represent set inclusion. We have $A \wedge B = A \cap B$ (intersection), $A \vee B = A \cup B$ (union) and $A^e = A^c$

(complement). The Newtonian mechanics of particle systems can be based on such a universe when X is a finite set.

(β) Let Ω' consist in all closures of open sets in an arbitrary topological space and let $\prec$ represent set inclusion. In this case the greatest common part of two bodies A, B is *not* always their intersection, but rather $A \wedge B = \overline{\mathring{A} \cap \mathring{B}}$ (a superimposed $\circ$ denotes the interior and a superimposed bar denotes the closure). The least envelope of a collection $\{A_i | i \in I\}$ is given by $\bigvee_{i \in I} A_i = \overline{\bigcup_{i \in I} \mathring{A}_i}$, *i.e.* by the closure of the union of the interiors of the sets in the collection. This least envelope is equal to the union $\bigcup_{i \in I} A_i$ if the collection is finite, but not necessarily if it is infinite. The exterior of a body is $A^e = \overline{A^c}$, *i.e.* the closure of the complement.

(γ) Let Ω consist in all finite unions of closed polyhedra and closed exteriors of polyhedra in a Euclidean space. Ω satisfies the axioms of a material universe when $\prec$ represents set inclusion. Greatest common parts, least envelopes, and exteriors are given by the same formulas as in example (β).

(δ) Let Ω consist in all closed regions with piecewise smooth boundaries, in a Euclidean space. If $\prec$ represents set inclusion, then the axioms (B1)–(B5) are satisfied, but (B6) is not. Since this example is of importance in conventional continuum mechanics, it may be desirable to develop the mechanics of material universes without postulating (B6). However, this leads to considerable technical difficulties, and the issue has not yet been resolved.

Acknowledgement. The research leading to this paper was supported by grants from the National Science Foundation.

References

1. Noll, W., The foundations of mechanics. Non-Linear Continuum Theories, C.I.M.E. Lectures, 1965, pp. 159–200, Rome (1966)
2. Noll, W., A new mathematical theory of simple materials. Archive for Rational Mech. Anal. **48**, 1–50 (1972)
3. Noll, W., Euclidean geometry and Minkowskian chronometry. American Mathematical Monthly **71**, 129–144 (1964)
4. Gurtin, M. E., & W. O. Williams, An axiomatic foundation for continuum thermodynamics. Archive for Rational Mech. Anal. **26**, 83–117 (1967)
5. Noll, W., The foundations of classical mechanics in the light of recent advances in continuum mechanics. Proceedings of the Berkeley Symposium on the Axiomatic Method, 226–281, Amsterdam, 1959
6. Gurtin, M. E., V. J. Mizel & W. O. Williams, A note on Cauchy's stress theorem. J. Math. Anal. Appl. **22**, 398–401 (1968)
7. Noll, W., La Mécanique Classique, Basée sur un Axiome d'Objectivité. Colloque International sur la Méthode Axiomatique dans les Mécaniques Classiques et Nouvelles (1959), pp. 47–56, Paris, 1963
8. Coleman, B. D., & W. Noll, On the thermostatics of continuous media. Archive for Rational Mech. Anal. **4**, 97–128 (1959)
9. Coleman, B. D., & W. Noll, The thermodynamics of elastic materials with heat conduction and viscosity. Archive for Rational Mech. Anal. **13**, 167–178 (1963)

Department of Mathematics
Carnegie-Mellon University
Pittsburgh

(Received June 7, 1973)

GPSR Compliance

The European Union's (EU) General Product Safety Regulation (GPSR) is a set of rules that requires consumer products to be safe and our obligations to ensure this.

If you have any concerns about our products, you can contact us on ProductSafety@springernature.com

In case Publisher is established outside the EU, the EU authorized representative is:

Springer Nature Customer Service Center GmbH
Europaplatz 3
69115 Heidelberg, Germany

Batch number: 10584556

Printed by Printforce, the Netherlands